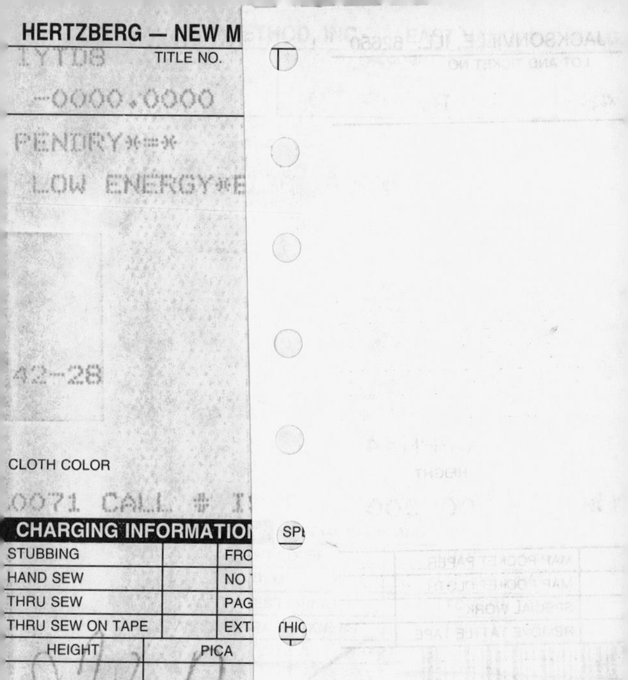

L ENERGY
N DIFFRACTION

its Application to
f Surface Structure

Techniques of Physics

Editors
G. K. T. Conn and K. R. Coleman

Techniques of physics find wide application in biology, medicine, engineering and technology generally. This series is devoted to techniques which have found and are finding application. The aim is to clarify the principles of each technique, to emphasize and illustrate the applications and to draw attention to new fields of possible employment.

1. D. C. Champeney: Fournier Transforms and their Physical Applications
2. J. B. Pendry: Low Energy Electron Diffraction

In preparation

R. E. Meads and F. W. D. Woodhams: Mossbauer Techniques.

W. Eccleston and R. A. Stuart: Physical Principles of MOS Devices.

G. H. Lunn: The Practice of High Speed Photography.

B. R. Garfield: Photo-Emissive Surfaces.

P. Iredale: Optical Image Detectors.

M. J. Goringe and C. J. Humphreys: Principles and Applications of High Voltage Electron Microscopy.

LOW ENERGY ELECTRON DIFFRACTION

The Theory and its Application to Determination of Surface Structure

J. B. PENDRY
The Cavendish Laboratory,
University of Cambridge,
Cambridge, England.

1974

ACADEMIC PRESS
London and New York
A Subsidiary of Harcourt Brace Jovanovich, Publishers

ACADEMIC PRESS INC. (LONDON) LTD.
24/28 Oval Road,
London NW1

United States Edition published by
ACADEMIC PRESS INC.
111 Fifth Avenue
New York, New York 10003

Library of Congress Catalog Card Number: 73–9472
ISBN: 0–12–550550–7

Filmset by Ramsay Typesetting Ltd. (London and Crawley)
and printed in Great Britain by T. & A. Constable Ltd., Edinburgh

Preface

It has long been realized that low energy electron diffraction (LEED) is a tool of great potential value in the study of surfaces. Confinement of electrons by strong scattering to the first few Angstroms of the crystal, combined with sensitivity to structural, electronic, and vibrational properties of a surface makes for wide application. Solid-state physicists, chemists, and metallurgists have all been involved in development of the technique. Yet LEED has by no means made its full potential contribution to surface studies because theoretical interpretation of experiments has proved to be a non-trivial problem. Some of the most interesting information has remained locked in the measurements.

The past five years have seen very satisfactory progress on the formerly lagging theoretical front, to the point where most workers would now agree that theoretical models give a good account of experiments. Even more recently LEED experiments have been used in conjunction with theory to deduce previously unknown surface structures of adsorbate systems.

The author believes that further application of the new theories will be a most profitable exercise for the understanding of surface phenomena. This book has been designed to give impetus to such practical applications. Theories are explained in detail starting from a knowledge of scattering theory such as most post-graduate students are equipped with at the beginning of their research. Every effort has been made to relate a topic that sometimes is necessarily mathematical to its experimental basis, and more abstract arguments are illustrated by analogies with familiar systems. The unifying aim throughout the book is towards applying the new theories with understanding to characterization of surfaces.

To assist the reader in future schemes he may have, a large appendix of computer programs has been supplied. All of these have been tested and reproduced directly from machine output, to minimize inaccuracies. The same programs can also find application in other fields, such as band structure and surface-state calculations.

Some acknowledgments are in order. Volker Heine drew my attention to LEED theory, guided my first steps, and has taken a continuing interest in the field. David Titterington has been responsible for a large part of the programs in the appendix, enduring with fortitude the pesterings of an impatient author. Downing College, Cambridge, U.K., supported me as a research fellow during the greater part of the writing of this book; Bell Laboratories, Murray Hill, New Jersey, U.S.A., took over the burden of my support for the final months of writing.

J. B. Pendry

Downing College
Cambridge
1973

Contents

PREFACE v

Chapter 1. INTRODUCTION

1A	Preamble	1
IB	Experimental apparatus	3
IC	Preparing the sample	7
ID	The clean surface	8
IE	The diffraction pattern	11
IF	Intensity measurements	17
IG	Incoherent and inelastic scattering	26

Chapter 2. SCATTERING PROCESSES FOR LOW ENERGY ELECTRONS

IIA	Experimental clues to the nature of electron scattering	31
IIB	The ion-core potential	37
IIC	Ion-core scattering	47
IID	The optical potential	57
IIE	Calculating the optical potential inside a solid	62
IIF	The solid-vacuum interface	70

Chapter 3. PRINCIPLES OF DIFFRACTION AT $T = 0°K$

IIIA	Kinematic theory	75
IIIB	The general case in one dimension-normal modes	84
IIIC	The general case in three dimensions	96
IIID	Bloch waves and character of LEED spectra	105
IIIE	Surface state resonances	109
IIIF	Peak widths	112
IIIG	Sensitivity of spectra	115

Chapter 4. SCHEMES OF CALCULATION

IVA	Introduction	121
IVB	Bloch waves	123
IVC	Multiple scattering within a plane of ion-cores	128
IVD	Kambe's method for planar scattering	135
IVE	Assembling planes into layers	138
IVF	The matrix doubling method	141
IVG	Pseudopotential methods for calculating Bloch waves	143
IVH	Simplifications brought about by symmetry	147
IVI	Critical review	151

Chapter 5. PERTURBATIVE METHODS AND RELATED TECHNIQUES
VA Introduction 153
VB Perturbation theory and planar scattering 157
VC Perturbation schemes for interplanar scattering 162
VD Renormalised forward scattering perturbation theory 168
VE Kinematic theory and the averaging postulate 174
VF Theory of averaged LEED data 178

Chapter 6. TEMPERATURE EFFECTS
VIA The experimental situation 186
VIB Lattice vibrations 193
VIC Theoretical treatment 201
VID Relationship of theory with experiment 207

Chapter 7. APPLICATIONS OF LEED TO SURFACE STRUCTURE ANALYSIS
VIIA Surface structures and LEED experiments 220
VIIB Formal description of surface structure 228
VIIC Scattering processes in the presence of overlayers 236
VIID Averaging methods and determination of structure using kinematic theory 240
VIIE Extension of dynamical methods to overlayers 245
VIIF Efficient calculation of intra-layer scattering 249
VIIG Dynamical methods and their sensitivity 251
VIIH Some determinations of surface structures 260

Chapter 8. APPENDIX
A Definitions of special functions 268
B Evaluation of the integral for kinematic scattering 273
C LEED computer programs 275

REFERENCES 395

AUTHORS INDEX 400

SUBJECT INDEX 403

Chapter 1

Introduction

IA Preamble

The early origins of low energy electron diffraction (LEED) are bound up with the development of atomic physics and quantum theory. Key developments came in 1897 when J. J. Thomson discovered beams of electrons, measuring e/m, and in 1927 when Davisson and Germer performed the first electron diffraction experiment, using an apparatus that was the prototype for the modern LEED system, to demonstrate the wavelike properties of electrons. Details of these early developments can be found in Ritchmeyer, Kennard and Lauritsen's (1955) book on atomic physics.

After the auspicious start that Davisson and Germer's experiment gave to the subject, interest began to move away from LEED. The greater ease of generating and controlling higher energy beams was partly responsible for this shift. Also there was the question of small penetration of the sample by low energy electrons (usually defined to have energies between 0 and 1000 eV) making experiments on transmission through even the thinnest films impossible. In reflection only a small fraction of the incident beam goes towards producing a diffraction pattern, raising problems of sensitivity. The state of the immediate surface region was found to modify strongly the diffraction process. In an era when high vacuum techniques were not available this meant that results reflected either the condition of an oxide layer on the sample or some other covering of foreign atoms. High energy electron diffraction with its greater powers of penetration removed these difficulties, and at the same time simplifications brought about in the effective scattering power of atoms by the high incident energies made theory much more tractable. At this stage high energy diffraction proved much the more interesting tool though some workers, notably H. E. Farnsworth, did continue to develop LEED techniques.

In the early 1960s interest in LEED revived. Development in techniques had by that time made it possible to control the composition of surfaces by cleaning with ion bombardment and flashing to high temperatures. Ultra high vacuum systems were available to reduce the background gases to a level where deposition of a monolayer of foreign atoms took hours rather than seconds. The post acceleration grid enabled weak beams reflected from the surface to be given enough energy to cause the screen on which they impinge to fluoresce efficiently, thus increasing sensitivity.

One of the original disadvantages was the cause of the revival: sensitivity to the immediate surface region. Few techniques are available for examining the immediate surface region in high resolution. Field ion microscopy can do this, but the high fields required interfere with many of the sensitive processes occurring at surfaces. By comparison the disturbing influence of LEED is very small. There follow some examples of areas in which LEED can make a contribution.

Even perfect, clean surfaces are not yet well understood from a theoretical standpoint. Part of the reason is the shortage of data about surfaces against which theories can be tested. For example the decay of the crystal potential in the region of transition between crystal and vacuum is a difficult quantity to calculate, but details of its structure can be probed by LEED. Even the precise geometry of the surface atoms is not known. It is believed from theoretical calculations that the surface layer dilates outwards by of the order of 5% of the bulk inter-layer spacing: a result that remains to be confirmed experimentally.

Then there is the whole question of surface dynamics. Bulk vibrations are understood in some detail thanks to neutron scattering experiments and pseudopotential calculations. Some theoretical work has been done for surfaces, but again there is a shortage of measurements with which to make comparison. LEED is sensitive to vibrations of atoms in a manner similar to X-ray diffraction where intensities are reduced by a Debye-Waller factor (James, 1962) and though interpretation is more complicated, intensity/temperature data can be used to give information about surface vibrations.

Some clean surfaces may not be 'perfect'. Semiconductor surfaces can undergo rearrangement and have a structure different from that obtained by truncation of a bulk crystal. Such a change in structure can influence surface states occurring on these materials, and other properties. Since many semiconductor devices depend on the surface region this is another cause of interest in LEED experiments.

The first few atomic layers of solids, constituting perhaps only one part in 10^6 of the mass of a macroscopic sample, play a disproportionate role in chemistry, especially the chemistry of catalysis. The reactants are concentrated there by adsorbtion and by interaction with the surface move to configurations favourable to reaction. An understanding of the processes involves knowing activation energies, and configurations involved. Theoretical approaches are complicated and must necessarily rely on certain parameters being measured directly. Trying to calculate activation energies for catalytic processes without detailed knowledge of where the atoms sit is like trying to calculate the band structure of a solid without knowing the size and contents of the unit cell. Corrosion is another field in which similar considerations hold, at any rate in the initial stages.

Much has been achieved already in some of these fields. Diffraction *patterns* as opposed to *intensities* are open to simple interpretation and give the size and shape of unit surface cell, in directions parallel to the surface. Contents of the unit cell, and spacings between overlayer and substrate are more difficult. Sometimes progress can be made from considerations of chemical bonding between adsorbate atoms and the substrate, bond lengths, and sheer physical size, but these estimates are crude, and sometimes either fail or give ambiguous answers. To extract such information by LEED it is necessary to measure diffracted intensities as functions of incident energy or angle.

A detailed theory of the diffraction process is required. What we need to be able to do is to postulate a model for the surface with atomic positions, vibrational amplitudes etc. specified, to calculate resulting LEED intensities and compare them with experiment. By a process of refinement detailed properties of the surface are to be determined. This is the goal towards which we shall be working.

The book sets out to describe the physical origin of diffracted beams and to provide the means to calculate them. It is hoped to provide an impetus to the interaction of theoretical and experimental LEED in order to further investigation of surface properties. To this practical end an appendix of computer programs has been provided as part of the book. So our aim is the practical one of setting LEED to work to unravel the surface problem.

IB Experimental apparatus

Figure 1.1 shows the basic arrangement of a LEED experiment.

The electrons are produced in a well collimated monoenergetic beam by an electron gun. Inside this gun is a heated cathode acting as an electron source. Temperatures are of the order of 2500°K if the cathode is made of tungsten, but more usually a bariated material is used that can give the currents required at a lower temperature of about 1000°K (Somorjai and Farrell,

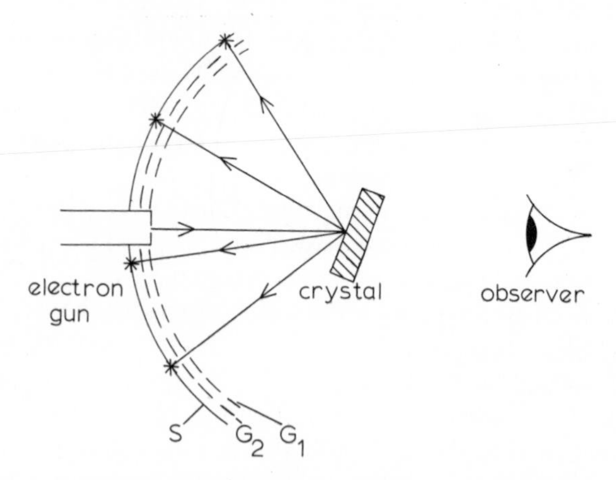

FIGURE 1.1. A schematic LEED experiment. S is a fluorescent screen; G_1 and G_2 are grids.

1971). It is this temperature that determines the basic energy resolution of the equipment because no energy filtering is applied to the emitted electrons. They have a thermal spread of energies of the order of $\frac{3}{2}$ kT or ± 0.1 eV at 1000°K and ± 0.3 eV at 2500°K.

Figure 1.2 shows how electrons leave the cathode. A grid that can be negatively biased by up to 30 volts controls current from the gun. There follow three anodes. At higher beam energies A1 and A3 are biased to the final energy of the electron beam whilst A3 takes a potential somewhere between that of A1 and the cathode, but adjustable so as to focus the electrons coming out of the gun. At low energies A2 and A3 are connected together, A1 being used for focusing. Using this configuration the gun can be made to deliver more current. Beams are parallel to within better than 1°.

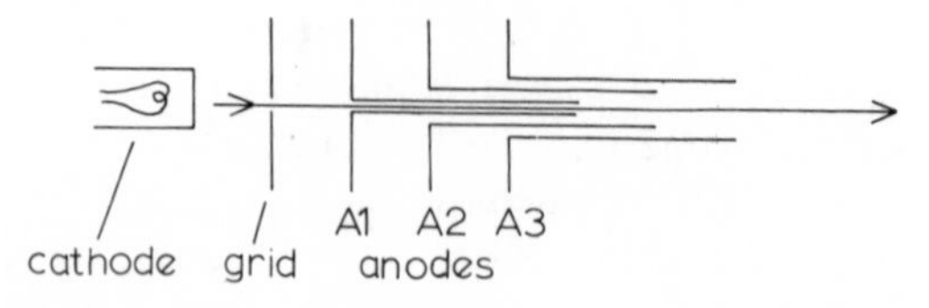

FIGURE 1.2. Schematic cross-section of an electron gun used for LEED experiments, taken through the axis of cylindrical symmetry.

The electron gun has dimensions of the order of 10 cms in length by 2 cms in diameter. The beam is about 0.1 cms in diameter and carries a current of the order 1 μA but both quantities vary with the energy range in which the gun is operated. More details can be had in an article by Heppel (1967).

Although in Fig. 1.1 the beam from the gun is shown as a thin pencil of electrons, on the microscopic scale on which diffraction operates a more appropriate picture is of an infinitely wide beam incident on the surface. An ideal beam of this nature would be described by a plane wave having amplitude at point **r**

$$A \exp(i\mathbf{k}\cdot\mathbf{r}), \tag{1.1}$$

which has a well defined energy, $\hbar^2|\mathbf{k}|^2/2m_e$ and a well defined direction given by the wave-vector **k**. In practice, as we have seen, neither the direction nor the energy are defined precisely and the beam really consists of a whole set of waves having slightly different directions and energies. The result is that whereas a simple beam described by expression (1.1) has a definite phase at all parts of the surface on which it is incident, our more complicated beam varies in phase in a manner that is not entirely predictable because of uncertainties in **k**. For variations of phase over a short distance on the crystal surface, these uncertainties are not important but over larger distances they become so. There is a characteristic length called the coherence length such that atoms in the surface within a coherence length of one another can be thought of as illuminated by the simple wave described by (1.1). Atoms further apart must be regarded as illuminated by waves whose phase-relationship is arbitrary. For waves scattered from points within the coherence length *amplitudes* are to be added, but waves scattered from points separated by more than a coherence length add in intensity because of the arbitrary phase factor. Thus no surface structure on a scale larger than the coherence length forms a diffraction pattern (Lander, 1965).

If E is the average energy of a beam, wave-vector **k**, variations in **k** parallel to **k** itself are given by

$$|\mathbf{k}|^{-2}\overline{(\Delta\mathbf{k}\cdot\mathbf{k})^2} = |\mathbf{k}|^2\overline{(\Delta E)^2}/(4E^2) \tag{1.2}$$

since

$$\hbar^2|\mathbf{k}|^2 = 2m_e E,$$

where $\overline{(\Delta E)^2}$ is the mean square spread in energy. Variations in **k** perpendicular to **k** are given by

$$|\mathbf{k}|^{-2}\,\overline{|\Delta\mathbf{k}\times\mathbf{k}|^2} = \overline{(\Delta\theta)^2}\,|\mathbf{k}|^2 \tag{1.3}$$

$\overline{(\Delta\theta)^2}$ being the mean square angular spread of the beam.

Two points in the surface on which the beam is incident, separated by distance ℓ, differ in phase on average by $\ell\cdot\mathbf{k}$ and the mean square deviation from this phase, $\overline{(\Delta\phi)^2}$, has contributions from both the angular and energy spread of the beam:

$$\overline{(\Delta\phi)^2} = \overline{(\Delta\theta)^2}\tfrac{1}{2}|\mathbf{k}|^2|\ell|^2\sin^2(\alpha) + \overline{(\Delta E)^2}(2E)^{-2}|\mathbf{k}|^2|\ell|^2\cos^2(\alpha). \tag{1.4}$$

α is the angle between ℓ and **k**. When the mean square error in phase becomes of the order of π^2, the phase relationship between the two points has vanished. Thus the coherence length is

$$\ell_c = \frac{2\pi|\mathbf{k}|^{-1}}{[2\sin^2(\alpha)\overline{(\Delta\theta)^2} + \cos^2(\alpha)\overline{(\Delta E/E)^2}]^{\frac{1}{2}}} \tag{1.5}$$

Choosing $\Delta\theta \simeq 0.001$ radians, $\Delta E \simeq 0.2\,\text{eV}$ at $E = 150\,\text{eV}$ and $\alpha = 45°$, gives

$$\ell_c \simeq 500\text{Å}. \tag{1.6}$$

To continue with our description of the equipment: in most sets of apparatus the crystal is mounted at the centre of a system of hemispherical grids (Fig. 1.1), and there are various arrangements that can be used. The simplest one is considered, where there are two grids. G_1 is connected to the final anode, A3, and to the crystal so that once electrons are inside the region enclosed by G_1 they are in a field-free region. Electrons are diffracted from the surface of the crystal and speed towards the grids. G_2 is normally biased relative to G_1 so that only electrons having lost less than 1–2 eV of their original energy can get through to the far side of G_2. All others are turned back and collected by G_1. Finally the screen is biased with a large accelerating potential of the order of $2k$ eV to give the electrons enough energy to excite the phosphor efficiently. More details can be had in the survey by Estrup (1970).

The diffraction pattern on the screen can be photographed to fix positions of features, or intensities of separate parts of the pattern can be measured with a spot photometer. In some systems a more direct way of measuring intensities is used: a device called a Faraday cup collects diffracted beams in the field-free region and in this way it is possible to obtain measurements of absolute intensities. (Farnsworth, 1964). Figure 1.3 shows a sketch of a Faraday cup.

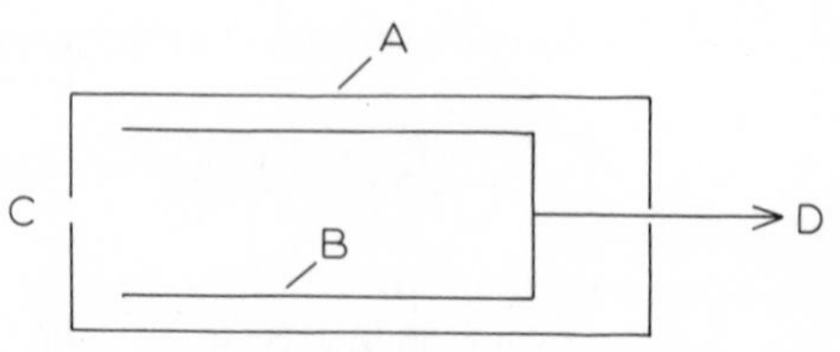

FIGURE 1.3. A Faraday cup. A: outer shield connected to G_1. B: inner conductor to collect electrons entering through a small hole: C. D: lead to a current measuring device.

The Faraday cup is capable of making accurate measurements both of absolute intensities and of angular position of structure which can be fixed to within $\frac{1}{2}°$, but it is much more clumsy to operate than the spot photometer. For this reason the Faraday cup is sometimes used to calibrate the spot photometer, subsequent work being done with the latter device when the less precise angular resolution of around 3° is not of importance.

The assembly must be mounted inside an ultra high vacuum system with pressure of the order of 10^{-9} to 10^{-10} Torr. Only at these pressures can the crystal surface be preserved from contamination by residual gases. Assuming a sticking probability of unity for molecules incident on the crystal surface, the kinetic theory of gases can be used to show that it takes of the order of a day to develop a monolayer of adsorbed gas in a vacuum of this quality.

The cleaning problem is a crucial one. Two facilities are provided for cleaning inside the chamber. The specimen can be heated or 'flashed' to a high temperature just below its melting point, as a means of removing impurities. Refractory materials are naturally cleaned more successfully by this technique. A mass spectrometer can be used to analyse gases flashed from the surface, giving an idea of impurities present.

In addition a source of 500 eV argon ions can be used to bombard the surface, providing a scouring action. In particularly intransigent cases the cleaning problem can sometimes be sidestepped by cleaving the specimen in high vacuum using a bellows-mounted chisel.

On the other hand, sometimes it is desired to introduce adsorbates deliberately. Usually this is done by providing ports in the vacuum chamber through which samples can be introduced in the gaseous form to give a small pressure in the chamber and hence at the crystal surface. Materials such as sodium with low partial pressures at room temperature can be introduced onto the crystal surface by generating a beam of ions directed at the surface.

Various other facilities are provided. The sample can always be rotated about at least one axis, to change angles of incidence and to turn the crystal to face one or other of the various guns in the chamber. The crystal itself can be heated, as already mentioned, and by mounting a thermocouple effects of temperature on diffraction can be investigated. Less common is provision of cooling for the specimen either by liquid nitrogen ($\simeq 72°$K) or by liquid helium ($< 10°$K).

IC Preparing the sample

The crystal surface on which the experiments are to be done, is usually cut outside the chamber, either by spark cutting or by mechanical grinding for harder materials. Orientation of the crystal is achieved by X-ray methods and is usually accurate to within 1°. Mechanical polishing with carbide powders and chemical etching follows, to make the surface optically smooth.

A mirror finish on the surface guarantees smoothness only on a scale of the order of the wavelength of light, say greater than 1000Å, which is not even less than the coherence length of the incident beam! When the crystal goes into the diffraction chamber it almost certainly has much surface damage on a scale comparable with the coherence length, and several layers of foreign molecules—oxygen, water etc.—adsorbed on the surface.

We have described how ion bombardment is used to strip away layers from the surface, followed by flashing to high temperatures to remove impurities, any of the bombarding ions embedded in the surface, and more gross mechanical imperfections. Finally the crystal is annealed to allow the atoms to arrange themselves into a well ordered surface.

During the process of preparation, it is important to be able to monitor the condition of the surface. In early experiments the diffraction pattern was used, and is useful for detecting mechanical imperfection. Disorder, faceting, impurities that form a new structure, can all be detected by the diffraction pattern but impurities that adsorb without changing the basic symmetry of the surface, or go down in a disordered manner, are not so easily detected in this way. The next sophistication is to monitor intensities of diffracted beams. The surface is cleaned until intensities become stable. Here the assumption is made that the cleaning process is having an impact on impurities, otherwise the process could be producing a constantly dirty surface.

It could happen that impurities diffuse to the surface from the bulk—an effectively infinite source. Thus repeating the cleaning cycle would not change the pattern. Difficulties of this nature require some direct monitoring of surface impurities and most modern equipment has an Auger facility which can detect and identify impurities on the surface. By using both the diffraction pattern, and the Auger facility it is possible to do experiments in which one is reasonably certain of an atomically clean surface, sufficiently well ordered that more than 90% of the surface contributes to the diffraction process.

ID The clean surface

Almost all specimens on which LEED studies are made have two-dimensional symmetry parallel to the surface. The more perfectly the specimen is prepared, the more perfect will be the symmetry. In many substances the symmetry is given simply by that of the perfect bulk crystal in a plane parallel to the surface we consider. In others, rearrangement of the surface layer can take place to form a structure with less symmetry than expected if the surface were formed by truncation of a perfect bulk crystal at a plane (it cannot have more symmetry than this).

Surfaces are usually named after the planes of the bulk crystal to which they are parallel, e.g. (100), (110), etc. Their two-dimensional structure can be described in terms of a unit cell, the smallest unit that can be repeated to build up the whole surface. The two sides of the unit cell will be designated by vectors **a** and **b** lying in the plane of the surface, shown in Fig. 1.4.

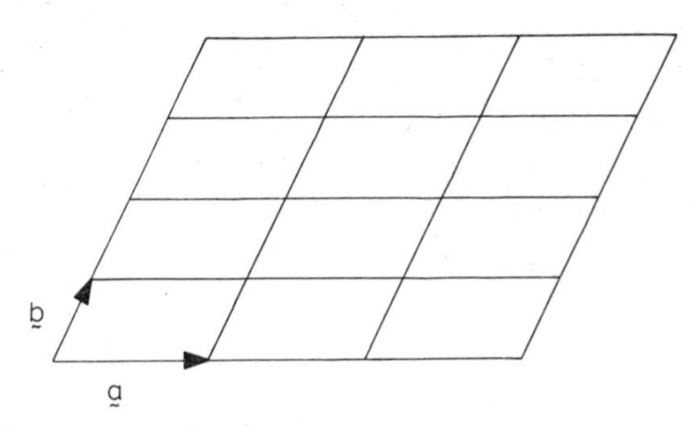

FIGURE 1.4. Two-dimensional array of unit cells.

The origin of any unit cell in Fig. 1.4 can be written

$$\ell\mathbf{a} + m\mathbf{b}, \tag{1.7}$$

relative to the origin of a reference unit cell. ℓ and m are integers. Any lattice of points, the positions of all of which are given by a formula of the type (1.7), and which contains points for all values of ℓ and m is called a Bravais lattice. In two dimensions there are four different kinds of Bravais lattice, or five if we allow the centred rectangular lattice which is sometimes useful for describing the hexagonal lattice in rectangular coordinates. The unit cells of the five lattices are given in Fig. 1.5.

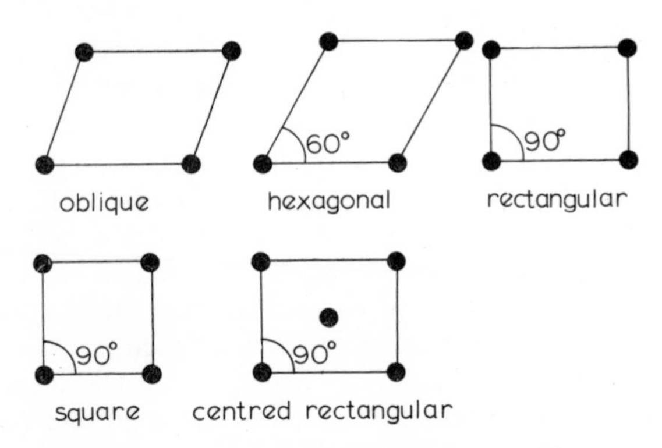

FIGURE 1.5. Five two-dimensional Bravais lattices.

Details of various conventions in surface crystallography, some of which we follow here, can be found in an article by Wood (1964).

The symbols x, y, z will always refer to a set of orthogonal right-handed axes. In this book the x and y axes are always taken to lie in the plane of the surface, and the z axis is the inward-pointing normal to the surface.

Deep inside the surface, the bulk crystal has symmetry in the z direction.

A full account of complexities of three-dimensional crystal structure can be found in volume 1 of the International Tables of X-ray Crystallography (ed. Lonsdale (1959), but for our purposes in LEED theory it suffices to observe that in the bulk, crystal structure can be built up of identical layers of atoms, the plane of each layer parallel to the surface. Each successive layer deeper in the crystal is displaced relative to the previous one by vector **c**. For some crystal surfaces, such as those of graphite or mica parallel to the cleavage planes, this is an obvious decomposition to make, but even in crystals not possessing an obviously layered structure such a decomposition can always be made for any of their surfaces.

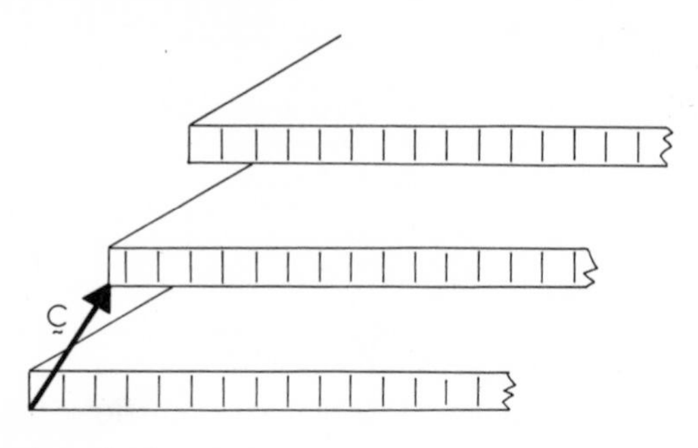

FIGURE 1.6. A crystal can be described as a stack of layers of atoms, the plane of each layer parallel to the surface in question.

Within each layer a further classification can be made into atoms all lying with their centres in the same plane, parallel to the surface.

Layers near the surface and planes of atoms within the layers can relax in their relative spacings in comparison with the bulk. Theoretical calculations by Burton and Jura (1967) indicate a dilation of the order of 10% for the top-most atomic layer. Less common is rearrangement of composition in a plane of atoms near the surface.

The following conventions will be used in connection with the layers. Planes of atoms making up each layer will have their positions relative to some origin in the layer, denoted by $\mathbf{d}_p$. Each plane can be decomposed into unit cells and the displacement of the jth unit cell relative to the one at the origin in the plane is $\mathbf{R}_j$. Coordinates of atoms within a unit cell of the plane relative to the origin of the unit cell are denoted by $\mathbf{r}_k$. $\mathbf{R}_j$ and $\mathbf{r}_k$ are both vectors which are parallel to the surface plane. Thus the position of the kth atom in the jth unit cell of the pth plane of a layer has coordinates given by

$$\mathbf{r}_k + \mathbf{R}_j + \mathbf{d}_p \tag{1.8}$$

relative to the origin of that layer. Figure 1.7 will help to make this rather complicated situation more clear.

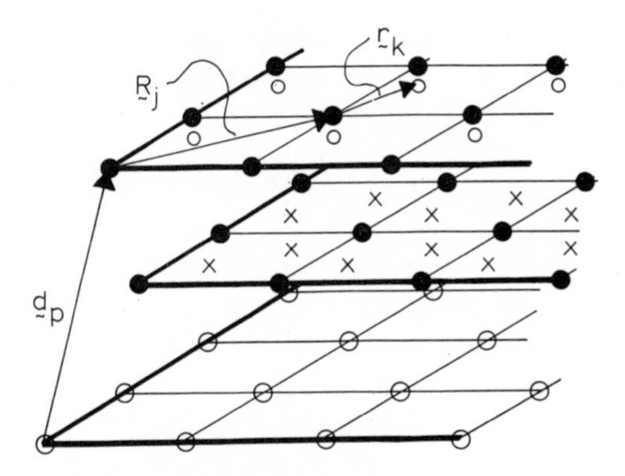

FIGURE 1.7. Structure of a layer composed of three planes of atoms. Atoms are denoted by ●, o or × and some possible values of $\mathbf{d}_p$, $\mathbf{R}_j$, and $\mathbf{r}_k$ are shown.

The orientation of the incident beam of electrons relative to the surface will be described by two angles: θ and ϕ. θ is the angle between the beam of electrons and the surface normal, ϕ the angle between the plane containing the surface normal and incident beam, and the x axis. Figure 1.8 clears up any ambiguities.

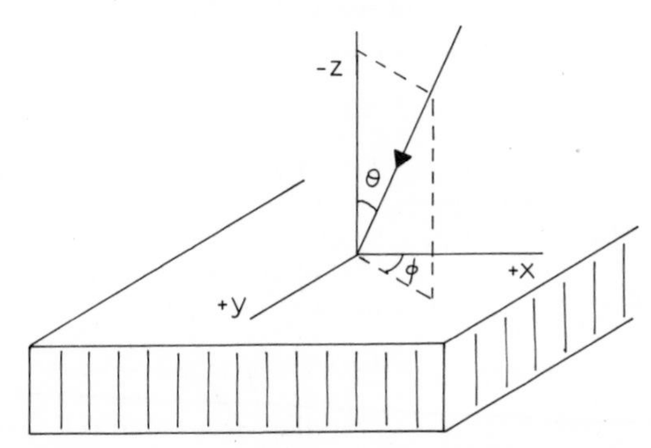

FIGURE 1.8. Conventions concerning the incidence of an electron beam on a surface.

For convenience the pair of angles (θ, ϕ) is often denoted by Ω, and $\Omega(\mathbf{k})$ will be taken to stand for the angular coordinates of a beam with wave-vector $\mathbf{k}$.

IE The diffraction pattern

We shall assume that the experiment has been sufficiently well prepared that we can consider a mono-energetic, collimated beam of electrons incident on a perfectly clean, well ordered surface. Our interest will centre on the elastically scattered electrons, because they produce almost all the structure in the diffraction pattern. The motion of these electrons will be taken to be determined by a Schrödinger equation of the form

$$-\frac{\hbar^2}{2m_e}\nabla^2\psi(\mathbf{r}) + V(\mathbf{r})\psi(\mathbf{r}) = E\psi(\mathbf{r}). \tag{1.9}$$

For convenience atomic units in which

$$\hbar^2 = m_e = e^2 = 1, \tag{1.10}$$

will be used in equations. Units of energy are Hartrees, and

$$1H = 27.2\,\text{eV} \tag{1.11}$$

The Bohr radius is the unit of length

$$1\ B.r.\,(\text{a.u.}) = 0.5292\ \text{Å} \tag{1.12}$$

The Schrödinger equation simplifies to

$$-\tfrac{1}{2}\nabla^2\psi + V\psi = E\psi. \tag{1.13}$$

Atomic units will be employed for formulae and mathematical working, but when discussing quantities that compare directly with experiment electron volts and Angströms will sometimes be used.

The incident electron has a wavefunction

$$\psi_i = B\exp(i\mathbf{k}_0\cdot\mathbf{r}), \tag{1.14}$$

$$\tfrac{1}{2}|\mathbf{k}_0|^2 = E. \tag{1.15}$$

We want to discover what properties the scattered electron's wavefunction, ψ_s, has in consequence of the surface symmetry. We know that the Schrödinger equation must be obeyed by the total wavefield:

$$-\tfrac{1}{2}\nabla^2[\psi_i(\mathbf{r}) + \psi_s(\mathbf{r})] + V(\mathbf{r})[\psi_i(\mathbf{r}) + \psi_s(\mathbf{r})] = E[\psi_i(\mathbf{r}) + \psi_s(\mathbf{r})]. \tag{1.16}$$

The surface has two-dimensional periodicity such that translating the whole crystal through an integral number of cell sides, $(\ell\mathbf{a} + m\mathbf{b})$, merely gives the same positions of unit cells in Fig. 1.4. Since the potential acting on the electron is due to the crystal, the same arguments of symmetry apply to V,

$$V(\mathbf{r} + \ell\mathbf{a} + m\mathbf{b}) = V(\mathbf{r}). \tag{1.17}$$

In equation (1.16) we can substitute for

$$\mathbf{r} = \mathbf{r}' + \ell\mathbf{a} + m\mathbf{b} \tag{1.18}$$

then make use of (1.17) and the fact that

$$\nabla^2\psi = \frac{\partial^2\psi}{\partial x^2} + \frac{\partial^2\psi}{\partial y^2} + \frac{\partial^2\psi}{\partial z^2}$$

$$= \frac{\partial^2\psi}{\partial x'^2} + \frac{\partial^2\psi}{\partial y'^2} + \frac{\partial^2\psi}{\partial z'^2} = \nabla'^2\psi, \tag{1.19}$$

(easily deduced from (1.18)) to write

$$\begin{aligned} &-\tfrac{1}{2}\nabla'^2[\psi_i(\mathbf{r}' + \ell\mathbf{a} + m\mathbf{b}) + \psi_s(\mathbf{r}' + \ell\mathbf{a} + m\mathbf{b})] \\ &+ V(\mathbf{r}')[\psi_i(\mathbf{r}' + \ell\mathbf{a} + m\mathbf{b}) + \psi_s(\mathbf{r}' + \ell\mathbf{a} + m\mathbf{b})] \\ &= E[\psi_i(\mathbf{r}' + \ell\mathbf{a} + m\mathbf{b}) + \psi_s(\mathbf{r}' + \ell\mathbf{a} + m\mathbf{b})]. \end{aligned} \tag{1.20}$$

In other words, $\psi(\mathbf{r}' + \ell\mathbf{a} + m\mathbf{b})$ obeys the same equation as $\psi(\mathbf{r})$. $\psi(\mathbf{r}' + \ell\mathbf{a} + m\mathbf{b})$ is very similar to $\psi(\mathbf{r})$ in the incident wave component except that instead of being given by (1.14) it is now

$$\begin{aligned} \psi_i(\mathbf{r}' + \ell\mathbf{a} + m\mathbf{b}) &= B\exp(i\mathbf{k}_0\cdot\mathbf{r}') \\ &\times \exp(i\ell\mathbf{k}_{0\parallel}\cdot\mathbf{a} + im\mathbf{k}_{0\parallel}\cdot\mathbf{b}) \\ &= \psi_i(\mathbf{r}')\exp(i\ell\mathbf{k}_{0\parallel}\cdot\mathbf{a} + im\mathbf{k}_{0\parallel}\cdot\mathbf{b}). \end{aligned} \tag{1.21}$$

$\mathbf{k}_{0\parallel}$ denotes the component of $\mathbf{k}_0$ parallel to the surface. In this new situation we are beaming onto the crystal an incident wave the same as before except that its amplitude differs by a phase factor. Now the amplitude of the reflected wave is always proportional to that of the incident wave, therefore the new scattered wave must differ from the old by the same factor as the new incident wave differs from the old.

$$\psi_s(\mathbf{r}' + \ell\mathbf{a} + m\mathbf{b}) = \psi_s(\mathbf{r}')\exp(i\ell\mathbf{k}_{0\parallel}\cdot\mathbf{a} + im\mathbf{k}_{0\parallel}\cdot\mathbf{b}). \tag{1.22}$$

Equation (1.22) shows that the scattered wave obeys what is known as a Bloch theorem.

ψ_s itself does not have the periodicity of the crystal. Rather, it obeys a more complicated law of translation. But by writing

$$\psi_s(\mathbf{r}) = \exp(i\mathbf{k}_{0\parallel}\cdot\mathbf{r}_\parallel)\chi_s(\mathbf{r}), \tag{1.23}$$

and substituting into equation (1.22), we can show that the function $\chi_s(\mathbf{r})$ does have the same simple periodic properties as the potential.

$$\chi_s(\mathbf{r} + \ell\mathbf{a} + m\mathbf{b}) = \chi_s(\mathbf{r}). \tag{1.24}$$

If we wish $\chi_s(\mathbf{r})$ can in virtue of (1.24) be written as a Fourier expansion

$$\chi_s(\mathbf{r}) = \sum_{\mathbf{g}} \alpha_{\mathbf{g}}(z)\exp(i\mathbf{g}\cdot\mathbf{r}_\parallel). \tag{1.25}$$

The quantities $\mathbf{g}$ are two dimensional vectors having only x and y components.

They have the property, in consequence of (1.24), that

$$\mathbf{g}\cdot\mathbf{a} = \text{integer} \times 2\pi, \tag{1.26a}$$

$$\mathbf{g}\cdot\mathbf{b} = \text{integer} \times 2\pi. \tag{1.26b}$$

a pair of equations that can be satisfied automatically by the following form of

$$\mathbf{g} = h\mathbf{A} + k\mathbf{B}, \tag{1.27}$$

$$\mathbf{A} = (A_x, A_y) = \frac{2\pi}{(a_x b_y - b_x a_y)}(b_y, -b_x), \tag{1.28a}$$

$$\mathbf{B} = (B_x, B_y) = \frac{2\pi}{(a_x b_y - b_x a_y)}(-a_y, a_x). \tag{1.28b}$$

h and k are integers.

Allowed values of $\mathbf{g}$ fall on the points of a Bravais lattice. This lattice is called the reciprocal lattice of the surface defined by $\mathbf{a}$ and $\mathbf{b}$. At this stage it appears as an abstract construction but we shall soon see that it has a very direct interpretation in LEED experiments. Figure 1.9 shows the unit cells of some real Bravais lattices and the unit cells of corresponding reciprocal lattices.

To return to the scattered wave: since χ_s can be expanded in a fourier series, a series expansion of ψ_s can be made from equation (1.23),

$$\psi_s(\mathbf{r}) = \sum_{\mathbf{g}} \alpha_{\mathbf{g}}(z)\exp[i(\mathbf{g} + \mathbf{k}_{0\parallel})\cdot\mathbf{r}_{\parallel}]. \tag{1.29}$$

Equation (1.29) holds quite generally, but outside the crystal an even simpler expression can be found, because we know that there the potential is zero. Substituting for the total wave-function, $\psi_s + \psi_i$, into equation (1.13) and rearranging, making use of equation (1.12),

$$\begin{aligned} &[|\mathbf{k}_0|^2 - 2E]B\exp(i\mathbf{k}_0\cdot\mathbf{r}) \\ &\quad + \sum_{\mathbf{g}}\left[(|\mathbf{k}_{0\parallel} + \mathbf{g}|^2 - 2E)\alpha_{\mathbf{g}} + \frac{d^2\alpha_{\mathbf{g}}}{dz^2}\right]\exp[i(\mathbf{k}_{0\parallel} + \mathbf{g})\cdot\mathbf{r}_{\parallel}] \\ &= 0 = \sum_{\mathbf{g}}\left[(|\mathbf{k}_{0\parallel} + \mathbf{g}|^2 - 2E)\alpha_{\mathbf{g}} + \frac{d^2\alpha_{\mathbf{g}}}{dz^2}\right]\exp[i(\mathbf{k}_{0\parallel} + \mathbf{g})\cdot\mathbf{r}_{\parallel}]. \end{aligned} \tag{1.30}$$

As it stands equation (1.30) is a sum over Fourier components, but by multiplying by $\exp[-i(\mathbf{k}_{0\parallel} + \mathbf{g}')\cdot\mathbf{r}_{\parallel}]$ and integrating over unit two dimen-

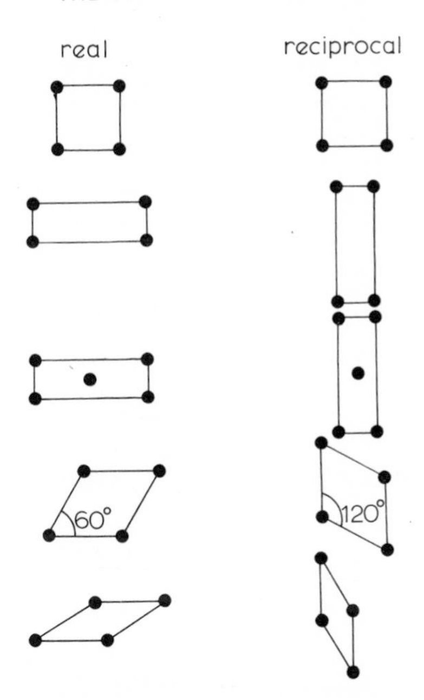

FIGURE 1.9. Unit cells of corresponding real and reciprocal two-dimensional lattices.

sional cell, it is easy to show that each Fourier component must obey its own separate equation

$$[|\mathbf{k}_{0\parallel} + \mathbf{g}|^2 - 2E]\alpha_{\mathbf{g}} + \frac{d^2\alpha_{\mathbf{g}}}{dz^2} = 0 \tag{1.31}$$

thus

$$\alpha_{\mathbf{g}}(z) = \beta_{\mathbf{g}} \exp[\pm i(2E - |\mathbf{k}_{0\parallel} + \mathbf{g}|^2)^{\frac{1}{2}} z], \tag{1.32}$$

where $\beta_{\mathbf{g}}$ is a constant. Since we are dealing with waves that have been scattered from the crystal, the minus sign must be chosen in equation (1.32) so that the waves do in fact travel away from the crystal. The scattered wave outside the crystal can now be written

$$\psi_s(\mathbf{r}) = \sum_{\mathbf{g}} \beta_{\mathbf{g}} \exp(i\mathbf{K}_{\mathbf{g}}^- \cdot \mathbf{r}) \tag{1.33}$$

$$\mathbf{K}_{\mathbf{g}}^- = [\mathbf{k}_{0x} + \mathbf{g}_x, \mathbf{k}_{0y} + \mathbf{g}_y, -(2E - |\mathbf{k}_{0\parallel} + \mathbf{g}|^2)^{\frac{1}{2}}]. \tag{1.34}$$

So it follows from quite general considerations, independent of detailed mechanisms of scattering, that the diffracted wavefield has the form of a series of discrete beams each with a different parallel component of momentum, $(\mathbf{k}_{0\parallel} + \mathbf{g})$. Figure 1.10 illustrates the situation.

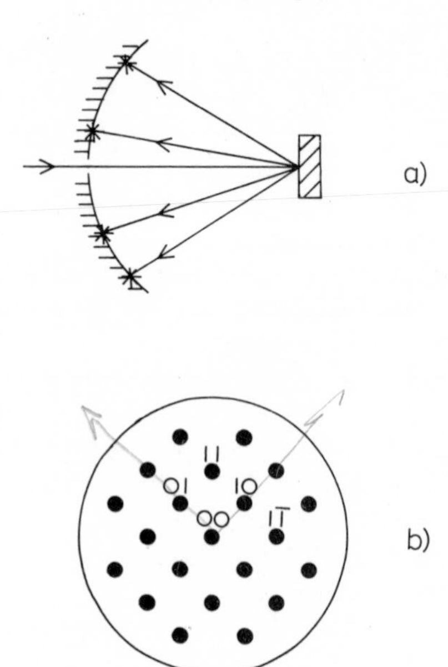

FIGURE 1.10. a) showing the diffraction of an incident beam into a series of discrete beams, b) a plan view of a typical pattern made on the fluorescent screen by these beams.

The directions of the beams are determined by $\mathbf{K}_\mathbf{g}^-$, and hence by $\mathbf{k}_{0\parallel}$, $\mathbf{g}$ and E, see Fig. 1.11.

Only a finite number of beams emerge from the crystal at finite incident energy. As $(\mathbf{k}_{0\parallel} + \mathbf{g})$ becomes larger, so the z component of momentum in equation (1.34) diminishes until beams for a certain value of $(\mathbf{k}_{0\parallel} + \mathbf{g})$ such that

$$|\mathbf{k}_{0\parallel} + \mathbf{g}|^2 \simeq 2E \tag{1.35}$$

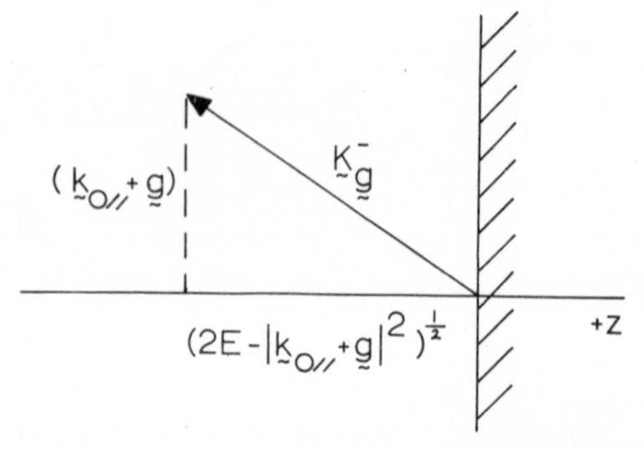

FIGURE 1.11. $(\mathbf{k}_{0\parallel} + \mathbf{g})$ and E determine the direction of a beam.

are travelling almost parallel to the surface. If values of $(\mathbf{k}_{0\parallel} + \mathbf{g})$ even larger than this critical value are chosen the nature of the beam changes as the z-exponent becomes complex implying that the beam becomes evanescent, dying away exponentially in amplitude away from the surface within a few atomic units. Such beams can never be observed on the screen. As the incident energy is increased equation (1.35) tells us that more and more Fourier components produce beams that can reach the screen, and from Fig. 1.11 it is clear that the pattern seen on the screen contracts to allow room for the larger number of spots. Figure 1.12 shows photographs taken of a diffraction pattern at increasing incident energies.

For a given $\mathbf{k}_{0\parallel}$ and E, $\mathbf{g}$ determines the diffraction pattern: one spot on the screen for every value of $\mathbf{g}$. In fact, by making the screen spherical, and placing the crystal at the centre of the sphere, the spot pattern provides us with a direct picture of the reciprocal lattice, or at any rate that part of it producing beams that emerge from the crystal. Use is made of this fact in naming the spots. The spot produced by a beam with parallel component of momentum $(\mathbf{k}_{0\parallel} + \mathbf{g})$, where

$$\mathbf{g} = h\mathbf{A} + k\mathbf{B}$$

as in equation (1.27), is referred to as the (hk) spot. Figure 1.10b shows this labelling applied to some spots in a pattern.

The diffraction pattern gives us the reciprocal lattice of the surface, and hence **A** and **B**, the sides of the unit cell of that lattice. Knowing **A** and **B** it is a simple matter to invert equations (1.28) to get the unit cell of the real lattice:

$$\mathbf{a} = (a_x, a_y) = \frac{2\pi}{A_x B_y - B_x A_y}(B_y, -B_x), \tag{1.36a}$$

$$\mathbf{b} = (b_x, b_y) = \frac{2\pi}{A_x B_y - B_x A_y}(-A_y, A_x). \tag{1.36b}$$

In this way much structural information about surfaces has been found. Chapter VII will investigate this problem in greater depth.

IF Intensity measurements

Locations of beams in the elastic diffraction pattern give the unit surface cell of the crystal, but no information about disposition of contents of the cell, such as separation between planes of atoms near the surface or location of adsorbates relative to the substrate. This information is contained in intensities of elastically scattered beams measured as functions of energy or angle of incidence. Intensity/energy curves are the usual form for presenting data.

a

b

FIGURE 1.12. Diffraction patterns taken by Huang and Estrup (1973) from a molybdenum (001) surface with an incident energy of (a) 32 eV (b) 62 eV (c) 76 eV (d) 100 eV (e) 167 eV (f) 267 eV.

d

c

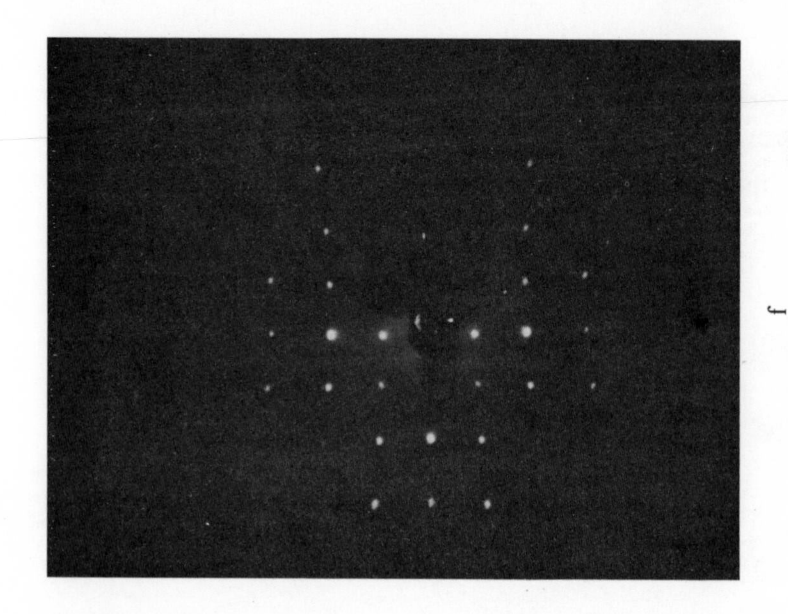

f

e

We reproduce in Fig. 1.13 the intensity as a function of energy of the specularly reflected (00) beam from a copper (001) surface, under conditions of near-normal incidence. The measurements are due to Andersson (1969). The first point to notice is that there is a great deal of structure in the curves, hence potentially much information.

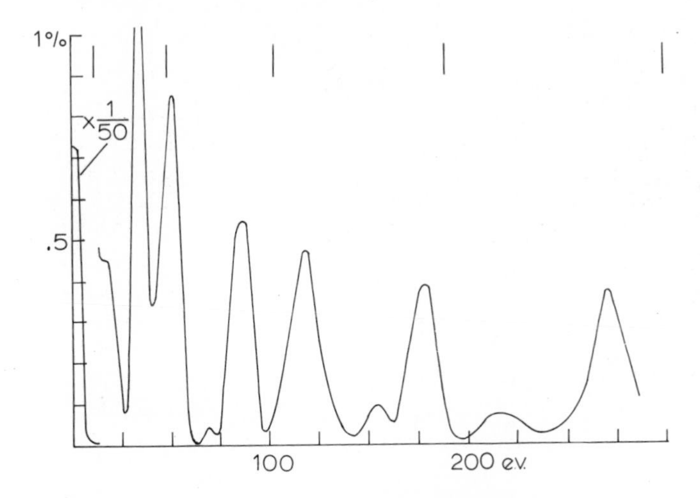

FIGURE 1.13. Intensity of the 00 beam from a copper (001) surface, measured as a function of energy. The crystal has been rotated about the x-axis by 3° to enable the 00 beam to be measured.

In discussing the curves it is convenient to divide energies into three ranges. Below 15 eV intensities are commonly in the 10–50% range at peak values. Structure in the curves is at its most detailed in this range and though it happens that no such structure appears in Fig. 1.13, peaks only 1–2 eV wide are commonly observed.

At intermediate energies between, say 15 eV, and 150 eV though with no sharp division at the upper end, intensities drop markedly to of the order of 1%. At the same time structure in the curves is less fine, being on a scale of 5–10 eV, but still leaving the curves quite complicated.

At high energies above 150 eV reflectivities fall even lower to an extent that depends much on the material. Structure at high energies appears as widely spaced peaks of irregular shape, rather than the jumble familiar at lower energies. Peak widths in this range are typically 20–30 eV.

Some more intensity energy curves are shown in Fig. 1.14. They are for graphite and were measured by McRae and Caldwell (1967). A logarithmic plot of intensities makes peaks appear less prominent than for copper, but the same general picture holds, of decreasing intensities and broadening structure with increasing energy. Variation of intensity with angle can be seen in this

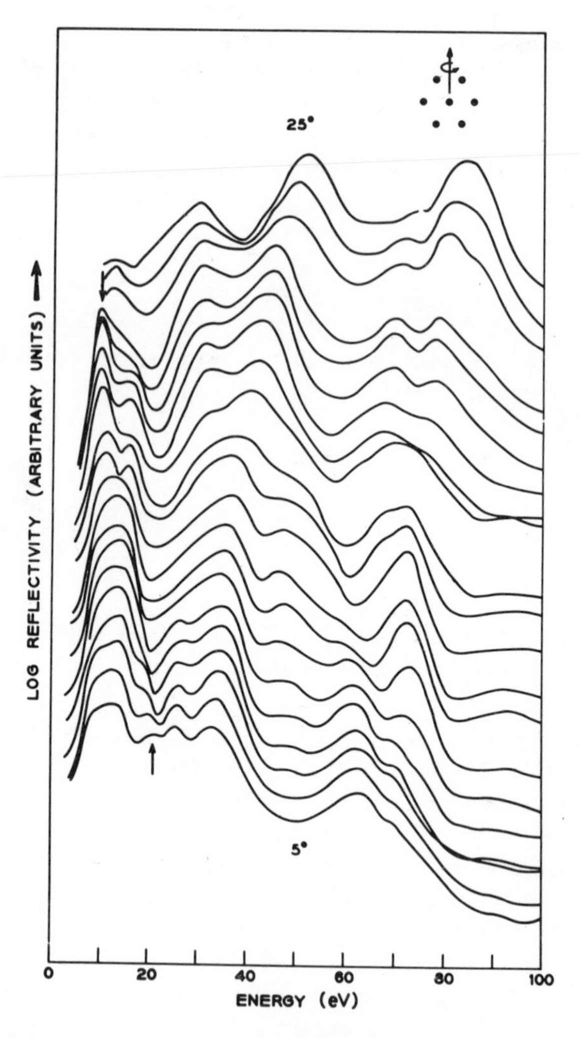

FIGURE 1.14. Intensity in arbitrary units of the 00 beam from a graphite (0001) surface measured as a function of energy. The curves are for various angles of incidence in 1° steps starting with 5° for the lowest curve. The axis of rotation of the crystal is indicated relative to the diffraction pattern at normal incidence in the figure inset. Successive curves are displaced upwards for clarity.

figure and implies that angles of incidence should be defined within 1° for accurate work with intensities.

A simple theory of peak formation runs as follows: although the electrons do not penetrate deeply into the crystal, we might suppose, especially at high energies, that they do go sufficiently deep to be sensible of the periodic structure of the crystal in the z-direction. In the case of X-rays, which penetrate much more efficiently, peaks in reflectivities are determined by conditions for Bragg reflection off planes of atoms. If we try to apply Bragg theory to LEED we deduce that peaks in the 00 beam are caused by reflection from planes of atoms parallel to the surface. If the spacing between these planes is c, the Bragg condition for a peak is

$$\frac{2\pi}{\lambda} = \text{integer} \times c, \tag{1.37}$$

ensuring that reflections from successive planes are in phase. λ, the wavelength of the electrons, will not be quite the same as it is in free space because of the 'inner potential' well inside the crystal: U_0,

$$\lambda = 2\pi(2E - U_0)^{-\frac{1}{2}}. \tag{1.38}$$

E, U_0 and λ are in atomic units. Substituting (1.38) into (1.37)

$$E = \tfrac{1}{2}c^2(\text{integer})^2 + U_0. \tag{1.39}$$

The bars on Fig. 1.13 mark energies at which 1.36 is satisfied for $U_0 = 0$. In the high energy range there is reasonable correlation with Bragg theory but in intermediate and low energy ranges there are more experimental peaks than Bragg theory predicts. Peaks that coincide most closely with the Bragg condition are called 'primary' peaks and all others 'secondary' peaks, but sometimes the complications are such that even this classification can be ambiguous.

Provided that peaks can be assigned to particular Bragg reflections, then plotting peak energies against (integer)^2 gives values for c and for U_0. This has been done by Andersson for his copper results and we reproduce this graph in Fig. 1.15. Values of inter-planar spacing, c, and inner potential, U_0, are calculated to be

$$c = 3.42 \pm 0.02 \text{ a.u.}, U_0 = -13.5 \text{ eV}. \tag{1.40}$$

The X-ray value of c is 3.41 a.u.

So by using this rather unsophisticated theory we have been able to extract from LEED intensities some structural information not available from the diffraction pattern. However we shall see that to obtain really useful information requires a more complete understanding of the diffraction process.

We turn now from peak positions to their shape. Finite penetration of the beam is implied by the width of primary peaks in intensity/energy curves. If only a finite number, N, of planes take part in the Bragg reflection process

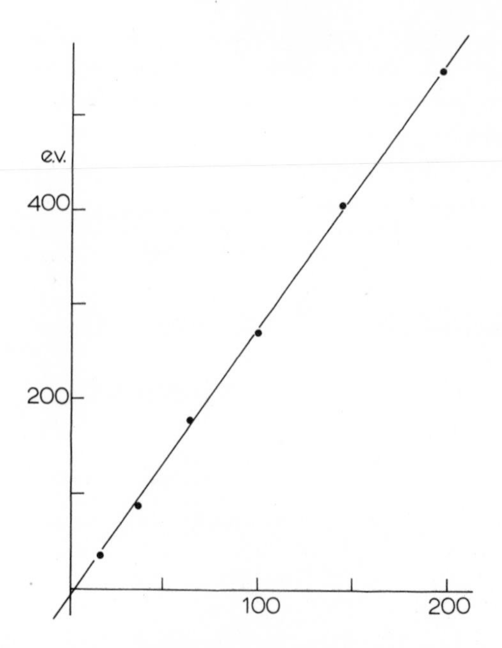

FIGURE 1.15. A plot of peak energies against square of peak order for the copper (001) results of Fig. 1.13. The E intercept gives U_0 and the slope the inter-planar spacing.

then from analogy with optical interference calculations, the peaks have widths in terms of wavelength

$$\Delta\left(\frac{2\pi}{\lambda}\right) \simeq \frac{\pi}{2Nc}, \tag{1.41}$$

implying via (1.38) an energy width of approximately

$$\Delta E \simeq \pi(2E)^{\frac{1}{2}}/(2Nc). \tag{1.42}$$

Substituting

$$\Delta E = 10\,\text{eV} = 0.37\ \text{Hartrees},$$

and

$$E = 100\,\text{eV} = 3.70\ \text{Hartrees}, \tag{1.43}$$

gives

$$Nc \simeq 12\ \text{a.u.} \simeq 6\text{Å}, \tag{1.44}$$

or $\simeq$ 4 atomic layers for most materials.

Other experimental estimates can be made. The observation that LEED patterns from the (111) surface of cubic close packed crystals show 3-fold rather than 6-fold symmetry implies that at least three layers must contribute to the patterns because the first two layers taken together have 6-fold symmetry. Palmberg and Rhodin (1968) have shown that when gold is grown epitaxially on a copper surface, the diffracted intensities closely approximate those obtained from a pure gold surface after 3–4 monolayers have been deposited.

The first few layers of atoms being dominant in determining intensities, changing the nature of the first layer, for example by adsorbing a layer of foreign atoms, can be expected to change intensities considerably. Figure 1.16 makes this point by comparing $I(E)$ spectra for a clean molybdenum surface and for the same surface with oxygen adsorbed. The measurements were made by Wilson (1971).

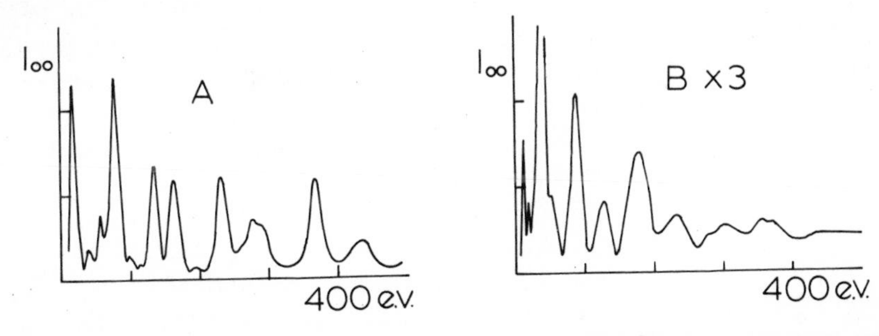

FIGURE 1.16. Intensity energy spectra from a molybdenum (001) surface. A: clean surface, B: oxygen covered surface. The curves have not been corrected for variation of incident current with energy.

LEED intensity spectra are strongly influenced by temperature. The effect is to reduce heights of peaks and is more marked at higher energies. Tabor and Wilson's (1970) results for niobium, reproduced in Fig. 1.17, show this

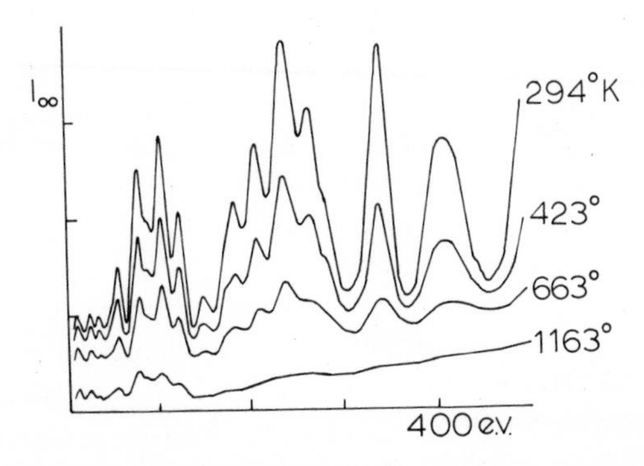

FIGURE 1.17. Intensity/energy spectra taken at 8° from normal incidence on a niobium (001) surface at various temperatures. For clarity each successive curve is displaced downwards by a small amount. Results have not been corrected for variation of incident current with energy.

Note that these curves taken with a spot photometer have an arbitrary intensity scale and have not been corrected for variation of incident current with energy.

The temperature dependence can be explained by another analogy with X-ray theory. Random displacements of atoms caused by finite temperature mean that phases of waves scattered by them do not quite add coherently even at Bragg peaks, and intensities are lowered. At shorter wavelengths (higher energies in the electron case) the phase differences introduced by a given displacement are greater, hence the greater sensitivity at higher energy.

It is clear that temperature has a marked effect even in a hard material such as niobium where amplitudes of vibration are smaller than in most materials. Temperature effects are generally of great importance in spectra taken above 150 eV, and are responsible for reflected intensities being so small at higher energies. In softer materials reduction of intensity can be so great as to make it impossible to observe spectra at high energies without considerable cooling of the sample.

Because of the information they contain additional to that supplied by the diffraction pattern, intensity spectra hold the key to several important problems in surface physics. Their interpretation has until recently proved elusive, but finally an understanding of their structure has been arrived at. The more complete theory will be presented in later chapters.

IG Incoherent and inelastic scattering

In speaking of the diffraction pattern and intensities of beams, we have specified that electrons considered be elastically scattered. As we have seen in the last section, intensities of elastically scattered electrons even at peak values typically amount to only $\simeq 2\%$ of the incident current. Many of the other electrons suffer inelastic collisions and cascade down in energy eventually to drain away via an earthing connection to the crystal. Some are inelastically scattered out of the crystal, back towards the fluorescent screen. In addition, energies comparable with those of the inelastically scattered electrons can be transferred to an electron originally resident in the crystal giving rise to so called 'secondary electrons'.

Normally electrons with energies more than about 1 eV less than that of the incident beam would be prevented from reaching the screen by energy selecting grids put there for that purpose, but it is possible by varying the retarding potential on the grids to admit all electrons with energies greater than E_ℓ. This number can then be differentiated electronically to give the number of electrons with energies between E_ℓ and $E_\ell + dE_\ell$:$N(E_\ell)dE_\ell$. Figure 1.18 shows a schematic picture of $N(E_\ell)$.

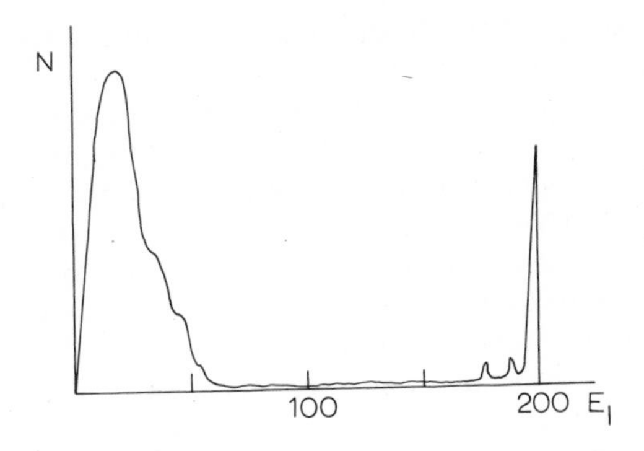

FIGURE 1.18. A schematic energy loss spectrum from a primary beam of 200 eV.

The elastic peak at E_p forms quite a small fraction of the distribution, even though it includes electrons having undergone loss processes involving energy changes too small to be resolved from the true elastic peak. These are the phonon or thermal losses and we shall return to them in a moment. Other losses involve electronic excitations. For $(E_p - E)$ less than 10–20 eV most of the spectrum comes from primary electrons having lost energy to individual electrons resident in the crystal. Of greater importance are the losses of primary beam energy in the creation of plasmons or charge density waves. Plasmons have energies in the 10–20 eV region and provide the dominant inelastic scattering mechanism for the primary electrons, though in some materials plasmon losses become tangled with single particle losses and no clear distinction can be made. In Fig. 1.18 we show peaks corresponding to one and two plasmon losses. Often peaks due to creation of surface plasmons can be seen too.

The dominant loss mechanism in most materials is by either plasmon creation or by excitation of some mixture of plasmon-single particle state such as occurs in copper where the plasmon peaks are not sharply defined. Other processes such as excitation of deep core levels play a minor role as can be shown theoretically (Ing, 1972) or by observing the weakness of any such characteristic losses in the $N(E)$ spectrum.

Before the electrons can excite plasmons they must have a certain minimum energy often around 15 eV. It is this minimum energy that led us to draw a distinction between energies about 15 eV and below. Above, strong attenuation of the elastic beam holds; below, the elastic beam is usually much less strongly affected.

Other contributions to $N(E_\ell)$ come from electrons that were not part of the primary beam but have been excited from below the Fermi level to higher energy states. They are called secondary electrons and together with primary electrons that have cascaded down in energy, contribute mainly to the large broad peak at low energies.

There is a small but important component of the secondary electron contribution, from Auger processes. If an electron in a deep state, energy E_1, is excited leaving behind an electron hole, after a certain length of time another electron in a higher energy state, E_2, will make a transition to the lower level. The energy difference can either be emitted as an X-ray, or more probably given up to a third electron having initially energy E_3. The energy balance means that finally the third electron has energy $(E_3 - E_1 + E_2)$, possibly sufficient to escape from the crystal. Figure 1.19 illustrates the process.

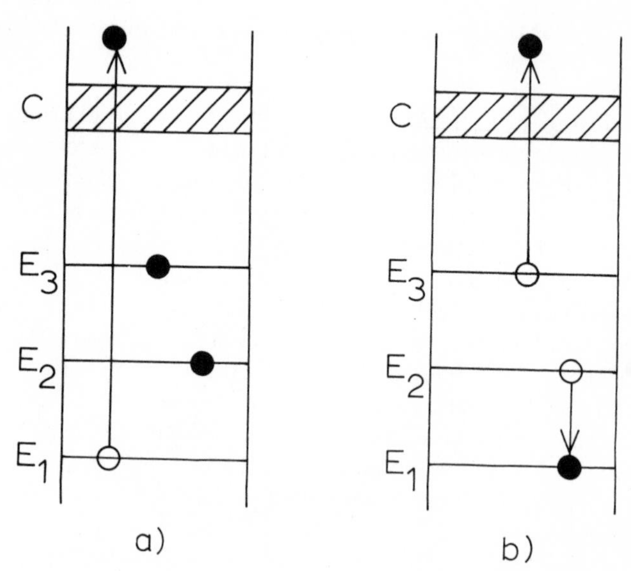

FIGURE 1.19. $E_1E_2E_3$: core levels in a solid. C conduction band. (a) creation of a deep hole, (b) Auger transition.

Auger electrons appear as tiny humps on the broad secondary electron peak in Fig. 1.18. They can be distinguished from other structure by the fact that their energies are not related to the incident energy, but given directly by $(E_3 - E_1 + E_2)$. Therefore changing the incident energy changes the relative position of Auger peaks in the spectrum.

Auger peaks are small in intensity, but narrow in width and are more easily observed against the large background by differentiating $N(E)$ to give an Auger spectrum one of which for a chromium surface taken by Wilson (1971) is shown in Fig. 1.20. Peaks are characteristic of energy levels in the atoms of the solid and can be used to identify atoms present. More-

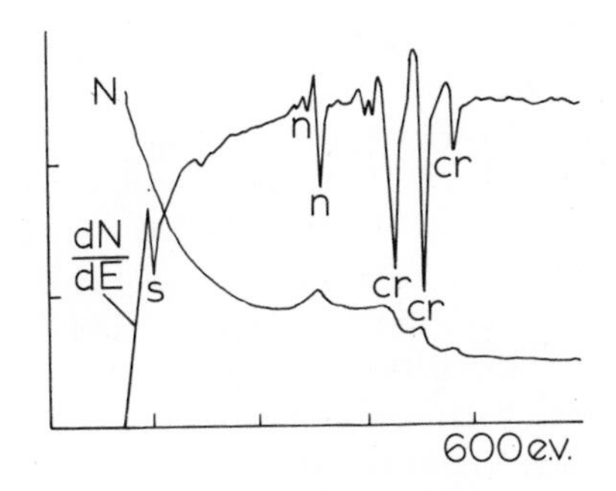

FIGURE 1.20. Auger spectrum for a 'clean' chromium surface. Both $N(E)$ and dN/dE are shown. Peaks are characteristic of atomic species, and this spectrum shows that some sulphur and some nitrogen is present on the surface.

over the technique is a useful complement to LEED because the Auger electrons come from about the same depth in the crystal as the primary LEED electrons penetrate to. Therefore they provide an analysis of just the region of the crystal which determines the diffraction pattern.

As well as losing energy to other electrons in the crystal the incident electron can lose energy to the crystal lattice by scattering off a phonon, giving up some energy and momentum. Energy losses to phonons are very small: of the order of $\hbar \times$ the phonon frequencies, or less than 0.1 eV. Electrons with such a small loss cannot be filtered out by the energy selecting grids and are usually called quasi-elastic. Sometimes it is possible when phonon energies are large e.g. for optical modes, and with high quality electron guns and analysing equipment, to resolve phonon losses. Figure 1.21 shows results by Ibach (1970) in which he measures intensities of electrons that have lost energy to an optical surface mode. So far no apparatus

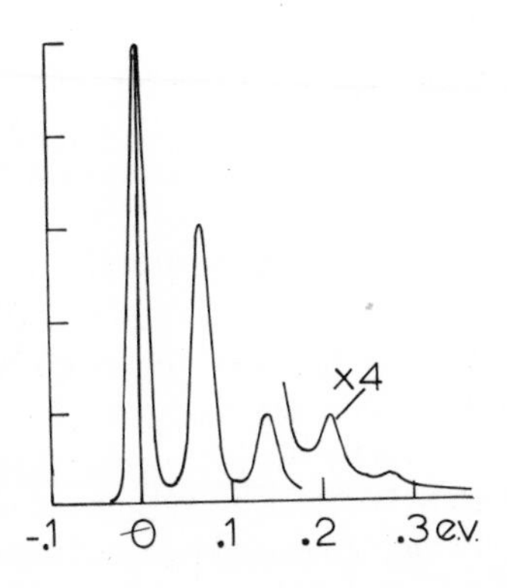

FIGURE 1.21. Energy loss peaks in electrons reflected from the $(1\bar{1}00)$ surface of a ZnO crystal at 45° angle of incidence. Temperature is 127° K, and incident energy 7.5 eV.

has been built with sufficient resolution to detect electrons having suffered energy loss to an acoustic phonon.

The effect of phonon losses on the elastically scattered electrons is rather different from electronic losses. The latter tend to exert their influence through a finite path length for elastic electrons in the crystal. Phonon losses also contribute to this path length (Stern and Howie, 1972) but their most dramatic effect is not path length limitation. Since phonons are merely motions of atoms in the crystal, those electrons interacting with phonons are in the process of scattering from an atom. Phonon scattering robs the elastically scattered beams of their intensities in this direct way, appearing to modify the strength of elastic scattering matrix elements by a Debye-Waller factor as we discussed in the section on intensities.

What sort of a pattern do inelastic electrons make on the fluorescent screen? For the elastic electrons we deduced the Bloch theorem: the electron can only be scattered into a discrete set of beams, as a consequence of crystal symmetry parallel to the surface. No such restriction holds for inelastically scattered electrons. They can leave behind a continuously variable amount of momentum, with either an electron, a plasmon, or a phonon. Thus in the absence of energy analysing grids the screen is illuminated not only by discrete beams of elastically scattered electrons, but also with a more uniform background of inelastically scattered electrons.

That is not to say that inelastically scattered electrons have no angular structure in their intensities. At very high temperatures when atoms have large displacements, each atom behaves as an independent scatterer, its position being uncorrelated with its neighbours'. Then the background due to phonon scattering reflects structure in the atomic scattering factor, and often at high temperatures a series of dark rings can be seen in the background intensity correlating with minima in the atomic scattering factors (Schilling and Webb, 1970).

At low temperatures only phonons with low momenta are excited and electrons scattering off them stay quite close in angle to the elastically scattered beams. Thus a halo of low-momentum-phonon-loss electrons forms around the elastic beams.

Angular structure of intensities scattered by electronic processes is present, but is less characteristic than that due to phonon scattering.

Finally we mention the Kikuchi effect. Inelastically scattered electrons on their way out of the crystal can be elastically scattered and it has been shown how much electrons tend to be scattered preferentially in certain directions, forming a pattern on the screen that reflects the symmetry of the surface. Kikuchi patterns are characteristic of diffraction experiments at higher energies and temperatures.

Chapter 2

Scattering Processes for Low Energy Electrons

IIA Experimental clues to the nature of electron scattering

What can happen to an electron in a crystal? Firstly it can be scattered by the strong potentials inside the surface. For convenience we are going to divide the scattering into two types: forward scattering where the angle of scatter is $< 90°$, and back scattering that reverses the momentum normal to the surface. In LEED we are dependent on the latter process for the electrons we observe. Secondly it can undergo a loss of energy. If the energy-selecting grids are in position such an energy loss is tantamount to absorption of the electron because it will never subsequently have enough energy to make an appearance on the fluorescent screen. Thirdly there is a uniform lowering of the electron's energy as it enters the potential well of the crystal. (The same well is responsible for keeping the conduction electrons in the crystal).

The strength of each of these processes can be gauged by a matrix element. Electrons are scattered at a rate determined by forward and backward scattering matrix elements due to each of the atoms comprising the crystal, T_f and T_b respectively. T is of course a continuous function of scattering angle and the division into T_f and T_b a conceptual convenience. T is also a function of incident energy.

Inelastic scatterings of all sorts we treat together in terms of the lifetime of an electron, τ. An electron with energy E would normally have temporal variation in wavefunction amplitude of $\exp(-iEt)$. By giving the energy an imaginary component of $+iV_{0i}$, the *intensity* of the wave function can be made to decay away in time as $\exp(+2V_{0i}t)$ and we can equate

$$V_{0i} = -\frac{1}{2\tau} \tag{2.1}$$

Attenuation of elastically scattered beams can be simulated by an imaginary component of the energy, i.e. by adding a constant imaginary potential, iV_{0i}, to the Schrödinger equation.

The potential well we take to have a depth V_{0r}, a quantity that can be identified with the so called 'inner potential' when scattering by T_f and T_b is weak.

Figure 2.1 shows the various processes schematically. In a sense, the back scattering is most important of all these processes; it enables us to get a picture on the screen. To scatter electrons with energies well above the Fermi level through the large angles necessary to reverse their direction requires a strong potential and only the potential deep inside the atom, in ion core regions is strong enough to do this. Yet even the ion cores do not provide enough scattering power to make T_b very large. For example, in Fig. 1.13 the elastically reflected intensities are as weak as 0.5% of the incident intensity at around 50–100 eV, and smaller still at higher energies showing that T_b is generally quite weak.

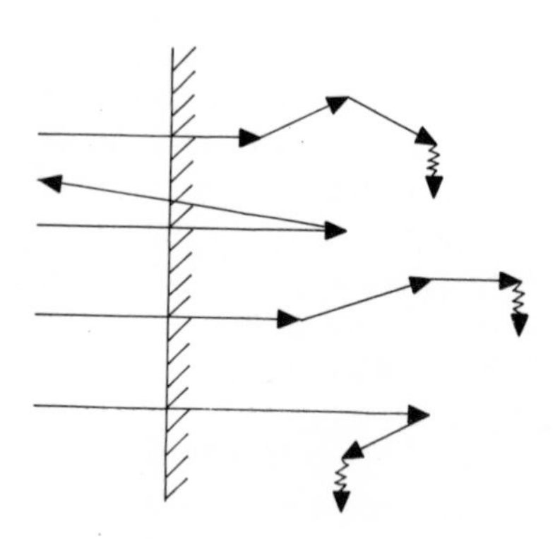

FIGURE 2.1. Once inside a crystal an electron can be scattered in both forward and backward directions. The electrons without energy loss (straight lines) terminate their paths either by exit from the surface or more usually by sustaining an energy loss (wavy lines).

The question immediately arises of 'weak in comparison with what?'. The answer lies in consideration of inelastic processes. These permanently remove flux from elastically scattered beams and one may imagine a competition developing. If V_{0i} is large and hence inelastic scattering is large, the electron loses some energy before it has a chance to be back scattered. Flux is removed from the incident beam at a rate dominated by V_{0i}, therefore the lifetime of an incident electron before inelastically scattering, τ, is given by equation (2.1). The elastically scattered beam grows in amplitude at a rate given by T_b, and its intensity after a time $\tau = -1/(2V_{0i})$ will be

$$I = \left| \frac{T_b}{2V_{0i}} \right|^2 \simeq 0.01 \tag{2.2}$$

We have inserted a typical experimentally observed value of I, but the precise figure is often strongly dependent on the precise direction of back-scattering. Figure 1.13 shows specular reflection through an angle of almost 180°, but in Fig. 2.2 other beams are shown that scatter through smaller angles.

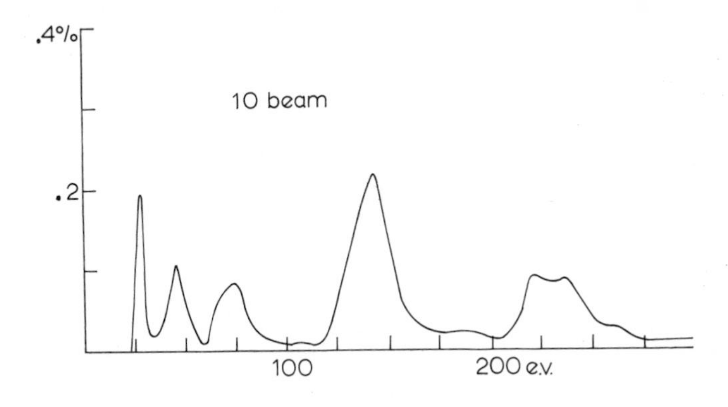

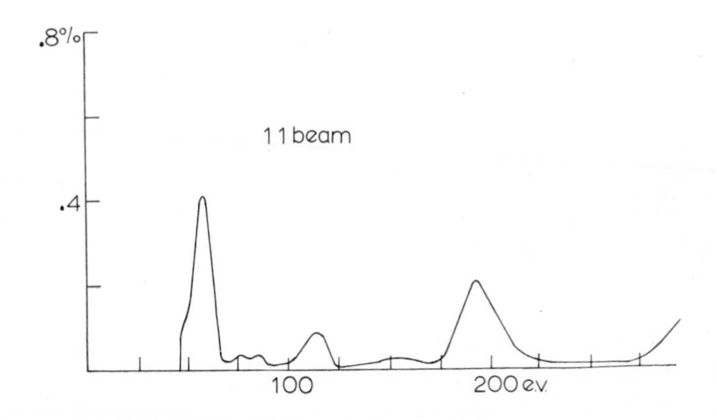

FIGURE 2.2. Intensities of the 10 and 11 beams from a copper (001) surface for normal incidence, measured as functions of energy. The data are due to Andersson (1969).

If we ignore the detailed structure as a function of energy which is due to the arrangement of atoms in the crystal, and concentrate on the overall intensity of the beams we see that the 10 beam shows intensities much lower than the 00 beam by a factor of about five, and the 11 beam is different again: perhaps only a factor of two weaker than the 00 beam.

The most striking evidence of structure in T_b as a function of angle is provided by scattering from liquid mercury. The disordered nature of a liquid means that elastically reflected electrons no longer emerge as discrete beams, but as a continuous distribution on the screen. To a large extent the total intensity is the sum of intensities scattered by the individual atoms and therefore the intensity distribution is not far removed from that given by the single atom matrix element, T_b. Figure 2.3 shows results of measurements by Schilling and Webb (1970) of angular distribution of intensity scattered from liquid mercury. A large amount of structure can be seen.

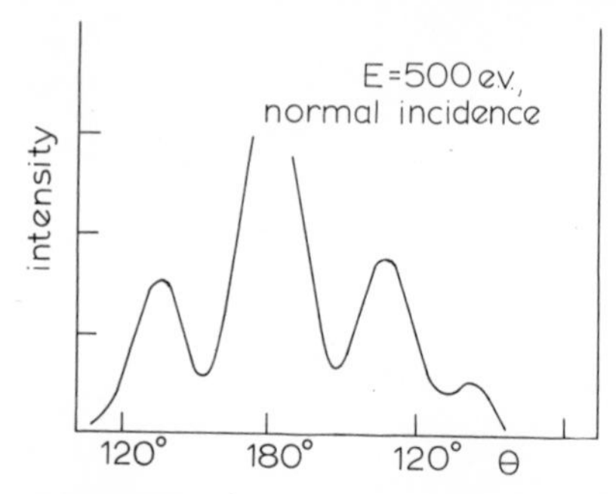

2.3. Angular structure in intensity of electrons elastically scattered from liquid mercury as a function of scattering angle, θ.

Forward scattering influences what appears on the screen much less directly. If it were weak, then it would have hardly any influence on the reflected intensities. On the other hand if T_f is large (again in comparison with V_{0i}) then it has the ability to produce several strong forward travelling beams in addition to the incident beam. The result of such a situation would be that, whereas in a simple picture we imagine peaks in reflected intensities as occuring when the incident beam satisfies a Bragg condition for reflection, there are now in effect several 'incident' beams any one of which can satisfy a Bragg condition. Many more peaks will be possible in a strong forward scattering situation, and any profusion of the number of peaks above the number expected from simple Bragg-law considerations must be taken as evidence of strong forward scattering. In fact a wide range of materials does show such a profusion. We can see that it is the case for the copper results in Fig. 1.13. It also holds for experiments on beryllium (Baker *et al.* 1969) one of which is presented in Fig. 2.4. More details can be found in Baker (1970).

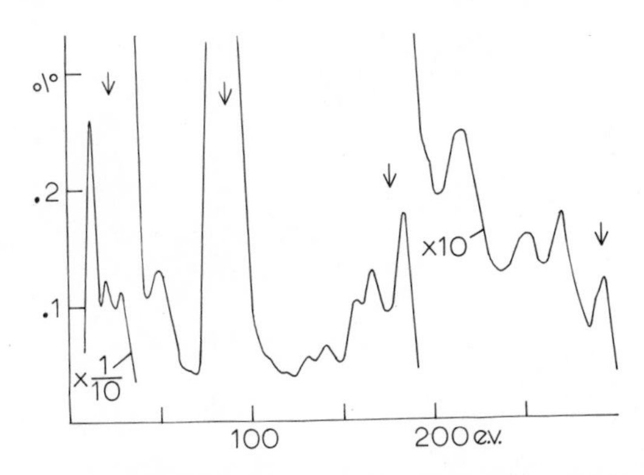

FIGURE 2.4. Intensity of the 00 beam from a beryllium (0001) surface, rotated away from normal incidence by 17.3° about the $[11\bar{2}0]$ axis. $T = 70°$ K. Arrows mark energies at which Bragg peaks are expected on a simple theory.

We are forced to conclude that T_f is at least as strong as V_{0i}. Yet it cannot be very much greater because, if it were, electrons would have time, before losing energy and being removed from the elastic system, to undergo many scatterings each in the forward direction but adding up to a net back scattering. If this did happen to any appreciable extent back reflection would be strong, not weak as observed. Hence T_f is not vastly greater than V_{0i}:

$$\overline{|T_f|^2} \gtrsim |V_{0i}|^2, \tag{2.3}$$

where the bar denotes an average value.

Next we consider the potential well of depth V_{0r}. Not only does it play a role in LEED, it also operates on conduction electrons and unless V_{0r} is a strong function of energy estimates made for conduction electrons should give some idea of V_{0r} for LEED calculations.

In metals such as aluminium with nearly-free-electron conduction bands the wavevector of electrons is given by the equation

$$\tfrac{1}{2}|\mathbf{k}|^2 = E - V_{0r} \tag{2.4}$$

(inelastic scattering of conduction electrons is small, therefore V_{0i} is neglected). The electron energy levels are filled up to the Fermi energy

$$\tfrac{1}{2}k_F^2 = E_F - V_{0r} \tag{2.5}$$

and the Fermi wavevector follows from the density of electrons, ρ,

$$\frac{2}{(2\pi)^3}\frac{4}{3}\pi k_F^3 = \rho \tag{2.6}$$

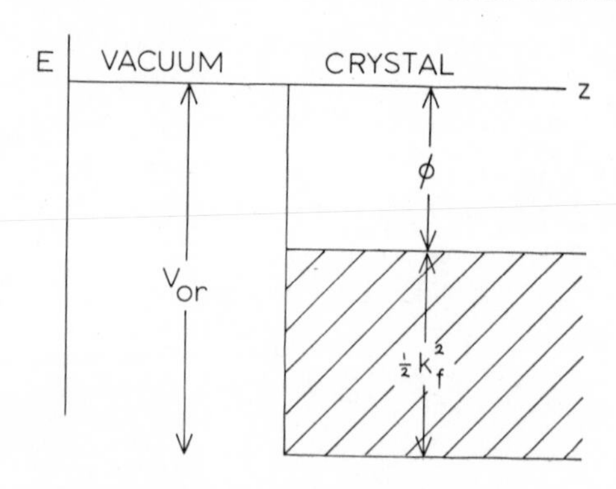

FIGURE 2.5. Diagram showing a crystal with band-width $\frac{1}{2}k_F^2$ and work function ϕ.

The situation is illustrated in Fig. 2.5. The electrons with greatest energy in the crystal are those at the Fermi energy, and therefore the work function is the energy needed to remove these electrons from the crystal. Experimentally we know the electron density for aluminium, therefore k_F, and the work function and we can deduce V_{0r}:

$$\tfrac{1}{2}k_F^2 \simeq 12.5\,\text{eV},$$

$$\phi = 3.7\,\text{eV}$$

$$V_{0r} = -(\tfrac{1}{2}k_F^2 + \phi) \simeq -16\,\text{eV}. \tag{2.7}$$

In the case of a metal with a conduction band that is not free electron like the calculation is more complicated, but Mattheiss (1964) has calculated for copper that the Fermi energy is given, apart from V_{0r}, by

$$E_F = 9.5\,\text{eV}$$

from band-structure calculations. Experiment gives for the work function

$$\phi = 4.5\,\text{eV}$$

To bring the electrons at the Fermi energy below the vacuum zero by an amount ϕ, requires that

$$V_{or} = -(9.5 + 4.5) = -15\,\text{eV}. \tag{2.8}$$

Equations (2.7) and (2.8) give an idea of values of V_{0r} expected at low incident energies. At higher energies another estimate can be made from our simple condition (1.39) for the appearance of a Bragg peak. The intercept of this expression with the energy axis in Fig. 1.15 gives V_{0r} for copper to be

$$V_{0r} = -13.5\,\text{eV} \tag{2.9}$$

using structure above 100 eV. Equation (2.9) is not a very reliable estimate because, as we have seen, the simple Bragg reflection picture implying weak T_f and T_b does not provide an accurate description of diffraction processes.

What is clear is that as far as can be judged from experiments interpreted according to simple theory, V_{0r} is not a strong function of incident energy in the range covered by LEED electrons, and takes values between -10 and -20 eV.

Finally there is V_{0i}, the matrix element for inelastic scattering. Under its influence the intensity of the elastic component of the wavefunction dwindles on a time scale $-1/(2\,V_{0i})$. That means elastic diffraction of electrons must take place within a time $-1/(2\,V_{0i})$, on average. Hence from the uncertainty principle any features in the diffracted intensities cannot be defined with a better half-width than $(\Delta E/2)$, given by

$$\left(\frac{\Delta E}{2}\right)^2 \left(\frac{1}{2V_{0i}}\right)^2 \geqslant \tfrac{1}{4}, \tag{2.10}$$

or

$$\Delta E \geqslant 2|V_{0i}|. \tag{2.11}$$

In particular, peaks in intensity/energy curves can never be more narrow than $2|V_{0i}|$. Examination of experimental curves such as those given in Fig. 1.13 and 2.2. shows that below 15 eV incident energy, features as narrow as 1–2 eV are sometimes observed, whereas in the range from around 15 eV–100 eV peaks are generally broader, say 5–10 eV wide. At higher energies peaks are broader still rising to perhaps 30–40 eV in width in the range 500–1000 eV but as we shall see, some doubt occurs as to whether the equality in equation (2.11) is applicable at energies higher than 100 eV. Our guesses from experiment based on equation (2.11) are

$$\begin{aligned} E &< 15\text{ eV}, & V_{0i} &\simeq -1\text{ eV}, \\ 15\text{ eV} < E &< 100\text{ eV}, & V_{0i} &\simeq -4\text{ eV}, \\ 100\text{ eV} < E & & ,\ |V_{0i}| &< 10\text{ eV}. \end{aligned} \tag{2.12}$$

From equations (2.12) for V_{0i} we also deduce that $T_b \simeq 1$ eV and T_f is of the order of, or a little greater than 6 eV in the 15–100 eV range. If T_b does not fall drastically in the $E < 15$ eV range, we might expect from equation (2.2) that the decrease in peak width in this range should be associated with large reflectivities; exactly what we have observed.

IIB The ion-core potential

At first sight it might seem that the potential scattering an electron incident on a crystal is a complicated entity. The placing of atoms in close proximity to one another inevitably distorts their electronic structure, and furthermore the distortions will not have spherical symmetry. Rather they will have the same symmetry as the unit cell. However experience has shown that for many materials the situation can be simplified by certain accurate approximations.

Near centres of atoms the potential will always be given by $-Z/r$ because of the nuclear charge. Further away from the nucleus, the potential is less strong and influence of neighbouring atoms more pronounced. If we wish we can draw spheres about the nuclei such that within the spheres there is to good accuracy a spherically symmetric potential, but outside these spheres non-sphericity is important but the potential is weaker.

Now we are going to make the approximation that, if we take the largest possible non-overlapping spheres drawn about each nucleus, the potential inside will be spherically symmetric, and the potential outside will be constant. This procedure has become known by the curious name of the 'muffin tin approximation'. Figure 2.6 gives a cross-section of a crystal that has been divided up into muffin tins. Figure 2.7 shows the potential in the solid plotted along the z-axis of Fig. 2.6.

The approximation has been extensively tested in band structure calculations and is known to be accurate for many materials, especially the close-packed metals.

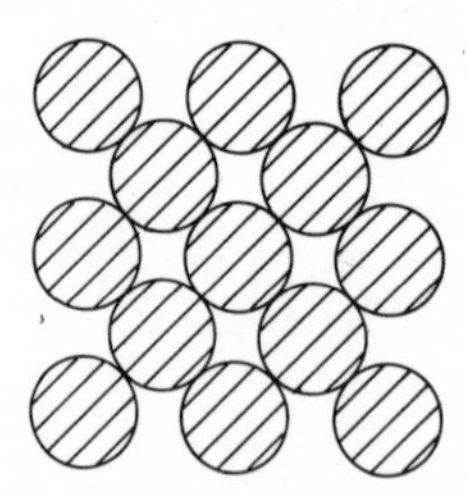

FIGURE 2.6. Cross-section of a crystal divided into muffin tins.

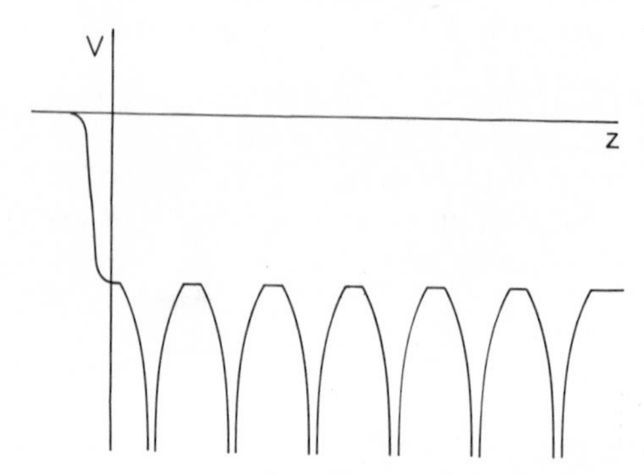

FIGURE 2.7. A cross-section of the potential in a crystal. Note the regions of constant potential between the ion cores, and the barrier at the surface.

The constant potential between spheres is due mainly to the incident electron pushing aside conduction or valence electrons to form a correlation hole. This effect will be discussed in Section D and for the time being we are going to measure all energies inside the solid relative to this constant potential. What our interest in this section centres upon is the ion-core potential inside the muffin tin sphere.

The ion-core potential is largely determined by tightly bound core-state electrons and the nuclear charge, modified a little by conduction/valence electron screening, less so by the proximity of neighbouring atoms. The core region not being very polarisable, scattering by the core does not change very much from an atomic to a crystalline environment. By the same argument, ion cores near the surface of a solid would not be expected to change their scattering powers appreciably. What *does* happen near the surface is that the potential outside the muffin tins ceases to be constant. The details of its variation near the surface will be discussed in Section F. We wish to calculate first the ion-core potential, then to find how it scatters the incident electron.

Important ingredients of our calculations will be wavefunctions for the core electrons. In the main these wavefunctions are sufficiently tightly bound that they can be taken to be unchanged from those of the free atom. Calculations of free atom wavefunctions have been made by various authors (Herman and Skillman, 1963; Clementi, 1965) and are readily available. Again the accuracy of using such wavefunctions to construct ion-core potentials has been extensively checked through band structure calculations. We refer to Snow (1968) as a good example amongst many.

Sometimes it happens that core electron wavefunctions overlap the muffin-tin sphere and interact with neighbouring atoms. This is true of the *d*-levels in transition and noble metals. Overlap is usually not large and the levels retain their atomic nature to a large extent. Various corrections have been suggested (see Loucks, 1967, for example) but the one we shall employ is to assume that the influence of neighbouring atoms acts so as to push the overlapping wavefunction back inside its own muffin tin sphere. The motivation behind this step is that although an ion core may lose charge because its own wavefunctions overlap the muffin-tin boundary, it gains charge from the overlap of wavefunctions from neighbouring atoms. We take this into account by cutting off the core state wavefunction outside the sphere and increasing its amplitude within to give the proper number of electrons in the level.

A further complications can occur where core levels do not occur in 'closed shells'. For a spherically symmetric potential core wavefunctions all have the form

$$\psi_c(\mathbf{r}, s) = \delta_{ss_c}\psi_{\ell_c}(|\mathbf{r}|)\, Y_{\ell_c m_c}(\Omega) \tag{2.13}$$

where s is the spin coordinate and can take values of $\pm\frac{1}{2}$; ψ_{ℓ_c} is the radial wavefunction; and $Y_{\ell_c m_c}$ is a spherical harmonic describing angular variation of the wavefunction. For a given value of angular momentum, ℓ_c, there are $(2\ell_c + 1)$ values that the z-component, m_c, can assume and two spin states for each. Unless all the $2(2\ell_c + 1)$ states are occupied with equal probability, the atom cannot be spherically symmetric because it would have a net spin, or component of angular momentum along some axis. For example, in nickel above the Curie temperature, band structure calculations tell us that the d-level holds not 10 electrons needed for a closed shell, but 9.4 electrons. To retain a spherically symmetric potential we multiply the d wavefunctions in nickel by $(0.94)^{\frac{1}{2}}$ then assume that each orbital is equally occupied. Now we have 0.94 of an electron in each of the 10 orbitals and since all m-values and spin states are equally occupied we regain spherical symmetry.

Finally there is the distribution of screening charge. The ion-core potential acts on conduction electrons influencing their motion, and therefore changes their charge density. The problem is a self-consistant one and indeed when accurate band structure calculations are needed an iterative procedure for calculating ion-core scattering, conduction electron density, and the ion core scattering again with the new screening charge, must be employed: a tedious procedure.

Fortunately LEED energies are much higher than those characteristic of conduction or valence electrons in crystals, and the situation is different. It will be argued in section C that cancellation of ion-core scattering to give a weak effective potential or pseudopotential, does not operate so effectively at higher energies and scattering by ion cores is much stronger. Therefore relatively speaking screening charge has a much greater effect on conduction electrons than it does on higher energy LEED electrons that are already being strongly scattered by the ion cores. For example, in copper the scattering cross-section at 100 eV is ten times its value just above the Fermi energy and in consequence screening is ten times less important. This figure is typical.

The main effect of screening charge as far as LEED electrons are concerned is to ensure electrical neutrality and most reasonable guesses at the screening charge density will give satisfactory results. We might choose to assume a uniform distribution of screening charge inside the muffin-tin sphere of the density required to give electrical neutrality of the whole sphere. Alternatively we could be a little more sophisticated and say that wavefunctions of electrons providing the screening charge are going to behave inside the atom not very differently from the outermost atomic orbital that went to form the conduction or valence band of the solid. Therefore a better approximation is to take the screening charge to be distributed within the muffin-tin sphere in the same way as in the outermost orbital of the free atom, again with the density normalised to give an electrically neutral muffin-tin.

Let us look at one of the muffin-tin spheres in isolation. It contains a modified neutral atom and we have established all the quantities required to calculate the potential of this modified atom. We are looking for a Schrödinger equation governing the motion of an incident electron, but initially it must be recognised that the problem is really a many-body one. We should be solving the complete Schrödinger equation that includes all electrons in the total wavefunction, Φ, for an atom with N electrons positions $\mathbf{r}_j$, spin coordinates s_j, plus an incident electron position $\mathbf{r}_0$ and spin s_0:

$$H\Phi = \left[\sum_{j=0}^{N} \left\{ -\tfrac{1}{2}\nabla_j^2 - \frac{Z}{|\mathbf{r}_j|} + v_s(\mathbf{r}_j) \right\} + \sum_{j=0}^{N} \sum_{i=j+1}^{N} \frac{1}{|\mathbf{r}_i - \mathbf{r}_j|} \right] \times \Phi(r_0, s_0; \ldots r_N, s_N) = E_t \Phi \tag{2.14}$$

E_t is the total energy and includes that of the core levels as well as that of the incident electron. v_s is the potential due to the screening charge, and z is the nuclear charge.

The difficulty in equation (2.14) is that electrons influence one-another via the Coulombic repulsion and the incident electron may distort wavefunctions of core electrons by its own electrostatic field, correlating their motion with its own and thus changing the effective potential seen by the incident electron. Such effects would be important if the core were easily polarised but fortunately, as we stated earlier and will show later, it is not.

So in this approximation we can write down Φ as a product of core-state wavefunctions, modified if necessary as described above. At the same time we must be sure to make the resulting wavefunction antisymmetric under exchange of particles because electrons are Fermions,

$$\Phi(\mathbf{r}_0, s_0; \ldots \mathbf{r}_N, s_N) = \sum_p \varepsilon_p \phi(\mathbf{r}_i, s_i)\psi_1(\mathbf{r}_j, s_j) \ldots \psi_N(\mathbf{r}_k, s_k). \tag{2.15}$$

ϕ is the incident electron's wavefunction.

The sum is over permutations of the particle coordinates and ε_p takes the value of $+1$ if permutation p can be achieved by exchanging an even number of particle coordinates, -1 for an odd number. There are $(N+1)!$ possible permutations.

Since the core state part of Φ is known already, we want to eliminate this part from our equations and concentrate on ϕ, the incident electron wavefunction. That is, we want to find a one-electron Schrödinger of the form

$$(-\tfrac{1}{2}\nabla^2 + V)\alpha(\mathbf{r}, s) = E\alpha(\mathbf{r}, s), \tag{2.16}$$

which is satisfied by all wavefunctions, both ϕ and the ψ_c's.

Assuming for the moment that the individual wavefunctions are orthonormal

$$\left.\begin{aligned}\sum_s \int \psi_i^*(\mathbf{r},s)\psi_j(\mathbf{r},s)d^3\mathbf{r} &= \delta_{ij},\\ \sum_s \int \psi_i^*(\mathbf{r},s)\phi\,(\mathbf{r},s)d^3\mathbf{r} &= 0,\end{aligned}\right\} \tag{2.17}$$

and defining a function

$$\chi(\mathbf{r}_1,s_1;\ldots\mathbf{r}_N,s_N) = \psi_1(\mathbf{r}_1,s_1)\ldots\psi_N(\mathbf{r}_N,s_N), \tag{2.18}$$

multiplying equation (2.14) by χ^*, integrating over space, and summing over spin coordinates of χ leaves an equation involving only $\mathbf{r}_0$ and s_0. On the right-hand side the only term remaining is that in which the coordinates have not been permuted, leaving just

$$E_t\phi(\mathbf{r}_0\,s_0)$$

after integration. The left-hand side can be split into three parts,

$$\sum_{s_1\ldots s_N}\int d^3\mathbf{r}_1\ldots d^3\mathbf{r}_N\chi^*\left\{\left[\sum_{j=1}^{N} -\tfrac{1}{2}\nabla_j^2 - \frac{Z}{|\mathbf{r}_j|} + v_s(\mathbf{r}_j) + \sum_{j=1}^{N}\sum_{i=j+1}^{N}\frac{1}{|\mathbf{r}_i-\mathbf{r}_j|}\right] + \left[-\tfrac{1}{2}\nabla_0^2 - \frac{Z}{|\mathbf{r}_0|} + v_s(\mathbf{r}_0)\right] + \left[\sum_{j=1}^{N}\frac{1}{|\mathbf{r}_0-\mathbf{r}_j|}\right]\right\}\Phi \tag{2.19}$$

The first term is zero unless the permutations in Φ are those leaving $\phi = \phi(\mathbf{r}_0, s_0)$, because it amounts to the left-hand side of the Schrödinger equation obeyed by the core-states alone, which is solved by the function

$$\sum_p \varepsilon_p\psi_1(\mathbf{r}_j,s_j)\ldots\psi_N(\mathbf{r}_k,s_k). \tag{2.20}$$

Hence from the first term we retrieve

$$E_{ct}\phi(\mathbf{r}_0\,s_0)$$

where E_{ct} is the total energy associated with the core states.

The second term has non-zero contributions only from the first term in Φ:

$$\left[-\tfrac{1}{2}\nabla_0^2 - \frac{Z}{|\mathbf{r}_0|} + v_s(\mathbf{r}_0)\right]\phi(\mathbf{r}_0,s_0).$$

Finally the third term has non-zero contributions only from the first component of Φ and from those permutations that exchange coordinates of $\phi(\mathbf{r}_0,s_0)$ with those of a core state $\psi_j(\mathbf{r}_j,s_j)$. Remembering to change the sign by -1 when the permutation is odd:

$$\sum_{j=1}^{N}\phi(\mathbf{r}_0,s_0)\int\sum_{s_j}\frac{\psi_j^*(\mathbf{r}_j,s_j)\psi_j(\mathbf{r}_j,s_j)}{|\mathbf{r}_0-\mathbf{r}_j|}d^3\mathbf{r}_j$$

$$- \sum_{j=1}^{N} \psi_j(\mathbf{r}_0, s_0) \int \sum_{s_j} \frac{\psi_j^*(\mathbf{r}_j, s_j)\phi(\mathbf{r}_j, s_j)}{|\mathbf{r}_0 - \mathbf{r}_j|} d^3\mathbf{r}_j$$

The complete equation is now

$$\left[-\tfrac{1}{2}\nabla_0^2 - \frac{Z}{|\mathbf{r}_0|} + v_s(\mathbf{r}_0) + \sum_j \int \sum_{s_j} \frac{|\psi_j(\mathbf{r}_j, s_j)|^2}{|\mathbf{r}_0 - r_j|} d^3\mathbf{r}_j \right] \phi(\mathbf{r}_0, s_0)$$

$$- \sum_j \left[\int \sum_{s_j} \frac{\psi_j^*(\mathbf{r}_j, s_j)\phi(\mathbf{r}_j, s_j)}{|\mathbf{r}_0 - \mathbf{r}_j|} d^3\mathbf{r}_j \right] \psi_j(\mathbf{r}_0, s_0)$$

$$= (E_t - E_{tc})\phi(\mathbf{r}_0, s_0)$$

$$= E\phi(\mathbf{r}_0, s_0). \tag{2.21}$$

Similar equations could be deduced for the core-states, from which the orthonormality relationships, (2.17) can be proven.

(2.21) has the form of a Schrödinger equation and neglecting the second term in brackets on the left-hand side, it has the simple interpretation that the electron moves in the electrostatic field of core electrons plus nucleus. The second term arises out of considerations of antisymmetry under exchange of particles, because no two electrons can be in the same place at the same time hence each electron is surrounded by a region depleted of other electrons and hence of lower potential. Equation (2.21) which includes exchange, is the Hartree-Fock approximation to the solution of our problem. Without exchange effects it would be the Hartree approximation.

It will be noticed that the exchange potential comes in the form of an operator, involving an integral with the wavefunction rather than a straightforward multiplication of $\phi(\mathbf{r}_0, s_0)$ by $V(\mathbf{r}_0)$, i.e. it is non-local. This property will introduce some complications.

We mention in passing that Slater (1967) has suggested that the exchange correction can be simplified considerably by (a) solving the problem of electrons with uniform density ρ in a large box, for which the exchange contribution to energies of particles can be found analytically in the Hartree Fock approximation (Dirac, 1930). (b) Averaging this exchange contribution for all electrons below the Fermi surface (c) assuming that the exchange potential in equation (2.21) is approximated by a local potential whose value at point $\mathbf{r}$ is set equal to the averaged exchange as evaluated in (a) and (b) with the density taken to be the local density of electrons. This gives

$$V_{ex}(\mathbf{r}) \simeq -3\left[\frac{3\rho(\mathbf{r})}{8\pi}\right]^{\frac{1}{3}}. \tag{2.22}$$

Equation (2.22) has been shown to work well when approximating the exchange interaction of conduction electrons with core states, but it is not certain how high in energy an expression derived from an average below the Fermi energy will continue to hold good. There is evidence that it often works well, from the work of Jepsen, Marcus and Jona (1971) who have made LEED calculations at energies up to 200 eV using an ion-core potential constructed using a local approximation to exchange, and get good agreement with experiment.

We shall continue to use the original expression for exchange given in equation (2.21), but a readable and practical account of a scheme based on equation (2.22) for exchange is given in Loucks' (1967) book. Several papers deal specifically with constructing ion-core potentials for LEED calculations: Capart (1971), Pendry (1969, 1971), Strozier and Jones (1971).

Equation (2.21) governs behaviour of the incident electron with wave-function ϕ, inside one of the muffin-tin spheres and of course, inside the other spheres it will obey a similar equation. If we wish, ϕ can be decomposed into its 'partial waves', each partial wave having a definite total, and z-component of angular momentum:

$$\phi(\mathbf{r}, s) = \sum_{\ell m} a_{\ell m} \delta_{ss'} \phi_\ell(|\mathbf{r}|) Y_{\ell m}(\Omega) \tag{2.23}$$

The potential acting on electrons inside the muffin tin is spherically symmetric and therefore the radial component of each partial wave, $\phi_\ell(|\mathbf{r}|)$, cannot depend on the z-component of angular momentum, m, only on total angular momentum, ℓ.

A further consequence of the spherical symmetry we have insisted upon, is that the ion-core cannot absorb any angular momentum, and if the electron happens to have angular momentum quantum numbers ℓ and m, these are conserved. Hence each partial wave in equation (2.23) behaves independently: each partial wave separately satisfies equation (2.21),

$$\left[-\tfrac{1}{2}\nabla^2 - \frac{Z}{|\mathbf{r}|} + v_s(\mathbf{r}) - E + \sum_{js} \int \frac{|\psi_j(\mathbf{r}', s)|^2}{|\mathbf{r} - \mathbf{r}'|} d^3\mathbf{r}'\right] \phi_\ell(|\mathbf{r}|) Y_{\ell m}(\Omega)$$
$$- \sum_j \psi_j(\mathbf{r}, s') \int \frac{\psi_j^*(\mathbf{r}', s') \phi_\ell(|\mathbf{r}'|) Y_{\ell m}(\Omega')}{|\mathbf{r} - \mathbf{r}'|} d^3\mathbf{r}' = 0. \tag{2.24}$$

(2.21) as it stands is a partial differential equation with r, θ, ϕ, as variables. It can be reduced to a differential equation with r as the variable.

The first term in (2.24) is the kinetic energy operator which can be divided into a contribution from radial motion

$$-\frac{1}{2}\frac{1}{r^2}\frac{d}{dr}\left[r^2 \frac{d}{dr}\phi_\ell(r)\right] Y_{\ell m}(\Omega), \tag{2.25}$$

(we have written r in place of $|\mathbf{r}|$) and a contribution from angular motion

$$\frac{-\phi_\ell}{2}\left[\frac{1}{r^2 \sin(\theta)}\frac{\partial}{\partial\theta}\left(\sin(\theta)\frac{\partial Y_{\ell m}(\Omega)}{\partial\theta}\right) + \frac{1}{r^2 \sin^2(\theta)}\frac{\partial^2 Y_{\ell m}(\Omega)}{\partial\phi^2}\right] \tag{2.26}$$

Since we know that the total angular momentum quantum number is ℓ, total angular kinetic energy must be the square of the angular momentum divided by twice the moment of inertia of a particle, mass unity, at point **r**:

$$\ell(\ell + 1)/2r^2$$

Hence (2.26) reduces to

$$[\ell(\ell + 1)/2r^2]\phi_\ell(r) Y_{\ell m}(\Omega). \tag{2.27}$$

The other terms in (2.24) are independent of angle except for the exchange term. Making use of equation (2.13) for ψ_c, the exchange term becomes

$$-\sum_{\ell' m'} \psi_{\ell'}(r) Y_{\ell' m'}(\Omega) \times \int \frac{\psi_\ell^{*\prime}(r')\phi_\ell(r')\, Y^*_{\ell' m'}(\Omega')\, Y_{\ell m}(\Omega')}{|\mathbf{r} - \mathbf{r}'|} d^3\mathbf{r}' \tag{2.28}$$

Next $1/|\mathbf{r} - \mathbf{r}'|$ is to be expanded about the origin in a power series,

$$\frac{1}{|\mathbf{r} - \mathbf{r}'|} = \sum_{\ell''=0}^{\infty} \frac{r_<^{\ell''}}{r_>^{\ell''} + 1} P_{\ell''}(\cos\theta''), \tag{2.29}$$

where $r_<$ and $r_>$ stand for the lesser and greater of $|\mathbf{r}|$ and $|\mathbf{r}'|$. θ'' is the angle between **r** and **r**$'$.

Substitution of (2.29) into (2.28) gives

$$-\sum_{\ell'\ell''} \psi_{\ell'}(r) \int \psi^*_{\ell'}(r')\phi_\ell(r') \frac{r_<^{\ell''}}{r_>^{\ell''+1}} \frac{(2\ell' + 1)}{4\pi}$$
$$\times P_{\ell'}(\cos\theta'') P_{\ell''}(\cos\theta'') Y_{\ell m}(\Omega') d^3\mathbf{r}'. \tag{2.30}$$

We have made use of the summation rule for the $Y_{\ell m}$'s,

$$\sum_{m'=-\ell'}^{+\ell'} Y^*_{\ell' m'}(\Omega) Y_{\ell' m'}(\Omega') = \frac{(2\ell' + 1)}{4\pi} P_{\ell'}(\cos\theta'') \tag{2.31}$$

and θ'' is the angle between directions Ω and Ω'.

Finally note that in the angular part of the integration over $d^3\mathbf{r}'$, instead of measuring θ' relative to the z axis, it can perfectly well be measured relative to direction **r**, because integration is over all angles. Thus

$$d^3\mathbf{r}' = r'^2 dr' \sin(\theta'') d\theta'' d\phi'' \tag{2.32}$$

$Y_{\ell m}(\Omega')$ must now be re-expressed in terms of the new variable $d\Omega''$. By merely changing our axis, we can change only the z-component of angular momentum; the total angular momentum must be unchanged. Therefore the angular variation of $Y_{\ell m}(\Omega')$ relative to another axis involves only $Y_{\ell M}(\Omega'')$ with the same total angular momentum:

$$Y_{\ell m}(\Omega') = \sum_{M=-\ell}^{+\ell} b_M Y_{\ell M}(\Omega'') \tag{2.33}$$

We are interested only in the $M = 0$ component of (2.23) because in (2.27) we find the integral

$$\int Y_{\ell M}(\Omega'')d\phi'' = 0, \text{ unless } M = 0, \tag{2.34}$$

because $Y_{\ell M}$ has a ϕ'' dependence of $\exp(iM\phi'')$.

Multiplying (2.33) on both sides by $Y^*_{\ell 0}(\Omega'')$ and integrating:

$$\begin{aligned} b_{M=0} &= \int Y_{\ell m}(\Omega')\, Y^*_{\ell 0}(\Omega'')\, d\Omega'' \\ &= \left[\frac{2\ell + 1}{4\pi}\right]^{\frac{1}{2}} \int Y_{\ell m}(\Omega')\, P_\ell(\cos\theta'')\, d\Omega'' \\ &= \left[\frac{4\pi}{2\ell + 1}\right]^{\frac{1}{2}} \sum_{M'} \int Y_{\ell m}(\Omega')\, Y_{\ell M'}(\Omega)\, Y^*_{\ell M'}(\Omega')\, d\Omega'' \end{aligned} \tag{2.35}$$

The various relationships made use of can be found in the appendix. Again, we can integrate over $d\Omega'$ rather than $d\Omega''$ and

$$\begin{aligned} b_{M=0} &= \left[\frac{4\pi}{2\ell + 1}\right]^{\frac{1}{2}} \sum_{M'} \int Y_{\ell m}(\Omega')\, Y^*_{\ell M'}(\Omega')\, d\Omega'\, Y_{\ell M'}(\Omega) \\ &= \left[\frac{4\pi}{2\ell + 1}\right]^{\frac{1}{2}} Y_{\ell m}(\Omega). \end{aligned} \tag{2.36}$$

Collecting together expressions (2.36), (2.33) and (2.32) and using them to simplify (2.30):

$$-\,Y_{\ell m}(\Omega) \sum_{\ell'\ell''} \psi_{\ell'}(r) \int \psi^*_{\ell'}(r')\phi_\ell(r') \frac{r_<^{\ell''}}{r_>^{\ell''+1}} r'^2 dr' \times C(\ell',\ell'',\ell), \tag{2.37}$$

$$C(\ell',\ell'',\ell) = \frac{2\ell' + 1}{2} \int_0^\pi P_{\ell'}(\cos\theta'')\, P_{\ell''}(\cos\theta'')\, P_\ell(\cos\theta'') \sin(\theta'')\, d\theta'' \tag{2.38}$$

Finally, by substituting expression (2.37) for the exchange term, and expressions (2.25) and (2.27) for the radial and angular components of the kinetic energy, (2.24) can be written

$$\begin{aligned} -\frac{1}{2}\frac{1}{r^2}\frac{d}{dr}\left[r^2 \frac{d\phi_\ell(r)}{dr}\right] + \frac{\ell(\ell + 1)}{2r^2}\phi_\ell(r) \\ + V_H(r)\phi_\ell(r) + \int V_{ex}^{(\ell)}(r,r')\phi_\ell(r')r'^2\, dr' = E\phi_\ell(r), \end{aligned} \tag{2.39}$$

$$V_H(r) = v_s(r) - \frac{Z}{r} + \sum_{js} \int \frac{|\psi_j(\mathbf{r}',s)|^2}{|\mathbf{r} - \mathbf{r}'|} d^3\mathbf{r}', \tag{2.40}$$

$$V_{ex}^{(\ell)}(r,r') = -\sum_{\ell'\ell''} \psi_{\ell'}(r)\psi_{\ell'}^{*}(r')\frac{r_{<}^{\ell''}}{r_{>}^{\ell''+1}}\,C(\ell',\ell'',\ell). \tag{2.41}$$

Equations (2.39), (2.40), (2.41) are the ones used in practice to calculate scattering by the ion-core potential. There are computer programs in the appendix based on these equations.

IIC Ion-core scattering

Equation (2.39) enables us to predict how incident electrons will move in the region of the ion core. In particular, in the region immediately outside the muffin-tin sphere the wavefunctions are very easily written down. V_H is zero because the sphere is electrically neutral, and V_{ex} is also zero because $\psi_{\ell'}(r')$ is zero for r' lying outside this sphere (see Fig. 2.8). Then (2.39) becomes

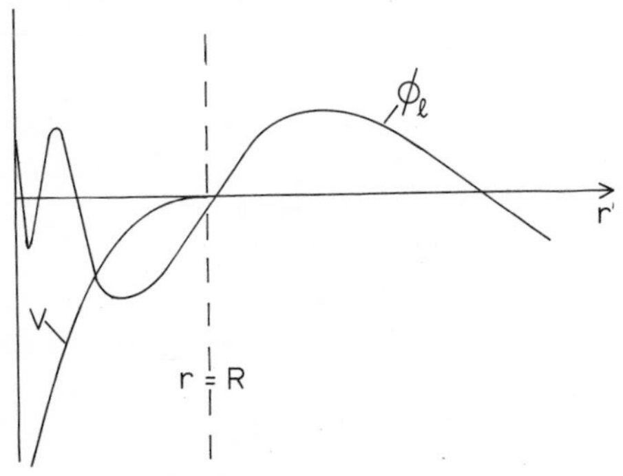

FIGURE 2.8. A schematic diagram showing the radial wave-function and ion-core potential V. $\phi_\ell(r)$ is calculated by outward integration from the origin to the muffin-tin radius, R.

$$-\frac{1}{2}\frac{1}{r^2}\frac{d}{dr}\left(r^2\frac{d\phi_\ell}{dr}\right) + \frac{\ell(\ell+1)}{2r^2}\phi_\ell = E\phi_\ell, \tag{2.42}$$

which has solutions

$$\phi_\ell = \alpha_\ell h_\ell^{(1)}(\kappa r) + \beta_\ell h_\ell^{(2)}(\kappa r), \tag{2.43}$$

$$\kappa = +(2E)^{\frac{1}{2}}. \tag{2.44}$$

α_ℓ and β_ℓ are constants. $h_\ell^{(1)}$ and $h_\ell^{(2)}$ are spherical Hankel functions of the first and second kind, order ℓ. $h_\ell^{(2)}(\kappa r)$ behaves for large values of r as

$$h_\ell^{(2)}(\kappa r)\underset{r\to\infty}{=} i^{(\ell+1)}\frac{\exp(-i\kappa r)}{\kappa r}, \tag{2.45}$$

which leads us to identify this part of the expression as an incoming wave. On the other hand

$$h_\ell^{(1)}(\kappa r) \underset{r\to\infty}{=} i^{-(\ell+1)} \frac{\exp(+i\kappa r)}{\kappa r} \tag{2.46}$$

and the second term is an outgoing wave.

A concise description of the wavefunction outside the muffin-tin sphere is possible, even when the full ion-core potential is operating inside. We have an incoming wave of amplitude β_ℓ and angular momentum quantum numbers (ℓ, m). As the electron moves away from the centre, since the ion-core potential conserves current, $h_\ell^{(1)}$ must have the same amplitude as the incident wave, but may differ in phase. Therefore outside the range of the potential the wave has the form

$$\phi_\ell = \beta_\ell[\exp(2i\delta_\ell)h_\ell^{(1)} + h_\ell^{(2)}]. \tag{2.47}$$

δ_ℓ is known as the phase shift for angular momentum ℓ.

Now if the potential happens to be zero inside as well as outside the sphere, i.e. the muffin tin is empty, then equation (2.47) holds inside the muffin tin, in particular at $r = 0$. $h_\ell^{(1)}$ and $h_\ell^{(2)}$ are singular at the origin; since a physical wavefunction cannot be singular we must choose $\delta_\ell = 0$ when there is no potential so that the two singularities in equation (2.47) cancel. If there is no potential there can be no scattering of electrons by the ion core, hence $\delta_\ell = 0$ is the no-scattering situation.

For $\delta_\ell \neq 0$ we can decompose (2.47) into unscattered, $\phi_\ell^{(0)}$, and scattered, $\phi_\ell^{(s)}$, components

$$\phi_\ell = \phi_\ell^{(0)} + \phi_\ell^{(s)} \tag{2.48}$$

$$\phi_\ell^{(s)} = \beta_\ell[\exp(2i\delta_\ell) - 1]h_\ell^{(1)} \tag{2.49}$$

$$\phi_\ell^{(0)} = \beta_\ell[h_\ell^{(1)} + h_\ell^{(2)}] = \beta_\ell \; 2j_\ell \tag{2.50}$$

j_ℓ is the spherical Bessel function of order ℓ.

The phase shifts can be calculated by integration of equation (2.39) starting at the origin, where we have the boundary condition that $\phi_\ell(r)$ must not be singular, and proceeding out to the boundary of the muffin-tin sphere. Once out of range of the potential we know that ϕ_ℓ has the form (2.47) and from details of the integration we calculate the logarithmic derivative at the boundary of the muffin tin sphere, radius R.

$$L_\ell(R) = \frac{\phi'_\ell(R)}{\phi_\ell(R)} = \frac{\exp(2i\delta_\ell)h_\ell^{(1)\prime}(\kappa R) + h_\ell^{(2)\prime}(\kappa R)}{\exp(2i\delta_\ell)h_\ell^{(1)}(\kappa R) + h_\ell^{(2)}(\kappa R)},$$

$$\exp(2i\delta_\ell) = \frac{L_\ell h_\ell^{(2)} - h_\ell^{(2)\prime}}{h_\ell^{(1)\prime} - L_\ell h_\ell^{(1)}}. \tag{2.51}$$

A prime denotes differentiation with respect to r.

Scattering by the ion core can be characterised by a set of phase shifts, one for every value of the angular quantum number, ℓ, of the incident electron. The phase shifts depend not only on the potential, but also on the energy of the incident electron.

Once the incident wave is given, the total wave function, incident plus scattered, is known outside the muffin tin spheres via equations (2.48)–(2.50). In fact this is all we shall need to know in the LEED problem as our experimental observations are always confined to the region outside the muffin-tin spheres! The muffin-tin spheres are to be thought of as typical "black boxes" in so far as knowing the phase shifts we can predict what will come out from what goes in, and that is enough for solving the LEED problem.

As an example, let us suppose that conditions in the solid were such that a plane wave would travel through the ion-core shown in Fig. 2.9 in the absence of the ion-core potential,

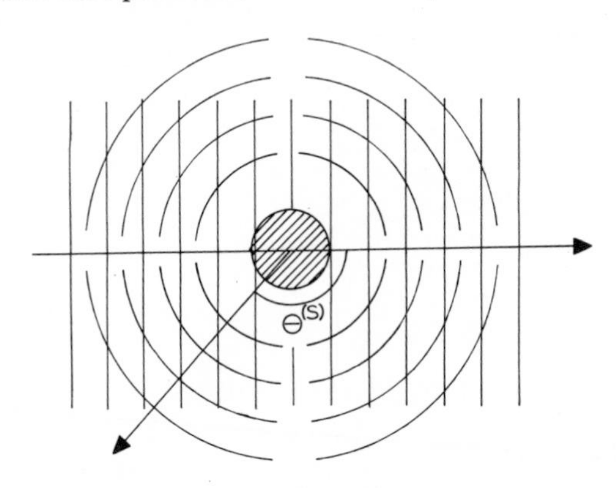

FIGURE 2.9. An ion-core immersed in a plane-wave. Scattered waves radiate from the ion-core, their amplitudes being functions of the scattering angle, $\theta^{(s)}$.

$$\begin{gathered}\exp(i\mathbf{k}\cdot\mathbf{r}),\\ \tfrac{1}{2}|\mathbf{k}|^2 = \tfrac{1}{2}\kappa^2 = E.\end{gathered} \tag{2.52}$$

To find how the ion-core scatters, the plane wave must be decomposed into spherical waves centred on the ion-core.

$$\exp(i\mathbf{k}\cdot\mathbf{r}) = \sum_{\ell=0}^{\infty} \sum_{m=-\ell}^{+\ell} 4\pi i^\ell j_\ell(\kappa r)\, Y_{\ell m}^*(\Omega(\mathbf{k}))\, Y_{\ell m}(\Omega(\mathbf{r})) \tag{2.53}$$

Therefore $\phi_\ell^{(0)}$ of equation (2.50) is

$$\phi_\ell^{(0)} = 2\pi i^\ell \, Y_{\ell m}^*(\Omega(\mathbf{k}))\cdot 2 j_\ell(\kappa r), \tag{2.54}$$

thus we identify

$$\beta_\ell = 2\pi i^\ell \, Y_{\ell m}^*(\Omega(\mathbf{k})),$$

and the scattered wave is

$$\phi_\ell^{(s)} = 2\pi i^\ell \, Y_{\ell m}^*(\Omega(\mathbf{k}))\left[\exp(2i\delta_\ell) - 1\right] h_\ell^{(1)}(\kappa r). \tag{2.55}$$

Adding together the various scattered spherical waves, the total is

$$\begin{aligned}\phi^{(s)}(\mathbf{r}) &= \sum_{\ell m} 2\pi i^\ell \left[\exp(2i\delta_\ell) - 1\right] Y_{\ell m}^*(\Omega(\mathbf{k})) \\ &\quad \times Y_{\ell m}(\Omega(\mathbf{r}))\, h_\ell^{(1)}(\kappa r) \\ &= \sum_\ell i^{\ell+1} \sin(\delta_\ell)\exp(i\delta_\ell)(2\ell + 1) \\ &\quad \times P_\ell(\cos(\theta^{(s)}))\, h_\ell^{(1)}(\kappa r),\end{aligned} \tag{2.56}$$

where $\theta^{(s)}$ is the angle through which the electron is scattered, i.e. the angle between $\mathbf{r}$ and $\mathbf{k}$.

Although in a solid the scattered wave does not have far to go before it meets another ion-core it is nevertheless instructive to see how equation (2.56) behaves for large values of r. In this case we can substitute the asymptotic form of $h_\ell^{(1)}$ from (2.46) to get a simpler picture of the scattered wave:

$$\phi^{(s)}(\mathbf{r}) \underset{r\to\infty}{=} t(E, \theta^{(s)})\left[-\frac{\exp(i\kappa r)}{2\pi r}\right], \tag{2.57}$$

$$t(E,\theta^{(s)}) = \frac{-2\pi}{\kappa}\sum_\ell (2\ell + 1)\sin(\delta_\ell)\exp(i\delta_\ell)\, P_\ell(\cos\theta^{(s)}). \tag{2.58}$$

It is evident that at large distances the scattered wave takes the form of a spherical wave, $\exp(i\kappa r)/r$), centred on the ion core, whose amplitude varies with angle. t is usually referred to as the "t matrix" of the ion core. Equation (2.58) expresses it in terms of phase shifts and is exact, but when scattering by the ion core potential can be treated in first order perturbation theory (which it never can in the LEED range of energies), the direct connection between t, and the Fourier transform of the ion-core potential is evident:

$$\begin{aligned} t(E, \theta^{(s)}) &\simeq \int\int \exp(-i\mathbf{k}'\cdot\mathbf{r}')\left[V_H(\mathbf{r})\delta(\mathbf{r} - \mathbf{r}') + V_{ex}(\mathbf{r}, \mathbf{r}')\right] \\ &\quad \times \exp(i\mathbf{k}\cdot\mathbf{r})\, d^3\mathbf{r}\, d^3\mathbf{r}', \end{aligned} \tag{2.59}$$

$$|\mathbf{k}'| = |\mathbf{k}| = \kappa,$$

$$\mathbf{k}'\cdot\mathbf{k} = \kappa^2\cos(\theta^{(s)}).$$

The incident plane wave given by (2.52) has a flux of κ per unit cross-sectional area. From (2.57) we deduce that the differential cross-section for scattering into solid angle $d\Omega$ is

$$\frac{d\sigma(E,\theta^{(s)})}{d\Omega} = \left|\frac{t(E,\theta^{(s)})}{2\pi}\right|^2. \tag{2.60}$$

The total flux scattered follows from integration of (2.60) over angles,

$$\sigma_T = \frac{4\pi}{\kappa^2}\sum_{\ell}(2\ell + 1)\sin^2(\delta_\ell). \tag{2.61}$$

An elementary treatment of scattering theory can be found in Schiff (1968), and a more advanced approach in Newton (1966).

The Legendre polynomials, $P_\ell(\cos\theta^{(s)})$, are oscillatory functions of angle. With increasing ℓ they oscillate more rapidly, therefore the greater the angular structure of scattering, the more phase shifts will be needed to describe it. Although the summation is in principle over all values of ℓ, in practice only a finite number of angular momenta are of importance in $\phi^{(s)}(\mathbf{r})$. A simple classical argument shows why. All of the potential is confined within a sphere of radius R. If there were a classical particle of angular momentum ℓ about the centre of the ion core, and linear momentum κ, the closest it would approach to the centre would be R_ℓ,

$$\kappa R_\ell = \ell, \tag{2.62}$$

and if R_ℓ were greater than R the classical particle would not see the potential at all. Figure 2.10 illustrates the situation.

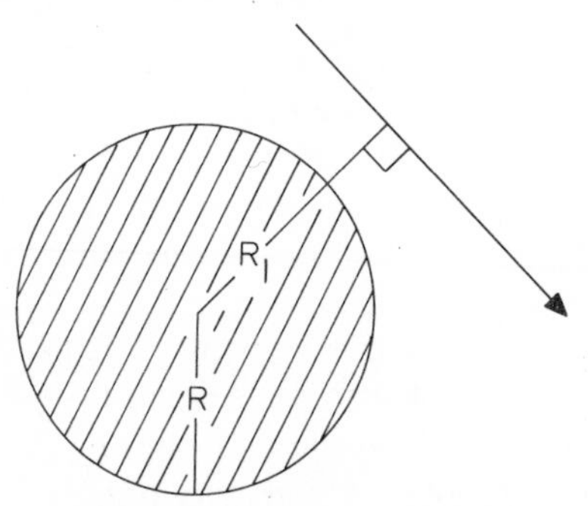

FIGURE 2.10. The classical trajectory of an electron approaches the ion-core no closer than a distance R_ℓ determined by the angular momentum. If R_ℓ is greater than the muffin-tin radius, R, the classical particle does not penetrate the potential.

A quantum mechanical particle can penetrate into classically forbidden regions, but the more the classical laws are violated by its doing so, the less probability there is of this happening. Therefore by the time R_ℓ has grown to R, the electron is seeing only the outermost parts of the ion-core, where the potential is weakest. Substituting typical values for a 50 eV electron we retrieve for the maximum value of ℓ that is going to be significant

$$\ell_m \simeq 4 \tag{2.63}$$

In Fig. 2.11 phase shifts for aluminium are shown, calculated according to the prescription we have outlined. The general picture is clear: of an increasing number of phase shifts with increasing energy, reflecting increasing complexity of angular structure in the scattering.

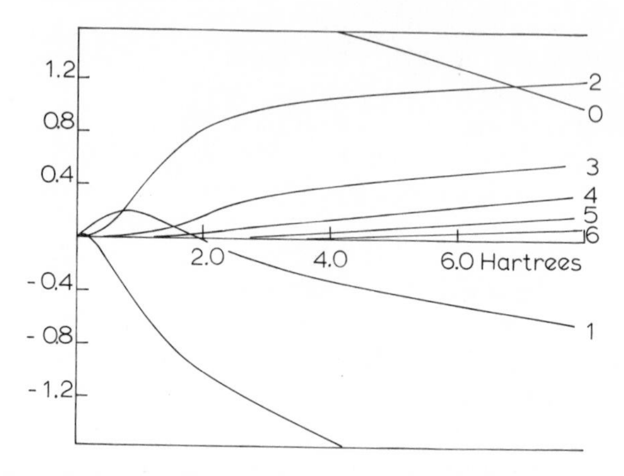

FIGURE 2.11. Phase shifts for aluminium as functions of energy plotted modulo π. The energy is measured relative to the constant potential between muffin tins.

At a typical energy for a LEED electron, there will be much more angular structure in the t-matrix than in the t-matrix experienced by a conduction-band electron, even though the same potential is responsible for the scattering. Figure 2.12 shows the t-matrix for aluminium (divided by the volume of the unit cell in the metal) for several energies. Increasing strength of scattering with increasing energy is apparent, the forward scattering peak growing in strength, back scattering becoming weaker at high energies. It is the sharply peaked nature of forward scattering that necessitates use of many phase shifts to describe high energy scattering. Also notice the detailed angular structure, even at 50 eV, both in amplitude and phase.

It is also instructive to examine how great differences between various ways of constructing potentials are. Figure 2.13 shows how the t-matrix for nickel changes when core-state wavefunctions as calculated by Clementi (1965) and Herman Skillman (1963) are employed, and when screening charge is distributed either as a uniform charge density inside the muffin-tin, or as following the density of the atomic 4s state. It would appear that there is not a great deal of difference between the various procedures, and in so far

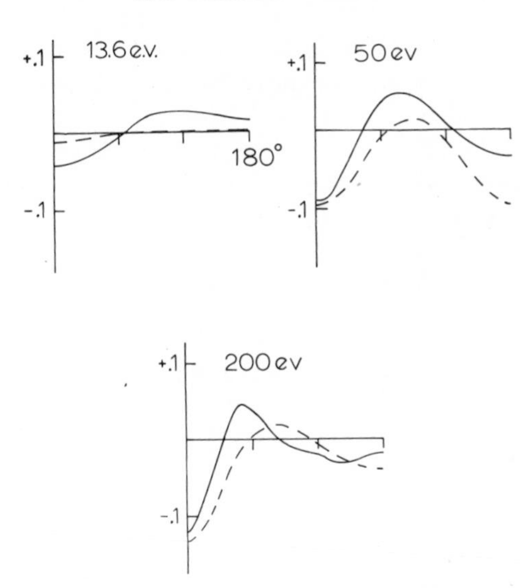

FIGURE 2.12. The t-matrix for aluminium divided by the volume of unit cell, showing the effect of increasing energy on scattering. ——— real part, ------ imaginary part.

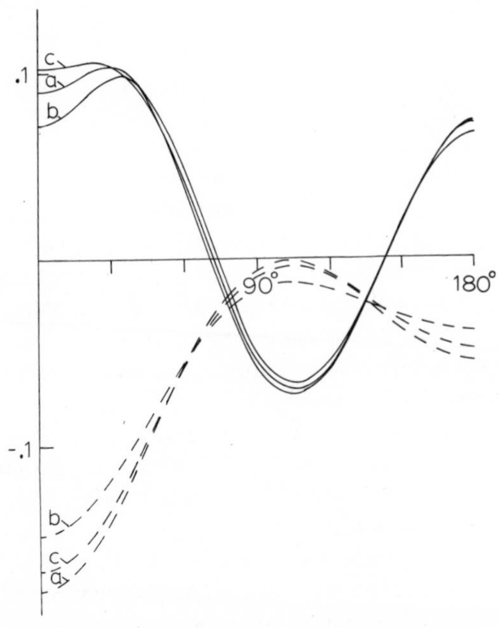

FIGURE 2.13. t-matrices for nickel/volume of unit cell, at 50 eV incident energy. (a) Herman-Skillman core-state wavefunctions, uniform screening charge. (b) Clementi core-states, uniform screening charge. (c) Clementi core-states, screening charge follows the 4s density. ——— real part, ------ imaginary part.

as differences are representative of overall errors, we can expect to calculate phase shifts by the methods given above. However the importance of including an adequate number of phase-shifts cannot be too strongly stressed. To drive home the point Fig. 2.14 shows how the t-matrix varies when successive phase shifts are omitted.

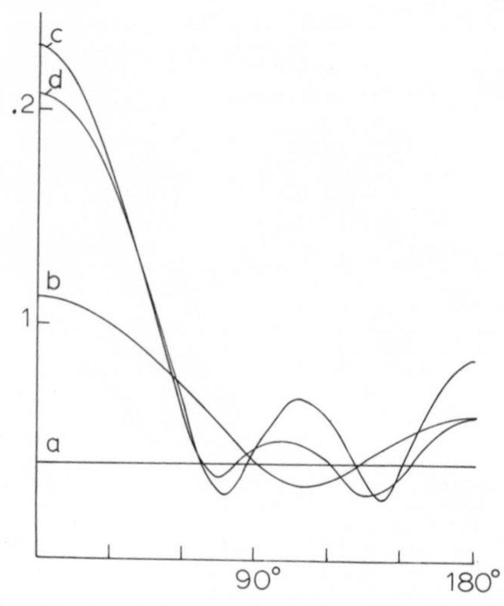

FIGURE 2.14. Modulus of the t-matrix for nickel/volume of unit cell, at 50 eV incident energy, including phase shifts (a) $\ell = 0$, (b) $\ell = 0,1$, (c) $\ell = 0,1,2$, (d) $\ell = 0,1,2,3,4$.

It is of interest to investigate why the t-matrix varies as it does with energy, and to find if there is any systematic variation from one material to another. The potential giving rise to scattering can be roughly characterised by atomic number, Z, of the ion-core and it might be supposed that the larger Z becomes, the stronger the scattering factor, because the potential is stronger. Such arguments are rudely shattered by comparing the almost free-electron-like conduction band of sodium, $Z = 11$, with the valence band of solid hydrogen where electrons are very strongly influenced in their behaviour by ion cores.

The answer to this paradox is to be found in pseudopotential theory (Herring, 1940, Phillips and Kleinman, 1959), and in particular the pseudopotential cancellation theorem (Cohen and Heine, 1961, Pendry, 1971). The theorem states that an attractive potential strong enough to have bound states behaves as if its scattering strength is sapped away by the effort of binding a state, and scatters electrons in the conduction band only weakly.

The proof of this statement involves several sophisticated concepts of formal scattering theory, which need not detain us here. Some idea of how the effect comes about can be gleaned from the following classical argument. If a particle moves in a weak potential, motion of the particle will not be much affected, and it samples the potential in a uniform manner. In a strong potential another effect comes into play: the velocity of the particle will increase in regions of attractive potential, and thus the particle will pass through these regions more rapidly than in the absence of the potential. Hence the influence of a strong attractive potential on a particle is less than one might anticipate because motion of the particle is changed so that it spends less time under the influence of strong potentials. Evidently the lower the incident energy of the electron, the greater will be the fractional increase in velocity in the core region, the less it will see of the potential there and the greater the cancellation effect. At higher energies the electron sees more and more of the core region.

Applying these arguments to ion-cores we can easily see why properties of conduction and valence bands are determined by structure of outer regions of the ion core. At LEED energies penetration is greater and scattering consequently stronger; this fact changes the balance of importance of quantities affecting electronic motion at LEED energies. It is of the first importance to calculate an accurate value of ion-core scattering for LEED energies because this is the dominant scattering process. In fact it is virtually the only mechanism that turns around electrons so that they are reflected from a crystal surface. Without ion-core scattering, LEED would not be possible!

Despite the fact that LEED electrons penetrate core-regions more than conduction electrons, for most elements the penetration is by no means complete. Elements with a higher value of Z are penetrated less than those with low Z and the result is that Z is no sure guide to scattering strength at LEED energies. Roughly speaking, all elements are equivalent in their scattering strength *per ion core*, though details of the t-matrix vary considerably. Of greater importance in determining scattering strength *per unit volume of solid* is the size of unit cell occupied by the ion. Thus solid xenon with a unit cell eight times as large as that of nickel might be expected to be a much weaker scatterer, the larger atomic number (54 as opposed to 28 for nickel) having relatively small influence on scattering powers of the ions.

A rough guide to penetration of a region of an ion-core, is the binding energy of core-states occupying that region. So it is only at the highest LEED energies that electrons penetrate the innermost core regions and scattering strength reflects size of atomic number. For higher energies such as are used in transmission electron diffraction it is usually the case that the core is completely penetrated, and in the kilovolt range Z is a good guide to scattering strength.

In the appendix there are programs for calculating phase shifts, and in Table 2.1 we reproduce some calculations made with these programs for various elements.

The phase shifts in Table 2.1 are shown as functions of energy measured relative to the constant potential between muffin-tins. Eventually we wish to refer energies to the vacuum level, and must correct for the constant potential between muffin-tins, V_0. If the imaginary part of V_0 were zero this would be a trivial shift in energy, but V_{0i} is not generally zero as we shall see

TABLE 2.1. Phase shifts for several elements as they occur in their normal crystalline form, given modulo π. Energy is calculated relative to the constant potential between muffin tins and is given in Hartrees.

element: aluminium

ℓ / E	0	1	2	3	4	5	6
0.50	−0.21	+0.18	+0.09	—	—	—	—
0.75	−0.40	+0.21	0.22	0.02	—	—	—
1.00	−0.55	+0.18	0.36	0.04	—	—	—
1.50	−0.80	+0.09	0.63	0.10	0.02	—	—
2.00	−1.00	−0.01	0.80	0.17	0.03	—	—
2.50	−1.17	−0.11	0.90	0.25	0.07	0.01	—
3.00	−1.31	−0.19	0.97	0.31	0.10	0.02	—
4.00	−1.52	−0.32	1.06	0.40	0.15	0.06	0.01
5.00	+1.44	−0.42	1.12	0.46	0.22	0.09	0.04
6.00	+1.30	−0.51	1.14	0.51	0.26	0.13	0.06
7.00	+1.16	−0.58	1.16	0.55	0.29	0.16	0.08
8.00	+1.02	−0.65	1.18	0.58	0.32	0.17	0.09

element: nickel

ℓ / E	0	1	2	3	4	5	6
0.50	−0.70	−0.12	−0.36	—	—	—	—
0.75	−0.90	−0.22	−0.40	+0.01	—	—	—
1.00	−1.07	−0.32	−0.45	0.03	—	—	—
1.50	−1.34	−0.52	−0.50	0.09	0.02	—	—
2.00	−1.55	−0.66	−0.53	0.15	0.03	—	—
2.50	+1.41	−0.79	−0.55	0.22	0.05	0.01	—
3.00	+1.27	−0.90	−0.56	0.28	0.07	0.02	0.01
4.00	+1.01	−1.09	−0.58	0.42	0.11	0.03	0.01
5.00	+0.81	−1.24	−0.60	0.54	0.15	0.05	0.02
6.00	+0.64	−1.38	−0.61	0.64	0.02	0.07	0.03
7.00	+0.49	−1.49	−0.62	0.73	0.24	0.09	0.04
8.00	+0.36	+1.55	−0.63	0.82	0.28	0.11	0.05

Table 2.1 – *continued*

element: xenon

ℓ / E	0	1	2	3	4	5	6	7	8
0.50	+0.96	+1.66	+0.93	+0.19	+0.02	—	—	—	—
0.75	+0.58	+1.33	+0.92	+0.55	0.06	—	—	—	—
1.00	+0.27	+1.05	+0.89	+0.89	0.11	0.02	—	—	—
1.50	−0.18	+0.68	+0.73	−1.23	0.22	0.06	0.02	—	—
2.00	−0.51	+0.40	+0.59	−0.69	0.35	0.11	0.04	—	—
2.50	−0.77	+0.17	+0.45	−0.43	0.48	0.16	0.06	—	—
3.00	−0.99	−0.01	+0.36	−0.29	0.59	0.22	0.08	—	—
4.00	−1.35	−0.32	+0.15	−0.13	0.81	0.33	0.14	0.01	—
5.00	+1.52	−0.55	−0.02	−0.05	0.97	0.43	0.21	0.07	0.03
6.00	+1.28	−0.76	−0.13	−0.01	1.12	0.53	0.27	0.15	0.08
7.00	+1.08	−0.93	−0.25	−0.02	1.23	0.61	0.33	0.19	0.11
8.00	+0.91	−1.09	−0.36	+0.05	1.33	0.67	0.38	0.23	0.13

in section *D*. The correction proceeds as follows. We are given $\delta_\ell(E')$ where E' is a real energy measured relative to the muffin-tin zero. We want δ_ℓ for a complex energy relative to the muffin-tin zero of $(E - V_{0r} - iV_{0i})$. Now V_{0i} is always small in comparison with $(E - V_{0r})$ and to a good approximation

$$\delta_\ell(E - V_{0r} - iV_{0i}) \simeq \delta_\ell(E - V_{0r}) - iV_{0i}\frac{d\delta_\ell(E - V_{0r})}{dE}. \qquad (2.64)$$

From equation (2.47) we see that an imaginary part of the phase shift implies that either more, or less flux is being absorbed inside the muffin-tin according to whether $d\delta_\ell/dE$ is positive or negative, than would be absorbed by V_{0i} in the absence of the ion core. The conclusion is that if $d\delta_\ell/dE$ is positive the electron spends longer in the region of the ion core, and has more chance of being inelastically scattered there. Note that for those values of ℓ that core states exist for, $d\delta_\ell/dE$ is negative, elegantly confirming our statement that the electron spends less time in the region of a potential strong enough to bind a state.

Usually $d\delta/dE$ is so small that the correction to δ_ℓ from the second term in (2.64) can be neglected. We mention it here for completeness' sake, and for the insight it gives of potential scattering theory.

IID The optical potential

We must now turn to consider the real and imaginary parts of V_0 (V_{0r} and V_{0i}) which are the non-ion-core parts of the crystal potential. Some rather drastic assumptions have been made, not least of which is that all other scattering can be summarised by V_{0r} and V_{0i}. In this section our aim is to establish that the remaining contributions to electron scattering can be expressed as a one electron potential acting on the incident electron, and then to discuss what sort of qualitative properties this potential has. Quantitative details are left until Section E.

There are two major contributions that we have omitted: from the conduction or valence electrons, and at finite temperatures from phonons. The latter contribution will be dealt with in a different manner later in the book. For the moment we assume $T = 0°K$ when the dominant term is that from conduction and valence electrons.

It has been remarked above that unlike core electrons conduction and valence electrons are readily polarisable. In consequence wavefunctions of these electrons are changed by the incident electron in a manner depending on the exact nature of that electron's motion. The result is that we have a many-body problem and the equations do not factorise into a set of single-electron equations as they did in the Hartree-Fock approximation where electrons of the solid were assumed unpolarisable.

In Section B we derived equation (2.21) describing motion of an incident electron under the influence of a single ion core. Except for the screening potential, v_s, influence of conduction electrons was neglected. When all ion-cores in the solid are included equation (2.21) becomes

$$-\tfrac{1}{2}\nabla^2\phi + v_s\phi + V_c\phi = E\phi, \tag{2.65}$$

$$V_c\phi = \left[\sum_{jst}\int d^3\mathbf{r}'\,\frac{|\psi_j(\mathbf{r}'-\mathbf{R}_t,s)|^2}{|\mathbf{r}-\mathbf{r}'|}\right]\phi(\mathbf{r},s) - \left[\sum_{jst}\int d^3\mathbf{r}'\,\frac{\psi_j^*(\mathbf{r}'-\mathbf{R}_t,s)\,\phi(\mathbf{r}',s)}{|\mathbf{r}-\mathbf{r}'|}\right]\psi_j(\mathbf{r}-\mathbf{R}_t,s). \tag{2.66}$$

The location of a given ion-core we have denoted by $\mathbf{R}_t$ and assumed for simplicity's sake that only one sort of ion core is present.

Now we must construct a more accurate equation. Conduction electrons as well as the incident electron obey an equation derived from (2.65) except that now we wish to improve upon our treatment of electron-electron interaction. If we write the total wavefunction for incident electron plus N conduction/valence electrons as

$$\Psi(\mathbf{r}_0, s_0; \cdots \mathbf{r}_N, s_N),$$

then Ψ obeys an equation similar to (2.65)

$$-\tfrac{1}{2}\sum_{i=0}^{N}\nabla_i^2\Psi + \sum_{i=0}^{N}V_c(\mathbf{r}_i)\Psi + \sum_{i=0}^{N}\sum_{j=i+1}^{N}\frac{\Psi}{|\mathbf{r}_i-\mathbf{r}_j|} = E_s\Psi \tag{2.67}$$

E_s is the total energy associated with the incident and conduction electrons.

If we insist on knowing all details of the incident electron's motion, we must solve equation (2.67) as it stands. But usually we do not need such detailed information. Usually the energy selecting grids are in position and the experiment is concerned only with elastically scattered electrons. Under these circumstances we can make use of a theorem from nuclear physics stating that the wavefunction of an elastically scattered particle obeys a Schrödinger equation with an effective or "optical" potential term. Unlike potentials in a conventional Schrödinger equation the optical potential is not in general Hermitian. The non-Hermitian character causes violation of flux conservation and is associated with loss of intensity through inelastic scattering processes.

To understand how this theorem comes about we observe that the total wavefunction Ψ can be written in terms of the ground and excited states of the conduction/valence band. Let those states of the conduction electrons alone be

$$\Gamma_k(\mathbf{r}_1, s_1; \cdots \mathbf{r}_N, s_N)$$

corresponding to excitation energies E_k. E_0 is the ground-state energy. Γ_k obeys equation (2.68)

$$\sum_{i=1}^{N}\left[-\tfrac{1}{2}\nabla_i^2 + V_c(\mathbf{r}_i) + \sum_{j=i+1}^{N}\frac{1}{|\mathbf{r}_i - \mathbf{r}_j|}\right]\Gamma_k = E_k\Gamma_k. \tag{2.68}$$

Interaction of an incident electron with a system originally in the ground state can be thought of as a set of excitations of conduction electrons

$$\Psi(\mathbf{r}_o, s_o; \cdots \mathbf{r}_N, s_N) = \sum_k \phi_k(\mathbf{r}_o, s_o)\Gamma_k(\mathbf{r}_1, s_1; \cdots \mathbf{r}_N, s_N). \tag{2.69}$$

ϕ_j is a state of the incident electron that has lost energy $(E_j - E_o)$ to conduction electrons. We assume for simplicity that these states are not degenerate, but degeneracy introduces only mathematical complications. It is understood that Ψ should be antisymmetrised for particle exchange, an exercise left to the reader.

Substituting for (2.69) into (2.67) we retrieve after making use of (2.68)

$$\sum_k\left[-\tfrac{1}{2}\nabla_o^2 + V_c(\mathbf{r}_o) + \sum_{j=1}^{N}\frac{1}{|\mathbf{r}_o - \mathbf{r}_j|}\right]\phi_k(\mathbf{r}_o, s_o)$$
$$\times\ \Gamma_k(\mathbf{r}_1, s_1; \cdots \mathbf{r}_N, s_N) = \sum_k (E_s - E_j)\,\phi_k\Gamma_k. \tag{2.70}$$

Multiplying by Γ_0^*, integrating over conduction electron coordinates and making use of the orthogonality of the Γ_k's gives us an equation for the elastically scattered component of the wavefunction

$$[-\tfrac{1}{2}\nabla_o^2 + V_c(\mathbf{r}_o)]\phi_o(\mathbf{r}, s_o) + v_{op}(\mathbf{r}_o, s_o)\phi(\mathbf{r}_o, s_o)$$
$$= (E_s - E_o)\phi_o = E\phi_o(\mathbf{r}_o, s_o), \tag{2.71}$$

$$v_{op}(\mathbf{r}_0, s_0) = \sum_{j=1}^{N} \sum_{k=0}^{\infty} \frac{1}{\phi(\mathbf{r}_0, s_0)} \sum_{s_1 \ldots s_N}$$
$$\times \int d^3\mathbf{r} \cdots d^3\mathbf{r}_N \frac{\Gamma_0^*(\mathbf{r}_1, s_1 \cdots \mathbf{r}_N, s_N)\Psi_k(\mathbf{r}_0, s_0, \cdots \mathbf{r}_N, s_N)}{|\mathbf{r}_0 - \mathbf{r}_j|}, \tag{2.72}$$

which is of the form we wish for. Note the absence of any indication of an Hermitian form to v_{op}.

The optical potential which acts on the elastic component of the wavefunction evidently contains contributions from excited states of conduction electrons. These fall into two classes: 'virtual' excitations that fall back into the ground state giving back energy after the incident electron has passed, and 'real' excitations that do not return energy. The first type corresponds to an elastic deformation or polarisation. Short lived virtual excitations are always allowed by the uncertainty principle but the second term can only be non-zero if the incident electron has enough energy to spare for creating a permanent excitation. It is the real excitations that are responsible for the non-Hermitian flux-absorbing part of v_{op} in (2.72).

The real part of the optical potential can be understood simply in terms of additional contributions to elastic electron scattering. It is in the non-Hermitian part that the optical potential differs from conventional potentials. As we have seen it produces absorption of the elastically scattered beam and thereby broadening of structure in intensity/energy curves of reflectives thus obscuring a lot of detailed structure that might otherwise be seen in these curves. At the same time strong absorption can bring about welcome simplifications in LEED calculations by reducing the length and therefore complexity of the path an electron can follow inside a crystal. The imaginary part of the optical potential is second only to the ion-core potentials in its importance in LEED calculations and we shall spend a little time discussing it in detail. A good elementary account of various excitations involved will be found in Pine's (1963) book and more detailed considerations in a review article by Hedin and Lundqvist (1969).

Although we shall show that once away from the immediate surface region spatial variation of v_{op} is small,

$$v_{op}(\mathbf{r}) \simeq V_{0r} + iV_{0i}, \tag{2.73}$$

the absorptive part, V_{0i}, has structure as a function of incident energy, at least in the lower energy range. No contribution to V_{0i} from a particular loss mechanism can come about until the incident electron has enough energy to excite this mechanism. Plasmons have energies around 15 eV and in metals with free-electron conduction bands the large contribution of plasmons to V_{0i} does not come into play below an incident energy of this value (measured relative to the Fermi-level, because that is as much energy as can be lost). Even in non-free electron metals there is a more general cluster of excitations usually at around the equivalent free-electron plasmon energy. At higher energies still the deep core levels can be excited and V_{0i} takes a further increase though, it will transpire, by very small amounts.

Just above the threshold energy, V_{0i} rises rapidly; at very high energies the speed of the electron reduces absorption and V_{0i} falls as $E^{-\frac{1}{2}} \ell n\, E$; between these extremes a broad maximum occurs in the contribution to V_{0i} from the excitation. In fact once the major part of V_{0i} has been established on crossing the plasmon threshold, there is a relatively small amount of structure in V_{0i} at higher energies.

Below the plasmon energy, single electrons can still be excited from the conduction band, though they produce a smaller V_{0i} than plasmons. As the incident electron's energy creeps closer to E_F the phase space for excitation of conduction electrons falls, with the result that V_{0i} falls to zero near the Fermi level as $(E - E_F)^2$.

Variations from one material to another are not spectacular above plasmon energies, say above 20–30 eV, most materials tending to give a V_{0i} of around -4 eV within a factor of two, i.e. a path length of around 5Å at 100 eV. Increase in electron density is usually accompanied by increasing 'stiffness' of the electrons, pushing back V_{0i} towards a slower variation with electron density.

Below the plasmon energy variations can be more interesting. In metals single electrons can be excited by an incident electron that has any energy

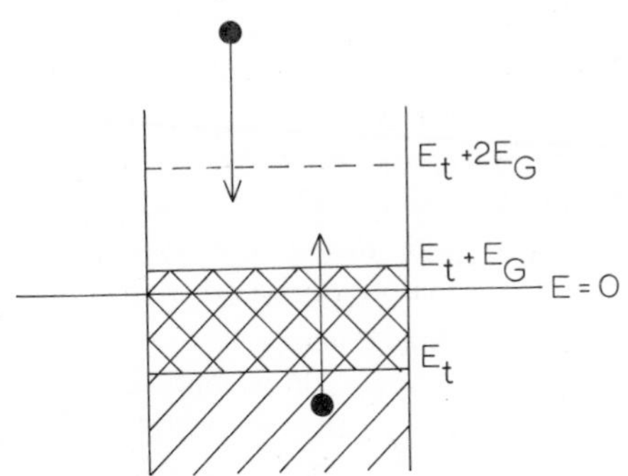

FIGURE 2.15. A schematic energy-level diagram for an insulator showing the top of a fully occupied valence band, E_T, above that a gap of width E_G where there are no allowed levels, and finally allowed but unoccupied states. The vacuum zero of energy is marked. Incident electrons can fall no further than the top of the gap, and excited electrons must have an energy of at least $E_T + E_G$, because of the exclusion principle.

above the Fermi-level, but in insulators the situation is different. Suppose we take a wide band-gap insulator such as magnesium oxide. Figure 2.15 shows allowed energy levels inside the crystal. Incident electrons can only fall down to the top of the forbidden energy range; valence electrons must receive at least enough energy to clear the gap before they can be excited. There is a range of incident energies

$$0 < E < E_T + 2E_G \tag{2.74}$$

where the incident electron either cannot penetrate the solid, or cannot fall far enough to excite a valence electron.

LEED from insulators can be expected to have a very low energy 'window' where electronic excitations are virtually switched off and only residual scattering by phonons and impurities remains, though little has been done to date in investigating path lengths for non-free-electron materials in this lower energy range, either theoretically or experimentally.

IIE Calculating the optical potential inside a solid

The amount of quantitative information available from theory is not large, being limited to free-electron-like bands or excitation of very tightly bound core-states. Nevertheless the theory is able to confirm our earlier estimates of V_{0i} for these free-electron materials, to show that core states make only small contributions as we have postulated, and to make semi-quantitative estimates of the spatial variation of the optical potential showing that it is small in the bulk of the solid for many materials.

Calculations are easier well away from the surface of a solid where the crystal can be regarded as having effectively full symmetry in the z, as well as x and y, directions. Therefore, sparse as it is, data for the bulk are more complete than for the less well understood case of the immediate surface region which will be examined in Section F.

Equation (2.72) enables us to understand how v_{op} comes about but as a means of calculation it is not so suitable. Calculations of v_{op} for cases of any generality have yet to be done, the most extensive investigations being for materials in which the ion-core potential in equation (2.67) can be neglected as far as incident and conduction electrons are concerned. For conduction electrons this is a good approximation in materials like the alkali metals or aluminium in which free-electron behaviour can be verified. It is almost never true of the incident electron and the best we can hope for is that since expression (2.72) for v_{op} involves a sum over several states in which the incident electron has lost energy, some averaging will take place in which the effect of ion-core scattering disappears.

Theory of the optical potential, or self-energy as it is sometimes called, has been developed in terms of perturbation theory and Feynman diagrams. A careful account will be found in Nozière (1964) and we shall quote some of his results. Also Quinn and Ferrell (1958) made an approximate numerical treatment of the problem we describe below.

The first order contribution to the optical potential involves the electron which we shall assume to have momentum $\mathbf{p}$, energy p_0, and spin s, interacting with one of the conduction electrons with momentum energy and spin, $\mathbf{p}'$ p_0' and s'. The Feynman picture is given in Fig. 2.16.

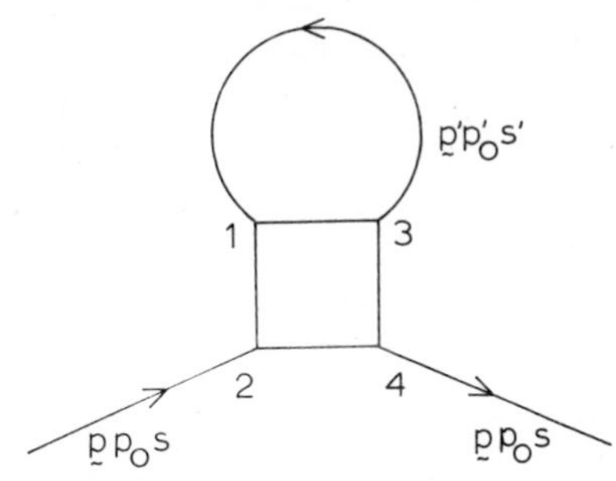

FIGURE 2.16. A Feynman diagram representing the first order term in the self-energy of an electron.

Lines represent electrons and the square represents an interaction. The formula derived for this contribution from the rules given in Nozières book is

$$v_{op} = \frac{-i}{(2\pi)^4} \sum_{s'} \int d^3\mathbf{p}' dp_0' \Big[v(q = 0, \omega = 0)$$

$$- \delta_{ss'} v(q = |\mathbf{p} - \mathbf{p}'|, \omega = (p_0 - p_0')) \Big] \frac{\exp(i\delta q_0')}{p_0' - \frac{1}{2}|\mathbf{p}'|^2 + i\eta}, \tag{2.75}$$

$$\delta = +0,$$

$$\eta = +0, \qquad |\mathbf{p}'| > k_F,$$

$$\eta = -0, \qquad |\mathbf{p}'| < k_F.$$

k_F is the Fermi momentum. The infinitesimals will be important in evaluation of the integral over p_0'. The brackets contain interaction terms, $v(|\mathbf{q}|)$ being the matrix element of the effective electron-electron interaction, scattering an electron from $\mathbf{k}$ to $\mathbf{k} + \mathbf{q}$. The first term inside the brackets is a direct interaction between incident and conduction electrons whereas the second term comes about from antisymmetry of the wavefunction on interchanging the two electrons, only occurring when the two electrons have parallel spins.

It will be noticed that v_{op} as given by (2.75) depends only on momentum and energy of the incident electron and is not a function of position in the crystal, a consequence of neglecting ion-core scattering and therefore assuming that once clear of the surface, the solid is translationally invariant. For this special case we can write

$$v_{op}(\mathbf{r}) = V_{0r} + iV_{0i}.$$

The electron-electron interaction is not simply a Coulomb interaction. Just as conduction electrons screen ion-cores, they also screen the bare Coulombic interaction between incident and conduction electrons. The bare Coulombic interaction is $4\pi/q^2$, but in a metal it is important to divide by the dielectric constant to take account of screening:

$$v(q, q_0) = \frac{4\pi}{q^2 \varepsilon(q, q_0)}. \tag{2.76}$$

The dielectric constant is a dynamical one, that is screening by conduction electrons depends on the frequency, q_0, of the interaction being screened.

The q_0' integration in (2.75) is performed first, by closing the contour of integration in the upper half of the complex q_0' plane (Fig. 2.17). The factor $\exp(i\delta q_0')$ ensures that the semicircle adds nothing to the integral. We must

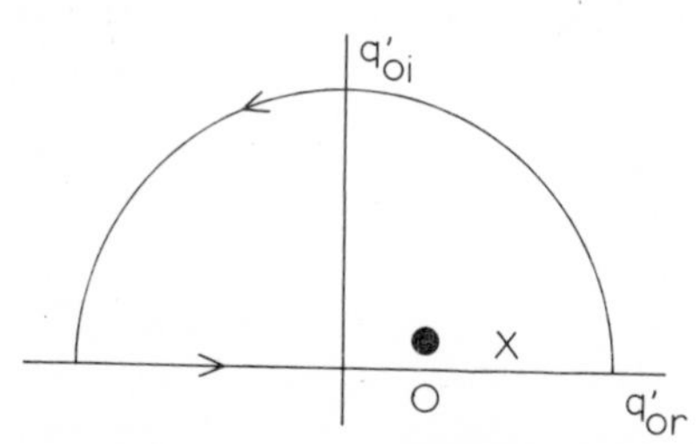

FIGURE 2.17. The contour for the q_0' integration in equation (2.75), × pole of the screened interaction, o pole of the last term in (2.75), $|\mathbf{p}'| > k_F$, ● ditto, $|\mathbf{p}'| < k_F$.

now count poles (and any cuts) trapped in the upper half plane. First we notice that the last term (2.75) gives rise to poles at

$$p_0' = \tfrac{1}{2}|\mathbf{p}|^2$$

which are caught inside the contour if $|\mathbf{p}'|$ is less than the Fermi momentum, contributing to the integral

$$\frac{1}{(2\pi)^3} \sum_{s'} \int d^3\mathbf{p}' \left[v(q = 0, \omega = 0) - \delta_{ss'} v(q = |\mathbf{p} - \mathbf{p}'|, \omega = p_0 - \tfrac{1}{2}|\mathbf{p}'|^2) \right]_{f|\mathbf{p}'|}$$

$$\begin{aligned} f|\mathbf{p}'| &= 1, \qquad |\mathbf{p}'| < k_F, \\ &= 0, \qquad |\mathbf{p}'| > k_F. \end{aligned} \tag{2.77}$$

The first term in (2.77) is the average field due to all electrons inside the Fermi sphere. In the more general case where ion-core interactions are included this term would have screened the ion-cores. Since we included screening in Sections B and C we shall now drop this term. The second is the exchange term and comes about because the region around the incident electron is denuded of parallel spin electrons by the exclusion principle and the potential it sees is not as repulsive as that due to a uniform distribution of electrons.

In addition to poles of the last term in equation (2.75), the screened interaction has a pole caused by a zero of the dielectric constant. Zero dielectric constant for given q and ω implies that the medium can sustain oscillations, wavelength q, frequency ω: the charge density executes plasma oscillations. This is how plasmons make their appearance in the formalism. The oscillations will always be decaying ones, i.e. the frequency has a negative imaginary component and the pole will be at a value of $(p_0 - p'_0)$ which has a small negative imaginary part. Hence the pole is in the upper half of the complex p'_0 plane and is caught in our contour.

There is other analytic structure of the screened interaction arising out of single particle excitations, but the plasmon pole is the dominant term in this expression. The remaining integration over p' contributes an imaginary component to v_{op} if a further pole occurs in the denominator

$$p'_0 - \tfrac{1}{2}|\mathbf{p}'| + i\eta$$

and the condition for this to happen is that the electron has enough energy to excite a real plasmon.

Lundqvist (1969) has carried out these calculations with the aid of a computer. In his results the energy in $v_{op}(p, p_0)$ is measured relative to the Fermi energy so the optical potential correction to a particle's energy is

$$E(k) = \tfrac{1}{2}k^2 + v_{op}(k, E(k) - E(k_F)) \simeq \tfrac{1}{2}k^2 + v_{op}(k, \tfrac{1}{2}k^2 - \tfrac{1}{2}k_F^2)$$
$$+ \left.\frac{\partial v_{op}(k\omega)}{\partial\omega}\right|_{\omega=\frac{1}{2}k^2-\frac{1}{2}k^2{}_F} \times \left[E(k) - E(k_F) - \tfrac{1}{2}k^2 + \tfrac{1}{2}k_F^2\right] \quad (2.78)$$

Also we have that

$$E(k_F) = \tfrac{1}{2}k_F^2 + v_{op}(k_F, 0), \quad (2.79)$$

therefore

$$E(k) \simeq \tfrac{1}{2}k^2 + Z(k)M(k) + M(k_F)[1 - Z(k)], \quad (2.80)$$

$$Z(k) = \left[1 - \left.\frac{\partial v_{op}(k,\omega)}{\partial\omega}\right|_{\omega=\frac{1}{2}k^2-\frac{1}{2}k^2{}_F}\right]^{-1} \quad (2.81)$$

$$M(k) = v_{op}(k, \tfrac{1}{2}k^2 - \tfrac{1}{2}k_F^2). \quad (2.82)$$

From equation (2.80) V_{0r} and V_{0i} can readily be recognised

$$V_{0r} + iV_{0i} = Z(k)M(k) + [1 - Z(k)]M(k_F). \tag{2.83}$$

In Figs 2.18–2.20 we reproduce Lundqvist's calculations of M and Z. Electron density is expressed in terms of r_s, the radius of a sphere containing one electron in a gas of that density. The Fermi momentum, k_F, is related to r_s by

$$k_F = 1.9192/r_s. \tag{2.84}$$

Lundqvist's results show numbers correlating with those we estimated from experiment in Chapter I and in Section A of this chapter. For $r_s = 2$, typical of metallic electron densities:

$$V_{0r} \simeq -0.5 \text{ Hartrees} = -13.6 \text{ eV}, \tag{2.85}$$

and an absorptive part

$$V_{0i} \simeq -0.2 \text{ Hartrees} = -5.5 \text{ eV}. \tag{2.86}$$

Fixing E at around 4.0 Hartrees (109 eV) and solving for the wave vector in equation (2.80) gives

$$\begin{aligned} \tfrac{1}{2}k^2 &= 4.0 + 0.5 + 0.2i, \\ k_r + ik_i &\simeq 3.0 + 0.067i. \end{aligned} \tag{2.87}$$

Suppose the wave is travelling normal to the surface

$$\exp(ikz) = \exp(i3.0z - 0.067z), \tag{2.88}$$

implying that the *intensity* of the wave decays by a factor e^{-1} after travelling 7.5 a.u. (4Å), consistent with our earlier estimates of path lengths.

We have asserted that the core states will not make a large contribution to correlation effects discussed in this section. Our conclusion was based on the argument that core states are tightly bound entities and will not be easily polarised or excited. Ing (1972) has confirmed this conclusion by direct calculation.

He started with a formula related to (2.75), but instead of letting the incident electron polarise the conduction electrons, core electrons were substituted. Now the free electron approximation will not do and the last term in (2.75) is more complicated. To regain simplicity he argued that for a tightly bound core state, the incident electron must pass very close by before it can polarise or ionise it. Only the short range part of the electron-electron interaction matters; screening is not important and the bare Coulombic interaction can be used.

We have seen that conduction electrons contribute a typical figure of -0.2 Hartrees to the imaginary part of the optical potential, whereas Ing calculates for aluminium that the core states add only a small fraction to this

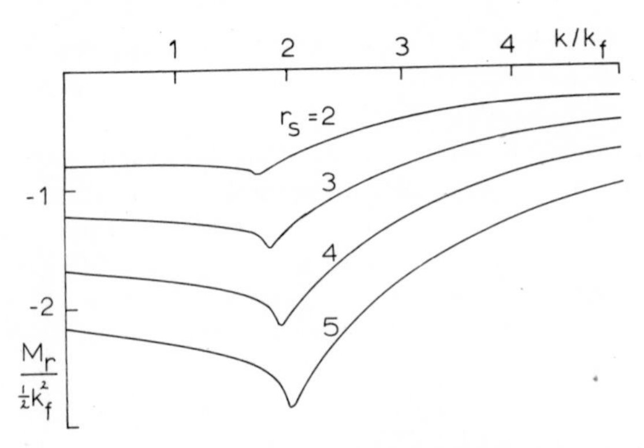

FIGURE 2.18. The real part of M plotted against k for various electron densities.

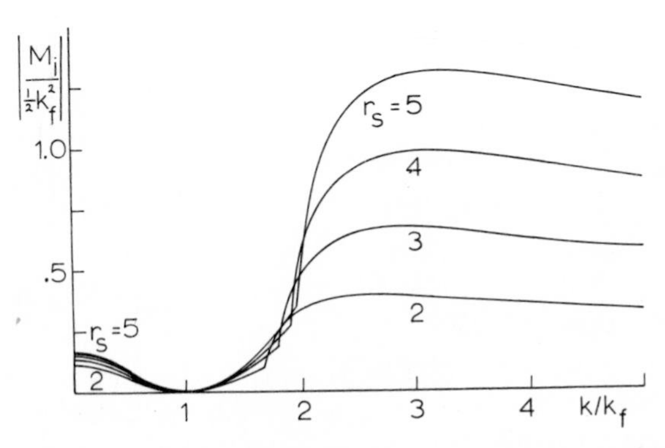

FIGURE 2.19. The imaginary part of M plotted against k for various electron densities.

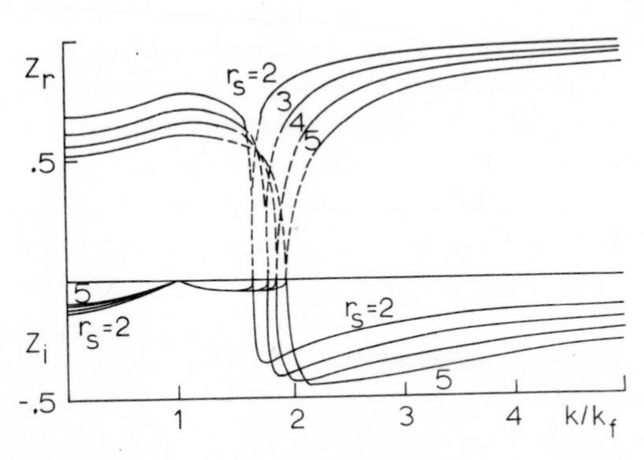

FIGURE 2.20. The real and imaginary parts of Z plotted against k for various electron densities.

number. Table 2.2 shows the maximum contribution to V_{0i} made by various core states in aluminium. Clearly contributions of core states can be neglected. A similar situation holds for the deep core states of other materials.

Materials whose electron states do not fit into the scheme of division between deep unpolarisable core states and free electron conduction electrons, are the transition and noble metals which are common subjects for LEED experiments. The *d*-bands are well localised about the ion-cores but their energies are often within a few electron volts of the Fermi surface. We might expect that these states have to be treated as contributing to the potential both as core states and as polarisable entities that affect the optical potential. Their localised nature makes them easily included in the calculation of the ion-core potential, but this fact excludes any simple treatment of the optical potential.

core state	ΔV_{oi} (Hartrees)
$A\ell$ 1s	−0.0003
$A\ell$ 2s	−0.0028
$A\ell$ 2p	−0.0030

TABLE 2.2. Showing the maximum contribution of various core states to the imaginary part of the optical potential.

The possibility of a major complication now occurs. It is that the highly non-uniform distribution of electrons in transition metals may lead to the optical potential being a strong function of position. Working against the effect of inhomogeneities in electron density there is the fact that the incident electron interacts with electrons in the *d*-levels even when it is outside their charge density. Therefore we expect the optical potential to be at least less inhomogeneous than the electron density causing it.

Ing has made upper estimates for inhomogeneities in the optical potential for copper and concludes that they probably amount to less than a 10% variation over the unit cell, a conclusion supported by the good agreement with experiment of calculations made for transition and noble metals on the assumption that the optical potential is uniform. In solids with larger, more complicated unit cells, inhomogeneity is probably of greater importance.

So much for theoretical estimates. Except for the nearly-free-electron metals, they mainly serve as a guide to the situation rather than provide accurate numbers for LEED calculations. For the more complicated transition and noble metals it is necessary to use empirical methods.

Firstly we can be confident from Ing's work that usually we need find only two quantities: V_{0r} and V_{0i}, because the optical potential is approximately homogeneous over the whole unit cell. Secondly, although both these numbers are in principle functions of the incident momentum, it has been found by comparing theory and experiment that the value of V_{0r} needed to fit experiments with an incident energy of 100 eV differs from that at 10 eV by only of the order of 10% (i.e. $\simeq$ 2 eV) in nickel, copper and molybdenum, conclusions that can be taken to apply to other transition metals. Hence we can fit V_{0r} to peak positions in intensity/energy spectra since to a first approximation its effect is to shift these uniformly. V_{0i} can be found by adjusting it to give correct peak widths because it is the principal agent in peak broadening. This procedure works well above the plasmon threshold where V_{0i} varies slowly, but sometimes it is not easy to find suitable peaks in the region where V_{0i} varies rapidly, near the plasmon threshold. However, it remains the only way of finding V_{0i} with good accuracy in these metals.

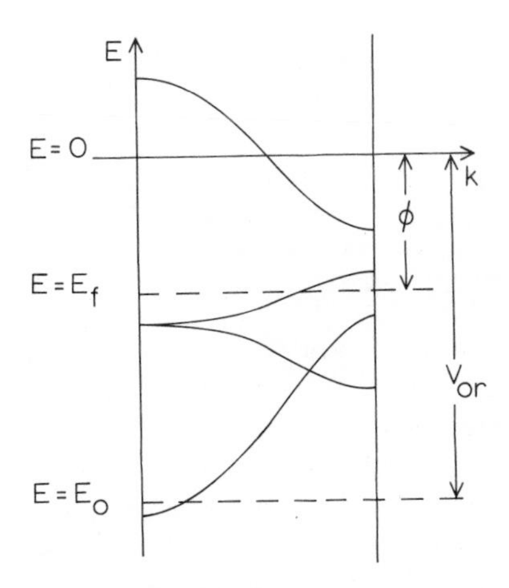

FIGURE 2.21. A schematic band structure showing the vacuum zero of energy ($E = 0$), the Fermi level ($E = E_F$) and the zero of energy between muffin tins ($E = E_0 = V_{0r}$). V_{0r} is to be adjusted to give the correct work function, ϕ.

There is an alternative empirical method for finding V_{0r} that is independent of LEED experiments. A band structure calculation is made for the solid, near the Fermi level using the ion-core potential previously constructed. V_{0i} is zero at the Fermi level, but V_{0r} has to be added to energies of bands, and takes the form of a uniform shift in energy. We know that the Fermi level must be lower than the vacuum zero of energy by the work-function, ϕ. Thus by using V_{0r} to adjust the Fermi level in our band structure to the work function, we can determine V_{0r} independent of LEED experiments. Figure 2.21 illustrates how the scheme works. The Fermi level itself is found simply by filling up the bands until the requisite number of electrons per atom have been used.

IIF The solid-vacuum interface

In the bulk of the crystal we have found that in addition to ion-core potentials there is another potential, V_0, acting due to correlation with the conduction/valence electrons. It was our good fortune that V_0 acted to produce a uniform lowering of the energy throughout the crystal, in many materials. This is true of the bulk, but what of the surface region? As we pass out of the solid and become more remote from the conduction/valence electron density to which V_0 owes its origin, so V_0 itself must diminish. We can even predict its asymptotic behaviour; Maxwell's equations tell us that at a large distance, d, from a metal a charge experiences an image potential $-1/4d$. Schematically V_0 must behave near the surface as in Fig. 2.22.

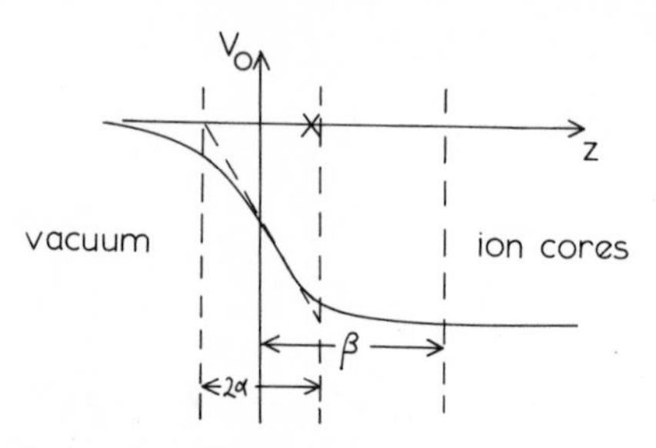

FIGURE 2.22. Schematic variation of V_0 near a surface. X represents the first layer of ion cores.

The barrier can be characterised by the distance from the centre of the last layer of ion cores, β, and by its steepness, α.

In addition to truncation of V_0 at the surface, other deviation from bulk behaviour can occur. Ion cores in layers near the surface may dilate away from the bulk. In extreme cases the geometric configuration at the surface may be completely reconstructured as happens for germanium and silicon. Reconstructured surface layers are more properly treated as overlayer problems and will be considered in a later chapter.

Relatively few calculations of the surface properties we are interested in exist, partly because of their theoretical difficulty, and partly because it is only recently that experiments directly sensitive to these quantities could be made and interpreted. It will probably be that their determination will come from experimental measurements rather than theoretical calculations. This gap in our knowledge of physical systems provides a spur to developing LEED as a probe of the surface region.

First we consider dilation of ion-cores near the surface. If forces between ion-cores are known, dilation can be calculated theoretically. In fact forces are generally not well characterised and are almost certainly dependent on the environment of the interacting ion-cores. Nevertheless Burton and Jura

(1967) in the approximation that forces remain unchanged for a surface environment, and are to be derived from a 'Morse potential',

$$D[\exp(-2\gamma(r - r_0)) - 2\exp(-\gamma(r - r_0))], \tag{2.76}$$

where D, γ and r_0 are parameters to be fitted to experimental data, have calculated dilation of surface layers in face centred cubic materials. Their results are given in Table 2.3. Tick and Witt (1971) have made similar calculations using instead of a Morse potential, the Lennard Jones potential, and their results are qualitatively similar.

The (001) surface		
solid	$d_1\%$	$d_2\%$
Ca	12.5	3.6
Ag	6.5	1.3
Al	11.0	3.0
Pb	5.5	1.0
Cu	9.7	2.4
Ni	9.1	2.2
Ar	2.6	0.6

The (110) surface		
solid	$d_1\%$	$d_2\%$
Ca	9.6	2.6
Ag	4.8	0.8
Al	8.4	2.1
Pb	4.1	0.6
Cu	7.3	1.7
Ni	6.9	1.5
Ar	1.8	0.4

The (111) surface		
solid	$d_1\%$	$d_2\%$
Ca	4.3	0.9
Ag	1.9	0.2
Al	3.7	0.7
Pb	1.6	0.2
Cu	3.1	0.5
Ni	2.9	0.5
Ar	0.8	0.2

TABLE 2.3. Dilation of surfaces in face-centred-cubic materials. d_1 is the percentage increase in spacing normal to the surface plane between the first and second layers. d_2 is the same quantity for the second and third layers.

According to the calculations increases in inter-planar spacings are typically of the order of 10% or between 0.1 and 0.2Å, for the first pair of planes, decreasing very rapidly into the bulk. Comparison of LEED calculations for dilated nickel (001) surfaces with experiment (Andersson and Pendry 1973) indicate that the distance between the first pair of planes does not increase by more than 10%. Taking account of so small a displacement leads to only small changes in intensity/energy LEED spectra.

Now we turn to the optical potential. Both its real and imaginary parts must go to zero beyond the surface. For conduction electrons the barrier formed by decay of the real part, V_{0r}, is of great importance; it stops them from running out of the crystal! It is important where surface states are concerned too, because it is across this barrier that bulk crystal wavefunctions are matched to decaying vacuum wave-functions. In LEED the influence is not so drastic, as the increase in potential changes the momentum of an electron crossing it by only a small percentage, except for very low energy electrons, and in those situations where electrons travelling nearly parallel to the surface are important. Waves reflected from the outer surface of the barrier add or subtract from the wave amplitude reflected by ion cores, according to phase. Reflection from the inner surface turns back electrons into the crystal and under certain circumstances, as we shall see, electrons can be trapped between the barrier and the rest of the crystal, changing $I(E)$ strongly.

The imaginary part of the optical potential has its own characteristic variation near the surface. At large distances from the surface V_0 behaves as an image potential and V_{or} dominates. V_{0i} must die away into the vacuum more quickly than V_{or}.

In the bulk V_{0i} came about because flux in the elastic wavefield was continually being lost to inelastic processes; in free-electron materials these were mainly plasmon excitations. Far outside the crystal no inelastic processes can be excited, of course, but at the surface itself other kinds of losses can take place. At an interface between a free electron metal and vacuum, a new sort of charge density oscillation is possible called a surface plasmon (Ritchie, 1957, Stern and Ferrell, 1960). It has been shown that the surface plasmon frequency, ω_{sp}, is related to the bulk plasmon frequency, ω_{bp}, by

$$\omega_{sp} = \frac{\omega_{bp}}{\sqrt{2}}. \tag{2.89}$$

Real excitations of surface plasmon contribute to V_{0i} in the surface region, and virtual excitations to V_{0r}. An interesting situation comes about if the incident electron has enough energy to excite a surface plasmon, but not a bulk plasmon. Loss of flux occurs only in the surface region and an absorbing barrier exists just inside the surface of the crystal.

A certain amount of evidence has been gleaned from comparison with experiment of LEED intensities calculated using different barriers. The conclusions reached so far are that the scale of variation of both V_{0r} and V_{0i} (i.e. α in Fig. 2.22) in the surface region is the dielectric screening length of the bulk material: if the barrier is much sharper than this, e.g. a step function, the barrier seems to give too strong a reflectivity. Even so V_{0i} must decay away quite rapidly because if it did not do so, beams emerging from the surface at shallow angles would have to travel through a vast length of absorbing medium and would be severely attenuated relative to specularly reflected beams. There is some evidence for such attenuation but not enough to imply that V_{0i} extends into the vacuum beyond a fraction of an interlayer spacing.

Several calculations of the surface barrier have been made: by Appelbaum and Hamann (1972) by Duke and Bennet (1967) by Lang and Kohn (1971), and by Gadzuk (1969). Perhaps the most complete theory so far is due to Inkson (1971). He determines the screened electron-electron interaction in the surface region of a free-electron material by assuming the role of the surface to consist of producing image forces as a consequence of classical requirements of continuity of potentials across a boundary. Having found how the incident electron interacts with conduction/valence electrons, he then proceeds to calculate the optical potential from an equation to (2.75). The model includes both real and virtual excitations of surface and bulk plasmons thus accounting for both V_{0i} and V_{0r} respectively.

In Fig. 2.23 we see his results for V_{0r}, with the screening length chosen to be equal to that found in copper ($\frac{1}{2}$Å). The $z = 0$ plane can be associated roughly with the edge of the conduction/valence charge density distribution. First note that the barrier in V_{0r} does coincide with the $z = 0$ plane and that the scale of variation is of the same order as the screening length confirming the less precise data obtained so far from experiment. Positions of ion core-layers have been drawn in and it is evident that V_{0r} has nearly obtained its

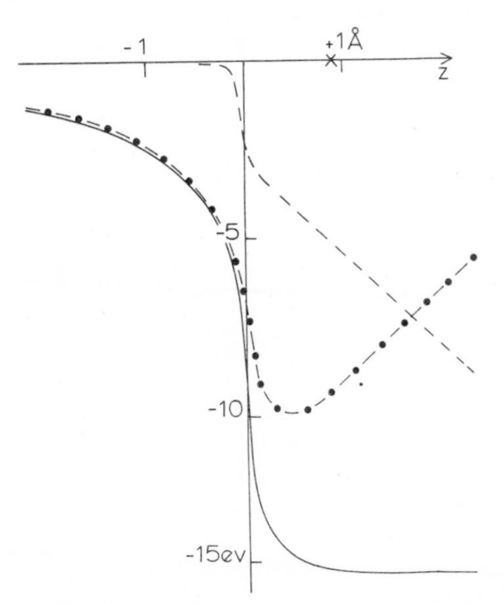

FIGURE 2.23. Calculations by Inkson of V_{0r} near a surface, — — — virtual bulk plasmon contribution, -•-•- virtual surface plasmon, ——— total. Locations of layers of ion cores are shown by crosses.

bulk value by the centre of the first layer. Calculations of V_{0r} were almost independent of energy in the 0–100 eV range.

Contributions to V_{0r} from virtual excitations of surface and bulk plasmons can be separated. The interesting point emerges that the role of surface plasmons seems to be to continue the optical potential out through the first layer of ion cores at a more or less constant value.

Figure 2.24 shows Inkson's calculations of contributions from real surface plasmon excitations to V_{0i} at an energy between surface and bulk plasmon thresholds.

Figure 2.24 confirms the rapid rise in V_{0i} near $z = 0$. A point of interest is that absorption by the surface plasmon is mainly on the inside surface of the crystal and is relatively small outside.

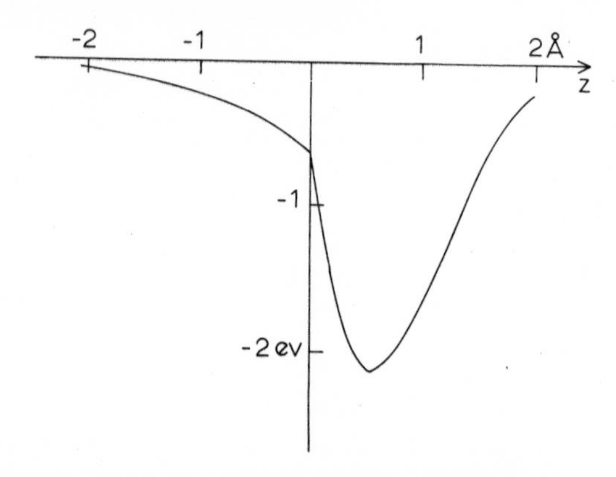

FIGURE 2.24. Calculations by Inkson of the surface plasmon contribution to V_{0i}. It is localised near the surface and concentrated on the inner surface of the crystal.

No complete account of electronic surface phenomena has been possible in this section, simply because much of the theory and many of the experiments have not been done. It is to be hoped that the combination of LEED experiments with theoretical analysis will remedy this situation.

Chapter 3

Principles of Diffraction at T = 0°K

IIIA Kinematic theory

We have discussed at some length components of the potential affecting behaviour of electrons and we know, for example, how to calculate scattering properties of individual ion-cores. Assembling ion-cores into the sort of structures on which LEED experiments are performed complicates the problem, even when the ions fall on a periodic lattice, and general solutions are not simple. At the same time an extremely simplified model is capable of providing qualitative understanding of many features of electron diffraction experiments and in some cases may even give quantitative results. The model is the kinematic theory and for many years was the only theory available for interpretation of diffracted intensities.

The assumption basic to kinematic theory is that scattering by ion-cores is weak in the sense that for one reason or another, only a small fraction of electrons is scattered from the path that would have been taken in the absence of ion-cores. The theory is closely akin to that used in interpretation of X-ray intensities. Other names for kinematic theory are 'first order perturbation' or 'single scattering' theory. Our condition that only a small number of electrons are scattered implies that a negligible number are scattered twice, hence the term 'single scattering'. In contrast other more complicated theories involve multiple scattering.

The assumption of weak ion-core scattering means that the zeroth order wavefunction inside the crystal is easily found. To zeroth order ion-core scattering can be omitted from the Schrödinger equation,

$$-\tfrac{1}{2}\nabla^2\phi + V_0\phi = E\phi, \tag{3.1}$$

V_0 being the optical potential due to conduction and valence electrons discussed in Sections IID and IIE. The incident electron is represented outside the crystal by a wavefunction of the form

$$\phi(\mathbf{r}) = B \exp(i\mathbf{k}_0 \cdot \mathbf{r}), \tag{3.2}$$
$$\tfrac{1}{2}|\mathbf{k}_0|^2 = E.$$

At the surface of the crystal the incident wave is partly reflected by the step down to the constant potential inside the crystal (Fig. 2.22), but mainly transmitted through the step into the crystal. Let the reflected wave be

$$R \cdot B \cdot \exp(i\mathbf{k}_{0\|} \cdot \mathbf{r}_\| - k_{0z}z), \tag{3.3}$$

and the wave transmitted into the crystal

$$T^+ \cdot B \exp(i\mathbf{k} \cdot \mathbf{r}), \tag{3.4}$$
$$\tfrac{1}{2}|\mathbf{k}|^2 = E - V_0, \qquad \mathbf{k}_\| = \mathbf{k}_{0\|}.$$

Since the step down in potential to V_0 exerts only forces normal to the surface, the parallel component of momentum, $\mathbf{k}_{0\|}$, is not affected by the step. The reflected wave will have small amplitude in most situations as has been indicated in Section IIF.

The zeroth order wave in kinematic LEED theory differs from that in X-ray theory in that extinction is already included in equation (3.4). Because of absorption represented by the imaginary component of V_0, k_z is complex and its imaginary part causes the wavefield to die away exponentially into the crystal even in the absence of ion-core scattering. If absorption is strong enough kinematic or single scattering theory is bound to work because multiple scattering processes are inhibited by the wavefield being strongly attenuated between scattering events.

Now ion-cores must be introduced into the problem. The crystal is to be divided into layers of ion-cores (Section ID), each layer having perfect symmetry parallel to the surface. The layer can be divided into two-dimensional unit cells parallel to the surface. Choosing one of these cells as the origin, if the jth unit cell is displaced from the origin by $\mathbf{R}_j$, the position of the sth atom in the jth unit cell can be written

$$\mathbf{R}_j + \mathbf{u}_s \tag{3.5}$$

where $\mathbf{u}_s$ is the position of the atom relative to its own unit cell. For the moment we do not need to subdivide the layer into planes as we did in Section ID, and vector $\mathbf{u}_s$ is given in terms of (1.8) by,

$$\mathbf{u}_s = \mathbf{r}_k + \mathbf{d}_p.$$

Figure 1.7 illustrates a possible structure. All the vectors $\mathbf{R}_j$ are parallel to the surface plane, but the vectors $\mathbf{u}_s$ need not be. If $\mathbf{a}$ and $\mathbf{b}$ represent adjacent sides of the unit cell at the origin, $\mathbf{R}_j$ can be written

$$\mathbf{R}_j = \ell\mathbf{a} + m\mathbf{b} \tag{3.6}$$

We already know the zeroth order wavefield and the next step is to find the contribution of a single layer to the scattered wavefield. Let the single layer be immersed in a uniform potential, V_0, with a single wave, amplitude $T^+(\mathbf{k})B$ incident upon it. Each ion-core scatters the incident wave in a manner that we have already investigated in Chapter II. The wave incident on the sth ion-core in the jth unit cell has amplitude at the centre of the ion-core

$$T^+(\mathbf{k})B\exp\left[i\mathbf{k}\cdot(\mathbf{R}_j + \mathbf{u}_s)\right], \tag{3.7}$$

and the amplitude of the scattered wave follows from equation (2.56)

$$T^+(\mathbf{k})B\exp\left[i\mathbf{k}\cdot(\mathbf{R}_j + \mathbf{u}_s)\right]\chi_s(\mathbf{r} - \mathbf{R}_j - \mathbf{u}_s), \tag{3.8}$$

$$\chi_s(\mathbf{r}) = \sum_\ell (2\ell + 1)P_\ell\left(\frac{\mathbf{k}\cdot\mathbf{r}}{|\mathbf{k}|\,|\mathbf{r}|}\right)\sin\left(\delta_\ell(s)\right)\exp\left(i\delta_\ell(s)\right)$$

$$\times\, i^{\ell+1}h_\ell^{(1)}(|\mathbf{k}|\,|\mathbf{r}|). \tag{3.9}$$

The total wavefield scattered by a layer is a sum of contributions from all ion cores within the layer

$$\phi_{scatt.} = \sum_{js} T^+(\mathbf{k})B\exp\left[i\mathbf{k}\cdot(\mathbf{R}_j + \mathbf{u}_s)\right]\chi_s(\mathbf{r} - \mathbf{R}_j - \mathbf{u}_s). \tag{3.10}$$

Because the ion cores are arranged in a periodic structure within the layer we know from general considerations advanced in Section IE that contributions to $\phi_{scatt.}$ from all ion cores add to give a set of diffracted beams. Descriptions of the wavefield in terms of spherical waves as in equation (3.10), or in terms of beams diffracted from the layer, are equivalent and we may use which is more convenient. Since experiment observes diffracted beams rather than spherical waves we choose to re-express equation (3.10) in terms of beams.

The beams differ from the incident beam in parallel components of momentum $\mathbf{k}'_{\|}$ by discrete amounts: the reciprocal lattice vectors, $\mathbf{g}$,

$$\mathbf{k}'_{\|} = \mathbf{k}_{\|} + \mathbf{g} \tag{3.11}$$

The reciprocal lattice vectors have been shown to have the property,

$$\mathbf{g}\cdot\mathbf{a} = 2\pi \times \text{integer}, \tag{3.12}$$

$$\mathbf{g}\cdot\mathbf{b} = 2\pi \times \text{integer}. \tag{3.13}$$

Equation (3.11) determines the z component of $\mathbf{k}'$ as well, because

$$\tfrac{1}{2}(|\mathbf{k}'_{\|}|^2 + \mathbf{k}'^2_z) + V_0 = E,$$

$$\mathbf{k}'_z = \pm(2E - 2V_0 - |\mathbf{k}'_{\|}|^2)^{\frac{1}{2}}. \tag{3.14}$$

Scattered beams always travel away from the layer and in equation (3.14) the positive sign is chosen for positive z and the negative sign for negative z. Figure 3.1 shows a Huyghen's construction (Jenkins and White, 1957) for a set of spherical waves adding to give a set of beams.

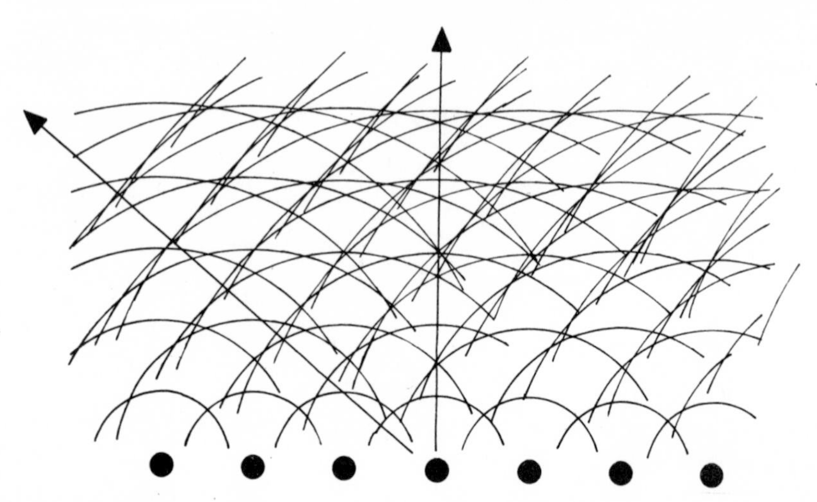

FIGURE 3.1. A Huyghen's construction showing how spherical waves add to give diffracted beams.

Writing $\phi_{scatt.}$ as a set of beams

$$\phi_{scatt.} = \sum_{\mathbf{k}'} b_{\mathbf{k}'} \exp(i\mathbf{k}' \cdot \mathbf{r}), \tag{3.15}$$

the coefficients $b_{\mathbf{k}''}$ can be found by multiplying by $\exp(-i\mathbf{k}'' \cdot \mathbf{r})$ and integrating over the whole area of layer illuminated by the incident beam,

$$\begin{aligned} \int \exp[i(\mathbf{k}'_{\|} - \mathbf{k}''_{\|}) \cdot \mathbf{r}]\, d^2\mathbf{r}_{\|} &\simeq 0 \qquad \text{if } \mathbf{k}'_{\|} \neq \mathbf{k}''_{\|}, \\ &= N^2 A \qquad \text{if } \mathbf{k}'_{\|} = \mathbf{k}''_{\|}. \end{aligned} \tag{3.16}$$

N^2 is the total number of unit cells, area A, illuminated by the incident beam. Therefore

$$b_{\mathbf{k}''} = \frac{1}{N^2 A} \int \phi_{scatt.}(\mathbf{r}) \exp(-i\mathbf{k}'' \cdot \mathbf{r})\, d^2\mathbf{r}_{\|}. \tag{3.17}$$

Substituting from equation (3.10) for $\phi_{scatt.}$ gives

$$b_{\mathbf{k}''} = \sum_{js} \frac{1}{N^2 A} \int T^+ B \exp[i\mathbf{k}\cdot(\mathbf{R}_j + \mathbf{u}_s)]$$

$$\times \chi_s(\mathbf{r} - \mathbf{R}_j - \mathbf{u}_s) \exp(-i\mathbf{k}''\cdot\mathbf{r}) d^2\mathbf{r}$$

$$= \frac{T^+ B}{A}\left[\frac{1}{N^2}\sum_j \exp[i(\mathbf{k} - \mathbf{k}'')\cdot\mathbf{R}_j]\right.$$

$$\times \sum_s \exp[i(\mathbf{k} - \mathbf{k}'')\cdot\mathbf{u}_s] \int \chi_s(\mathbf{r}') \exp(-i\mathbf{k}''\cdot\mathbf{r}') d^2\mathbf{r}'_{\parallel}, \tag{3.18}$$

where $\mathbf{r}' = \mathbf{r} - \mathbf{R}_j - \mathbf{u}_s$

Firstly consider the summation over unit cells in equation (3.18). $\mathbf{R}_j$ being a vector describing displacements of successive unit cells, is always parallel to the x–y plane and therefore from (3.11),

$$S(\mathbf{g}'') = \frac{1}{N^2}\sum_j \exp[i(\mathbf{k} - \mathbf{k}'')\cdot\mathbf{R}_j] = \frac{1}{N^2}\sum_j \exp[i\mathbf{g}''\cdot\mathbf{R}_j]. \tag{3.19}$$

S is known as a structure factor and it can be evaluated by rewriting $\mathbf{R}_j$ in terms of $\mathbf{a}$ and $\mathbf{b}$ of equation (3.6)

$$S(\mathbf{q}) = \frac{1}{N^2} \sum_{n_a=0}^{N-1} \sum_{n_b=0}^{N-1} \exp(in_a\mathbf{q}\cdot\mathbf{a}) \exp(in_b\mathbf{q}\cdot\mathbf{b})$$

$$= \frac{1}{N^2} \frac{1 - \exp(iN\mathbf{q}\cdot\mathbf{a})}{1 - \exp(i\mathbf{q}\cdot\mathbf{a})} \frac{1 - \exp(iN\mathbf{q}\cdot\mathbf{b})}{1 - \exp(i\mathbf{q}\cdot\mathbf{b})}. \tag{3.20}$$

When $\mathbf{q}$ is equal to a reciprocal lattice vector, $\mathbf{g}$,

$$S(\mathbf{q} = \mathbf{g}) = 1, \tag{3.21}$$

otherwise $S(\mathbf{q})$ is of order $1/N^2$, a very small number. Thus our assumption that diffracted beams have parallel components of momentum of the form $(\mathbf{k} + \mathbf{g})$ is born out because the form of S prevents other values of $\mathbf{k}'$ from contributing. Figure 3.2 shows S plotted against q_x for a simple example.

For a perfectly regular array of unit cells, S forbids scattering changing $\mathbf{k}_{\parallel}$ other than in units of $\mathbf{g}$, but for an imperfect array that might result from thermal motion disturbing the array of ions, $S(\mathbf{q})$ no longer excludes such changes of $\mathbf{k}_{\parallel}$. Chapter VI will elaborate on this point.

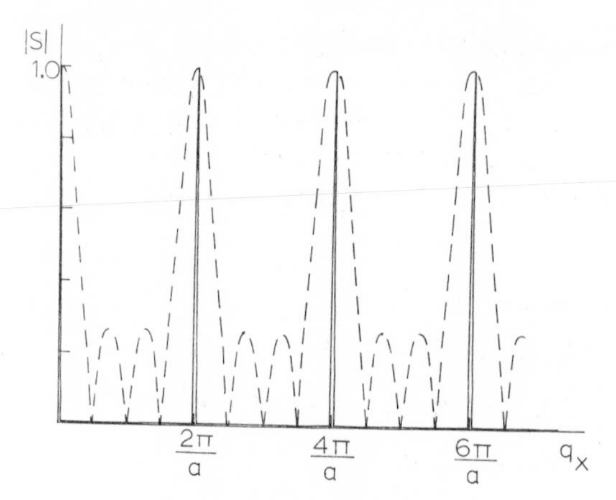

FIGURE 3.2. The structure factor $S(\mathbf{q})$, plotted as a function of q_x for a square lattice, cell side a. – – – – $N = 4$, ——— $= N$ effectively infinite.

Secondly there is the integral in equation (3.18). We quote the result derived in Appendix B:

$$\int \chi_s(\mathbf{r}') \exp(-i\mathbf{k}''\cdot\mathbf{r}')\, d^2\mathbf{r}'_{\|} = \frac{2\pi i}{|\mathbf{k}|\,|k''_z|} \sum_{\ell} (2\ell + 1) \sin(\delta_\ell(s)) \exp(i\delta_\ell(s)) \times P_\ell\left(\frac{\mathbf{k}\cdot\mathbf{k}''}{|\mathbf{k}|\,|\mathbf{k}''|}\right). \tag{A 53}$$

Substitution of results (A 53) and (3.21) into (3.18) gives

$$b_{\mathbf{k}''} = M(\mathbf{k}'',\mathbf{k})\, T^{+}(\mathbf{k}) B, \tag{3.22}$$

$$M(\mathbf{k}'',\mathbf{k}) = S(\mathbf{k}''_{\|} - \mathbf{k}_{\|}) \frac{2\pi i}{A|\mathbf{k}|\,|k''_z|} \sum_{\ell s} (2\ell + 1) \sin(\delta_\ell(s)) \exp(i\delta_\ell(s)) \times P_\ell\left(\frac{\mathbf{k}\cdot\mathbf{k}''}{|\mathbf{k}|\,|\mathbf{k}''|}\right) \exp[i(\mathbf{k} - \mathbf{k}'')\cdot\mathbf{u}_s]. \tag{3.23}$$

We have now re-expressed the sum of spherical scattered waves about each ion-core in the layer, as a sum of discrete beams diffracted from the layer. Within the kinematic approximation the variation in intensity of these beams, as functions of energy, for example, is all contained in the variation of the t-matrix for a unit cell

$$-\frac{2\pi}{|\mathbf{k}|} \sum_{\ell s} (2\ell + 1) \sin(\delta_\ell(s)) \exp(i\delta_\ell(s)) \times P_\ell\left(\frac{\mathbf{k}\cdot\mathbf{k}''}{|\mathbf{k}|\,|\mathbf{k}''|}\right) \exp[i(\mathbf{k} - \mathbf{k}'')\cdot\, \mathbf{u}_s]. \tag{3.24}$$

Most of the structure in LEED intensities is brought about not through (3.24), but by interference between scattering from different layers. By way of example take a crystal composed of equally spaced identical layers. The nth layer being displaced relative to the 0th layer by $n\,\mathbf{c}$. The incident wave expressed relative to the origin of the nth layer is

$$\exp(in\mathbf{k}\cdot\mathbf{c})\,T^{+}(\mathbf{k})\,B\exp[i\mathbf{k}\cdot(\mathbf{r}-n\mathbf{c})], \tag{3.25}$$

and the scattered wave is

$$\begin{aligned} &M(\mathbf{k}',\mathbf{k})\exp(in\mathbf{k}\cdot\mathbf{c})\,T^{+}(\mathbf{k})\,B\exp[i\mathbf{k}'\cdot(\mathbf{r}-n\mathbf{c})] \\ &=T^{+}(\mathbf{k})\,B\exp[in(\mathbf{k}-\mathbf{k}')\cdot\mathbf{c}]\,M(\mathbf{k}',\mathbf{k})\exp(i\mathbf{k}'\cdot\mathbf{r}). \end{aligned} \tag{3.26}$$

Therefore the total scattered wave is

$$T^{+}(\mathbf{k})\cdot B\cdot M(\mathbf{k}',\mathbf{k})\,\frac{\exp(i\mathbf{k}'\cdot\mathbf{r})}{1-\exp[i(\mathbf{k}-\mathbf{k}')\cdot\mathbf{c}]} \tag{3.27}$$

The total scattered wave is transmitted through the step in potential at the surface, amplitude transmission coefficient $T^{-}(\mathbf{k}')$, and there emerge from the crystal waves of the form

$$\begin{aligned} &\sum_{\mathbf{k}'} B\left\{R\delta_{\mathbf{k}'_{\|}k_{\|}}+T^{+}(\mathbf{k})\,T^{-}(\mathbf{k}')\,\frac{M(\mathbf{k}',\mathbf{k})}{1-\exp[i(\mathbf{k}-\mathbf{k}')\cdot\mathbf{c}]}\right\}\times\exp(i\mathbf{k}'_0\cdot\mathbf{r}), \\ &= F.B.\exp(i\mathbf{k}'_0\cdot\mathbf{r}). \end{aligned} \tag{3.28}$$

Expression (3.28) is the kinematic formula for amplitudes of beams diffracted from a crystal surface.

Experiments measure total electron flux in a beam, which is easily found from the amplitude of that beam. The current flowing parallel to z is proportional to the modulus squared of the beam's amplitude, times the area of crystal illuminated, times the component of momentum normal to the surface. Therefore reflected intensity in a beam, wave vector $\mathbf{k}'_0$, expressed as current for unit incident current, is

$$|F|^2\left|\frac{k'_{0z}}{k_{0z}}\right|$$

Neglecting structure in reflected intensities arising from $M(\mathbf{k}',\mathbf{k})$, the main variation in reflected intensity in the kinematic approximation comes from the factor

$$D(\boldsymbol{\Delta}\cdot\mathbf{c})=(1-\exp[i\boldsymbol{\Delta}\cdot\mathbf{c}])^{-1}, \tag{3.29}$$

$$\boldsymbol{\Delta}=\mathbf{k}-\mathbf{k}'=\boldsymbol{\Delta}_r+i\boldsymbol{\Delta}_i \tag{3.30}$$

a function that has maxima when

$$\mathbf{\Delta}_r \cdot \mathbf{c} = 2\pi \times \text{integer} \tag{3.31}$$

There is strong similarity between equation (3.29) and the structure factor of equation (3.20). Only the imaginary part of $\mathbf{\Delta}$ prevents D being a sharply peaked function of $\mathbf{\Delta}_r \cdot \mathbf{c}$. In other words absorption limits penetration and relaxes the 'third Laue condition' in X-ray terminology. Expression (3.29) for D is plotted in Fig. 3.3 for a crystal where

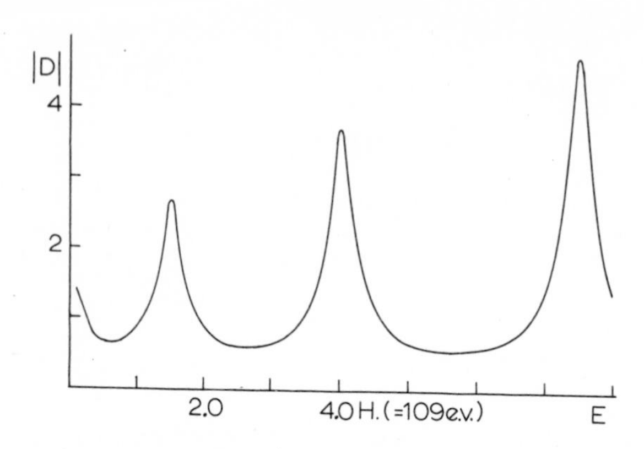

FIGURE 3.3. Function D plotted against energy for normal incidence on a simple cubic crystal showing strong peaks near the Bragg reflection conditions.

$$V_{0r} = -0.5 \text{ a.u.}, \quad V_{0i} = -0.15 \text{ a.u.}, \quad \mathbf{c} = (0, 0, \pi) \text{ a.u.},$$

typical values of these parameters. The incident wave has $\mathbf{k}_0$ normal to the surface and $\mathbf{k}'_0$ is parallel to the outward normal. Then

$$|\mathbf{k}|^2 = 2E - 2V_0, \qquad k_z = +(2E - 2V_0)^{\frac{1}{2}},$$

$$|\mathbf{k}'|^2 = 2E - 2V_0, \qquad k'_z = -(2E - 2V_0)^{\frac{1}{2}},$$

and equation (3.29 becomes

$$D = (1 - \exp[i\mathbf{\Delta} \cdot \mathbf{c}])^{-1} = (1 - \exp[i2(2E - 2V_0)^{\frac{1}{2}}\pi])^{-1}. \tag{3.32}$$

Figure 3.3 also illustrates the great importance of absorption in LEED. It has the effect of reducing intensities and without any absorption at all, the kinematic theory would predict infinite reflected intensities at the Bragg condition. Absorption also broadens the peaks. We saw in Chapter II that there was a limit to the narrowness of structure, imposed by the uncertainty principle:

$$\Delta E \geqslant 2|V_{0i}|$$

In these curves it is evident that the minimum width at half-peak height is achieved at normal incidence in kinematic theory.

Figure 3.3 is also characteristic of structure given by kinematic theory in reflected amplitudes. The slow variation of the other energy dependent factor, $M(\mathbf{k}',\mathbf{k})$ leads to different peak intensities rather than introducing any additional structure.

Some materials do seem to satisfy the conditions for kinematic theory to hold, namely that all elastic scattering should be weak in comparison with absorption. In the terminology of Section IIA:

$$T_f \ll V_{0i}, \tag{3.33}$$

$$T_b \ll V_{0i}. \tag{3.34}$$

It has been observed above that ion core scattering is at its weakest relative to unit volume of crystal, when the unit cell is large, a condition that is best satisfied in crystals of inert gases. Figure 3.4 shows intensity-energy curves for xenon and these curves behave in a kinematic manner: *all* peak positions are determined by the Bragg condition; all peaks have the simple shape predicted by kinematic theory as can be seen by comparing them with curves calculated from kinematic theory using scattering factors fitted to experiment. The spectra were taken from Ignatjevs, Pendry and Rhodin (1971).

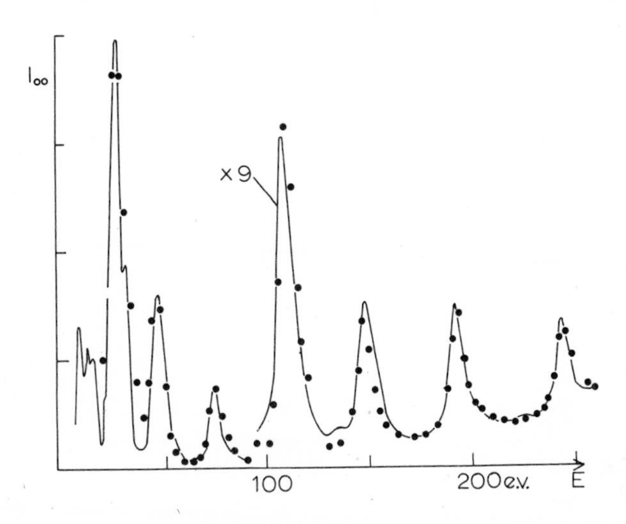

FIGURE 3.4. Intensity/energy spectrum taken near normal incidence from a xenon (100) surface. Black dots show points calculated in the kinematic approximation, with scattering fitted to experiment. Close agreement of peak shape and position indicates a kinematic experimental spectrum.

On the other hand, xenon is a most favourable candidate for kinematic theory with its enormous 232Å^3 unit cell. Most materials commonly used in LEED experiments are substances having strong ion-core scattering powers in relation to absorption. Some results for a nickel (001) surface are shown in Fig. 3.5, putting in accurate ion-core scattering factors and values of V_{0r} and V_{0i} estimated by methods related in the last chapter. Agreement with experiment (Andersson and Kasemo, 1971) is not good and a similar picture holds for other transition metals.

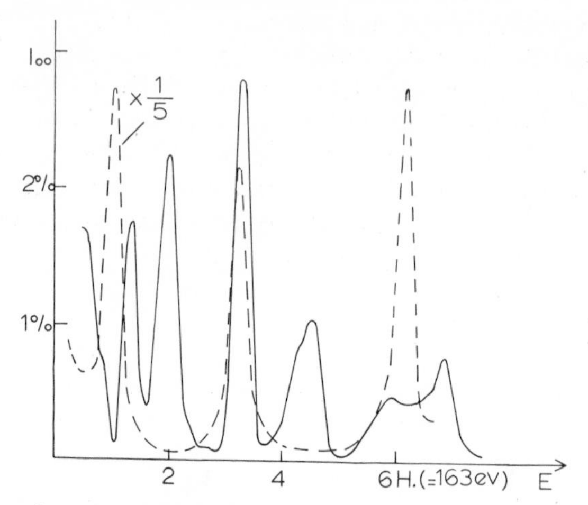

FIGURE 3.5. Comparison with experiment (———) of kinematical (— — —) intensities reflected from a (001) nickel surface at normal incidence, 00 beam.

Conditions (3.33) and (3.34) for the accuracy of kinematic theory cannot hold good. In fact we have seen already in Section IIA that they do not do so for most materials. It is the forward elastic scattering that is primarily responsible for spoiling the theory. No account is ever taken of forward scattering in kinematic theory.

The imbalance between forward and backward scattering is a fact playing an important role in LEED theory. It is responsible for what at first appears as a paradox: that LEED intensities are generally very weak, but at the same time their energy variation does not follow kinematic theory, indicating the presence of strong scattering. It will transpire that by keeping in mind this division in magnitude, simplified theories of LEED intensities can be developed.

IIIB The general case in one dimension-normal modes

Kinematic theory depends for its validity on ion-core scattering being weak, an assumption that does not hold for many materials. Some sort of multiple scattering theory must then be used, and to illustrate the ideas behind the so-called 'normal mode' or 'band structure' approach to multiple scattering, it is simplest to start in one dimension. Then many of the formulae are so simple that analytic expressions for them exist and at the same time the essence of the method is conveyed. Generalisation to three dimensions will be left until the principles have been established.

One-dimensional theory has little application beyond providing insights of the three-dimensional case, but there are some situations where it is applicable in practice. To be valid, it is necessary that only scattering into the direct forward or specularly reflected beams takes place. Thus if many diffracted beams are present as is the case at all but the lowest energies, beams other than the specular beam may be of importance if ion-core scattering is strong. We are led to the conclusion that practical applications can be found only in the very low energy (10 eV) spectra of nearly-free-electron materials.

To make our one-dimensional theory reasonably analogous to the real situation, we adopt a model illustrated in Fig. 3.6.

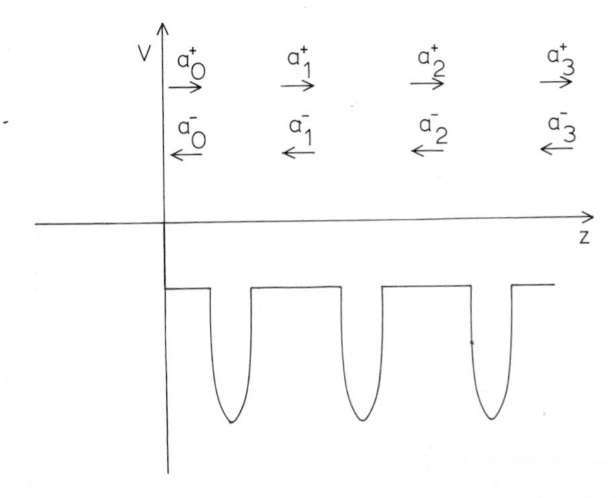

FIGURE 3.6. One dimensional potential in which the electron moves. Amplitudes of forward travelling waves between scatterers are indicated.

The one-dimensional crystal is taken to consist of a regular array of scattering potentials. Each scatterer is the analogue of a layer of ion cores. Between scattering potentials, is a uniform background potential, V_0, just as we had in the three-dimensional case. At the surface, taken to be a distance $\frac{1}{2}c$ from the last scatterer, V_0 steps up to the zero of the vacuum. Such a step is in fact a bad approximation to the true variation of V_0 at the surface which is much smoother but it has been adopted here for simplicity. V_0 can have an imaginary component to take account of absorption.

In the vacuum outside the crystal, waves obey the equation

$$-\frac{1}{2}\frac{d^2\phi}{dz^2} = E, \tag{3.35}$$

and therefore incident and reflected waves are respectively

$$\exp(ik_0 z),\ C\exp(-ik_0 z),\ \tfrac{1}{2}k_0^2 = E. \tag{3.36}$$

Inside the crystal in the region of constant potential between scatterers the equation for the wavefunction is

$$-\frac{1}{2}\frac{d^2\phi}{dz^2} + V_0\phi = E\phi. \tag{3.37}$$

Between each pair of scatterers there will be in general both forward and backward travelling waves. The wavefunction in the region of constant potential between the nth and $(n + 1)$th scatterers will be written

$$a_n^+ \exp[ik(z - nc)] + a_n^- \exp[-ik(z - nc)], \tag{3.38}$$
$$k^2 = 2E - 2V_0.$$

In a multiple scattering theory both forward and backward travelling waves can have large amplitudes. This generalisation causes trouble, for not only does the wave coming in from the surface contribute in amplitude to a_n^+, but also there are contributions caused, for example, by a backward travelling wave amplitude a_n^-, being reflected at the nth layer. This contribution is not small as it was assumed to be in kinematic theory.

The properties of the scattering potential can be summarised in two numbers. t we shall take to be the amplitude with which the scatterers transmit waves, and r the amplitude with which they reflect waves. Evidently if

$$t^*t + r^*r = 1, \tag{3.39}$$

no flux is lost by absorption within the scatterers. The choice

$$t = \text{real number}, \quad r = i(1 - t^2)^{\frac{1}{2}}, \tag{3.40}$$

satisfies equation (3.39) and is sufficiently general for our purposes.

We can now write down some equations for a_n^+ and a_n^-. There are two contributions to a_n^+. Firstly the forward travelling wave between the $(n - 1)$th and nth scatterers, having amplitude at the nth scatterer

$$a_{n-1}^+ \exp(ik\tfrac{1}{2}c)$$

can be transmitted by the scatterer to give a forward travelling wave of amplitude

$$t\,a_{n-1}^+ \exp(ik\tfrac{1}{2}c) \tag{3.41}$$

between the next pair of layers. Expression (3.41) is the amplitude of the wave referred to an origin at the centre of the nth scatterer. Referred to the point mid way between the nth and $(n + 1)$th scatterers it has amplitude

$$t a_{n-1}^{+} \exp(ikc). \tag{3.42}$$

The second contribution to a_n^+ comes about by the backward travelling wave between the nth and $(n + 1)$th scatterers, which has amplitude at the nth scatterer of

$$a_n^- \exp[-ik(-\tfrac{1}{2}c)],$$

being reflected by the nth scatterer to give a wave of amplitude

$$r a_n^- \exp(ik\tfrac{1}{2}c). \tag{3.43}$$

Referring this second contribution to an origin half way between layers gives

$$r a_n^- \exp(ikc). \tag{3.44}$$

Therefore in total

$$a_n^+ = (r a_n^- + t a_{n-1}^+) \exp(ikc) = (r a_n^- + t a_{n-1}^+)\alpha^{-1} \tag{3.45a}$$

In a similar way, a_{n-1}^- can be expressed as

$$a_{n-1}^- = (r a_{n-1}^+ + t a_n^-) \exp(ikc) = (r a_{n-1}^+ + t a_n^-)\alpha^{-1} \tag{3.45b}$$

Amplitudes of waves before the first scatterer are special because they must in addition to equations such as (3.45) satisfy the requirement that at the surface step, waves inside and outside the crystal should match in amplitude and derivative. From (3.36) and (3.38)

$$a_0^+ + a_0^- = 1 + C \tag{3.46a}$$

$$a_0^+ - a_0^- = (1 - C)\frac{k_0}{k} \tag{3.46b}$$

Equations (3.45) and (3.46) determine the problem mathematically. They correspond closely with equations we shall derive for the full three-dimensional theory.

Between each pair of scatterers there is a forward and backward travelling wave. The reflection and transmission coefficients complicate the problem by coupling together forward to backward waves, and waves between successive pairs of scatterers. Systems in which motion in one part is coupled to motion in another part of the system have been widely studied, and we can profit by drawing an analogy with one of the simplest of such systems. Figure 3.7 shows two pendulums suspended so that the motion of one affects that of the other. If the pendulums are set to swing in an arbitrary fashion, the motion will be complicated as first one pendulum, then the other swings with large amplitude. But there are some displacements for which simple behaviour is observed. If both pendulums are given equal displacements then the whole assembly will swing as one pendulum with a single period. Another simple case occurs when the displacements are equal but opposite. In the second case the pendulums swing anti-symmetrically but again have a single period of oscillation. These simple cases of motion in which motions of separate parts of the system differ in amplitude only by a constant factor are known as the normal modes (Goldstein, 1950). There are two of them for the simple system shown in Fig. 3.7. In any system of coupled motions the normal modes

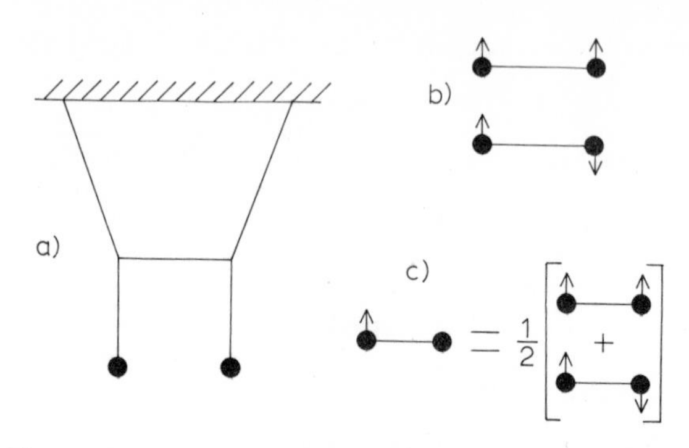

FIGURE 3.7. (a) Two coupled pendulums. (b) Their normal modes. (c) Expansion of a general displacement in normal modes.

have the useful property that an arbitrary motion can be written as a superposition of the normal modes and the simplicity of structure of a normal mode means that such a decomposition provides physical insight of the general motions. For example, in the two-pendulum case transfer of motion that takes place between pendulums can be ascribed to the beating together of the different frequencies of the two normal modes. A further consequence of the simplicity of the normal modes is that they are often more easily calculated than is a general motion, and sometimes afford the only way of solving the coupled motion problem.

Applying the concept of normal modes to the problem of electron waves between scatterers leads to a definition of normal modes as wavefunctions in which amplitudes describing the waves between the $(n-2)$th and $(n-1)$th scatterers differ from those between the $(n-1)$th and (n)th scatterers by a constant factor

$$a_n^+ = \beta a_{n-1}^+ \tag{3.47a}$$

$$a_n^- = \beta a_{n-1}^- \tag{3.47b}$$

Since in our example the layers are evenly spaced, the $(n+1)$th scatterer sits in the same relationship to the nth scatterer as the nth does to the $(n-1)$th.

Hence the same factor, β, must be employed:

$$a_{n+1}^{+} = \beta a_{n}^{+} = \beta^{2} a_{n-1}^{+} = \dots \tag{3.48a}$$

$$a_{n+1}^{-} = \beta a_{n}^{-} = \beta^{2} a_{n-1}^{-} = \dots \tag{3.48b}$$

The LEED problem can be solved by finding the normal modes from equations (3.45) and (3.47), then using equations (3.46) to find out what amplitudes must be given to the normal modes to correspond with the situation of a wave of unit amplitude incident on the surface.

Using equations (3.47) to eliminate $a_{n}^{\pm}$ from (3.45):

$$(t - \beta\alpha) a_{n-1}^{+} + \beta r a_{n-1}^{-} = 0 \tag{3.49a}$$

$$r a_{n-1}^{+} + (\beta t - \alpha) a_{n-1}^{-} = 0. \tag{3.49b}$$

Equations (3.49) have a solution provided that β satisfies the condition

$$(t - \beta\alpha)(\beta t - \alpha) - \beta r^{2} = 0$$

or

$$\alpha t \beta^{2} - (t^{2} + \alpha^{2} - r^{2})\beta + \alpha t = 0. \tag{3.50}$$

For our choice of t and r, (3.40), $t^{2} - r^{2}$ is unity, hence on solving for β,

$$\beta = \frac{(1 + \alpha^{2}) \pm [(1 + \alpha^{2})^{2} - 4\alpha^{2} t^{2}]^{\frac{1}{2}}}{2\alpha t}. \tag{3.51}$$

There are two normal modes. The ratios of a_{n-1}^{+} to a_{n-1}^{-} are

$$\frac{a_{n-1}^{+}}{a_{n-1}^{-}} = -\frac{(\beta t - \alpha)}{r} = \frac{i}{(1 - t^{2})^{\frac{1}{2}}} \frac{1}{2\alpha} (1 - \alpha^{2} \pm [(1 + \alpha^{2})^{2} - 4\alpha^{2} t^{2}]^{\frac{1}{2}}). \tag{3.52}$$

β is more commonly written

$$\beta = \exp(iKc). \tag{3.53}$$

Solid state physicists will recognise K to be the band structure of a one-dimensional crystal, and the normal modes to be Bloch waves. In fact the normal mode condition, (3.47), is the same as the so-called Bloch condition. *Normal modes of the LEED problem are Bloch waves of the crystal.* This is the key message of the chapter.

When the kinematic condition of weak scattering ($r \simeq 0$, $t \simeq 1$) is fulfilled, it follows from (3.51) and definitions of α and β, that

$$\exp(iKc) \simeq \exp(\pm ikc). \tag{3.54}$$

In kinematic theory k played a crucial role because the size of k determined the kinematic Bragg condition for peaks in diffracted intensities. The Bragg condition in one dimension is

$$k - (-k) = \frac{2\pi}{c} \times \text{integer},$$

i.e.

$$k = \frac{\pi}{c} \times \text{integer}. \tag{3.55}$$

When scattering is not small k no longer approximates K, but we shall see that we can still apply a Bragg condition, but now to K, the true band structure.

First of all we wish to investigate how strong scattering changes properties of K. Figure 3.8 shows K as a function of energy when both V_0 and r are zero:

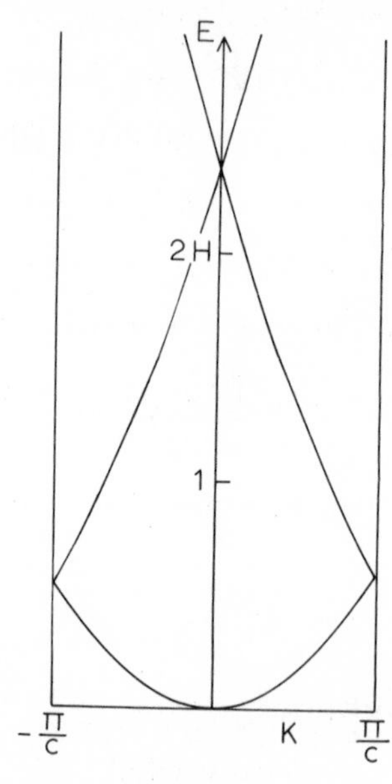

FIGURE 3.8. Free electron bands, $K(E)$, in one dimension. K is defined only to within modulo $2\pi/c$.

the free electron situation K is defined by equation (3.53) only to within modulo $2\pi/c$, and is generally plotted in the range $-\pi/c$ to $+\pi/c$. For $t < 1$ it is convenient to rewrite (3.51) substituting

$$\beta = \exp(iKc) \quad \text{and} \quad \alpha = \exp(-ikc)$$

$$\exp(iKc) = t^{-1}(\cos(kc) \pm i[t^2 - \cos^2(kc)]^{\frac{1}{2}}), \tag{3.56a}$$

if

$$t^2 > \cos^2(kc),$$

$$\exp(iKc) = t^{-1}(\cos(kc) \pm [\cos^2(kc) - t^2]^{\frac{1}{2}}), \tag{3.56b}$$

if

$$t^2 < \cos^2(kc).$$

In case (3.56a) $\exp(iKc)$ has modulus unity and therefore K is real. In case (3.56b) $\exp(iKc)$ does not have unit modulus but is real, and in these gaps in the band structure it must be the case that

$$K_r c = \pi \times \text{integer}, \tag{3.57}$$

where K_r is the real part of K. Figure 3.9 shows K when $t = 0.9$ but with V_0 still zero. In the band gaps the imaginary part of K, K_i, varies whilst K_r holds one of the constant values given by (3.57). Figure 3.9b shows variation in the imaginary part of K. More detailed information on band structure can be had in Ziman's (1964) book, for real K, and in papers by Blount (1962) and Heine (1963) for complex K.

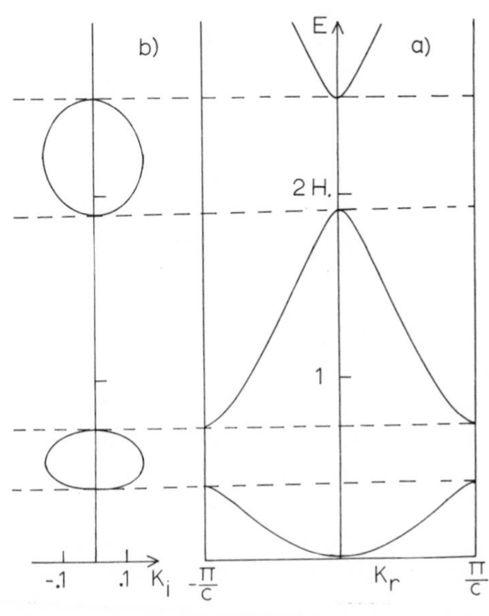

FIGURE 3.9. Bands for the case $t = 0.9$, $V_0 = 0$ (a) real part of K, (b) imaginary part of K. The lattice constant $c = 3.0$ a.u. (1.59Å).

Structure in $K(E)$ is complementary to structure in composition of the associated normal mode: substituting (3.56) into (3.52),

$$\frac{a_{n-1}^{+}}{a_{n-1}^{-}} = \frac{i}{(1-t^2)^{\frac{1}{2}}}(+i\sin(kc) \pm [t^2 - \cos^2(kc)]^{\frac{1}{2}}), \tag{3.58a}$$

if

$$t^2 > \cos^2(kc)$$

$$= \frac{i}{(1-t^2)^{\frac{1}{2}}}(+i\sin(kc) \pm i[\cos^2(kc) - t^2]^{\frac{1}{2}}) \tag{3.58b}$$

if

$$t^2 < \cos^2(kc)$$

In other words, when K is real (3.58a) holds and one normal mode has $a_{n-1}^{+} > a_{n-1}^{-}$, the other $a_{n-1}^{+} < a_{n-1}^{-}$, the first carrying net flux into the crystal, the second net flux out of the crystal. In the gap where K becomes complex (3.58b) holds and it is evident that

$$|a_{n-1}^{+}|^2 = |a_{n-1}^{-}|^2$$

For the special case we are considering of no absorption (in consequence of $V_0 = 0$) modes with complex values of K carry no current. If K is complex, then $|\beta| \neq 1$ and our arguments no longer hold. When (3.58b) is appropriate, that is to say in the band gap, forward and backward travelling components of the normal modes are equal. Therefore it is in this situation that strong reflectivities are to be expected because the total solution must reflect behaviour of the component normal modes. (3.57) is the condition for strong reflectivity and is to be compared with the kinematic condition, (3.55). (3.57) is a multiple scattering generalisation in one dimension of the Bragg condition.

Absorption played an important role within the kinematic theory, and so it does in the strong scattering case. If V_0 is now given an imaginary component all our careful arguments about the reality of K do not hold because k becomes complex, via equation (3.38). Distinctions between regions of real and complex K are blurred, although it is still true that where condition (3.57) is satisfied, then the imaginary component, K_i, tends to be large and strong mixing of forward and backward travelling components of the normal mode takes place.

In Fig. 3.10 we show band structure for a finite value of V_{0i}. All structure in the bands tends to melt away reflecting influence of the uncertainty principle; structure in energy is not discernible on a scale finer than 2 V_{0i}. No clearly defined band gaps remain and in their place we see crossings of real parts of two branches of K_r. In the presence of absorption it is no contradiction for normal modes to carry current and at the same time decay away into the crystal.

There are two normal modes of the one-dimensional system at any energy, and in the absence of absorption either one carries net current into the crystal and the other out of the crystal or in the band gaps one decays exponentially into the crystal, the other increases exponentially. Now we wish to turn to the question of how to express the total wave function in terms of normal modes. We can see immediately that only one of the normal modes is excited in a LEED experiment because, unless there is an electron gun on the reverse side of the crystal, the normal mode with net flux out of the crystal cannot be excited, and in the second case the exponentially increasing solution leads to the unphysical result that the wavefunction gets large deeper inside the crystal. When absorption is present the modes are always of increasing or decreasing nature and we must always choose the decreasing mode.

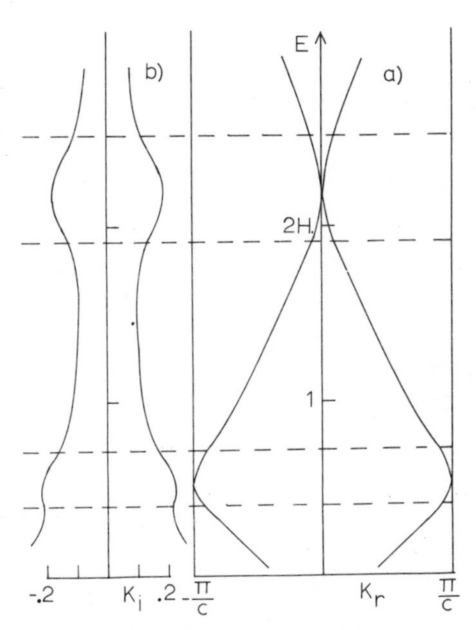

FIGURE 3.10. Bands for the case $t = 0.9$, $V_0 = -0.15i$. (a) real part of K (b) imaginary part of K. Dashed lines show locations of band-edges when $V_0 = 0$. The lattice constant $c = 3.0$ a.u. (1.59Å).

Knowing which normal mode to use, the wave function immediately inside the crystal is given by

$$a_0^+ \exp(ikr) + a_0^- \exp(-ikr),$$

and a_0^+/a_0^- is known from equations (3.58). The mode with correct behaviour outlined above is to be selected. Equations (3.46) expressing the fact that a wave of unit amplitude is incident on the crystal surface can be used to determine the amplitude of the reflected wave.

$$C = \frac{k_0(1 + a_0^+/a_0^-) + k(1 - a_0^+/a_0^-)}{k_0(1 + a_0^+/a_0^-) - k(1 - a_0^+/a_0^-)} \tag{3.59}$$

Reflected intensities, $|c|^2$, are plotted in Fig. 3.11 for the same values of parameters t and V_0 as in the band structures of Figs 3.9 and 3.10. Similar calculations can be found in papers by Molière (1939) and by Slater (1937).

Reflectivities for the case of zero absorption can be predicted from the band structure, for then we have seen that normal modes in band-gaps carry no current. Therefore as no absorption is taking place current cannot escape into the crystal; it can only be reflected. Regions of 100% reflectivity correspond precisely with the band gap. In the presence of absorption it is still

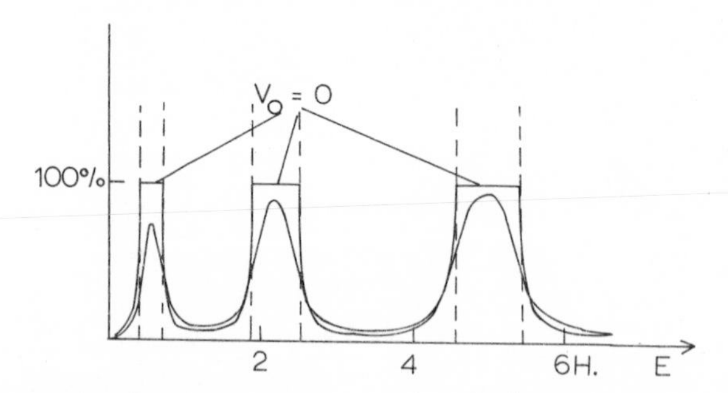

FIGURE 3.11. Reflected intensities from a one dimensional crystal. $t = 0.9$, $V_0 = 0$, $-0.15i$. Band edges in the case $V_0 = 0$ are marked by dashed lines. The lattice constant $c = 3.0$ a.u. (1.59Å).

true that peaks occur near to where band gaps were, in fact where 'crossings' of real parts, K_r, for two bands occur.

Absorption is seen to play the same two important roles it did in kinematic theory: reduction of intensities and broadening of peaks. However, in the strong scattering case it is seen from Fig. 3.11 that peaks are broader in general than the minimum width, $2|V_{0i}|$, dictated by the uncertainty principle. In particular, when V_{0i} is zero, peaks still have finite width.

In discussing kinematic theory strength of absorptive processes proved to be of great importance. The rate at which an electron is absorbed limits the extent of multiple scattering that can take place. The kinematic formula, (3.28), can be applied to one dimension simply by noting correspondence between the scattering matrix for a layer in three dimensions, M, and the one-dimensional reflectivity of scatterers, r.

$$C = R + \frac{T^+T^-r}{1 - \exp[i(k - (-k))c]} = R + \frac{T^+T^-r}{1 - \exp[2ikc]}. \qquad (3.60)$$

R and $T^\pm$ are reflectivities and transmissivities of the step at the surface, given by the well-known formula for reflectivity of square steps.

$$R = (k_0 - k)/(k_0 + k), \qquad (3.61a)$$

$$T^+ = [2k_0/(k_0 + k)] \times \exp(ik\tfrac{1}{2}c), \qquad (3.61b)$$

$$T^- = [2k/(k_0 + k)] \times \exp(ik\tfrac{1}{2}c). \qquad (3.61c)$$

There is a phase factor because the barrier is $\frac{1}{2}c$ from the first layer.

For large values of V_{0i} the kinematic formula gives the same result as the exact normal mode method. Figure (3.12) shows some examples.

The value of one dimensional theory lies mainly in the simplified picture it gives of the LEED problem. A few examples where reflectivities show qualitatively similar structure to the one dimensional results, occur at low

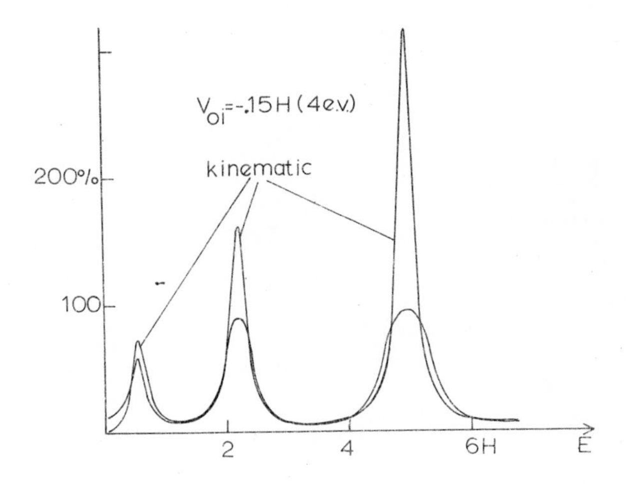

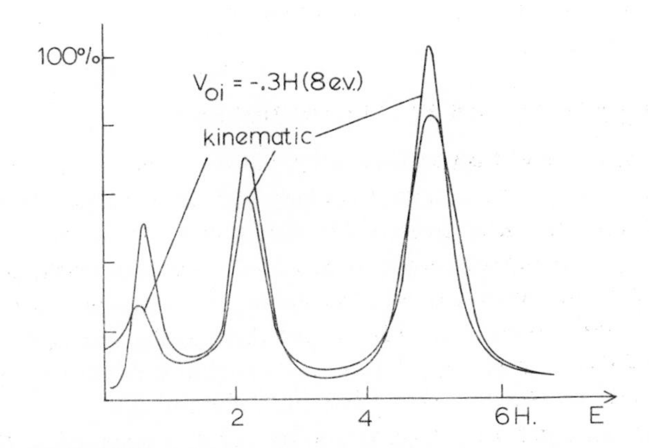

FIGURE 3.12. Comparison of normal mode (exact) and kinematic formulae for reflectivities of a one dimensional crystal, $t = 0.9$, $V = -0.15i$, $-0.3i$. The lattice constant $c = 3.0$ a.u. When absorption is large the kinematic formula is accurate.

energies. For example there is the strong peak in reflectivity near the zero of energy for normal incidence on a copper (100) surface. At low energies absorption is too weak for kinematic theory to be feasible and reflectivities rise to around 30%. Figure 3.13 shows Andersson and Kasemo's (1970) results compared with a band structure calculation by Burdick (1963). The analogy with curves shown in Fig. 3.11 is obvious. The number of such examples encountered is limited and confined to very low energies.

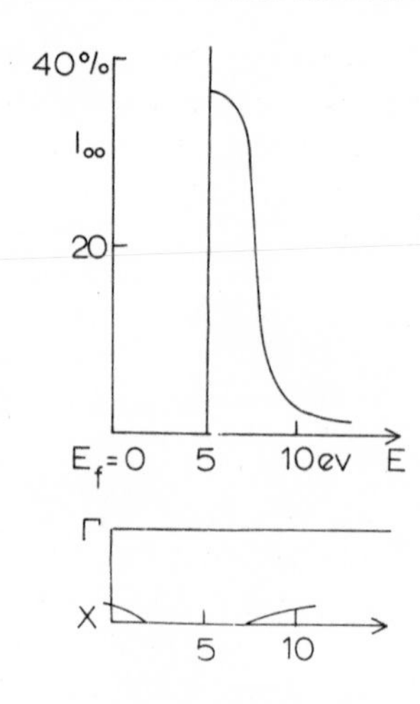

FIGURE 3.13. The top panel shows intensity specularly reflected from a copper (100) surface, near normal incidence. The lower panel shows band structure normal to the surface. Note correspondence of the reflectivity peak with the band gap.

IIIC The general case in three dimensions

In section A we saw that kinematic theory with its simple structure and easily visualised processes does not provide an accurate description of LEED processes in many crystals. On the other hand the model of Section B, complete in one dimension as it is, does not often correspond with the real situation. What we must do now is to start again using the one dimensional case as a guide and build a fully fledged three dimensional theory. Other accounts of theory presented here can be found in papers by Jepsen and Marcus (1971), McRae (1968) and Pendry (1971a, b).

In one dimension the electron was diffracted by a series of scatterers. The three dimensional analogy is of a series of layers parallel to the surface, each layer being composed of many ion-cores. Details of the structure will be left until the next chapter; all that we need to know here is that each layer is assumed periodic in directions parallel to the surface. Each layer being a component of the crystal, it has at least the symmetry of the surface in question. Once we pass the immediate surface region, in the bulk the crystal can be thought of as composed of identical layers stacked on top of one-another, each layer deeper inside the crystal displaced relative to its neighbour by a vector **c** as described in Section ID and illustrated in Fig. 1.6.

In regions close to the surface layers may have spacings different from those of the bulk, and in extreme cases may have different composition. The spacing between the nth and $(n + 1)$th layers, where it differs from the bulk spacing, will be referred to as $\mathbf{c}_n$. The step in potential at the surface can be thought of as an additional layer with different scattering properties placed on top of the crystal. There are not usually more than two layers, including the surface barrier, that are untypical of the bulk.

Between each pair of layers, away from the strongly scattering ion-cores of layers inside the surface, or the variation in potential at the surface barrier, the Schrödinger equation becomes

$$-\tfrac{1}{2}\nabla^2\psi + V_0\psi = E\psi, \tag{3.62}$$

an equation that has plane-wave solutions. In one dimension there were only two plane waves at a given energy: forward and backward travelling, but as we saw in Section A, in three dimensions other beams can be formed, one for each Fourier component parallel to the surface of the wavefield, corresponding to different parallel components of momentum. Because of symmetry of the surface only discrete values of parallel momentum are allowed: $\mathbf{k}_{0\|}$ the parallel component of the original incident beam, and $\mathbf{k}_{0\|} + \mathbf{g}_j$ where $\mathbf{g}_j$ is a reciprocal lattice vector of the surface defined in Section IE. The z-component of mementum of each beam is determined by substituting in (3.62):

$$\mathbf{K}_{\mathbf{g}}^{\pm} = [(\mathbf{k}_{0\|} + \mathbf{g})_x, (\mathbf{k}_{0\|} + \mathbf{g})_y, \pm (2E - 2V_0 - |\mathbf{k}_{0\|} + \mathbf{g}|^2)^{\frac{1}{2}}]. \tag{3.63}$$

Profusion of beams is the main difference from the one-dimensional case and (3.38) describing the one dimensional wavefunction between scatterers must be replaced by a sum over several forward and backward travelling beams,

$$\sum_{\mathbf{g}} a_{n\mathbf{g}}^{+} \exp\left[i\mathbf{K}_{\mathbf{g}}^{+} \cdot (\mathbf{r} - n\mathbf{c})\right] + a_{n\mathbf{g}}^{-} \exp\left[i\mathbf{K}_{\mathbf{g}}^{-} (\mathbf{r} - n\mathbf{c})\right]. \tag{3.64}$$

Figure 3.14 illustrates the new situation

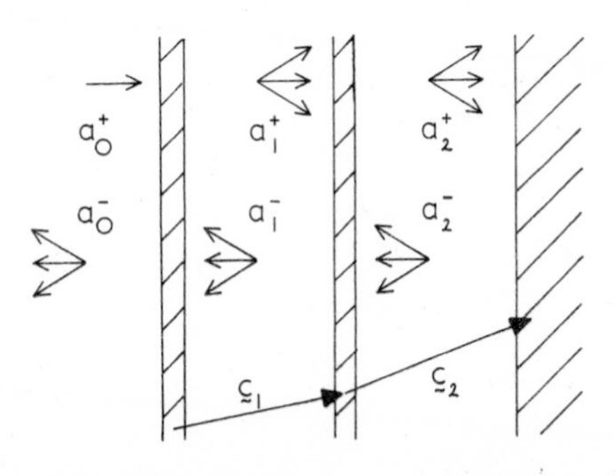

FIGURE 3.14. The wavefield between layers of a crystal is decomposed into sets of forward and backward-travelling beams. After the first few layers the configuration of further layers including interlayer spacings, $\mathbf{c}_n$, becomes typical of the bulk.

Expansion (3.64) holds only in the region of constant potential between layers and it may be thought that this restricts applicability of the approach to materials where a finite extent of such region exists. In fact it does not because the region can always be shrunk to an infinitesimally thin sheet winding its way between two layers. We can always postulate the existence of an infinitesimally thin sheet without loss of generality. We continue to speak of a finite width of region only because it makes the situation easier for the reader to visualise.

Any one of the beams incident on a layer can be diffracted forwards or backwards into a whole set of beams. In other words, whereas in one dimension two numbers, r and t, sufficed to describe scattering, the layers we encounter in three dimensions require matrices for description of scattering. Following through the same reasoning as in the one dimensional case: the forward travelling waves between the $(n-1)$th and nth layers have amplitudes at the nth layer

$$a_{n-1\mathbf{g}}^{+} \exp[i\mathbf{K}_{\mathbf{g}}^{+} \cdot \tfrac{1}{2}\mathbf{c}_{n-1}]. \tag{3.65}$$

They can be transmitted or reflected, and we shall take the general case in which the layer is not symmetric, and has reflection and transmission coefficients that depend upon which side of the layer the wave is incident. Transmitted and reflected waves have amplitudes at the nth layer respectively,

$$\sum_{\mathbf{g}} (I_{\mathbf{g}'\mathbf{g}} + M_{\mathbf{g}'\mathbf{g}}^{++}(n)) a_{n-1\mathbf{g}}^{+} \exp[i\mathbf{K}_{\mathbf{g}}^{+} \cdot \tfrac{1}{2}\mathbf{c}_{n-1}], \tag{3.66a}$$

and

$$\sum_{\mathbf{g}} M_{\mathbf{g}'\mathbf{g}}^{-+}(n) a_{n-1\mathbf{g}}^{+} \exp[i\mathbf{K}_{\mathbf{g}}^{+} \cdot \tfrac{1}{2}\mathbf{c}_{n-1}]. \tag{3.66b}$$

The matrices $M^{-+}(n)$ and $M^{+-}(n)$ are the reflection coefficients for the nth layer referred to above. There is an obvious convention for the $\pm$ signs. I is the unit matrix; $(I + M^{++})$ and $(I + M^{--})$ are transmission coefficients.

In a similar manner waves between the nth and $(n+1)$th layers travelling in the back direction reach the nth layer and are transmitted and reflected. The following matrix equations replace (3.45),

$$\begin{aligned}
a_{n\mathbf{g}'}^{+} &= \sum_{\mathbf{g}} (I_{\mathbf{g}'\mathbf{g}} + M_{\mathbf{g}'\mathbf{g}}^{++}(n)) \exp[i(\mathbf{K}_{\mathbf{g}}^{+} \cdot \tfrac{1}{2}\mathbf{c}_{n-1} + \mathbf{K}_{\mathbf{g}'}^{+} \cdot \tfrac{1}{2}\mathbf{c}_{n})] a_{n-1\mathbf{g}}^{+} \\
&\quad + M_{\mathbf{g}'\mathbf{g}}^{+-}(n) \exp[i(-\mathbf{K}_{\mathbf{g}}^{-} \cdot \tfrac{1}{2}\mathbf{c}_{n} + \mathbf{K}_{\mathbf{g}'}^{+} \cdot \tfrac{1}{2}\mathbf{c}_{n})] a_{n\mathbf{g}}^{-} \\
&= \sum_{\mathbf{g}} Q_{\mathbf{g}'\mathbf{g}}^{I}(n) a_{n-1\mathbf{g}}^{+} + Q_{\mathbf{g}'\mathbf{g}}^{II}(n) a_{n\mathbf{g}}^{-},
\end{aligned} \tag{3.67a}$$

$$\begin{aligned}
a_{n-1\mathbf{g}'}^{-} &= \sum_{\mathbf{g}} (I_{\mathbf{g}'\mathbf{g}} + M_{\mathbf{g}'\mathbf{g}}^{--}(n)) \exp[i(-\mathbf{K}_{\mathbf{g}}^{-} \cdot \tfrac{1}{2}\mathbf{c}_{n} - \mathbf{K}_{\mathbf{g}'}^{-} \cdot \tfrac{1}{2}\mathbf{c}_{n-1})] a_{n\mathbf{g}}^{-} \\
&\quad + M_{\mathbf{g}'\mathbf{g}}^{-+}(n) \exp[i(-\mathbf{K}_{\mathbf{g}'}^{-} \cdot \tfrac{1}{2}\mathbf{c}_{n-1} + \mathbf{K}_{\mathbf{g}}^{+} \cdot \tfrac{1}{2}\mathbf{c}_{n-1})] a_{n-1\mathbf{g}}^{+}
\end{aligned}$$

$$= \sum_{\mathbf{g}} Q^{\mathrm{IV}}_{\mathbf{g}'\mathbf{g}}(n) a^{-}_{n\mathbf{g}} + Q^{\mathrm{III}}_{\mathbf{g}'\mathbf{g}}(n) a^{+}_{n-1\mathbf{g}}. \tag{3.67b}$$

when layer n is typical of the bulk we shall drop subscripts on $\mathbf{c}$, M, and Q.

Expressions have already been derived for $M^{\pm\pm}$ in the case that scattering within a single layer is kinematic.

$$M^{\pm\pm}_{\mathbf{g}'\mathbf{g}} \simeq M(\mathbf{K}^{\pm}_{\mathbf{g}'}, \mathbf{K}^{\pm}_{\mathbf{g}}), \tag{3.68}$$

and the right hand side is defined in equation (3.23). In most cases multiple scattering within a layer is important and in Chapter IV we shall see how (3.68) is modified to take account of multiple scattering.

The last layer is special because on the crystal side of the barrier, waves obey equation (3.62) whereas on the vacuum side waves obey the vacuum Schrödinger equation. Therefore outside the crystal, waves of amplitudes $a^{\pm}_{0\mathbf{g}}$ have wave-vectors given by

$$\mathbf{K}^{\pm}_{0\mathbf{g}} = [(\mathbf{k}_{0\|} + \mathbf{g})_x, (\mathbf{k}_{0\|} + \mathbf{g})_y, \pm(2E - |\mathbf{k}_{0\|} + \mathbf{g}|^2)^{\frac{1}{2}}]. \tag{3.69}$$

Reflectivities and transmissivities of the surface barrier, $M^{\pm\pm}(1)$, can be found by numerical integration of the Schrödinger integration through the variation in potential from V_0 in the crystal to zero outside. $M^{\pm\pm}(1)$ has only diagonal matrix elements if the surface barrier has no structure parallel to the surface.

What we are seeking is the reflected in terms of the incident amplitudes

$$a^{-}_{0\mathbf{g}'} = \sum_{\mathbf{g}} R_{\mathbf{g}'\mathbf{g}}(0)\, a^{+}_{0\mathbf{g}}. \tag{3.70}$$

We cannot immediately apply the normal mode method to find $R(0)$ because we have assumed the first few layers to be atypical of the bulk. What we can do is express $R(0)$ in terms of the reflectivity of the whole crystal minus the first layer:

$$a^{-}_{1\mathbf{g}'} = \sum_{\mathbf{g}} R_{\mathbf{g}'\mathbf{g}}(1)\, a^{+}_{1\mathbf{g}},$$

or on the understanding that a^{+}_{1} and a^{-}_{1} are vectors and $R(1)$ is a matrix,

$$a^{-}_{1} = R(1) a^{+}_{1} \tag{3.71}$$

From (3.67) we have

$$a^{+}_{1} = Q^{\mathrm{I}}(1) a^{+}_{0} + Q^{\mathrm{II}}(1) a^{-}_{1}, \tag{3.72a}$$

$$a^{-}_{0} = Q^{\mathrm{III}}(1) a^{+}_{0} + Q^{\mathrm{IV}}(1) a^{-}_{1}. \tag{3.72b}$$

Given a^{+}_{0} as the incident wave we can solve the last three equations for the three unknowns, and in particular,

$$a_0^- = [Q^{III}(1) + Q^{IV}(1)[1 - R(1)Q^{II}(1)]^{-1}R(1)Q^{I}(1)]a_0^+$$
$$R(0) = [Q^{III}(1) + Q^{IV}(1)[1 - R(1)Q^{II}(1)]^{-1}R(1)Q^{I}(1)]. \tag{3.73}$$

The first term in brackets is direct reflection from the first layer; the second term results from transmission through the first layer followed by multiple reflections between the first layer and the rest of the crystal.

The process can be repeated to calculate $R(1)$ in terms of $R(2)$ and $Q(2)$. In practice after the surface barrier and the first layer of ion cores have been stripped away the crystal remaining has to a very good accuracy the ideal arrangement of identical layers that occurs in the bulk, and the reflectivity matrix can be calculated by the normal mode technique as we now show.

For the bulk of the perfect crystal equations (3.67) hold with the matrices $Q(n = \infty)$ and inter-layer spacings $\mathbf{c}_{n=\infty}$ substituted. Proceeding by analogy with the one dimensional case, we wish to find amplitudes $b_{n\mathbf{g}}^{\pm}$ which obey equations (3.67) and also satisfy *the normal mode condition* which can be written

$$b_{n+1\mathbf{g}}^{+}(\mathbf{k}) = \exp(i\mathbf{k}\cdot\mathbf{c})\,b_{n\mathbf{g}}^{+}(\mathbf{k}), \tag{3.74a}$$

$$b_{n+1\mathbf{g}}^{-}(\mathbf{k}) = \exp(i\mathbf{k}\cdot\mathbf{c})\,b_{n\mathbf{g}}^{-}(\mathbf{k}), \tag{3.74b}$$

$$\mathbf{k}_{\parallel} = \mathbf{k}_{0\parallel}, \tag{3.75}$$

k_z is to be found.

Solution of (3.67) and (3.74) is more complicated than in one dimension and as it involves mathematical rather than physical principles, will be left until the next chapter. For the time being it will be assumed that the equations can in principle be solved to give amplitudes for the normal modes, $b_{n\mathbf{g}}^{\pm}(\mathbf{k})$.

If there are $2n$ beams included in the problem (n forward travelling and n backward travelling) then, we shall show rigorously in the next chapter, there must be $2n$ normal modes: as many as there are beams. This is obvious in the zero scattering case because then the normal modes are the beams themselves. Further, n of the normal modes will either grow exponentially into the crystal or carry net flux out of the crystal. These are to be discarded because they can be excited only by a source of electrons inside the crystal. The other n beams either decay away into the crystal or carry flux into the crystal, and these normal modes have the correct behaviour for excitation by incident beams.

We have now to use normal modes to find the reflectivity of an ideal crystal surface, all the irregularities having been dealt with previously (Fig. 3.15). Expanding the wavefield in terms of normal modes, we can determine their amplitudes by requiring that at the surface they add to give the correct incident waves. Outside the last layer of perfect crystal

$$a_{\mathbf{g}}^{+} = \sum_{\mathbf{k}} A(\mathbf{k}) b_{0\mathbf{g}}^{+}(\mathbf{k}), \tag{3.76}$$

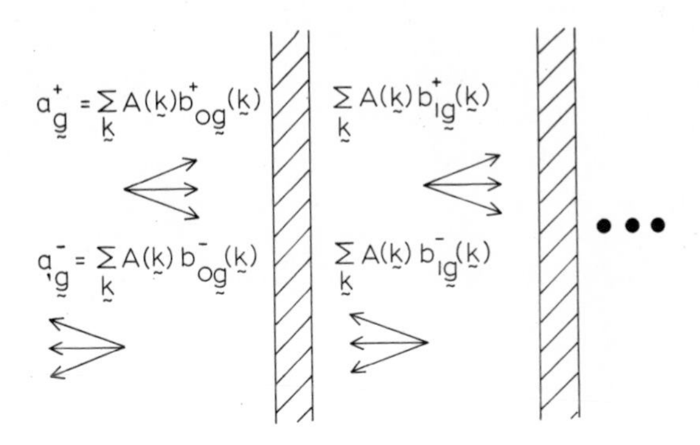

FIGURE 3.15. The sets of beams associated with the normal modes added together with proper amplitudes combine to give the correct amplitude of incident waves, wave amplitudes between each pair of layers, and the reflected wave amplitudes.

the summation being over n normal modes each designated by $\mathbf{k}$, for the n beams denoted by $\mathbf{g}$. $b_{0\mathbf{g}}^{+}(\mathbf{k})$ can be thought of as an $n \times n$ matrix with components

$$B_{\mathbf{gk}} = b_{0\mathbf{g}}^{+}(\mathbf{k}),$$

hence amplitudes of normal modes are

$$A(\mathbf{k}) = \sum_{\mathbf{g}} (B^{-1})_{\mathbf{kg}} a_{\mathbf{g}}^{+}. \tag{3.77}$$

Now that normal mode amplitudes are known, reflected beam amplitudes can be found by adding together contributions from each mode:

$$\begin{aligned} a_{\mathbf{g}'}^{-} &= \sum_{\mathbf{k}} A(\mathbf{k}) b_{0\mathbf{g}'}^{-}(\mathbf{k}) \\ &= \sum_{\mathbf{k}} \sum_{\mathbf{g}} b_{0\mathbf{g}'}^{-}(\mathbf{k}) (B^{-1})_{\mathbf{kg}} a_{\mathbf{g}}^{+}. \end{aligned} \tag{3.78}$$

Therefore the reflection matrix of the ideal crystal is given by

$$R_{\mathbf{g}'\mathbf{g}}(\text{ideal}) = \sum_{\mathbf{k}} b_{0\mathbf{g}}^{+}(\mathbf{k}) (B^{-1})_{\mathbf{kg}}. \tag{3.79}$$

Equations (3.73) and (3.79) solve the LEED problem under very general conditions. Some computer programs for calculating various quantities involved have been provided in the appendix.

Some calculations have been made using this method, for the (001) surface of copper. Phase shifts and values of V_0 calculated as specified in Chapter II have been used. The surface barrier was assumed so smooth that it reflected no flux at all, a much better approximation to the true barrier than a square step in V_0. No account was taken of displacement of surface layers.

First of all, as prescribed, the normal modes were calculated, and accompanying values of k_z, under conditions of normal incidence. Inclusion of 13 beams proved sufficient to give convergence: the Fourier expansion converges rapidly once $|\mathbf{g}|$ is large enough to give $(\mathbf{K}_\mathbf{g}^{\pm})_z$ a large imaginary component so that amplitudes decay between layers. The faster the associated beam decays, the less important is that Fourier component. Five phase shifts were used as was suggested in Section IIC.

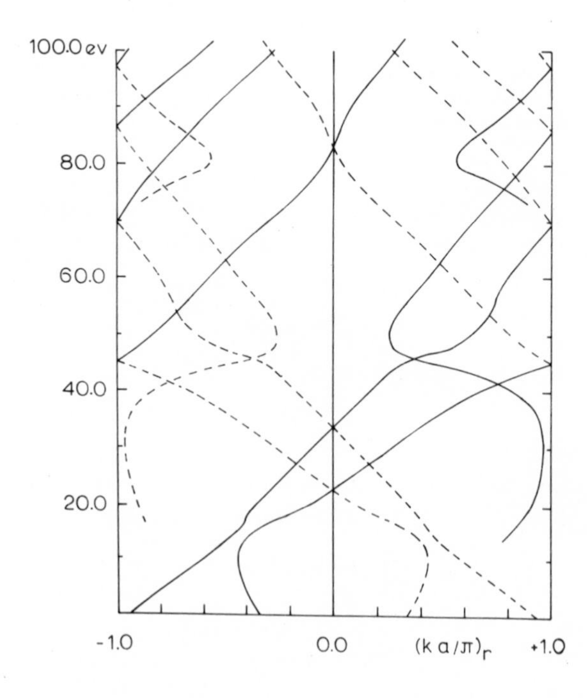

FIGURE 3.16. Band structure for copper normal to the (001) surface. Bands excited are shown in full lines, and those bands coupling only to sources inside the crystal in dashed lines. Only the real part of k_z is shown. Excited bands acquire some of the outward-flux of dashed bands when they are close-by.

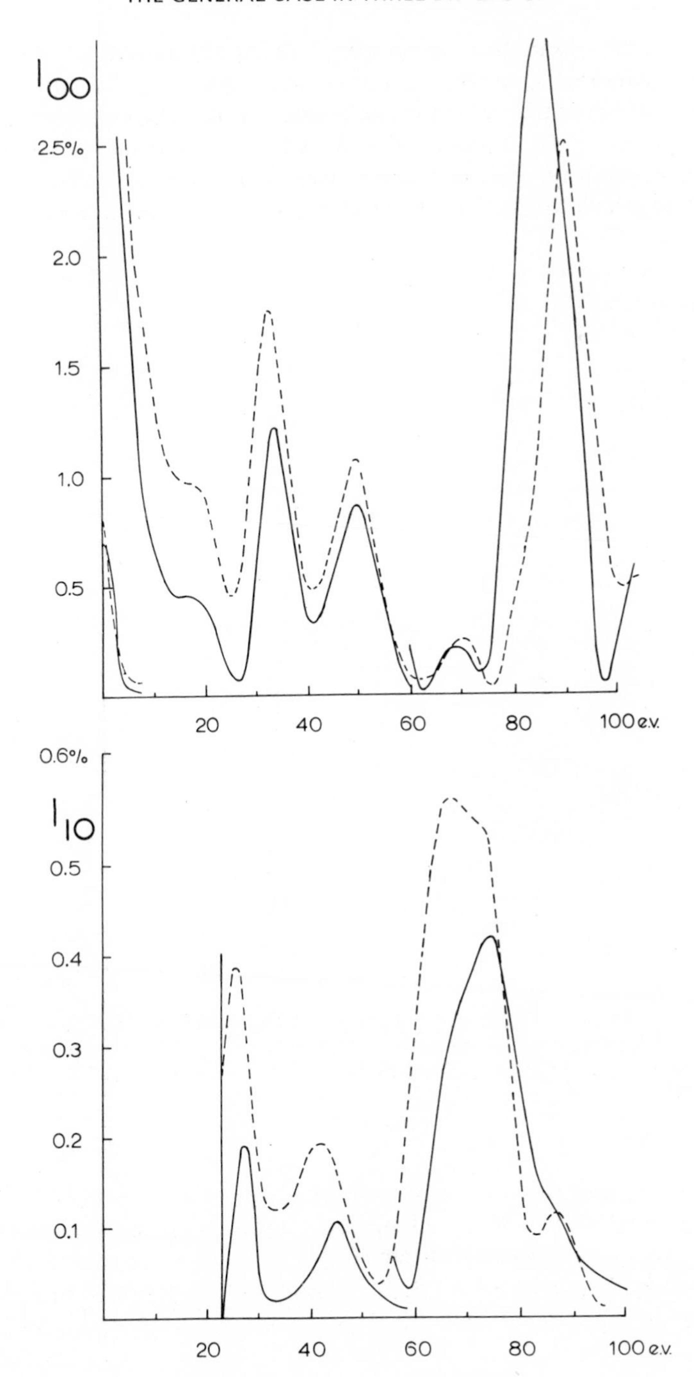
I00
2.5%
2.0
1.5
1.0
0.5
20
40
60
80
100 e.v.
0.6%
I10
0.5
0.4
0.3
0.2
0.1
20
40
60
80
100 e.v.

In Fig. 3.16 k_z is plotted against energy for those modes that are strongly excited. An increasing number of normal modes become important at higher energies, reflecting the fact that more beams crowd into the picture. Because of absorption all the normal modes excited decay away into the crystal, as was discussed in the one dimensional case. Equations (3.73) and (3.79) can be used to calculate reflectivities of our model and Fig. 3.17 shows the results

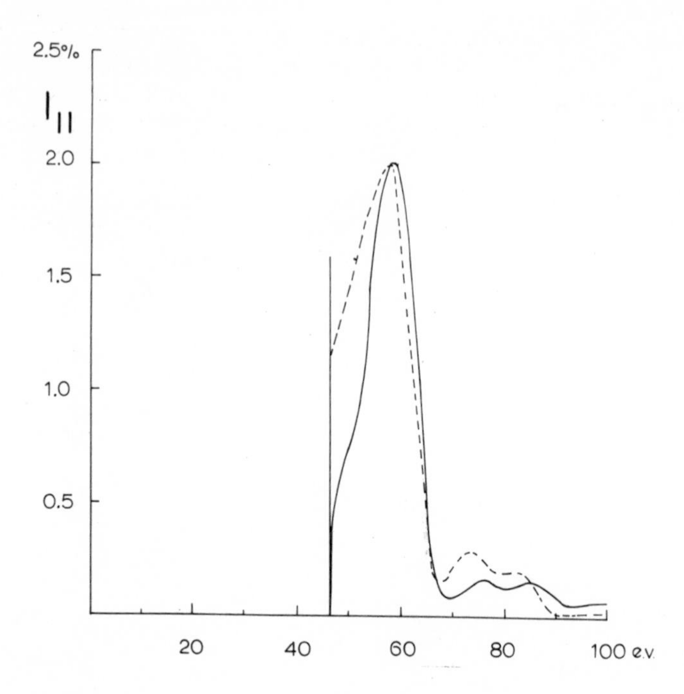

FIGURE 3.17. Reflectivities of a copper (001) surface, 00 beam taken with the crystal rotated 3° about the (010) axis, 10 and 11 beams taken at normal incidence. Full lines are experimental results by Andersson (1969), dashed lines our theory. Experimental intensities × 5 above 60 eV; both intensities × 1/50th below 10 eV.

It is clear from the good agreement, that the model we have adopted in applying the theory takes account of most of the important features of LEED. There are some remaining discrepancies, the largest of which involves intensities at high energies, a consequence of neglecting temperature effects. We shall see in Section E that other approximations have smaller effects on results.

IIID Bloch waves and character of LEED spectra

We saw how in one dimension band structure corresponding to normal modes excited, was useful in interpreting spectra, and so it will prove in three dimensions. The main properties of band structure as applied to LEED have been outlined for the one-dimensional case. The chief complication in three dimensions is that there is more than one Bloch wave (normal mode) to be excited at a given energy, a consequence of there being more than one diffracted beam.

The band structure appearing in Fig. (3.16) is not much like that normally encountered at energies near the Fermi energy. The reason is that we have included absorptive effects which are always present in LEED, and the Bloch waves have a decaying nature. Those excited by the incident wave decay away from the surface into the crystal and have a positive imaginary part to k_z, those that can only be excited by sources inside the crystal decay towards the surface and have negative imaginary parts to k_z. The magnitude of the imaginary part is typically 0.1 to 0.2 a.u. in the range 0–100 eV, but getting

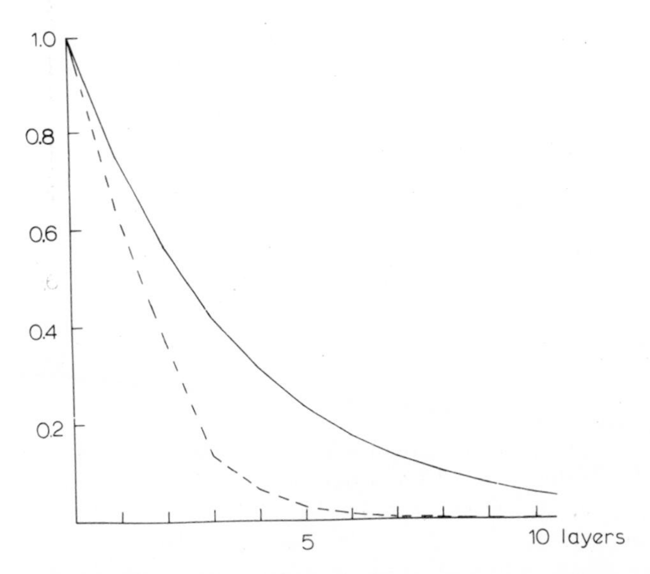

FIGURE 3.18. Decay of incident electron density plotted against number of layers. The full line is a Bloch wave calculation and the dashed line a kinematic calculation, both for normal incidence on a copper (001) surface at 150 eV. Thermal properties were included appropriate to a temperature of $T = 300°K$ (see Chapter VI).

smaller at higher energies because of greater penetration.

The decaying nature of Bloch waves is of course reflected in decay of the total wavefunction. In Fig. 3.18 we have plotted

$$A_n = \left[\sum_{\mathbf{g}} |a_{n\mathbf{g}}^+|^2 + |a_{n\mathbf{g}}^-|^2 \right] \Big/ \left[\sum_{\mathbf{g}} |a_{0\mathbf{g}}^+|^2 + |a_{0\mathbf{g}}^-|^2 \right], \tag{3.80}$$

a measure of the number of electrons between the nth and $(n + 1)$th layers. A_n is given both for a kinematic calculation and for a Bloch wave calculation.

In the more accurate normal mode calculation, density decreases more rapidly than in the kinematic case. This is a consequence of multiple scattering. Figure 3.19 illustrates how as a consequence of multiple scattering the

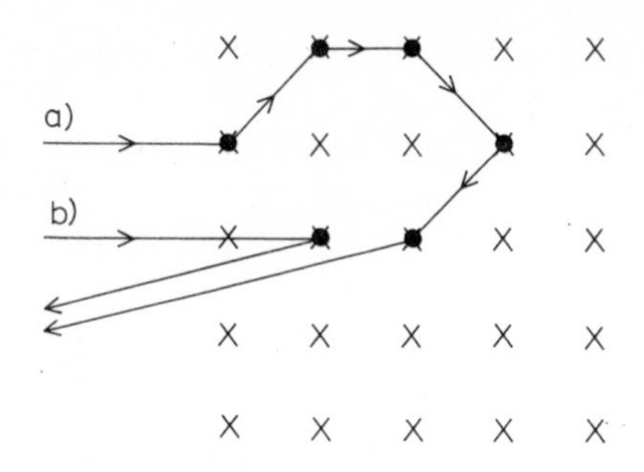

FIGURE 3.19. In multiple scattering processes, (a), the electron follows a more contorted path than in kinematic processes (b). As a result multiple scattering inhibits penetration of the crystal.

electron pursues a more tortuous path in the crystal than in single scattering theory. Since the path length is fixed by V_{0i}, actual distance travelled normal to the surface is less, and multiple scattering increases the influence of absorptive processes on restriction of penetration and reduction of reflected intensity.

The objects of prime interest in LEED intensity/energy spectra are the peaks. In kinematic theory peaks came about when a Bragg condition was satisfied by the incident wave causing successive layers to diffract in-phase. The three dimensional strong scattering process is more complicated than the one-dimensional case but can be understood by reference to the band structure. When two bands with the same **k**-vector are close together in energy they interact strongly, an effect that can be understood qualitatively from perturbation theory. If two states ϕ_1 and ϕ_2 with the same **k**-vector are coupled together by a matrix element $<\phi_1|V|\phi_2>$ they will mix with amplitude

$$\frac{<\phi_1|V|\phi_2>}{E_1 - E_2} \tag{3.81}$$

and if E_1 is close to E_2, strong mixing occurs. Thus if one of the Bloch waves we have excited in the crystal (allowed because it decays into the crystal and has net flux into the crystal) comes close in the band structure diagrams to another Bloch wave (this one not excited because it has net flux out of the crystal), the allowed Bloch wave will acquire some of the character of the forbidden one. That is to say it will acquire a component of back-reflected flux and peaks in reflected intensities will be seen. Which beams become strong depends on the exact composition of the Bloch wave.

Thus we expect strong reflected intensities where two Bloch waves with the same **k**-vector come close in energy, or alternatively two Bloch waves with the same energy are close in **k**. Wave-vectors of Bloch waves are defined by equation (3.74) only to within $2\pi/c_z$ in their z-components. Since $\mathbf{k}_{\parallel}$ is fixed for all Bloch waves by (3.75) our condition is

$$k_z^{(1)} - k_z^{(2)} = (2\pi/c_z) \times \text{integer}, \tag{3.82}$$

which is a Bragg condition on the Bloch wave-vectors.

It is important not to confuse the two sorts of decomposition that can be made of the wavefield inside a crystal. Between pairs of layers a decomposition into plane wave beams tells us how electrons move in this space. A decomposition into Bloch waves or normal modes tells us how the wavefunctions between consecutive pairs of layers are related because we know from (3.74) how to transfer each Bloch wave from one space to the next. Since peaks arise by constructive interference between waves scattered from successive layers, we are using a Bloch wave decomposition to simplify the problem.

In our simple Bragg condition (3.31) the free-electron wave-vector must be replaced by a Bloch wave-vector to generalise the condition. As a rule more than one Bloch wave is excited in the crystal by the incident beam and there may be as many as a dozen at higher energies and general angles of incidence, each one with a different wave-vector. Any one of these Bloch waves can Bragg-reflect and it is the assortment of wave-vectors introduced into the LEED problem by multiple scattering that gives rise to the variety of structure seen in spectra. Figure 3.20 shows how the single reflection for every integer in the Bragg reflection condition seen in kinematic theory, is replaced by several peaks in the multiple scattering case.

Numerous illustrations of this effect can be found in the band structure of Fig. 3.16 and spectrum of Fig. 3.17. For example, near the zero of energy, the crossing of bands at $k_{zr} = \pi/c_z$ gives rise to the very strong peak in the 00 beam at 00 eV. The peak in the 10 beam at 27.0 eV can be identified with the crossing at $k_{zr} = 0.1 \times \pi/c_z$, $E = 27.0$ eV, the peak in the 00 beam at 33.0 eV with the crossing at $k_{zr} = 0.0$, $E = 33.0$ eV, and so on. The peak in the 00 beam at 90.0 eV is of interest because it consists of several peaks all overlapping to form a conglomerate; a number of crossings are found in the band structure near 90 eV.

The coincidence of the energy of a peak with that of a crossing need not be exact because if at the same time as a Bloch wave undergoing a peak in back-scattering, its amplitude of excitation is diminishing, the peak observed will be shifted to lower energies and the shift can be of the same order as the peak width.

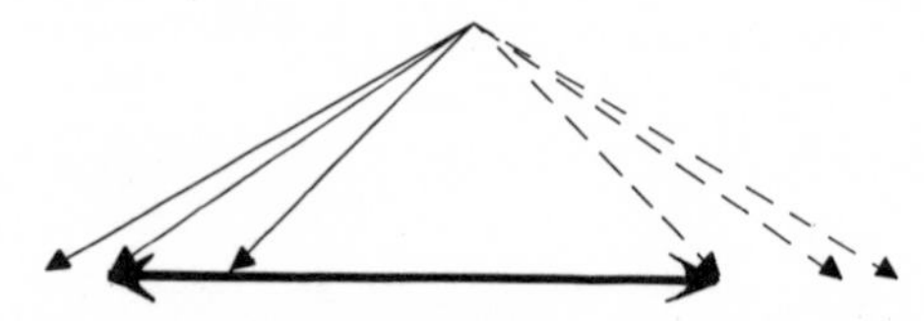

FIGURE 3.20. Real parts of several hypothetical Bloch wave-vectors at a given energy. As in Fig. 3.16 those drawn with full lines are excited by the incident beam and those with dashed lines are not excited. Whenever the wave-vectors of a pair of allowed and forbidden Bloch waves differ by a multiple of $2\pi/c_z$ the allowed Bloch wave acquires a component of flux reflected out of the crystal. Contrast the kinematic case where only one of the allowed Bloch waves is important giving much fewer peaks in reflectivity.

With the aid of band structure we can associate peaks in the LEED spectrum with Bloch waves. Thus by concentrating our attention on the normal mode concerned in a particular peak, we examine that part of the wave-function directly responsible for the peak and in this way some insight of the nature of the peak can be obtained. For example, if the Bloch wave happens to decay away into the crystal very strongly and so is strongly localised near the surface, we can deduce that changes in surface conditions, perhaps by contamination, will affect such a Bloch wave strongly and the associated peak will be sensitive to surface conditions.

We have seen already that multiple scattering causes stronger decay of the wavefield into the crystal as well as causing peaks away from positions predicted by single-scattering theory. Thus those peaks of a strong multiple scattering origin, far from the kinematic Bragg condition correspond to Bloch waves that decay more strongly than average. Hence peaks not close to the kinematic Bragg condition can be expected to show greater sensitivity to surface conditions, a fact that has been known experimentally for some years.

The full multiple scattering theory that usually has to be used to interpret LEED spectra involves complicated computer calculations and hence output of calculations is remote from the simple physical principles that were the input. The advantage of the Bloch wave picture is that it provides simple physical concepts in terms of which we can interpret our calculations, and enables us to bring to bear some physical insight, giving calculations meaning beyond a sequence of numbers.

One of the earliest theories of electron diffraction was cast in terms of Bloch waves, Bethe (1928) and subsequent authors have added to the theory, Boudreaux and Heine (1967), and used it to interpret spectra in detail, Capart (1969, 1971), Gersten and McRae (1972), Jennings and McRae (1970), Jepsen, Marcus and Jona (1972), Marcus and Jepsen (1968), Pendry (1969b, 1971b).

IIIE Surface state resonances

Peaks in LEED spectra can arise through a Bloch wave with relatively constant amplitude of excitation having a wave-vector that satisfies the Bragg condition. This is the most common mechanism by which structure arises, but there is another way. It can happen that Bloch waves, each producing a relatively constant back-reflected intensity, undergo a suddenly increased amplitude of excitation. The result is a peak in reflected intensity but the mechanism of excitation is very different from that of Bragg reflection. We are looking for a mechanism that in equation (3.76) will produce maxima in $A(\mathbf{k})$ rather than in $b_{0\mathbf{g}}^{-}(\mathbf{k})$. The effect was first noted by McRae (1966) and correctly interpreted and developed in greater detail in a second paper, McRae (1971).

The explanation is to be found in terms of surface state theory, and we shall need to digress a little to say what surface states are. To give a simple picture of a surface state we shall use the one-dimensional model of Section B and neglect for the moment influence of absorption. We saw in Fig. 3.9 that even when there is no absorption, multiple scattering can have the effect of making the wavefunction die away exponentially into the crystal. Ranges of energy where this happens are called band gaps.

In Fig. 3.9 there is a band gap, which by choosing V_{0r} to be sufficiently negative, we could lower below the zero of energy so that within this gap both waves inside and outside the crystal decay away from the surface. The question is 'can we match amplitudes and derivatives of these waves inside and outside across the surface to give a solution of the Schrödinger equation that is localised near the surface?' The situation is shown in Fig. 3.21. Whether or not the wavefunctions do have the correct form at some energy inside the band gap is dependent on the nature of the potential inside the crystal, but sometimes it does happen, and a solution called a surface state occurs.

The mathematical condition is that the wave outside the crystal

$$C\exp(-ik_0 z), \tag{3.83}$$

should match to the Bloch wave inside

$$a_0^{-}\left[\exp(-ikz) + (a_0^{+}/a_0^{-})\exp(ikz)\right] \tag{3.84}$$

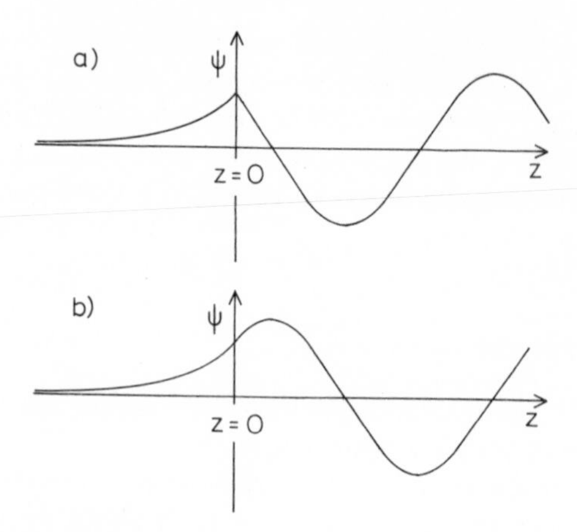

FIGURE 3.21. Wavefunctions inside and outside a surface at $z = 0$ that (a) do not match, (b) that do match and give a surface state.

at the surface. a_0^+/a_0^- is known from expression (3.52). Therefore if there is to be a surface state with continuous amplitude and derivative across the surface plane, the following equations must be satisfied,

$$a_0^- + a_0^+ = C, \tag{3.85a}$$

$$-a_0^- + a_0^+ = -C\frac{k_0}{k}. \tag{3.85b}$$

Note that these equations are the same as those we wrote down for the LEED problem (3.46) except that in (3.85) the incident wave is missing and if equations (3.85) are satisfied it implies that equations (3.46) can have finite solutions for zero incident amplitude! If the incident wave had finite amplitude and there was a surface state at that energy, then C and a_0^- would be infinite; the excitation amplitudes become singular. When absorption is introduced it takes away the singularity, but the excitation amplitudes still have a peak.

In three dimensions the condition for a surface state is again that there should be a finite amplitude of waves localised near the surface, even when there is no incident wave. The condition for this to happen is that the reflection matrix for the crystal surface, $R(0)$, must be infinite. In other words

$$\det\left[R_{\mathbf{gg}'}^{(0)}\right] = \infty. \tag{3.86}$$

We mention in passing that the whole formalism developed for LEED calculations, including many of the computer programs, can be taken over via equation (3.86) and used to calculate details of surface states.

A non-mathematical interpretation can be found for the surface state. Suppose that an electron wave is travelling in the region between the surface barrier and the crystal, Fig. 3.22. If it is travelling towards the barrier, it will be reflected because it does not have sufficient energy to escape into the vacuum. Travelling towards the crystal it will be reflected again, because in a band gap only decaying waves exist and these (providing no absorption is allowed) carry no flux away from the surface. Our picture is of an electron being continually scattered between the bulk and the surface barrier. When this happens in a self-consistent way there is a surface state.

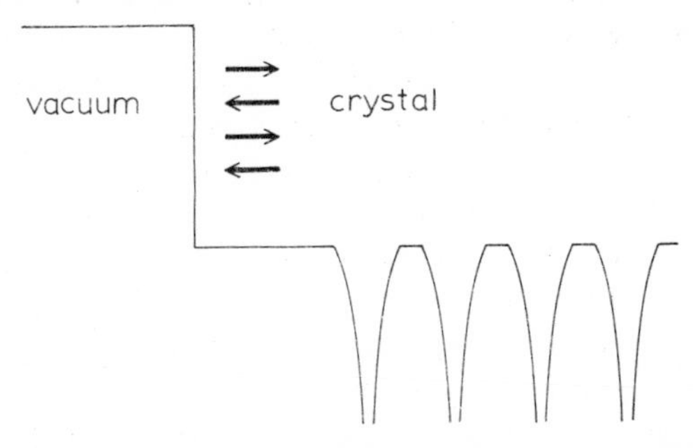

FIGURE 3.22. Multiple reflection of a wave between the surface barrier and the bulk of a crystal.

LEED takes place at positive energies, but still this analogy can help us in interpreting the surface state resonance peaks. We must look for energies where a beam can be strongly reflected by both the surface barrier and by the substrate. Energies just below the emergence condition for a new beam are good places because then the beam about to emerge is incident on the inside of the surface barrier at a shallow angle, and there is a large reflectivity; see Fig. 3.23. If the bulk also reflects strongly there is the possibility that a large amplitude will build up that will be our resonance. One difficulty comes at positive energies: it is that in addition to reflecting the beam in a specular

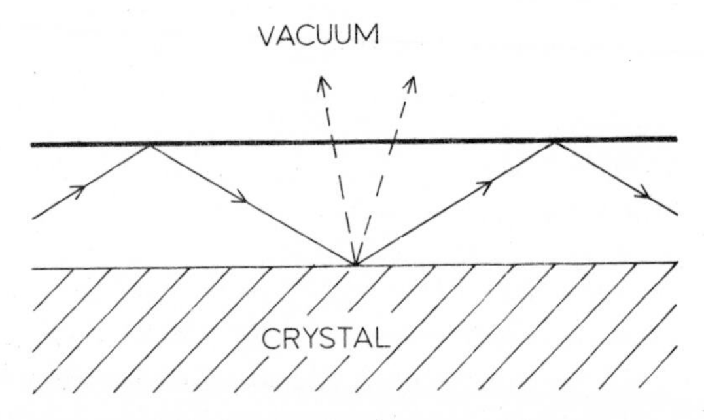

FIGURE 3.23. A surface state resonance in three dimensions, some beams shown by full lines are trapped, but others shown by dashed lines have enough normal component of momentum to escape from the surface.

fashion the substrate can also diffract it into another beam which is not incident on the barrier at a shallow angle, and can escape from the surface into the vacuum. So there is always some flux lost even in the absence of absorption, and the electron is never completely trapped in a surface state at positive energies. The formal expression is that the surface state is a resonance and not a bound state.

The reflection between surface barrier and substrate can build up a large amplitude of electrons at some energies which as we have seen, are to be expected near emergence conditions. As amplitude of the resonance increases, so do amplitudes of beams that escape and a maximum reflectivity is seen in one or other beam depending on the nature of the Bloch wave in the substrate.

An example of a resonance occurs in Fig. 3.17, where the 10 beam is seen to have a peak in intensity at 44.0 eV, just below the emergence energy of the 11 beam. Other examples are to be had in papers by McRae and Caldwell (1967) and Andersson (1970).

It will be appreciated that surface state resonance peaks are more sensitive to the form of the surface barrier than are other peaks. They are also strongly affected by surface contaminants. It is this great sensitivity of surface state peaks that is partly responsible for the generally worse agreement seen between theory and experiment away from normal incidence. At normal incidence—a symmetry direction—several of the beams emerge together and thus the number of energies at which a surface-state resonance appears is smaller than for a general angle of incidence. Thus inaccurate representation of the conditions for a surface state resonance is likely to affect the curves for normal incidence less.

IIIF Peak widths

When inelastic scattering is well approximated by uniform absorption of elastically scattered electrons inside the crystal, the uncertainty principle implies (Chapter II) that peak widths obey the inequality,

$$\Delta E \geqslant 2|V_{0i}| \tag{3.87}$$

When kinematic theory is appropriate the equality is obeyed, giving the narrowest peaks possible; all other effects can act only so as to broaden them. On the other hand, peaks can reach their limiting width in other circumstances too and observation of narrow peaks need not imply kinematic conditions. Equation (3.87) provides a useful way of estimating V_{0i}. Peak widths can be measured independently of a complete theory of LEED spectra and so information about inelastic processes is available directly from experiment. Provided only that enough peaks in the spectrum come close to the limiting width, a reasonable estimate of V_{0i} can be made. Several authors have made use of this fact and we reproduce the results of Andersson and Kasemo (1970) in Fig. 3.24. Copper and nickel are metals in which *d*-bands are important and it is difficult to make an accurate many electron calculation, hence estimates of V_{0i} from experiment are particularly useful.

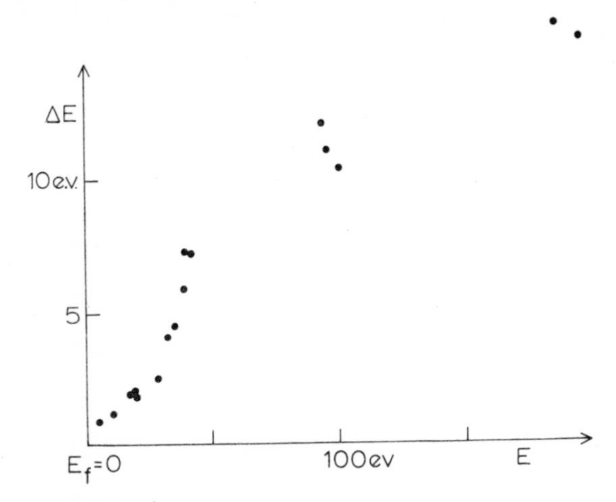

FIGURE 3.24. Peak widths at half height plotted against peak energy to determine an upper bound to $2V_{0i}$. Both copper and nickel spectra have been used to compile this plot.

There are circumstances in which misleading results can be obtained from such plots. For example, examining the calculated curve for copper in Fig. 3.17 where we know the value of $2|V_{0i}|$ to be 2 eV for energies less than 10 eV, because that is the number we inserted in the calculations, we can see that near $E = 0$ in the 00 beam there is a broad peak which is almost 5 eV wide, giving a very inaccurate estimate of V_{0i}. The reason for the enhanced width is that the layers of ion cores happen to reflect the electron beam strongly at this energy.

Section IIIB showed that when strong reflection occurs peak widths are increased. For example in the extreme case of zero absorption peaks become square topped as we have seen in Fig. 3.11. Widths of peaks are given not by $2|V_{0i}|$ which would imply zero width, but are determined instead by the width of the band-gap in Fig. 3.10. When absorption is finite but small, widths of peaks are determined partly by V_{0i} and partly by reflecting powers of layers.

Widening of peaks by strong elastic scattering is a process that tends to be confined to low energies where back-scattering properties of layers combine with small values of V_{0i} to make conditions favourable for this effect. In one dimension broadening can occur only via the 'strong reflectivity' mechanism, but in three dimensions another process is possible. In three dimensions several Bloch waves are excited in the crystal and any one of these can Bragg-reflect to give an intensity peak. There is nothing to stop these peaks from overlapping in energy to form a conglomerate whose width is determined not so much by the intrinsic width of its components but by the difference between energies of highest and lowest components.

An example is furnished in Fig. 3.17 by the peak in the 00 beam at 90 eV which is wider than $2|V_{0i}| = 8$ eV because of several overlapping components. In the 10 beam at 70 eV is another example, but in this case the two component structure is obvious. Overlap effects become increasingly common at higher energies as more Bloch waves are excited.

Conglomerate peaks tend to cluster around the kinematic Bragg condition: the stronger elastic scattering matrix elements, the further they can push peaks away from the Bragg condition. At high energies where there is a large separation between positions of kinematic peaks, a huddle of peaks appears around each kinematic condition and the density of peaks in this huddle is usually such that they overlap. Some evidence of underlying structure in high energy peaks is furnished by their often curious shape caused by varying intensities of component peaks. Figure 3.25 shows some measurements on nickel by Feinstein (1970) illustrating irregularity and width of high energy peaks.

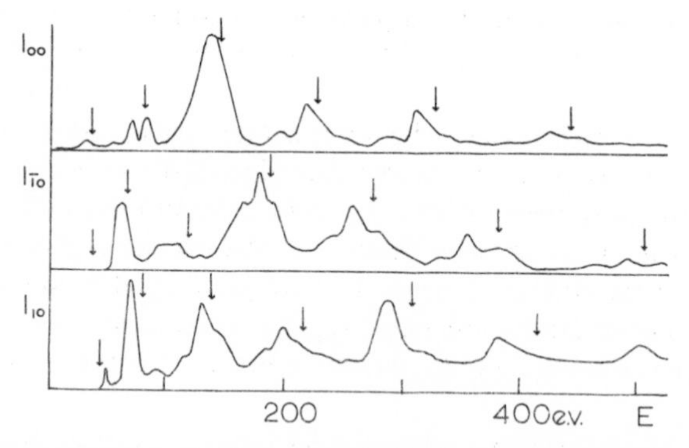

FIGURE 3.25. Spectra taken from a nickel (111) surface near normal incidence, uncorrected for varying incident current. Arrows indicate kinematic Bragg conditions. Note the irregular shape and enormous widths of higher energy peaks.

Because what appear to be single peaks occur near to the kinematic condition at high energies, it is sometimes said that kinematic behaviour is retrieved in the high energy regime. Theoretical calculations show that this is not so. Elastic scattering matrix elements increase as we have seen in Chapter II, tending to a constant value, and inelastic scattering processes according to all the estimates we have, do not in general increase sufficiently rapidly to inhibit the multiple scattering nature of the problem. Even at 100 k.eV many-beam theory is still required to interpret diffraction (Hirsch *et al.*, 1965) and recent work with 1 M.eV electron-microscopes shows even stronger multiple scattering effects.

Any attempt to apply kinematic theory to the problem immediately involves interpreting high energy peak widths in terms of $2|V_{0i}|$ leading to gross overestimates of this quantity.

IIIG Sensitivity of spectra

In Chapter II detailed information was given on how various ingredients of the potential inside a crystal and at the surface could be calculated. In this section we investigate to what extent changes or errors in various ingredients of the potential affect spectra.

The strongest scattering component of the potential is that due to ion-cores and it is this component that determines the character of the diffraction problem. Ion-core scattering is responsible for back reflection of electrons; it results in multiple scattering playing an important role in the theory and in a sense, other effects can be thought of as modifying the basic structure established by ion-core scattering. This being the case it might be anticipated that changing the ion-core scattering interferes with the diffraction process at a deep level, and as we have seen that process is a complicated and delicate one.

Errors in our calculations of ion-core potentials are expected to occur in peripheral regions of the atom populated by polarisable outer electrons of the core and subject to the influence of neighbouring atoms. By experimentation with various methods of constructing ion-core potentials it has been established that errors are of the order of 1 or 2 eV in this outer region. To estimate the effect of such errors a component of the form

$$\frac{3}{10}\left(\frac{3\rho(\mathbf{r})}{8\pi}\right)^{\frac{1}{3}} \tag{3.88}$$

will be added to the ion-core potential for copper to simulate errors. ρ is the charge density and taking the $\frac{1}{3}$ power weights the error potential to the outer regions of the atom in approximately the way we want it to. In outer regions (3.88) introduces errors of the order of 2 eV, and makes changes inside the core that are insignificant in comparison with the already strong potential there.

In Fig. 3.26 the change in the t-matrix is shown. Only very small changes are involved, mainly in forward scattering matrix elements sensitive to the periphery of the core. Figure 3.17 is recalculated and Fig. 3.27 shows the 00 beam calculated for what is believed to be the true potential and also using the erroneous potential. Changes in intensity are disproportionately large relative to the small changes in scattering shown in Fig. 3.26 and indicate the truth of our suspicion that the diffraction process is sensitive to ion-core

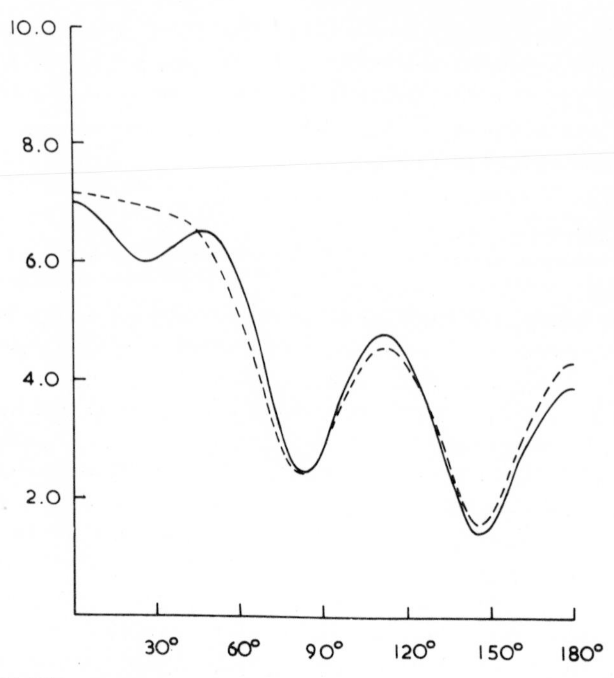

FIGURE 3.26. $|t(\theta)|$ for copper, $E = 50$ eV. Full curve calculated for a potential including the 'error potential', the broken curve omits the 'error potential'.

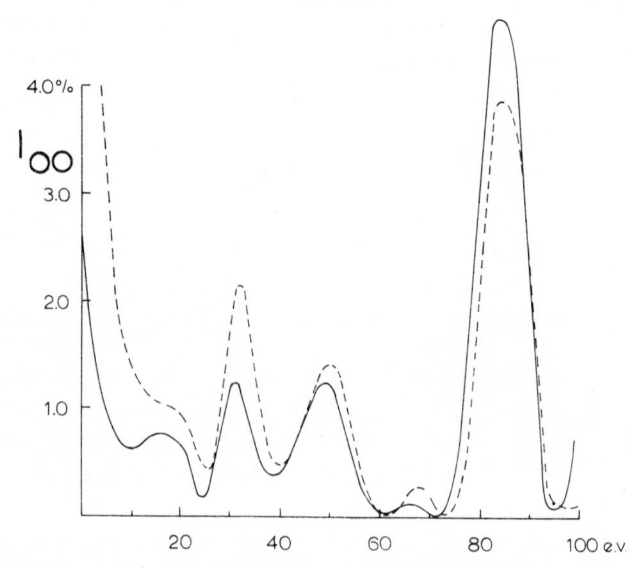

FIGURE 3.27. The 00 beam, normal incidence on a copper 100 surface, calculated using the erroneous potential, full curve, and what is believed to be the true potential, broken curve.

scattering. In contrast the kinematic formula, (3.28), says that changes in intensity are proportional to changes in $|t|^2$. Reference to Fig. 3.26 shows that back scattering is hardly changed by the sort of errors we expect to occur. Sensitivity is therefore a direct consequence of multiple scattering: the multiple scattering process is a delicate one.

Peak positions, as opposed to intensities seem to be less strongly affected than intensities.

It is clear that calculations are intolerant of errors in the ion-core potential and every effort must be made to use as accurate values as possible for this component of the total potential. Any attempt to make gross approximations either in construction of the ion-core potential, or by omitting some of the phase shifts will produce gross errors in calculated intensities.

The other important part of the potential is the constant potential, V_0, between muffin-tin spheres. Its real part, V_{0r}, lowers the energy of the electron in the crystal and in the absence of reflection at the surface barrier (to be dealt with later) errors in V_{0r} simply move calculated spectra up or down in energy. Fortunately V_{0r} appears to be constant to within 2 eV in the 0–100 eV range so it can be easily fixed.

The imaginary component, V_{0i}, representing absorption, has more complicated consequences. As we have seen, it places a lower limit of $2|V_{0i}|$ on peak widths, reduces intensities, and interferes with the multiple scattering process. In the kinematic case equation (3.28) tells us that when the condition for a peak is satisfied,

$$(\mathbf{k}_r - \mathbf{k}_r'')\cdot\mathbf{c} = 2\pi \times \text{integer}$$

then the intensity is proportional to

$$|1/[1 - \exp(-\Delta_{zi}c_z)]|^2, \tag{3.89}$$

$$\Delta_{zi} = (\mathbf{k}_i - \mathbf{k}_i'')_z.$$

Provided that absorption is not so strong as to obliterate the peak completely, (3.89) is approximated by

$$[\Delta_{zi}c_z]^{-1} \tag{3.90}$$

From equation (3.14) we have that

$$k_{zi} = \pm[(2E - 2V_0 - |\mathbf{k}_\parallel|^2)^{\frac{1}{2}}]_i \simeq \pm\frac{1}{2}\frac{-2V_{0i}}{(2E - 2V_{0r} - |\mathbf{k}_\parallel|^2)^{\frac{1}{2}}}, \tag{3.91}$$

if

$$2V_{0i} \ll 2E - 2V_{0r} - |\mathbf{k}_\parallel|^2. \tag{3.92}$$

Bearing in mind that k_{zi} being part of the incoming wave-vector takes the '+' sign in (3.91) and k_{zi}'' being part of the outgoing wave-vector takes the '–' sign, we deduce that Δ_{zi} is proportional to V_{0i} and from (3.90) that

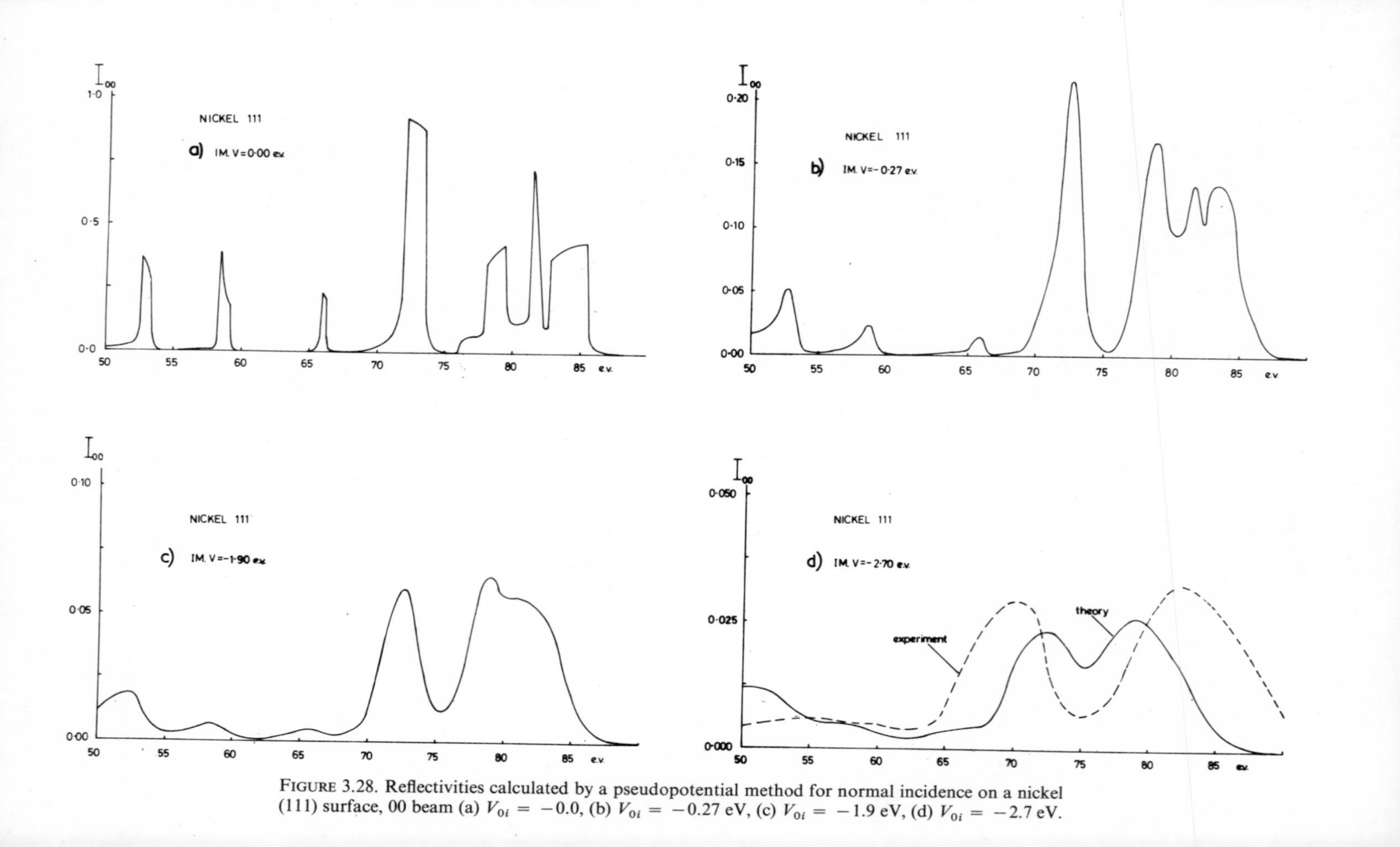

FIGURE 3.28. Reflectivities calculated by a pseudopotential method for normal incidence on a nickel (111) surface, 00 beam (a) $V_{0i} = -0.0$, (b) $V_{0i} = -0.27$ eV, (c) $V_{0i} = -1.9$ eV, (d) $V_{0i} = -2.7$ eV.

$$I \alpha |V_{0i}|^2 \tag{3.93}$$

where I is the approximate kinematic intensity.

Kinematic theory is not always a good guide to sensitivity and therefore we must put various values of V_{0i} into a multiple scattering calculation. In Fig. 3.28 this has been done for nickel.

In this multiple scattering calculation it will be seen that no excessive sensitivity either of widths or intensities occurs. Increasing V_{0i} results in a fairly uniform 'melting' of structure, with a scaling of intensities approximately according to the kinematic result, (3.93).

An ingredient that has been missing so far from our calculations of spectra, is reflectivity of the surface barrier. The potential between ion cores in the crystal must rise up to zero in the vacuum and the rise will result in reflection of electrons on both sides of the barrier. From the good agreement with experiment that we see for copper in Fig. 3.17 where barrier reflectivity is neglected, we might surmise that its influence is small.

If a step is used to approximate the barrier the calculation can be made easily using techniques developed in Section IIIC and the results of such a calculation for the (001) surface of nickel are shown in Fig. 3.29. Gross changes are seen between curves with no barrier reflectivity, and with a square step barrier because a step is a bad approximation to the true shape of the barrier. We know that V_0, being caused almost entirely by conduction or valence electrons, cannot die away on a scale less than the Fermi-Thomas screening wavelength which has a value of about 1 a.u. ($\frac{1}{2}$Å) for nickel. A third calculation is shown in Fig. 3.29 with a barrier chosen to vary on about the same scale as the Fermi-Thomas wavelength,

$$\frac{V_0}{1 + \exp(-2z)}. \tag{3.94}$$

The centre of the barrier, $z = 0$, is chosen to lie in a plane tangential to the muffin tins of the last layer of nickel atoms. No claim is made for accuracy of the model, simply that it has the essential physical properties of smooth variation on a scale of the Fermi-Thomas screening wavelength, and is in more or less the correct position indicated by calculations.

This more realistic barrier makes much less pronounced changes than the square-step barrier because being smoother it has smaller reflectivities. Again, changes are mainly confined to intensities, some of which are altered by of the order of 50% by introducing reflection coefficients of a realistic barrier. One of the most sentitive features is the peak at 46 eV in the 10 beam which is known to be caused by a surface state resonance.

Other processes that might change spectra are dilation of the first few layers of the crystal away from the bulk and temperature effects, including enhanced vibration of surface layers. Dilation of surface layers is to be investi-

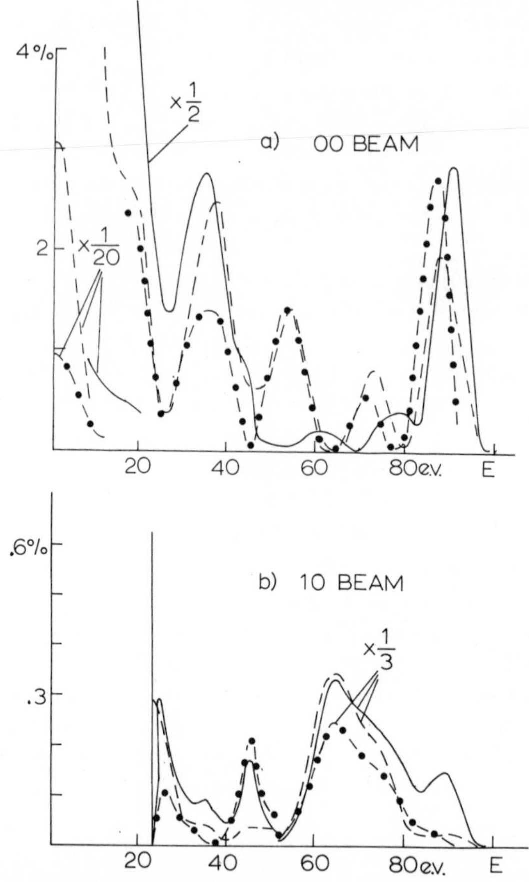

FIGURE 3.29. (a) the 00 beam and (b) the 10 beam for normal incidence on the (001) surface of nickel. Full line: surface barrier approximated by a step, broken line: with the reflectivity of the surface barrier neglected, dotted broken line: with a smooth barrier.

gated in Chapter VII and we shall see there that the main change is a slight shift in peak positions (by about 2 eV) when plausible dilations of the order of 10% are introduced. Temperature is more important and the whole of Chapter VI deals with this question.

Our conclusions are that spectra are sensitive to ion-core scattering factors and these must be calculated accurately. Small errors change intensities, but larger errors move peak positions. The constant potential between muffin tins, V_0, is less critical, V_{0r} shifting peak positions linearly, peak widths depending linearly on V_{0i} and intensities scaling as $|V_{0i}|^{-2}$. Spectra are not very sensitive to a surface barrier of any physically reasonable form, variations in intensities being the main results of changes in surface barriers.

Chapter 4

Schemes of Calculation

IVA Introduction

In the previous chapter the aim was to give an understanding of the nature of the diffraction process. In concentrating on making principles clear many mathematical or inessential details were not developed if they had not much understanding of the central processes to give. So the present chapter will be concerned with a large amount of heavy mathematical details, essential for a complete understanding of the problem, but which can if desired be omitted, on a first reading, or by those whose concern is to grasp essentials and not details. If the reader will be content with a few results quoted from this chapter, the rest of the book is to a large degree independent.

The previous chapter postponed derivation of several results. In Section B of this chapter we show how equations (3.67) and (3.74) for the normal modes can be manipulated into a standard mathematical form of a matrix eigenvalue equation. The eigenvalues give the band structure, and the eigenvectors the normal modes. Section C solves the multiple scattering problem within a plane of atoms and Section D gives details of an alternative and more general solution, but one which is more complicated. Section E shows how to build more complicated layers out of simple planes of atoms where scattering properties are known already.

There is more than one way to solve the LEED problem. Certainly the normal mode method is the most general one, and gives most detailed information, but other methods exist and it will be logical to describe some of them, those based on approximating the infinite crystal by a slab of finite thickness, at the end of Section E because a finite slab is nothing more than a thick layer. Those methods that can be thought of as perturbation-theory approaches will be considered in Chapter V. Generally speaking they do not involve complicated mathematics, and have physical interpretations of their own which merit separate consideration.

Even within the framework of a normal mode method, there are many variations, some in the use of the normal modes (Watts, 1968) to calculate intensities, but more usually in the method of calculating the normal modes. The methods most radically different from the layer method are those that Fourier expand each normal mode in the z-direction as well as in directions parallel to the surface. These latter methods are the ones most commonly adopted in calculations near the Fermi energy and Section G is devoted to explaining their relationship with layer methods, and describing the use of pseudopotentials in calculating normal modes.

In the past a great amount of expertise has been built up in calculating motion of conduction and valence electrons. Band structure calculations have been made by a variety of methods and numerous kinds of perturbation theory have been applied. To some extent LEED workers have been able to draw on this expertise, but it has proved that no techniques could be taken over without at least some modification. We have seen in Chapter II how different aspects of the potential inside the crystal are important at higher energies and the pseudopotential cancellation theorem does not operate as efficiently. Partly because of the stronger scattering and partly because the electron has shorter wave length and therefore picks out more details of the crystal potential, wavefunctions are generally more complicated. They need more Fourier components to describe them, and more phase shifts are of importance. Another difficulty is due to the increased accuracy needed. In dealing with conduction electrons it is usually the energy levels that are important, but in LEED the wavefunctions themselves are measured and at that the back reflected beams, the smallest components, containing around 1 % of the flux. As any variational principle reveals, a calculation that gives an inaccurate wavefunction can often be relied upon to give energy levels to much better accuracy. The much greater demands of accuracy and complexity in LEED usually prove too much for methods developed to deal with simpler situations.

Conversely, the methods developed in connection with LEED are very powerful in application to calculations of conventional band structure near the Fermi surface. They are particularly relevant to layered materials such as graphite or molybdenum disulphide. In these materials the large lattice spacing normal to the layers makes a Fourier expansion in this direction not very convergent. The layer method of calculating band structure is actually more easily applied to materials with such large spacings. Another field where the new techniques can be expected to make a contribution is surface state theory (Davison and Levine, 1970).

Of the several schemes used in LEED calculations it has been those making the conceptual division of the crystal into layers that have been most successful both from the point of view of reduced complexity, and reduced computation times. Yet even using these techniques it is sometimes desirable to reduce times even further, for example if calculations are to be pushed to very high incident energies around 500 eV. This is possible if experiments are done with the incident beam at normal incidence or in a mirror plane, so that the situation has a symmetry. Then those parts of the wavefunction having symmetry that does not conform with the incidence conditions can be discarded before the calculations starts. Considerable advantages are to be gained in this way. Section H shows how symmetry can be made use of.

Finally Section I makes comparison of methods described so far, comparing them in computing times, simplicity of execution, and information provided about scattering processes.

IVB Bloch waves

In Section IIIC we made use of normal modes, or Bloch waves in solving the multiple scattering problem. They had the simple property that wave amplitudes between each pair of layers differed only by a constant factor from those between the previous pair of layers. We saw in Chapter I that in the bulk a crystal structure can always be described as a regular array of identical layers each successive layer displaced by a vector **c** relative to its neighbour (Fig. 1.6). The layers may be simple in structure, as in the case of the (111) surface of a copper crystal where the layers are planes of close-packed atoms, or more involved as in crystals of mica where each layer is a complicated sandwich of atoms. Frequently the displacement, **c**, is not perpendicular to the layers. Successive layers parallel to the (111) surface in copper have displacements parallel as well as vertical to previous layers.

In Chapter III we wrote the wave field in the region of constant potential between the ith and $(i + 1)$th layers as a sum over beams.

$$\sum_{\mathbf{g}} b_{i\mathbf{g}}^{+} \exp(i\mathbf{K}_{\mathbf{g}}^{+} \cdot \mathbf{r}) + b_{i\mathbf{g}}^{-} \exp(i\mathbf{K}_{\mathbf{g}}^{-} \cdot \mathbf{r}), \tag{4.1}$$

$$\mathbf{K}_{\mathbf{g}}^{\pm} = [(\mathbf{k}_{0\|} + \mathbf{g})_x, (\mathbf{k}_{0\|} + \mathbf{g})_y, \pm (2E - 2V_0 - |\mathbf{k}_{0\|} + \mathbf{g}|^2)^{\frac{1}{2}}], \tag{4.2a}$$

if

$$2E - 2V_0 - |\mathbf{k}_{0\|} + \mathbf{g}|^2 > 0$$

$$= [(\mathbf{k}_{0\|} + \mathbf{g})_x, (\mathbf{k}_{0\|} + \mathbf{g})_y, \pm i(|\mathbf{k}_{0\|} + \mathbf{g}|^2 - 2E + 2V_0)^{\frac{1}{2}}], \tag{4.2b}$$

if

$$2E - 2V_0 - |\mathbf{k}_{0\|} + \mathbf{g}|^2 < 0.$$

In equation (4.1) it is assumed that the waves are referred to an origin half way between layers. In principle there are an infinite number of beams, but in practice convergent calculations can be made with a finite number, n, because if $|\mathbf{k}_{0\parallel} + \mathbf{g}|$ is large enough to make $K^{\pm}_{\mathbf{g}z}$ imaginary, that beam dies away exponentially between layers. The larger is $|\mathbf{k}_{0\parallel} + \mathbf{g}|$ the smaller amplitude that beam has at the $(i + 1)$th layer. In practice it has been found sufficient to include all those values of $\mathbf{g}$ for which $K^{\pm}_{\mathbf{g}z}$ is real, plus the first few with imaginary $K^{\pm}_{\mathbf{g}z}$. For most surfaces of simple metals about 10–20 beams are needed in the 0–100 eV range of incident energies.

Bloch waves were introduced to reduce the multiple scattering problem to two easier steps: calculation of Bloch waves followed by expansion of the wavefield in terms of Bloch waves. The latter part has been dealt with; now we must solve equations (3.67),

$$b_i^+ = Q^{\mathrm{I}} b_{i-1}^+ + Q^{\mathrm{II}} b_i^-$$

$$b_{i-1}^- = Q^{\mathrm{III}} b_{i-1}^+ + Q^{\mathrm{IV}} b_i^-$$

showing how beams scatter between layers and equations (3.74),

$$b_i^+(\mathbf{k}) = \exp(i\mathbf{k}\cdot\mathbf{c}) b_{i-1}^+(\mathbf{k}),$$

$$b_i^-(\mathbf{k}) = \exp(i\mathbf{k}\cdot\mathbf{c}) b_{i-1}^-(\mathbf{k}),$$

$$\mathbf{k}_{\parallel} = \mathbf{k}_{0\parallel}$$

specifying that we are calculating a Bloch wave. Subscripts $\mathbf{g}$ and $\mathbf{g}'$ have been omitted for simplicity, it being understood that we are dealing with vectors, $b_i^{\pm}$, and matrices, Q.

Derivations of equation (4.6) from which both band structure and Bloch waves can be found, have been given by McRae (1968) and by Jepsen and Marcus (1971).

Firstly equations (3.74) are used to eliminate b_{i-1}^- and b_{i-1}^+ from equations (3.67)

$$\exp(i\mathbf{k}\cdot\mathbf{c}) b_{i-1}^+ = Q^{\mathrm{I}} b_{i-1}^+ + Q^{\mathrm{II}} b_i^-, \tag{4.3a}$$

$$\exp(-i\mathbf{k}\cdot\mathbf{c}) b_i^- = Q^{\mathrm{III}} b_{i-1}^+ + Q^{\mathrm{IV}} b_i^-. \tag{4.3b}$$

Further simplification can be achieved by constructing a single vector of twice the size from b_{i-1}^+ and b_i^-. Equations (4.3) can now be written

$$\exp(i\mathbf{k}\cdot\mathbf{c}) \begin{bmatrix} I & 0 \\ Q^{\mathrm{III}} & Q^{\mathrm{IV}} \end{bmatrix} \begin{bmatrix} b_{i-1}^+ \\ b_i^- \end{bmatrix} = \begin{bmatrix} Q^{\mathrm{I}} & Q^{\mathrm{II}} \\ 0 & I \end{bmatrix} \begin{bmatrix} b_{i-1}^+ \\ b_i^- \end{bmatrix} \tag{4.4}$$

'I' and '0' represent unit and zero $n \times n$ matrices respectively. It can be shown by direct multiplication that the inverse of the matrix on the left of equation (4.4) is

$$\begin{bmatrix} I\,, & 0 \\ -[Q^{\mathrm{IV}}]^{-1} Q^{\mathrm{III}}, & [Q^{\mathrm{IV}}]^{-1} \end{bmatrix}. \tag{4.5}$$

Therefore multiplying (4.4) by (4.5),

$$\begin{bmatrix} Q^{\mathrm{I}} , & Q^{\mathrm{II}} \\ -[Q^{\mathrm{IV}}]^{-1}Q^{\mathrm{III}}Q^{\mathrm{I}}, & [Q^{\mathrm{IV}}]^{-1}[I - Q^{\mathrm{III}}Q^{\mathrm{II}}] \end{bmatrix} \begin{bmatrix} b_{i-1}^{+} \\ b_i^{-} \end{bmatrix} = \exp(i\mathbf{k}\cdot\mathbf{c}) \begin{bmatrix} b_{i-1}^{+} \\ b_i^{-} \end{bmatrix} \quad (4.6)$$

The importance of (4.6) is that it gives a prescription for finding band structure and Bloch waves by standard numerical procedures. The eigenvalues of the matrix on the left of equation (4.6) give the band structure, and the eigenvectors supply the Bloch wave amplitudes (Wilkinson, 1965). Equation (4.6) is of especial interest for LEED theory and surface problems in general because having fixed the energy, and $\mathbf{k}_{0\parallel}$ (i.e. the angle of incidence in a LEED problem), the eigenvalues give all the band structure we need at that energy, $k_z(E)$. Most eigenvalue methods in band structure calculations give E having fixed k_z, $E(k_z)$, and it is necessary to proceed by inserting a trial k_z into the equations, testing to see if it corresponds to the correct E. Therefore (4.6) is a great practical convenience.

Since the Bloch waves are derived from eigenvectors of a $2n \times 2n$ matrix, there are always $2n$ of them, just as many as we need to match amplitudes and derivatives of n Fourier components across a surface. It can be proved that half of the eigenvalues have modulus greater than unity, the other half less than unity. So we always have n Bloch waves with the correct decaying nature to be excited by an incident electron, to insert into our matching equations.

Under certain circumstances it is possible to reduce equation (4.6) to one involving a matrix only half the size (Pendry, 1971a). Since times for solving a matrix eigenvalue problem scale as n^3 this may afford a useful saving of computing effort. For crystal structures having a mirror plane parallel to the surface in question, if we know one eigenvalue

$$\exp(i\mathbf{k}\cdot\mathbf{c})$$

and eigenvector, we can from symmetry predict that another eigenvalue is

$$\exp(-ik_z c_z + i\mathbf{k}_{\parallel}\cdot\mathbf{c}_{\parallel})$$

and write down the corresponding eigenvector.

For example suppose that each layer has mirror plane symmetry about its centre and that the layers are stacked so that this mirror plane is also a symmetry of the crystal. In equations (3.67) describing scattering by a layer we have chosen origins half way between layers. The symmetry is more evident if origins are moved to the centre of the layer in question, for all waves.

$$b'^{+}_{i\mathbf{g}} = b^{+}_{i\mathbf{g}} \exp(-i\tfrac{1}{2}\mathbf{K}^{+}_{\mathbf{g}} \cdot \mathbf{c}), \tag{4.7a}$$

$$b'^{+}_{i-1\mathbf{g}} = b^{+}_{i-1\mathbf{g}} \exp(+i\tfrac{1}{2}\mathbf{K}^{+}_{\mathbf{g}} \cdot \mathbf{c}), \tag{4.7b}$$

$$b'^{-}_{i\mathbf{g}} = b^{-}_{i\mathbf{g}} \exp(-i\tfrac{1}{2}\mathbf{K}^{-}_{\mathbf{g}} \cdot \mathbf{c}), \tag{4.7c}$$

$$b'^{-}_{i-1\mathbf{g}} = b^{-}_{i-1\mathbf{g}} \exp(+i\tfrac{1}{2}\mathbf{K}^{-}_{\mathbf{g}} \cdot \mathbf{c}). \tag{4.7d}$$

A prime denotes an origin taken at the centre of the ith layer. The new wave amplitudes scatter as follows, from (3.67) and (4.7),

$$b'^{+}_{i} = Q'^{\mathrm{I}} b'^{+}_{i-1} + Q'^{\mathrm{II}} b'^{-}_{i}, \tag{4.8a}$$

$$b'^{-}_{i-1} = Q'^{\mathrm{III}} b'^{+}_{i-1} + Q'^{\mathrm{IV}} b'^{-}_{i}, \tag{4.8b}$$

where we have simplified the equations by writing

$$Q'^{\mathrm{I}}_{\mathbf{g}\mathbf{g}'} = Q^{\mathrm{I}}_{\mathbf{g}\mathbf{g}'} \exp(-i\tfrac{1}{2}\mathbf{K}^{+}_{\mathbf{g}} \cdot \mathbf{c} - i\tfrac{1}{2}\mathbf{K}^{+}_{\mathbf{g}'} \cdot \mathbf{c}), \tag{4.9a}$$

$$Q'^{\mathrm{II}}_{\mathbf{g}\mathbf{g}'} = Q^{\mathrm{II}}_{\mathbf{g}\mathbf{g}'} \exp(-i\tfrac{1}{2}\mathbf{K}^{+}_{\mathbf{g}} \cdot \mathbf{c} + i\tfrac{1}{2}\mathbf{K}^{-}_{\mathbf{g}'} \cdot \mathbf{c}), \tag{4.9b}$$

$$Q'^{\mathrm{III}}_{\mathbf{g}\mathbf{g}'} = Q^{\mathrm{III}}_{\mathbf{g}\mathbf{g}'} \exp(+i\tfrac{1}{2}\mathbf{K}^{-}_{\mathbf{g}} \cdot \mathbf{c} - i\tfrac{1}{2}\mathbf{K}^{+}_{\mathbf{g}'} \cdot \mathbf{c}), \tag{4.9c}$$

$$Q'^{\mathrm{IV}}_{\mathbf{g}\mathbf{g}'} = Q^{\mathrm{IV}}_{\mathbf{g}\mathbf{g}'} \exp(+i\tfrac{1}{2}\mathbf{K}^{-}_{\mathbf{g}} \cdot \mathbf{c} + i\tfrac{1}{2}\mathbf{K}^{-}_{\mathbf{g}'} \cdot \mathbf{c}).$$

Because the layer has mirror plane symmetry its reflection and transmission coefficients are independent of which side of the layer a wave is incident upon, we write

$$Q'^{\mathrm{I}} = Q'^{\mathrm{IV}} = T \tag{4.10a}$$

$$Q'^{\mathrm{II}} = Q'^{\mathrm{III}} = R. \tag{4.10b}$$

The Bloch wave condition (3.74) becomes

$$b'^{+}_{i\mathbf{g}}(\mathbf{k}) = \exp(i\mathbf{k} \cdot \mathbf{c}) \exp(-i\mathbf{K}^{+}_{\mathbf{g}} \cdot \mathbf{c}) b'^{+}_{i-1\mathbf{g}}(\mathbf{k}), \tag{4.11a}$$

$$b'^{-}_{i\mathbf{g}}(\mathbf{k}) = \exp(i\mathbf{k} \cdot \mathbf{c}) \exp(-i\mathbf{K}^{-}_{\mathbf{g}} \cdot \mathbf{c}) b'^{-}_{i-1\mathbf{g}}(\mathbf{k}). \tag{4.11b}$$

Equations (4.11) can be simplified a little. Consider the $(i - 1)$th and $(i + 1)$th planes; they are related by the mirror plane in the ith plane and cannot have any relative parallel displacement, modulo a lattice vector. Since $\mathbf{c}$ is the displacement of consecutive planes

$$2\mathbf{c}_{\parallel} \cdot \mathbf{g} = \text{integer} \times 2\pi, \tag{4.12}$$

from the properties of real and reciprocal lattice vectors. Remembering that

$$\mathbf{K}^{\pm}_{\mathbf{g}\parallel} = (\mathbf{k}_{0\parallel} + \mathbf{g}),$$

equations (4.11) can be rewritten

$$b'^{+}_{i} = \exp(ik_z c_z) Y^{-1} X b'^{+}_{i-1}, \tag{4.13a}$$

$$b_i'^- = \exp(ik_z c_z) YX b_{i-1}'^-, \tag{4.13b}$$

$$X_{\mathbf{gg}'} = X_{\mathbf{gg}'}^{-1} = \exp(i\mathbf{g}\cdot\mathbf{c}_{\parallel})\delta_{\mathbf{gg}'} = \pm\delta_{\mathbf{gg}'}, \tag{4.14}$$

$$Y_{\mathbf{gg}'} = \exp(iK_{\mathbf{g}z}^+ c_z)\delta_{\mathbf{gg}'} = \exp(-iK_{\mathbf{g}z}^- c_z)\delta_{\mathbf{gg}'}. \tag{4.15}$$

Substituting from (4.13) into (4.8) gives

$$\exp(ik_z c_z) Y^{-1} X b_{i-1}'^+ = T b_{i-1}'^+ + R b_i'^-, \tag{4.16a}$$

$$\exp(-ik_z c_z) Y^{-1} X b_i'^- = R b_{i-1}'^+ + T b_i'^-. \tag{4.16b}$$

In the same manner as we derived equation (4.6) from equations (4.3) we deduce from (4.16)

$$\begin{bmatrix} YXT, & YXR \\ -T^{-1}RYXT, & T^{-1}[Y^{-1}X - RYXR] \end{bmatrix} \begin{bmatrix} b_{i-1}'^+ \\ b_i'^- \end{bmatrix} = \exp(ik_z c_z) \begin{bmatrix} b_{i-1}'^+ \\ b_i'^- \end{bmatrix} \tag{4.17}$$

It is evident from (4.16) that given one solution of (4.19),

$$[b_{i-1}'^+, \quad b_i'^-],$$

a second is immediately available,

$$[b_i'^-, \quad b_{i-1}'^+] \tag{4.18}$$

but with k_z reversed, as we predicted. Substituting solution (4.18) into (4.17) reversing k_z and rearranging,

$$\begin{bmatrix} T^{-1}[Y^{-1}X - RYXR], & -T^{-1}RYXT \\ YXR, & YXT \end{bmatrix} \begin{bmatrix} b_{i-1}'^+ \\ b_i'^- \end{bmatrix} = \exp(-ik_z c_z) \begin{bmatrix} b_{i-1}'^+ \\ b_i'^- \end{bmatrix} \tag{4.19}$$

By adding (4.19) to (4.17) and writing the result in terms of $n \times n$ matrices,

$$(YXT + T^{-1}[Y^{-1}X - RYXR]) b_{i-1}'^+ + (YXR - T^{-1}RYXT) b_i'^- = 2\cos(k_z c_z) b_{i-1}'^+, \tag{4.20a}$$

$$(YXR - T^{-1}RYXT) b_{i-1}'^+ + (YXT + T^{-1}[Y^{-1}X - RYXR]) b_i'^- = 2\cos(k_z c_z) b_i'^-, \tag{4.20b}$$

and adding (4.20a) to (4.20b)

$$(YXT + T^{-1}[Y^{-1}X - RYXR] + YXR - T^{-1}RYXT)(b_{i-1}'^+ + b_i'^-) = 2\cos(k_z c_z)(b_{i-1}'^+ + b_i'^-). \tag{4.21}$$

(4.21) is the equation we seek, an $n \times n$ matrix eigenvalue equation. Using equations (4.16) b'^{+}_{i-1} and b'^{-}_{i} can be found separately, and hence so can b^{+}_{i-1} and b^{-}_{i}.

The trick can be made to work for other locations of the mirror plane though detailed working may be somewhat more complicated when the layers themselves do not have mirror plane symmetry.

IVC Multiple scattering within a plane of ion-cores

In Chapter III and Sections A and B of this chapter the theory of LEED was developed by dividing the crystal into a regular stack of identical layers. Transmission and reflection coefficients of the layers were assumed to be known, and now to complete the theory we must delve into details of electron scattering within each layer, and relate transmission and reflection coefficients for the complete layer to scattering properties of component atoms.

The layers can be complicated entities as we saw in Section ID and we propose further to divide a layer into component planes. Within a plane of atoms all the ion cores are taken to have their centres on the same plane parallel to the surface. We shall choose the origin to lie at the origin of a given unit cell within the plane and following notation of Section ID the position of the kth atom within the jth unit cell is given by

$$\mathbf{R}_{jk} = \mathbf{R}_j + \mathbf{r}_k. \tag{4.22}$$

Figure 1.7 shows structure within a layer and within planes. In this section we consider scattering by a single plane. Later, planes will be assembled into layers.

In calculating scattering by a plane of ion-cores we shall once again make the muffin-tin approximation introduced in Chapter II: that the ion core potentials can be contained within non-overlapping spheres drawn about their centres, and between these spheres is a constant potential V_0. The ion core potentials are assumed to be spherically symmetric though this assumption is not essential to the working in this section.

It is the ion cores that are responsible for scattering taking place within the plane. So far we have made use of plane-waves in describing motion of electrons through regions of constant potential between layers. There may be incident on the plane a set of plane waves,

$$\sum_{\mathbf{g}} U^{+}_{\mathbf{g}} \exp(iK^{+}_{\mathbf{g}} \cdot \mathbf{r}). \tag{4.23}$$

referred to an origin in the plane. In situations of spherical symmetry it is more convenient to re-express a plane wave as a sum over spherical waves. A spherical wave expansion of (4.23) about the kth ion core in the unit cell at the origin, reads

$$\sum_{\mathbf{g}} U_{\mathbf{g}}^{+} \exp(i\mathbf{K}_{\mathbf{g}}^{+} \cdot \mathbf{r})$$

$$= \sum_{\ell m} A_{\ell m k}^{(0)} j_{\ell}(\kappa|\mathbf{r} - \mathbf{R}_{0k}|) Y_{\ell m}(\Omega(\mathbf{r} - \mathbf{R}_{0k}))$$

$$= \sum_{\ell m} A_{\ell m k}^{(0)} \tfrac{1}{2}[h_{\ell}^{(1)}(\kappa|\mathbf{r} - \mathbf{R}_{0k}|) + h_{\ell}^{(2)}(\kappa|\mathbf{r} - \mathbf{R}_{0k}|)] \times Y_{\ell m}(\Omega(\mathbf{r} - \mathbf{R}_{0k})), \quad (4.24)$$

$$\kappa = |\mathbf{K}_{\mathbf{g}}^{+}| = +(2E - 2V_0)^{\frac{1}{2}}. \quad (4.25)$$

$\Omega(\mathbf{r} - \mathbf{R}_{0k})$ stands for the angular coordinates of $(\mathbf{r} - \mathbf{R}_{0k})$; $h_{\ell}^{(1)}$ and $h_{\ell}^{(2)}$ are outgoing and incoming partial waves; the constant $A_{\ell m k}^{(0)}$ is given by

$$A_{\ell m k}^{(0)} = \sum_{\mathbf{g}} U_{\mathbf{g}}^{+} \exp(i\mathbf{K}_{\mathbf{g}}^{+} \cdot \mathbf{R}_{0k}) 4\pi i^{\ell}(-1)^{m} Y_{\ell -m}(\Omega(\mathbf{K}_{\mathbf{g}}^{+})). \quad (4.26)$$

Spherical wave expansions about the kth atom in other unit cells can easily be found from $A_{\ell m k}^{(0)}$. $(\mathbf{R}_{jk} - \mathbf{R}_{0k})$ is a lattice vector of the plane hence

$$\mathbf{K}_{\mathbf{g}}^{+} \cdot (\mathbf{R}_{jk} - \mathbf{R}_{0k}) = (\mathbf{k}_{0\|} + \mathbf{g}) \cdot (\mathbf{R}_{jk} - \mathbf{R}_{0k})$$
$$= 2\pi \times \text{integer} + \mathbf{k}_{0\|} \cdot (\mathbf{R}_{jk} - \mathbf{R}_{0k}). \quad (4.27)$$

Using this result we deduce from (4.23) that the wavefunction at $\mathbf{R}_{jk}$ is identical with the wavefunction at $\mathbf{R}_{0k}$ except for a phase factor

$$\exp[i\mathbf{k}_{0\|} \cdot (\mathbf{R}_{jk} - \mathbf{R}_{0k})],$$

and amplitudes in the partial wave expansion differ by the same factor

$$A_{\ell m k}^{(0)}(j\text{th cell}) = A_{\ell m k}^{(0)}(0\text{th cell}) \exp[i\mathbf{k}_{0\|} \cdot (\mathbf{R}_{jk} - \mathbf{R}_{0k})]$$
$$= A_{\ell m k}^{(0)}(0\text{th cell}) \exp[i\mathbf{k}_{0\|} \cdot \mathbf{R}_{j}]. \quad (4.28)$$

Scattering by an ion core can be expressed as a shift in phase of outgoing partial waves, and we saw in Section IIC that in the presence of a scatterer, outside the muffin tins the following scattered wave must be present in addition to the incident wave:

$$\sum_{\ell m} A_{\ell m k}^{(0)} \tfrac{1}{2}(\exp[2i\delta_{\ell}(k)] - 1) h_{\ell}^{(1)}(\kappa|\mathbf{r} - \mathbf{R}_{jk}|)$$
$$\times \exp[i\mathbf{k}_{0\|} \cdot (\mathbf{R}_{jk} - \mathbf{R}_{0k})] Y_{\ell m}(\Omega(\mathbf{r} - \mathbf{R}_{jk})). \quad (4.29)$$

The phase shift depends only on k, not on j, because atoms in different unit cells are related by symmetry.

In contrast to Section IIC, we are dealing with more than one ion core and a difficulty arises. It is that flux scattered by one atom can impinge on a second atom scattering again, and so on. If scattering were sufficiently weak for multiple scattering processes to be neglected the total scattered flux would be

$$\psi_{s}^{(0)} = \sum_{\ell m j k} A_{\ell m k}^{(0)} \tfrac{1}{2}(\exp[2i\delta_{\ell}(k)] - 1) h_{\ell}^{(1)}(\kappa|\mathbf{r} - \mathbf{R}_{jk}|)$$

$$\times \exp\left[i\mathbf{k}_{0\parallel}\cdot(\mathbf{R}_{jk}-\mathbf{R}_{0k})\right] Y_{\ell m}(\Omega(\mathbf{r}-\mathbf{R}_{jk}))$$
$$= \sum_{\ell jk}\sum_{\mathbf{g}} U_{\mathbf{g}}^{+} \exp(i\mathbf{K}_{\mathbf{g}}^{+}\cdot\mathbf{R}_{jk}) i^{\ell}(2\ell+1)\tfrac{1}{2}(\exp[2i\delta_{\ell}(k)]-1)$$
$$\times h_{\ell}^{(1)}(\kappa|\mathbf{r}-\mathbf{R}_{jk}|) P_{\ell}(\mathbf{K}_{\mathbf{g}}^{+}\cdot(\mathbf{r}-\mathbf{R}_{jk})/|\mathbf{K}_{\mathbf{g}}^{+}|\,|\mathbf{r}-\mathbf{R}_{jk}|). \quad (4.30)$$

The symmetry of the plane being at least as great as that of the surface as a whole the beams in the diffraction pattern of the layer (that is what equation (4.30) represents) also occur in the diffraction pattern of the surface as a whole and we can use the same set of wave-vectors to re-express the zeroth expression for flux scattered by the plane (4.30) in terms of beams:

$$\psi_{s}^{(0)} = \sum_{\mathbf{g}'} V_{\mathbf{g}'}^{(0)+} \exp(i\mathbf{K}_{\mathbf{g}'}^{+}\cdot\mathbf{r}), \qquad z>0, \quad (4.31a)$$

$$= \sum_{\mathbf{g}'} V_{\mathbf{g}'}^{(0)-} \exp(i\mathbf{K}_{\mathbf{g}'}^{-}\cdot\mathbf{r}), \qquad z<0. \quad (4.31b)$$

A different expression must be used on each side of the plane, because the scatterers prevent analytic continuation from one side to the other. In Chapter III an expression was calculated, (3.23), for $V_{\mathbf{g}'}^{(0)\pm}$ in terms of the incident plane wave amplitudes

$$V_{\mathbf{g}'}^{(0)\pm} = \sum_{\mathbf{g}} M_{\mathbf{g}'\mathbf{g}}^{\pm+(0)} U_{\mathbf{g}}^{+} \quad (4.32)$$

$$M_{\mathbf{g}'\mathbf{g}}^{\pm+(0)} = \frac{2\pi i}{|\mathbf{K}_{\mathbf{g}}^{+}| A K_{\mathbf{g}'z}^{+}} \sum_{\ell=0}^{\infty}\sum_{k=1}^{n} (2\ell+1)\sin[\delta_{\ell}(k)]\exp[i\delta_{\ell}(k)]$$
$$\times P_{\ell}(\mathbf{K}_{\mathbf{g}'}^{\pm}\cdot\mathbf{K}_{\mathbf{g}}^{+}/|\mathbf{K}_{\mathbf{g}'}^{\pm}|\,|\mathbf{K}_{\mathbf{g}}^{+}|)\exp[i(\mathbf{K}_{\mathbf{g}}^{+}-\mathbf{K}_{\mathbf{g}'}^{\pm})\cdot\mathbf{r}_{k}]. \quad (4.33)$$

Only a summation over atoms in a single unit cell remains. A is the area of unit cell.

Equivalently $V_{\mathbf{g}'}^{(0)\pm}$ can be expressed in terms of incident spherical wave amplitudes, $A_{\ell mk}^{(0)}$ by substituting (4.33) and (4.26) into (4.32) and expanding P_{ℓ}, the Legendre polynomials in terms of $Y_{\ell m}$ the spherical harmonics,

$$V_{\mathbf{g}'}^{(0)\pm} = \frac{2\pi i}{|\mathbf{K}_{\mathbf{g}}^{+}| A K_{\mathbf{g}'z}^{+}} \sum_{\ell mk} A_{\ell mk}^{(0)} i^{-\ell} \sin[\delta_{\ell}(k)]\exp[i\delta_{\ell}(k)]$$
$$\times Y_{\ell m}(\Omega(\mathbf{K}_{\mathbf{g}'}^{\pm}))\exp(-i\mathbf{K}_{\mathbf{g}'}^{\pm}\cdot\mathbf{r}_{k}). \quad (4.34)$$

In cases where scattering is strong there are other large contributions to the spherical waves incident on an ion core apart from $A_{\ell mk}^{(0)}$, due to flux arriving from neighbouring atoms. Beeby (1968), Kambe (1967a, b, 1968), and McRae (1966) have all shown how to take this into account. Our treatment will be closest to that of Beeby, with some improvements suggested by Pendry (1971a).

The total amplitude of waves incident on the kth ion core in the unit cell at the origin can be written

$$A_{\ell mk} = A^{(s)}_{\ell mk} + A^{(0)}_{\ell mk} \tag{4.35}$$

$A^{(s)}_{\ell mk}$ is the amplitude scattered from other ion cores and itself depends on $A_{\ell mk}$. It must be determined self-consistently.

Equation (4.28) tells us that the original incident wavefield transforms from one unit cell to another by a phase factor

$$\exp\left[i\mathbf{k}_{0\parallel}\cdot(\mathbf{R}_j - \mathbf{R}_{j'})\right].$$

Therefore the once-scattered wave transforms in this way as can be confirmed from (4.31), and so must the twice scattered wave, etc. Hence the total incident wave amplitudes in the jth unit cell are

$$A_{\ell mk}\exp(i\mathbf{k}_{0\parallel}\cdot\mathbf{R}_j). \tag{4.36}$$

Knowing the waves incident on the kth ion core in the jth unit cell we can calculate the scattered wave, and in particular we can calculate the total wavefield at the sth ion core in the unit cell at the origin, due to all the *other* ion cores:

$$\sum_{\ell m}\sum_{jk}{}' A_{\ell mk}\tfrac{1}{2}\left(\exp\left[2i\delta_\ell(k)\right] - 1\right)\exp(i\mathbf{k}_{0\parallel}\cdot\mathbf{R}_j) \times h^{(1)}_\ell(\kappa|\mathbf{r} - \mathbf{R}_{jk}|)\,Y_{\ell m}(\Omega(\mathbf{r} - \mathbf{R}_{jk})) \tag{4.37}$$

The prime on the summation over j and k denotes that the sth ion core in the unit cell at the origin has been omitted.

The summation over j and k in (4.37) converges because absorption of electron flux causes $h^{(1)}_\ell$ to decay exponentially for large values of $|\mathbf{r} - \mathbf{R}_{jk}|$. Duke and Tucker (1969) have pointed out that in LEED calculations absorption is usually large enough to make (4.37) converge rapidly without any transformations of the summation such as those introduced by Kambe.

As it stands (4.37) is a mixture of partial wave expansions about many centres. To be able to identify $A^{(s)}_{\ell ms}$ all terms must be written as expansions about the sth atom in the 0th unit cell. Fortunately there is a convenient formula to transfer an expansion about the (jk)th site to the $(0s)$th site,

$$\begin{aligned} h^{(1)}_\ell(\kappa|\mathbf{r} - \mathbf{R}_{jk}|)\,Y_{\ell m}(\Omega(\mathbf{r} - \mathbf{R}_{jk})) &= \sum_{\ell''m''} G_{\ell m,\,\ell''m''}(\mathbf{R}_{0s} - \mathbf{R}_{jk})\,j_{\ell''}(\kappa|\mathbf{r} - \mathbf{R}_{0s}|) \\ &\quad\times Y_{\ell''m''}(\Omega(\mathbf{r} - \mathbf{R}_{0s})), \end{aligned} \tag{4.38}$$

$$G_{\ell m,\,\ell''m''}(\mathbf{R}_{0s} - \mathbf{R}_{jk}) = \sum_{\ell'm'} 4\pi(-1)^{\frac{1}{2}(\ell-\ell'-\ell'')}(-1)^{m'+m''}$$

$$\times\, h^{(1)}_{\ell'}(\kappa|\mathbf{R}_{0s}-\mathbf{R}_{jk}|)\,Y_{\ell'-m'}(\Omega(\mathbf{R}_{0s}-\mathbf{R}_{jk}))B^{\ell}(\ell'm',\ell''m''), \tag{4.39}$$

$$B^{\ell}(\ell'm';\ell''m'') = \int Y_{\ell m}(\Omega)\,Y_{\ell'm'}(\Omega)\,Y_{\ell''-m''}(\Omega)d\Omega. \tag{4.40}$$

Explicit formulae for (4.40) can be found in the appendix. Substituting expansion (4.38) into expression (4.37) for the sum of scattered waves gives,

$$\sum_{jk}{}' \sum_{\ell m} A_{\ell mk}\tfrac{1}{2}(\exp[2i\delta_\ell(k)]-1)\exp(i\mathbf{k}_{0\|}\cdot\mathbf{R}_j)$$

$$\times \sum_{\ell''m''} G_{\ell m,\,\ell''m''}(\mathbf{R}_{0s}-\mathbf{R}_{jk})j_{\ell''}(\kappa|\mathbf{r}-\mathbf{R}_{0s}|)\,Y_{\ell''m''}(\Omega(\mathbf{r}-\mathbf{R}_{0s})) \tag{4.41}$$

From equation (4.41) for the scattered wavefield incident on the sth atom, amplitudes of partial waves can immediately be identified

$$A^{(s)}_{\ell''m''s} = \sum_{\ell mk} A_{\ell mk}X_{\ell mk,\,\ell''m''s} \tag{4.42}$$

$$X_{\ell mk,\,\ell''m''s} = \sum_j{}'\tfrac{1}{2}(\exp[2i\delta_\ell(k)]-1)\exp(i\mathbf{k}_{0\|}\cdot\mathbf{R}_j)$$

$$\times\, G_{\ell m,\,\ell''m''}(\mathbf{R}_{0s}-\mathbf{R}_{jk}) \tag{4.43}$$

Equation (4.42) enables us to solve (4.35) for $A_{\ell mk}$,

$$A_{\ell mk} = \sum_{\ell'm'k'} A^{(0)}_{\ell'm'k'}(1-X)^{-1}_{\ell'm'k',\,\ell mk} \tag{4.44}$$

Equations (4.39), (4.43) defining X can be simplified by introducing

$$D_{\ell'm'}(ks) = -\kappa i(4\pi)^{-\frac{1}{2}}\delta_{m'0}\delta_{\ell'0} - i\kappa(-1)^{\ell'+m'}$$

$$\times \sum_j{}' h^{(1)}_{\ell'}(\kappa|\mathbf{R}_{0s}-\mathbf{R}_{jk}|)\,Y_{\ell'-m'}(\Omega(\mathbf{R}_{0s}-\mathbf{R}_{jk}))\exp(i\mathbf{k}_{0\|}\cdot\mathbf{R}_j). \tag{4.45}$$

(No confusion should be caused between δ_ℓ the conventional symbol for a phase shift, and $\delta_{\ell\ell'}$ the conventional symbol which is unity if the subscripts are equal, zero otherwise.) All summations over lattice points being made in (4.45), X can be expressed in terms of D

$$X_{\ell mk,\,\ell''m''s} = -\exp[i\delta_\ell(k)]\sin[\delta_\ell(k)]\sum_{\ell'm'} 4\pi\kappa^{-1}$$

$$(-1)^{\frac{1}{2}(\ell''-\ell-\ell')}(-1)^{m''}B^{\ell}(\ell'm',\ell''m'')\left\{\frac{\kappa i\delta_{m'0}\delta_{\ell'0}}{(4\pi)^{\frac{1}{2}}} + D_{\ell'm'}(ks)\right\} \tag{4.46}$$

The particular choice of D in equations (4.45) has been made so that our definition of D coincides with Kambe's.

Equation (4.42) enables us to solve (4.35) for $A_{\ell mk}$,

$$A_{\ell mk} = \sum_{\ell'm'k'} A^{(0)}_{\ell'm'k'}(1-X)^{-1}_{\ell'm'k',\,\ell mk}. \tag{4.47}$$

We are now in possession of the total wave amplitudes incident on any ion core in the plane. To find the total scattered wave we substitute the proper value of $A_{\ell mk}$ instead of the approximate $A^{(0)}_{\ell mk}$, in equation (4.34), then making use of (4.26),

$$V^{\pm}_{\mathbf{g}'} = \sum_{\mathbf{g}} M^{\pm +}_{\mathbf{g}'\mathbf{g}} U^{+}_{\mathbf{g}}, \tag{4.48}$$

$$\begin{aligned} M^{\pm\pm}_{\mathbf{g}'\mathbf{g}} = {} & \frac{8\pi^2 i}{|\mathbf{K}^{\pm}_{\mathbf{g}}| A K^{+}_{\mathbf{g}'z}} \sum_{\substack{\ell' m' k' \\ \ell m k}} \exp(i\mathbf{K}^{\pm}_{\mathbf{g}} \cdot \mathbf{r}_k - i\mathbf{K}^{\pm}_{\mathbf{g}'} \cdot \mathbf{r}_{k'}) \\ & \times [i^{\ell}(-1)^m Y_{\ell -m}(\Omega(\mathbf{K}^{\pm}_{\mathbf{g}}))] (1 - X)^{-1}_{\ell mk, \ell' m' k'} [i^{-\ell'} Y_{\ell' m'}(\Omega(\mathbf{K}^{\pm}_{\mathbf{g}'}))] \\ & \times \exp[i\delta_{\ell'}(k')] \sin[\delta_{\ell'}(k')]. \end{aligned} \tag{4.49}$$

The negative sign above $\mathbf{g}$ is to take account of situations where waves of amplitude $U^{-}_{\mathbf{g}}$ are incident on the plane from the positive z side. This generalization follows easily by tracing the negative sign through the working. Total contributions to $V^{+}_{\mathbf{g}'}$ including unscattered waves passing straight through the plane are

$$V^{+}_{\mathbf{g}'} = \sum_{\mathbf{g}} (I_{\mathbf{g}'\mathbf{g}} + M^{++}_{\mathbf{g}'\mathbf{g}}) U^{+}_{\mathbf{g}} + M^{+-}_{\mathbf{g}'\mathbf{g}} U^{-}_{\mathbf{g}} \tag{4.50a}$$

$$V^{-}_{\mathbf{g}'} = \sum_{\mathbf{g}} M^{-+}_{\mathbf{g}'\mathbf{g}} U^{+}_{\mathbf{g}} + (I_{\mathbf{g}'\mathbf{g}} + M^{--}_{\mathbf{g}'\mathbf{g}}) U^{-}_{\mathbf{g}} \tag{4.50b}$$

All amplitudes of waves are referred to the origin of the 0th unit cell in the plane.

Although we have specified that all ion-cores have their centres in a single plane we have been careful to keep the derivation of equation (4.49) general so that if we wish ion core positions need not be in the plane, i.e. $\mathbf{r}_k$ may have a z-component for some values of k. In fact, if we had wished, the whole layer could be treated by (4.49) and associated equations. Sometimes this approach is forced upon us, as we shall see in Section IVE, but it is not the optimum method, because the dimensions of matrix X are proportional to the number of different ion cores in the unit cell. Times to assemble X scale as the square of the dimensions, and times to invert $(1 - X)$ scale as the cube of the dimensions. In Section IVE a method of compounding planes into layers will be given, where times scale only linearly with number of planes.

Quite apart from dividing the calculation into more manageable pieces, treatment of planes of ion-cores produces some worthwhile simplifications in formulae for $M^{\pm\pm}$. First of all, since the ion-cores are assumed spherically symmetric, and their centres are in a plane, the whole system of ion cores is symmetric about the plane, therefore transmission and reflection coefficients are independent of which side of the plane waves are incident:

$$M^{+-} = M^{-+} = M^{-} \tag{4.51a}$$

$$M^{++} = M^{--} = M^{+}. \tag{4.51b}$$

Secondly in equation (4.39) for G, because all vectors $\mathbf{R}_{0s}$ and $\mathbf{R}_{jk}$ lie in the plane, perpendicular to the z axis,

$$Y_{\ell'-m'}(\Omega(\mathbf{R}_{0s}-\mathbf{R}_{jk})) = Y_{\ell'-m'}(\theta=\tfrac{1}{2}\pi, \phi(\mathbf{R}_{0s}-\mathbf{R}_{jk}))$$
$$= Y_{\ell'-m'}(\theta=\tfrac{1}{2}\pi, \phi=0)(-1)^{m'}\exp(-im'\phi(\mathbf{R}_{jk}-\mathbf{R}_{0s})). \tag{4.52}$$

Spherical harmonics have the property

$$Y_{\ell'-m'}(\theta=\tfrac{1}{2}\pi, \phi)=0, \qquad (\ell+m)=\text{odd} \tag{4.53}$$

Another simplification follows by introducing the quantity

$$F_{\ell'm'}(ks) = \sum_j{}' ih^{(1)}_{\ell'}(\kappa|\mathbf{R}_{0s}-\mathbf{R}_{jk}|)\exp[-im'\phi(\mathbf{R}_{jk}-\mathbf{R}_{0s})]$$
$$\times(-1)^{m'}\exp(i\mathbf{k}_{0\|}\cdot\mathbf{R}_j). \tag{4.54}$$

Note the prime on the summation excluding

$$\mathbf{R}_{jk}=\mathbf{R}_{0s}$$

Then we can simplify equations (4.39), and (4.43) defining X, to

$$X_{\ell mk,\ell''m''s} = \sum_{\ell'+m'=\text{even}} C^{\ell}(\ell'm',\ell''m'')F_{\ell'm'}(ks)\exp[i\delta_\ell(k)]\sin[\delta_\ell(k)] \tag{4.55}$$

$$C^{\ell}(\ell'm',\ell''m'') = 4\pi(-1)^{\frac{1}{2}(\ell-\ell'-\ell'')}(-1)^{m'+m''}Y_{\ell'-m'}(\tfrac{1}{2}\pi,0)B^{\ell}(\ell'm',\ell''m''). \tag{4.56}$$

One more simplification is open to us. Equation (4.40) shows that constants C are zero unless

$$\ell'+m'=\text{even},$$

and inspection of equation (4.40) for B shows that if B hence C and hence X are not to be zero,

$$\ell+m+\ell''+m''=\text{even} \tag{4.57}$$

i.e. $(\ell+m)$ and $(\ell''+m'')$ must be both odd or both even, if X is to be non zero. X factorises into two sub matrices sketched in Fig. 4.1.

If phase shifts are used up to a maximum value ℓ_m of ℓ, X has dimensions

$$(\ell_m+1)^2. \tag{4.58}$$

The even-even submatrix has dimensions

$$\tfrac{1}{2}(\ell_m+1)(\ell_m+2) \tag{4.59}$$

and the odd–odd submatrix

$$\tfrac{1}{2}\ell_m(\ell_m+1) \tag{4.60}$$

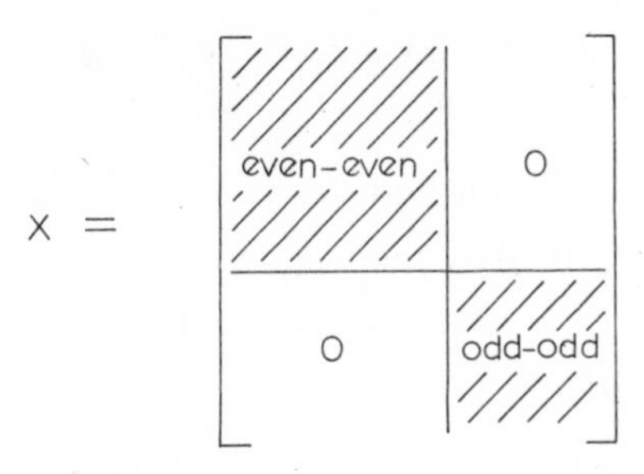

FIGURE 4.1. Matrix X factorises into two submatrices in consequence of all ion cores lying in a plane.

Equation (4.40) for B determines that

$$\tfrac{1}{2}(2\ell_m + 1)(2\ell_m + 2) \tag{4.61}$$

different values of ℓ' and m' in $F_{\ell'm'}(ks)$ make non-zero contributions to X.

In the computing appendix some programs are provided to calculate X when there is only one atom in the unit cell.

IVD Kambe's method for planar scattering

We have been able to correct for multiple scattering effects in a plane of ion-cores by adding to the original wave incident on an ion core, all the waves scattered by other ion cores. At some stage we always have to make a summation over ion cores, for example in formula (4.46) for matrix X, the quantity D is given by (4.45)

$$D_{\ell'm'}(ks) = -\kappa i(4\pi)^{-\frac{1}{2}}\delta_{\ell'0}\delta_{m'0} - i\kappa(-1)^{\ell'+m'}$$
$$\times \sum{}' h_{\ell'}^{(1)}(\kappa|\mathbf{R}_{0s} - \mathbf{R}_{jk}|)\, Y_{\ell'-m'}(\Omega(\mathbf{R}_{0s} - \mathbf{R}_{jk}))\exp(i\mathbf{k}_{0\|}\cdot\mathbf{R}_j).$$

For large values of $|\mathbf{R}_{0s} - \mathbf{R}_{jk}|$, the second two expressions inside the summation are of order unity,

$$|Y_{\ell'-m'}(\Omega(\mathbf{R}_{0s} - \mathbf{R}_{jk}))\exp(i\mathbf{k}_{0\|}\cdot\mathbf{R}_j)| = 0(1), \tag{4.62}$$

and the first term is

$$|h_{\ell'}^{(1)}(\kappa R)| \underset{R\to\infty}{=} \frac{\exp(-\kappa_i R)}{|\kappa|R}, \tag{4.63}$$

κ_i is the imaginary part of κ, caused by absorption in equation (4.25). Clearly if absorption is strong, good convergence is to be had. On the other hand if the method is to be applied where absorption is weak, and calculations of band structure near the Fermi energy usually neglect absorptive processes altogether, convergence is not good. In fact the series diverges when absorption is zero because we are summing the function $1/R$ over a two-dimensional lattice.

To understand how the difficulty can be overcome we shall take a simple example of a sum over lattice points.

$$S(\mathbf{r}) = \sum_j{}' H(\mathbf{r} - \mathbf{R}_j) \tag{4.64}$$

Function H might be

$$H(\mathbf{r} - \mathbf{R}_j) = h_{\ell'}^{(1)}(\kappa|\mathbf{r} - \mathbf{R}_j|)\, Y_{\ell' - m'}(\Omega(\mathbf{r} - \mathbf{R}_j)). \tag{4.65}$$

Poor convergence is obtained because $H(\mathbf{r})$ does not become rapidly smaller with increasing $|\mathbf{r}|$. Perhaps we can find another series with better properties. Except for the exclusion of $\mathbf{R}_j = 0$ in summation (4.64), $S(\mathbf{r})$ would be a periodic function of $\mathbf{r}$, with the same symmetry as the two-dimensional lattice. Therefore we can make a Fourier expansion of

$$S(\mathbf{r}) + H(\mathbf{r}) = \sum_{\mathbf{g}} s_{\mathbf{g}} \exp(i\mathbf{g}\cdot\mathbf{r}), \tag{4.66}$$

$$s_{\mathbf{g}} = \frac{1}{A}\int H(\mathbf{r}) \exp(-i\mathbf{g}\cdot\mathbf{r})\, d^2\mathbf{r}. \tag{4.67}$$

A is the area of unit cell, and the integration in (4.67) is over the whole two dimensional plane.

(4.66) gives us a new series for $S(\mathbf{r})$: a sum in reciprocal space rather than real space. Further we can hope that the very property of $H(\mathbf{r})$ preventing the real space sum (4.64) from converging rapidly, the long-range nature, will result in coefficients $s_{\mathbf{g}}$ that grow rapidly smaller with increasing $|\mathbf{g}|$. For many functions this is true, but if H has the form (4.65), $h_{\ell'}^{(1)}$ has highly singular behaviour near the origin, which requires many Fourier coefficients to describe it, and we do no better in reciprocal space for the sort of functions we are interested in. A similar problem arises when calculating the electrostatic energy of an ionic lattice; the function to be summed there is $|\mathbf{R}_j|^{-1}$.

Ewald (1921) first saw how to resolve the difficulty. Since $H(\mathbf{r})$ has troublesome behaviour for both large and small $|\mathbf{r}|$ we divide it into two functions

$$H = H_1 + H_2$$

$$H_1 = H(\mathbf{r}) \exp(-\eta|\mathbf{r}|^2), \qquad H_2 = [1 - \exp(-\eta|\mathbf{r}|^2)]H(\mathbf{r}), \tag{4.68}$$

one of which, H_1, is chosen to decrease rapidly for large $|\mathbf{r}|$, the other, H_2, is chosen to have the singular behaviour at small $|\mathbf{r}|$ removed, so that it will have a very convergent Fourier expansion. Contributions to $S(\mathbf{r})$ from the H_1 part of the series can be summed directly. Contributions from H_2 are found by transforming the H_2 series to reciprocal space. We have chosen a

Gaussian function in equation (4.68) to make the division, but other functions could have been chosen. The parameter η determines how much of the series is apportioned to H_1 and H_2.

Kambe has made an analogous division of the summation for D in equation (4.45). Mathematical details of the derivation are complicated and we shall not describe them here; they can be found in papers by Kambe (1967a, b, 1968). We shall quote his results for the case of all ion cores in a plane, corrected for a few printer's errors that occur in the papers and generalised to include V_{0i}. D is divided into two parts, plus a term to correct for adding in the point at the origin:

$$D_{\ell m} = D^{(1)}_{\ell m} + D^{(2)}_{\ell m} + D^{(3)}_{\ell m} \tag{4.69}$$

$D^{(1)}_{\ell m}$ is the term obtained by reciprocal-space summation,

$$\begin{aligned} D^{(1)}_{\ell m}(ks) = & -\frac{1}{A\kappa}\frac{i^{1-m}}{2^{\ell}}[(2\ell+1)(\ell+m)!(\ell-m)!]^{\frac{1}{2}}\sum_{\mathbf{g}}\exp[-im\phi(\mathbf{k}_{0\|}+\mathbf{g})] \\ & \times \exp[i(\mathbf{k}_{0\|}+\mathbf{g})\cdot(\mathbf{R}_{0s}-\mathbf{R}_{0k})]\sum_{n=0}^{(\ell-|m|)/2}\frac{(|\mathbf{k}_{0\|}+\mathbf{g}|/\kappa)^{\ell-2n}\times(K^{+}_{\mathbf{g}z}/\kappa)^{2n-1}}{n![\frac{1}{2}(\ell-m-2n)]![\frac{1}{2}(\ell+m-2n]!} \\ & \times \Gamma[(1-2n)/2, \exp(-i\pi)\alpha(K^{+}_{\mathbf{g}z}/\kappa)^2], \end{aligned} \tag{4.70}$$

$$\begin{aligned} D^{(2)}_{\ell m}(ks) = & -\frac{\kappa}{4\pi}\frac{(-1)^{\ell}(-1)^{\frac{1}{2}(\ell+m)}}{2^{\ell}[\frac{1}{2}(\ell-m)]![\frac{1}{2}(\ell+m)]!}[(2\ell+1)(\ell-m)!(\ell+m)!]^{\frac{1}{2}} \\ & \times \sum_{j}{}' \exp[-i\mathbf{k}_{0\|}\cdot\mathbf{R}_j - im\phi(\mathbf{R}_{0s}-\mathbf{R}_{jk})]\left[\frac{\kappa|\mathbf{R}_{jk}-\mathbf{R}_{0s}|}{2}\right]^{\ell} \\ & \times \int_0^{\alpha} u^{-\frac{3}{2}-\ell}\exp\left[u - \frac{(\kappa|\mathbf{R}_{jk}-\mathbf{R}_{0s}|)^2}{4u}\right]du, \end{aligned} \tag{4.71}$$

$$D^{(3)}_{\ell m}(ks) = -\delta_{\ell 0}\delta_{m0}(2\pi)^{-1}\kappa[2\textstyle\int_0^{\alpha^{1/2}}\exp(t^2)dt - \alpha^{-\frac{1}{2}}\exp(\alpha)]\delta_{ks}. \tag{4.72}$$

As previously $\phi(\mathbf{k}_{0\|}+\mathbf{g})$ represents the azimuthal angle of vector $\mathbf{k}_{0\|}+\mathbf{g}$. α is the parameter that determines division of the series between real and reciprocal space. Kambe shows that

$$\alpha = \left|\frac{\kappa^2 A}{4\pi}\right| \tag{4.73}$$

is the proper choice of α to give equally good convergence in real and reciprocal space.

The Γ function in (4.70) can be derived from the complex error function

$$\omega(z) = \exp(-z^2)[1 + 2i\pi^{-\frac{1}{2}}\textstyle\int_0^z \exp(t^2)\,dt], \tag{4.74}$$

via the formula,

$$\Gamma(\tfrac{1}{2}, x) = \pi^{\frac{1}{2}} \exp(-x)\omega(|x|^{\frac{1}{2}} \exp[i(arg(x) + \pi)/2]. \tag{4.75}$$

and recurrence relationships

$$n\Gamma(n, x) = \Gamma(n + 1, x) - x^n \exp(-x) \tag{4.76}$$

The integral

$$I_\ell = \int_0^\alpha u^{-\frac{3}{2}-\ell} \exp[u - \kappa^2|\mathbf{R}_{jk} - \mathbf{R}_{0s}|^2/(4u)]\, du,$$

can be calculated from

$$\begin{aligned} I_0 &= \pi^{\frac{1}{2}}[\tfrac{1}{2}\kappa|\mathbf{R}_{jk} - \mathbf{R}_{0s}|]^{-1} \exp[\alpha - \kappa^2|\mathbf{R}_{jk} - \mathbf{R}_{0s}|^2/(4\alpha)] \\ &\times \tfrac{1}{2}(\omega[\alpha^{\frac{1}{2}} + i\kappa|\mathbf{R}_{jk} - \mathbf{R}_{0s}|/(2\alpha^{\frac{1}{2}})] + \omega[-\alpha^{\frac{1}{2}} + i\kappa|\mathbf{R}_{jk} - \mathbf{R}_{0s}|/(2\alpha^{\frac{1}{2}})]), \end{aligned} \tag{4.77}$$

$$\begin{aligned} I_{-1} &= \pi^{\frac{1}{2}} \exp[\alpha - \kappa^2|\mathbf{R}_{jk} - \mathbf{R}_{0s}|^2/(4\alpha)]\,(2i)^{-1} \\ &\times (\omega[\alpha^{\frac{1}{2}} + i\kappa|\mathbf{R}_{jk} - \mathbf{R}_{0s}|/(2\alpha^{\frac{1}{2}})] - \omega[-\alpha^{\frac{1}{2}} + i\kappa|\mathbf{R}_{jk} - \mathbf{R}_{0s}|/(2\alpha^{\frac{1}{2}})]). \end{aligned} \tag{4.78}$$

and the recurrence relationship

$$\begin{aligned} \tfrac{1}{4}\kappa^2|\mathbf{R}_{jk} - \mathbf{R}_{0s}|^2 I_{\ell+1} &= \tfrac{1}{2}(2\ell + 1)I_\ell - I_{\ell-1} \\ &+ \alpha^{-\ell-\frac{1}{2}} \exp[\alpha - \kappa^2|\mathbf{R}_{jk} - \mathbf{R}_{0s}|^2/(4\alpha)]. \end{aligned} \tag{4.79}$$

The complex error function, ω, can either be taken from tabulated values (Fadeeva and Terent'ev, 1961) or, constructed from a power series expansion, or asymptotic formula, according to the argument. Details are to be had in the book by Abramowitz and Stegun (1964). Computer programs for the case of one ion core per unit cell can be found in the appendix.

The Kambe formula must be used wherever V_{0i} is very small at positive $(E - V_{0r})$. For negative $(E - V_{0r})$ the original series converges for any negative value of V_{0i} and can still be used when V_{0i} is zero.

The formulae derived by Kambe and quoted here are applicable to non-relativistic situations, such as are most commonly encountered in LEED calculations. Jennings (1970) and Jennings and Sim (1972) have shown how to include relativistic effects in the formalism and show that under special circumstances they can be important.

IVE Assembling planes into layers

To calculate Bloch waves we must have in our possession transmission and reflection matrices for the complete layer that repeats to form the crystal. In the simplest examples each layer is a single plane of ion-cores and the Q-matrices from which the Bloch waves are calculated can be found from equations (4.50) by change of origin from the centre of the layer to points half-way between layers. Waves on the $+z$ side of the layer referred to the new origin are

$$\sum_{\mathbf{g}} V_{\mathbf{g}}^{+} \exp[\tfrac{1}{2}i\mathbf{K}_{\mathbf{g}}^{+}\cdot\mathbf{c}] \exp[i\mathbf{K}_{\mathbf{g}}^{+}\cdot(\mathbf{r}-\tfrac{1}{2}\mathbf{c})] + U_{\mathbf{g}}^{-} \exp[\tfrac{1}{2}i\mathbf{K}_{\mathbf{g}}^{-}\cdot\mathbf{c}] \exp[i\mathbf{K}_{\mathbf{g}}^{-}\cdot(\mathbf{r}-\tfrac{1}{2}\mathbf{c})], \tag{4.80a}$$

and on the $-z$ side of the layer

$$\sum_{\mathbf{g}} U_{\mathbf{g}}^{+} \exp[-\tfrac{1}{2}i\mathbf{K}_{\mathbf{g}}^{+}\cdot\mathbf{c}] \exp[i\mathbf{K}_{\mathbf{g}}^{+}\cdot(\mathbf{r}+\tfrac{1}{2}\mathbf{c})] + V_{\mathbf{g}}^{-} \exp[-\tfrac{1}{2}i\mathbf{K}_{\mathbf{g}}^{-}\cdot\mathbf{c}] \exp[i\mathbf{K}_{\mathbf{g}}^{-}\cdot(\mathbf{r}+\tfrac{1}{2}\mathbf{c})] \tag{4.80b}$$

Now using equations (4.50), the Q-matrices defined in (3.67) relating amplitudes referred to origins half way between layers can be found.

$$Q_{\mathbf{gg}'}^{\mathrm{I}} = (I_{\mathbf{gg}'} + M_{\mathbf{gg}'}^{++}) \exp(i\mathbf{K}_{\mathbf{g}}^{+}\cdot\tfrac{1}{2}\mathbf{c} + i\mathbf{K}_{\mathbf{g}'}^{+}\cdot\tfrac{1}{2}\mathbf{c}), \tag{4.81a}$$

$$Q_{\mathbf{gg}'}^{\mathrm{II}} = M_{\mathbf{gg}'}^{+-} \exp(i\mathbf{K}_{\mathbf{g}}^{+}\cdot\tfrac{1}{2}\mathbf{c} - i\mathbf{K}_{\mathbf{g}'}^{-}\cdot\tfrac{1}{2}\mathbf{c}), \tag{4.81b}$$

$$Q_{\mathbf{gg}'}^{\mathrm{III}} = M_{\mathbf{gg}'}^{-+} \exp(-i\mathbf{K}_{\mathbf{g}}^{-}\cdot\tfrac{1}{2}\mathbf{c} + i\mathbf{K}_{\mathbf{g}'}^{+}\cdot\tfrac{1}{2}\mathbf{c}), \tag{4.81c}$$

$$Q_{\mathbf{gg}'}^{\mathrm{IV}} = (I_{\mathbf{gg}'} + M_{\mathbf{gg}'}^{--}) \exp(-i\mathbf{K}_{\mathbf{g}}^{-}\cdot\tfrac{1}{2}\mathbf{c} - i\mathbf{K}_{\mathbf{g}'}^{-}\cdot\tfrac{1}{2}\mathbf{c}) \tag{4.81d}$$

When a layer consists of two or more planes of atoms, scattering is not to be found by adding together amplitudes from each of the planes separately, it is necessary to consider processes whereby an electron can be scattered first by one plane, then by another, and so on. Figure 4.2 shows a possible

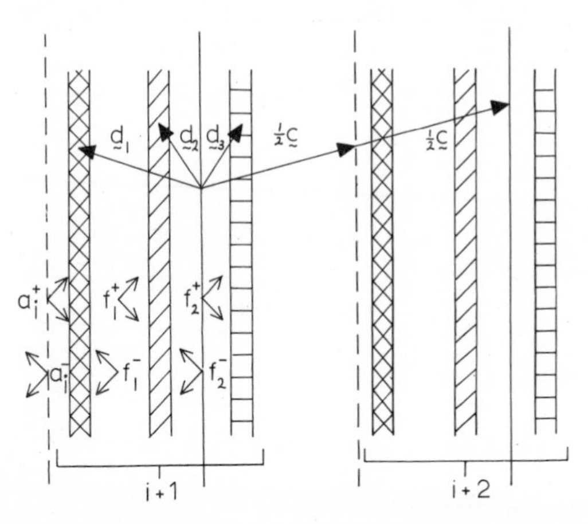

FIGURE 4.2. Two layers, each containing three planes of ion cores. Displacements of planes relative to the centre of the layer, and relative displacement of layers are shown. Amplitudes of waves between planes are shown referred to origins half-way between planes. Dashed lines denote origins half-way between layers.

arrangement of three planes per layer. Amplitudes of waves between planes are best referred to origins half way between the planes, except for the space between the last plane and the first in the succeeding layer, where we choose the origin half way between layers, at $\frac{1}{2}\mathbf{c}$ from the centre of the first layer.

The advantage of choosing origins in this way is that waves emerging from plane 1 have amplitudes referred to the correct origin for describing their transmission and reflection by plane 2, and we avoid the confusing algebra necessary to change origins.

We already know how to calculate transmission and reflection by a plane of atoms; they are described by matrices $M^{\pm\pm}$. With the new choice of origin we must modify $M^{\pm\pm}$ to give Q-matrices analogous to those of equations (4.81). For plane 1,

$$Q^{\mathrm{I}}_{\mathbf{gg}'}(1) = \left(I_{\mathbf{gg}'} + M^{++}_{\mathbf{gg}'}(1)\right)\exp\left[i\mathbf{K}^{+}_{\mathbf{g}}\cdot\tfrac{1}{2}(\mathbf{d}_2 - \mathbf{d}_1) + i\mathbf{K}^{+}_{\mathbf{g}'}\cdot(\tfrac{1}{2}\mathbf{c} + \mathbf{d}_1)\right], \quad (4.82\text{a})$$

$$Q^{\mathrm{II}}_{\mathbf{gg}'}(1) = M^{+-}_{\mathbf{gg}'}(1)\exp\left[i\mathbf{K}^{+}_{\mathbf{g}}\cdot\tfrac{1}{2}(\mathbf{d}_2 - \mathbf{d}_1) - i\mathbf{K}^{-}_{\mathbf{g}'}\cdot\tfrac{1}{2}(\mathbf{d}_2 - \mathbf{d}_1)\right], \quad (4.82\text{b})$$

$$Q^{\mathrm{III}}_{\mathbf{gg}'}(1) = M^{-+}_{\mathbf{gg}'}(1)\exp\left[-i\mathbf{K}^{-}_{\mathbf{g}}(\tfrac{1}{2}\mathbf{c} + \mathbf{d}_1) + i\mathbf{K}^{+}_{\mathbf{g}'}\cdot(\tfrac{1}{2}\mathbf{c} + \mathbf{d}_1)\right], \quad (4.82\text{c})$$

$$Q^{\mathrm{IV}}_{\mathbf{gg}'}(1) = \left(I_{\mathbf{gg}'} + M^{--}_{\mathbf{gg}'}(1)\right)\exp\left[-i\mathbf{K}^{-}_{\mathbf{g}}\cdot(\tfrac{1}{2}\mathbf{c} + \mathbf{d}_1) - i\mathbf{K}^{-}_{\mathbf{g}'}\cdot(\mathbf{d}_2 - \mathbf{d}_1)\right]. \quad (4.82\text{d})$$

with similar expressions for $Q(2)$ and $Q(3)$.

We take the first two planes as a pair and calculate the Q-matrices, $Q(12)$ for transmission and reflection by both planes together, i.e. we relate $a_i^{\pm}$ to $f_2^{\pm}$. We have four simultaneous equations describing transmission and reflection by planes 1 and 2, from which $f_1^{\pm}$ can be eliminated, leaving the two equations we desire. A little more insight is to be had by solving the problem in a multiple scattering way: a_i^{+} is incident on plane 1 and is transmitted (factor $Q^{\mathrm{I}}(1)$), the wave is then incident on plane 2; part is transmitted contributing to f_2^{+}

$$Q^{\mathrm{I}}(2)\,Q^{\mathrm{I}}(1)\,a_i^{+},$$

part is reflected by plane 2 $(Q^{\mathrm{III}}(2))$ again by plane 1 $(Q^{\mathrm{II}}(1))$ and transmitted by plane 2 $(Q^{\mathrm{I}}(2))$ contributing

$$Q^{\mathrm{I}}(2)\,Q^{\mathrm{II}}(1)\,Q^{\mathrm{III}}(2)\,Q^{\mathrm{I}}(1)\,a_i^{+}$$

subsequent terms give a geometric series.

$$Q^{\mathrm{I}}(2)\sum_{n=0}^{\infty}\left[Q^{\mathrm{II}}(1)\,Q^{\mathrm{III}}(2)\right]^n Q^{\mathrm{I}}(1)\,a_i^{+} = Q^{\mathrm{I}}(2)\left[I - Q^{\mathrm{II}}(1)\,Q^{\mathrm{III}}(2)\right]^{-1} Q^{\mathrm{I}}(1)\,a_i^{+}$$

which is the total contribution to f_2^{+} from a_i^{+} and leads us to identify

$$Q^{\mathrm{I}}(12) = Q^{\mathrm{I}}(2)\left[I - Q^{\mathrm{II}}(1)\,Q^{\mathrm{III}}(2)\right]^{-1} Q^{\mathrm{I}}(1), \quad (4.83\text{a})$$

similar arguments show that

$$Q^{\mathrm{II}}(12) = Q^{\mathrm{II}}(2) + Q^{\mathrm{I}}(2)Q^{\mathrm{II}}(1)[I - Q^{\mathrm{III}}(2)Q^{\mathrm{II}}(1)]^{-1}Q^{\mathrm{IV}}(2), \tag{4.83b}$$

$$Q^{\mathrm{III}}(12) = Q^{\mathrm{III}}(1) + Q^{\mathrm{IV}}(1)Q^{\mathrm{III}}(2)[I - Q^{\mathrm{II}}(1)Q^{\mathrm{III}}(2)]^{-1}Q^{\mathrm{I}}(1), \tag{4.83c}$$

$$Q^{\mathrm{IV}}(12) = Q^{\mathrm{IV}}(1)[I - Q^{\mathrm{III}}(2)Q^{\mathrm{II}}(1)]^{-1}Q^{\mathrm{IV}}(2). \tag{4.83d}$$

and we can now write,

$$f_2^+ = Q^{\mathrm{I}}(12)a_i^+ + Q^{\mathrm{II}}(12)f_2^-, \tag{4.84a}$$

$$a_i^- = Q^{\mathrm{III}}(12)a_i^+ + Q^{\mathrm{IV}}(12)f_2^-. \tag{4.84b}$$

To add the third plane to the system, we regard the first two planes as a single entity and repeat the procedure to eliminate $f_2^\pm$ and find the Q-matrices for the whole layer connecting $a_i^\pm$ and $a_{i+1}^\pm$.

Any number of planes can be treated in this way and it is evident that it takes the same amount of computing effort to add an extra plane to a layer whether there are three or thirty planes there already. Times scale linearly with the number of planes rather than as the cube of the number of planes if we were to treat the whole layer at once by the method of Section C.

Some difficulties can arise when planes are too close to one another. In principle we need an infinite number of Fourier components to describe the waves between planes, but in equations (4.82) we can see that Fourier components with very large $(\mathbf{k}_{0\parallel} + \mathbf{g})$ have a large imaginary $K_{\mathbf{g}z}^\pm$, and decay away rapidly from one plane to the next. The closer planes are the more Fourier components we must take before the ones we leave out have decayed sufficiently to be neglected. When planes are closer than a certain spacing (say, 0.3Å to give a rough average figure) it becomes more economical to treat the two planes as a single entity by the methods of Section C with the generalisation that not all ion cores are exactly in the same plane (equation (4.46)).

IVF The matrix doubling method

The only method of solving the LEED problem working for all values of various parameters that go into calculations is the normal mode method. Yet, partly because of its generality and completeness of information it provides, calculations can be time-consuming. There is a hierarchy of methods for making calculations, each making more restrictive assumptions about parameters, and gaining more speed in calculation.

The fastest methods are to be found amongst perturbation techniques described in Chapter V. Here we describe a method that for its validity requires only that finite absorption is present, i.e.

$$V_{0i} < 0. \tag{4.85}$$

In that case it is always possible to approximate the semi-infinite crystal by a slab of finite thickness, because all electron wave functions die away into the crystal, and electrons incident on one surface cannot sense the presence of the second surface if the thickness is much greater than an extinction distance.

The slab being finite in thickness can be treated as a thick layer and its scattering properties calculated as prescribed in the last section: first transmission and reflection by planes are calculated, then planes are built into the basic layer unit. Identical layers are stacked to form the slab by considering first a pair of layers then a pair of pairs, applying equations (4.83) each time to build up 2, 4, 8, . . . layers. Eight layers are usually sufficient to approximate the semi-infinite crystal, but occasionally sixteen are needed.

We shall refer to this as the layer doubling method of calculating reflectivities. It involves matrix multiplication and matrix inversion, much less time-consuming procedures than matrix eigenvalue problems. The number of processes scales as $\ell n(L)$ where L is the number of layers in the slab.

There are other methods of calculating reflectivities of a slab. One way would be to return to Section C, and calculate multiple scattering between all ion-cores in the slab at once, taking a two dimensional unit cell with many ion-cores in it. This is a most uneconomical way of proceeding because the number of ion cores in the unit cell is proportional to L, the number of layers in the slab, and computing times scale as the cube of the number of atoms. L^3 is a much more rapid increase than $\ell n(L)$ and the method is not competitive.

Yet another approach which has been used in some published calculations is what we shall call the giant matrix method. First transmission and reflection coefficients are calculated for layers. The unknowns are wave amplitudes between each pair of layers in the slab, and of course the outgoing wave at the surface and the wave transmitted through the far side of the slab, assumed very small in amplitude. We define a giant vector containing all the unknown amplitudes of waves set end to end

$$W = [a_0^-, a_1^+, a_1^-, a_2^+, a_2^-, \cdots a_{N-1}^+, a_{N-1}^-, a_N^+] \tag{4.86}$$

W has $n \times (L + 1)$ components where n is the number of beams and L the number of layers. We have equations (3.67) connecting $a_i^\pm$ to $a_{i+1}^\pm$, and the boundary conditions of an incident wave fixed to be a_0^+, and no waves incident on the far side of the crystal

$$a_N^- = 0. \tag{4.87}$$

These equations can be summarised by one giant matrix equation

$$AW = S \tag{4.88}$$

where S is a giant vector, the same size as W, having components

$$S = [Q^{\mathrm{III}} a_0^+, Q^{\mathrm{I}} a_0^+, 0, 0, \ldots], \tag{4.89}$$

and A is a giant matrix, dimensions $n(L+1) \times n(L+1)$,

$$A = \begin{bmatrix} I, & O, -Q^{\mathrm{IV}}, & O, & O, O, & O, O, & \cdot & \cdot & & \\ O, & I, -Q^{\mathrm{II}}, & O, & O, O, & O, O, & \cdot & \cdot & & \\ O, -Q^{\mathrm{III}}, & I, & O, -Q^{\mathrm{IV}}, O, & & O, O & \cdot & \cdot & & \\ O, -Q^{\mathrm{I}}, & O, & I, -Q^{\mathrm{II}}, O, & & O, O & \cdot & \cdot & & \\ O, & O, & O, -Q^{\mathrm{III}}, & I, O, -Q^{\mathrm{IV}}, O & & & \cdot & \cdot & & \\ O, & O, & O, -Q^{\mathrm{I}}, & O, I, -Q^{\mathrm{II}}, O & & & \cdot & \cdot & & \\ \cdot & \cdot & \cdot \quad \cdot & \cdot \quad \cdot & \cdot \quad \cdot & \cdot & \cdot & & \\ \cdot & \cdot & \cdot \quad \cdot & \cdot \quad \cdot & \cdot \quad \cdot & \cdot & \cdot & & \\ & & & & & \cdot & \cdot O, -Q^{\mathrm{III}}, & I, O \\ & & & & & \cdot & \cdot O, -Q^{\mathrm{I}}, & O, I \end{bmatrix} \tag{4.90}$$

S is given by the incident wave amplitudes, and at one fell swoop all the unknowns, contained in W, can be found by inverting matrix A in (4.88). This matrix inversion scales as L^3 in time for large L and therefore the method is not a practically viable one and has been superseded by matrix doubling methods.

Finally we mention that we have in this section chosen to describe the wave between layers in terms of amplitudes of Fourier components, and our matrices have subscripts **g** referring to beams. It has been noted that an alternative expansion of the wave is in terms of spherical waves about an ion-core. Equation (4.24) shows how to make such an expansion. Some workers have preferred such a spherical wave expansion, and their matrices have subscripts ℓm, giving the appearance of a different method, whereas in fact it is only the representation that differs.

IVG Pseudopotential methods for calculating Bloch waves

The method described in Section B is not the only way of calculating Bloch waves. It has been adopted as the prime method here because of its elegance and close relationship with non-Bloch wave approaches to the LEED problem. Of the wide variety of methods the best known, because of its application to calculation of conduction and valence bands, is the pseudopotential method and in this section we trace the relationship between the approach already described and the perhaps more familiar pseudopotential approach.

An account of relationships between various approaches to LEED can be found in Shen (1971), and a review of pseudo-potential methods applied to calculations of Bloch waves near the Fermi energy has been written by Ziman (1971). Accounts of its application to LEED are to be found in articles by Boudreaux and Heine (1967), by Capart (1969), by Feder (1972) and Sondhi (1972).

It will be remembered that Bloch waves satisfy the special condition that amplitudes of beams between consecutive pairs of layers differ by a constant factor:

$$b_{i\mathbf{g}}^{+} = \exp(i\mathbf{k}\cdot\mathbf{c})\, b_{i-1\mathbf{g}}^{+}, \tag{4.91a}$$

$$b_{i\mathbf{g}}^{-} = \exp(i\mathbf{k}\cdot\mathbf{c})\, b_{i-1\mathbf{g}}^{-}, \tag{4.91b}$$

or expressed in terms of the complete wavefunction between layers i and $(i+1)$.

$$\psi_{\mathbf{k}}(\mathbf{r}) = \sum_{\mathbf{g}} b_{i\mathbf{g}}^{+}(\mathbf{k}) \exp(i\mathbf{K}_{\mathbf{g}}^{+}\cdot\mathbf{r}) + b_{i\mathbf{g}}^{-}(\mathbf{k}) \exp(i\mathbf{K}_{\mathbf{g}}^{-}\cdot\mathbf{r}), \tag{4.92}$$

$$\psi_{\mathbf{k}}(\mathbf{r} + n\mathbf{c}) = \exp(in\mathbf{k}\cdot\mathbf{c})\,\psi_{\mathbf{k}}(\mathbf{r}). \tag{4.93}$$

So far $\psi_{\mathbf{k}}$ has been given only in the region of constant potential between layers of ion cores where equation (4.92) is valid. Of course the wavefunction exists inside the ion cores but we have not explicitly calculated its value.

One thing we do know about the complete wavefunction is that it still obeys equation (4.93) even when $\mathbf{r}$ is inside an ion core. Take two ion cores in equivalent positions in consecutive layers; we know the wavefunction surrounding the ion cores obeys (4.93) and differs by only a constant factor from one core to the other. Hence the wavefunction inside the ion cores also differs by only a constant factor because it is completely determined by integrating the Schrödinger equation into the core region.

We have already made use of periodicity parallel to the layers of the function

$$\psi_{\mathbf{k}} \times \exp(-i\mathbf{k}\cdot\mathbf{r}) \tag{4.94}$$

to make a Fourier expansion between layers, (4.92). For the special case of a Bloch wave, equation (4.93) tells us that (4.94) is also periodic in the direction of $\mathbf{c}$, and a Fourier decomposition can be made in the z direction.

$$\exp(-i\mathbf{k}\cdot\mathbf{r})\,\psi_{\mathbf{k}}(\mathbf{r}) = \sum_{G_z} \psi_{\mathbf{k}G_z}(\mathbf{r}_{\parallel}) \exp(iG_z z), \tag{4.95}$$

$$G_z = \frac{2\pi}{c_z} \times \text{integer}. \tag{4.96}$$

Then making the usually Fourier expansion parallel to the layers as well,

$$\exp(-i\mathbf{k}\cdot\mathbf{r})\,\psi_{\mathbf{k}}(\mathbf{r}) = \sum_{\mathbf{G}} \psi_{\mathbf{k}\mathbf{G}} \exp(i\mathbf{G}\cdot\mathbf{r}), \tag{4.97}$$

$$\mathbf{G} = (g_x, g_y, G_z)$$

Since condition (4.93) holds for any value of **r**, so does (4.97), inside ion cores as well as outside.

If the coefficients $\psi_{\mathbf{kG}}$ can be found, the Bloch wave is determined everywhere. Substitution of (4.96) into the Schrödinger equation gives

$$\sum_{\mathbf{G}} [\tfrac{1}{2}|\mathbf{k} + \mathbf{G}|^2 + V - E]\psi_{\mathbf{kG}} \exp[i(\mathbf{k} + \mathbf{G})\cdot\mathbf{r}] = 0. \tag{4.98}$$

The ion-core potentials and the optical potential have been included in V. The modulus bars on $|\mathbf{k} + \mathbf{G}|^2$ stand for vector modulus *not*

$$(\mathbf{k} + \mathbf{G})\cdot(\mathbf{k}^* + \mathbf{G}^*).$$

Multiplying (4.98) from the left by

$$\Omega^{-1} \exp[-i(\mathbf{k} + \mathbf{G}')\cdot\mathbf{r}], \tag{4.99}$$

and integrating with respect to **r** over the volume of the crystal, Ω,

$$\sum_{\mathbf{G}} \{[\tfrac{1}{2}|\mathbf{k} + \mathbf{G}|^2 - E]\delta_{\mathbf{G}'\mathbf{G}} + V_{\mathbf{G}'\mathbf{G}}\}\, \psi_{\mathbf{kG}} = 0, \tag{4.100}$$

$$V_{\mathbf{G}'\mathbf{G}} = \Omega^{-1} \int V \exp[i(\mathbf{G} - \mathbf{G}')\cdot\mathbf{r}]\, d^3\mathbf{r}. \tag{4.101}$$

(4.100) is a matrix equation which we must solve by fixing the energy, E, and the parallel components of momentum to be what the incident electron dictates, and solve for k_z. The elegant eigenvalue techniques which served us in the layer method of finding k_z are not directly applicable. (4.100) can be forced into an eigenvalue form (Wilkinson, 1965), provided that $V_{\mathbf{GG}'}$ is independent of **k**, but unfortunately this is usually not true, and we must solve by inserting trial values of k_z and testing for

$$\det M = 0, \tag{4.102}$$

$$M_{\mathbf{G}'\mathbf{G}} = [\tfrac{1}{2}|\mathbf{k} + \mathbf{G}|^2 - E]\delta_{\mathbf{G}'\mathbf{G}} + V_{\mathbf{G}'\mathbf{G}}. \tag{4.103}$$

Once we have solved (4.100) and found $\psi_{\mathbf{kG}}$, the Bloch wave can be re-expressed in the more familiar beam-expansion between layers, (4.92) by noting that the total amplitude of the **g**th Fourier component parallel to the surface, half-way between planes, is a sum over contributions from the G_z Fourier components.

$$b_{i\mathbf{g}}^{+}(\mathbf{k}) + b_{i\mathbf{g}}^{-}(\mathbf{k}) = \sum_{G_z} \psi_{\mathbf{kG}} \tag{4.104}$$

similarly for the derivative in the z-direction

$$K_{\mathbf{g}z}^{+} b_{i\mathbf{g}}^{+}(\mathbf{k}) + K_{\mathbf{g}z}^{-} b_{i\mathbf{g}}^{-}(\mathbf{k}) = \sum_{G_z} (k_z + G_z)\psi_{\mathbf{kG}}. \tag{4.105}$$

From (4.104) and (4.105) the two unknowns $b_{ig}^{\pm}$ can be found and theory of LEED reflectivities goes through as before.

Equation (4.100) is much used in solving diffraction problems at high energies (Hirsch *et al.*, 1965), but at lower energies below a few kilovolts certain practical difficulties occur. They are due to the fact that to make an accurate picture of $\psi_{\mathbf{k}}$, many Fourier components, $\psi_{\mathbf{kG}}$, are needed. It is the part of $\psi_{\mathbf{k}}$ inside the ion core which needs many components for an accurate description because of strong influence exerted by the ion core potential on the wavefunction. Figure 4.3 shows a sketch of the variation in the core

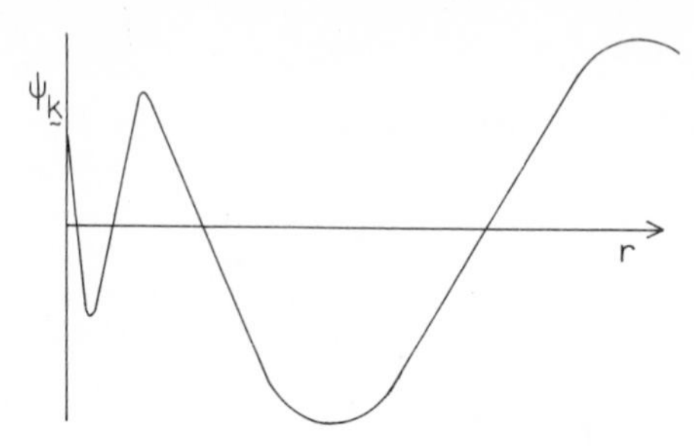

FIGURE 4.3. Schematic variation of a wavefunction near the centre of an ion-core. The strong potential causes many oscillations in $\psi_{\mathbf{k}}$ that need many Fourier components for an accurate description.

region. So difficult is this part of the wavefunction to expand in a Fourier series that even with a large computer it is not feasible to include enough components. To combat this difficulty pseudopotential theory was developed.

Behind theory of pseudopotentials is a simple observation we have made already; it is that LEED reflectivities can be found without at any time calculating wavefunctions inside ion cores. We need to know only the incident and scattered waves outside the ion core; detailed mechanisms of scattering inside the core are irrelevant because we observe electrons only in places remote from ion-cores. This statement is also true of many other types of experiment. Therefore provided that our calculation reproduces what happens outside muffin tins surrounding ion cores, it can put any wavefunction it pleases inside the muffin tins. In particular instead of a very strong ion-core potential with its associated wavefunction so difficult to expand in a Fourier series, we can choose another potential inside the cores, a pseudopotential, having an associated wavefunction more to our liking. The condition that two potentials have the same scattering properties at a given energy is not a restrictive one, and leaves a wide choice of possible pseudopotentials.

The concept has been widely used near the Fermi energy, and has been generalised to higher energies (Pendry, 1968, Pendry and Capart 1969). By choosing the correct sort of pseudopotential, Fourier expansions of wavefunctions can be made much more convergent and calculations based on equation (4.100) with $V_{\mathbf{G}'\mathbf{G}}$ replaced by $V_{ps\mathbf{G}'\mathbf{G}}$, are practicable especially at lower energies, below 50 eV.

IVH Simplifications brought about by symmetry

Schemes we describe in other parts of the book are mostly of a general nature. For crystals with a simple surface structure several of these schemes are practical methods of calculating, but the range of complexity possible in surface structures is such that for the most complicated of them even the fastest methods take too much time.

There is one trick left open to us, and that is to take account of symmetry. Most surface structures possess symmetry of some sort, even if it is only a mirror plane, and often there is an axis of symmetry perpendicular to the surface. If the incident beam is made parallel to the mirror plane, or to the axis of symmetry so that it does not spoil the symmetry, beams diffracted from this configuration must have the same symmetry as the surface.

Suppose that a surface has a fourfold axis; then in the case of a normally incident beam to find the complete diffraction pattern it is necessary to know only beam intensities in one quadrant and dimensions of the problem have been reduced by a factor of four. Times for methods described in this chapter scale as n^3, n being the number of beams. Thus a saving of a factor of 64 in times is achieved, and even in the fast perturbation schemes times scale as n^2 and a factor of 16 divides times.

Application of symmetry to quantum mechanical problems is a complex problem in group theory. A full exposition would occupy the whole book and in any case accounts by several authors are already available: Falicov (1964), Heine (1960), Tinkham (1964) to name only three. In our case the symmetry available to us is that of a plane; the surface prevents any three-dimensional symmetry. Also the fact that the incident beam is always a plane-wave further reduces complexity of the problem and we are able to derive the results we need without becoming involved in more sophisticated group-theoretical techniques.

We begin by examining implications of symmetry for the wavefield between layers. A slight change of convention will be introduced to make clear discussion of symmetry. Instead of representing amplitudes of beams between the $(i-1)$th and ith layers, as $U^{\pm}_{i\mathbf{g}}$, they will be represented by $U^{\pm}_{i}(\mathbf{g}_{jk})$. All $\mathbf{g}_{jk}$'s with the same subscript j are generated by the symmetry operations acting on one of the set. Subscript k labels members of the symmetrically related set. For example, if the surface had perpendicular to it a mirror plane, $U^{\pm}_{i}(\mathbf{g}_{j1})$ and $U^{\pm}_{i}(\mathbf{g}_{j2})$ would be amplitudes of beams symmetrically disposed about the mirror plane. Subscript j refers to different pairs of beams. The problem is simplified by choosing the origin of $x - y$ coordinates to coincide with the position of the axis or mirror plane. With this choice of origin, symmetry of the wavefield,

$$\sum_{jk} U_i^+(\mathbf{g}_{jk}) \exp[i\mathbf{K}^+(\mathbf{g}_{jk})\cdot\mathbf{r}] + U_i^-(\mathbf{g}_{jk}) \exp[i\mathbf{K}^-(\mathbf{g}_{jk})\cdot\mathbf{r}], \quad (4.106)$$

implies that

$$U_i^\pm(\mathbf{g}_{jk}) = U_i^\pm(\mathbf{g}_{jk'}) \quad (4.107)$$

for any k, k'. Therefore one symbol will serve for all of the symmetrically related set. If there are m_j members of the jth set, define

$$U_i^\pm(\mathbf{g}_{jk}) = V_{ij}^\pm/(m_j)^{\frac{1}{2}} \quad (4.108)$$

The smaller vectors $V_{ij}^\pm$ store all information contained in the larger vectors.

At present scattering by layers is described by equations (3.67) in terms of $U_i^\pm(\mathbf{g}_{jk})$.

$$U_{i+1}^+(\mathbf{g}_{jk}) = \sum_{j'k'} Q^{\mathrm{I}}(\mathbf{g}_{jk}, \mathbf{g}_{j'k'})\, U_i^+(\mathbf{g}_{j'k'}) + Q^{\mathrm{II}}(\mathbf{g}_{jk}, \mathbf{g}_{j'k'})\, U_{i+1}^-(\mathbf{g}_{j'k'}), \quad (4.109a)$$

$$U_i^-(\mathbf{g}_{jk}) = \sum_{j'k'} Q^{\mathrm{III}}(\mathbf{g}_{jk}, \mathbf{g}_{j'k'})\, U_i^+(\mathbf{g}_{j'k'}) + Q^{\mathrm{IV}}(\mathbf{g}_{jk}, \mathbf{g}_{j'k'})\, U_{i+1}^-(\mathbf{g}_{j'k'}) \quad (4.109b)$$

We wish to re-express equations (4.109) in terms of the smaller vectors. First make a summation over k in (4.109), then substitute for U from (4.108),

$$V_{i+1j}^+ = \sum_{j'} q_{jj'}^{\mathrm{I}} V_{ij'}^+ + q_{jj'}^{\mathrm{II}} V_{i+1j'}^-, \quad (4.110a)$$

$$V_{ij}^- = \sum_{j'} q_{jj'}^{\mathrm{III}} V_{ij'}^+ + q_{jj'}^{\mathrm{IV}} V_{i+1j'}^-, \quad (4.110b)$$

$$q_{jj'} = \sum_{kk'} Q(\mathbf{g}_{jk}, \mathbf{g}_{j'k'})\, m_j^{-\frac{1}{2}} m_{j'}^{-\frac{1}{2}}. \quad (4.111)$$

A similar reduction can be made in any matrix equation involving U. The reduced equations such as (4.110) have the same form as the original equation, implying that all methods used to calculate Bloch waves, or scattering by assemblies of planes, etc., can be adapted to the smaller vectors, V, simply by replacing Q with q.

A complication arises in some cases where more than one sort of symmetry is present. If a surface has both an axis of symmetry and a mirror plane, the axis of symmetry need not coincide with the mirror plane, and the origin of coordinates cannot coincide with both the plane and the axis. If the origin is chosen in the mirror plane, those values of k and k' connected by the mirror plane have

$$U_i^{\pm}(\mathbf{g}_{jk}) = U_i^{\pm}(\mathbf{g}_{jk'}) \tag{4.112}$$

as before, but for values of k and k' connected by the axis it is amplitudes about the *axis* that are related. If the axis of symmetry has coordinate relative to the origin of $\mathbf{r}_{a\|}$, then

$$U_i^{\pm}(\mathbf{g}_{jk}) \exp(i\mathbf{g}_{jk}\cdot\mathbf{r}_{a\|}) = U_i^{\pm}(\mathbf{g}_{jk'}) \exp(i\mathbf{g}_{jk'}\cdot\mathbf{r}_{a\|}), \tag{4.113}$$

and all the members of the set can again be expressed in terms of a single variable $V_{ij}^{\pm}$. The generalisation of equation (4.111) is easily found.

Symmetry can also be of assistance in calculating scattering within a plane of ion cores. In that case the wavefield incident on an ion core is expanded in spherical waves about the ion core's centre,

$$\sum_{\ell m} \mathrm{A}_{\ell m} \mathrm{j}_\ell(\kappa|\mathbf{r}|)\, \mathrm{Y}_{\ell m}(\Omega(\mathbf{r})) \tag{4.114}$$

Suppose that a p-fold axis of symmetry passes through the atom's centre, then for normally incident electrons the wavefield is p-fold symmetric about the axis. The angular variation of (4.114) is contained in

$$Y_{\ell m}(\Omega) = P_\ell^{|m|}(\cos\theta) \exp(im\phi) \operatorname{sign}(m) \tag{4.115}$$

and only when

$$m = p \times \text{integer} \tag{4.116}$$

does $Y_{\ell m}$ behave with correct symmetry about the axis. This argument applies both to the total wavefield at the ion core described by $A_{\ell mk}$ in equation (4.35) and the wavefield coming from other atoms, $A_{\ell mk}^{(s)}$. Matrix X connects these two quantities in equation (4.42)

$$A_{\ell mk}^{(s)} = \sum_{\ell' m' k'} A_{\ell' m' k'} X_{\ell' m' k', \ell mk} \tag{4.42}$$

Columns of X in which k' represents an atom through which a p-fold axis of symmetry passes, we know, multiply elements of $A_{\ell' m' k'}$ that are zero, because of arguments above, and we may as well delete all those columns wherein m' does not satisfy (4.116). Similarly, rows of X where m does not satisfy (4.116) correspond to zero values of $A_{\ell mk}^{(s)}$ and can also be deleted.

Once again symmetry reduces dimensions of matrices, with corresponding savings in calculations.

Not only is X reduced in size, but also, because only elements with m and m' satisfying (4.116) are needed, $F_{\ell m}(k'\,k)$ involved in the calculation is reduced in size. If both atoms k and k' have a p-fold axis passing through them, the occurence of

$$\int Y_{\ell m}\, Y_{\ell' m'}\, Y_{\ell'' -m''}\, d\Omega$$

in summation (4.55) dictates that only for values of m satisfying (4.116) is $F_{\ell m}(k'k)$ needed. A further reduction of computation is brought about because in equation (4.54) where F is calculated from a lattice summation,

$$\mathbf{k}_{0\|} = 0, \tag{4.117}$$

eliminating the final factor, and because of the axis of symmetry, for every

$$\phi(\mathbf{R}_{jk'} - \mathbf{R}_{0k})$$

there is another value of j, such that

$$\phi(\mathbf{R}_{j'k'} - \mathbf{R}_{0k}) = \phi(\mathbf{R}_{jk'} - \mathbf{R}_{0k}) + \frac{2\pi}{p} \times \text{integer} \tag{4.118}$$

$$|\mathbf{R}_{j'k'} - \mathbf{R}| = |\mathbf{R}_{jk'} - \mathbf{R}_{0k}| \tag{4.119}$$

and only summations in a segment $1/p$th of the plane need be made.

A mirror plane has more complicated consequences. If the mirror plane passes through the centre of the ion core, operation of the mirror plane corresponds to reversing the azimuthal angle ϕ, provided that we choose $\phi = 0$ to lie in the mirror plane:

$$\begin{aligned}\sum_{\ell m} & A_{\ell mk} j_\ell(\kappa|\mathbf{r}|)\, Y_{\ell m}(\theta, \phi) \\ &= \sum_{\ell m} A_{\ell mk} j_\ell(\kappa|\mathbf{r}|)\, Y_{\ell m}(\theta, -\phi) \\ &= \sum_{\ell m} A_{\ell mk} j_\ell(\kappa|\mathbf{r}|)\, Y_{\ell -m}(\theta, \phi)(-1)^m \end{aligned} \tag{4.120}$$

the last step following from properties of spherical harmonics. Therefore the mirror plane implies

$$A_{\ell mk} = A_{\ell -mk}(-1)^m \tag{4.121}$$

Defining new vectors for $m \neq 0$,

$$A_{\ell |m| k} = a_{\ell |m| k} 2^{-\frac{1}{2}}, \tag{4.122a}$$

$$A_{\ell -|m| k} = a_{\ell |m| k}(-1)^m 2^{-\frac{1}{2}}. \tag{4.122b}$$

otherwise

$$A_{\ell 0k} = a_{\ell 0k} \tag{4.123}$$

Substituting into equation (4.42) and summing over $\pm m$, $\pm m'$ gives

$$a^{(s)}_{\ell |m| k} = \sum_{\ell' |m'| k'} a_{\ell' |m'| k'} X_{\ell' |m'| k',\, \ell |m| k}, \tag{4.124}$$

$$x_{\ell'|m'|k', \ell|m|k} = \sum_{\pm m, \pm m'} X_{\ell'm'k', \ell mk} \varepsilon_{mm'} i^{(|m| - m + |m'| - m')}, \quad (4.125)$$

$$\begin{aligned} \varepsilon_{mm'} &= 1 && \text{if} \quad m = 0, \quad m' = 0 \\ &= 2^{-\frac{1}{2}} && \text{if} \quad m = 0, \quad m' \neq 0 \\ & && \text{or} \quad m \neq 0, \quad m' = 0, \\ &= 2^{-1} && \text{if} \quad m \neq 0, \quad m' \neq 0. \end{aligned} \quad (4.126)$$

IVI Critical review

In this chapter we have dealt with details of some of the more general methods for calculating LEED spectra. All methods considered above converge to the exact solution of our model in the limit of a large number of Fourier components and phase shifts, provided only that absorption is finite. (This statement is not generally true of perturbative and approximate methods in the next chapter). We wish to compare methods given above in speed of execution and usefulness in understanding the LEED process.

Only the Bloch wave method produces information additional to diffracted intensities themselves, that is of much interpretational value. We saw how Bloch waves are the normal modes of the process and a knowledge of band structure generated automatically by the Bloch wave method, is essential where a detailed understanding is required.

Two distinct ways of calculating Bloch waves were described: the layer method, and the pseudopotential method. Both have received considerable attention and it now seems clear that in LEED calculations the layer method has distinct advantages both in speed of execution and in simplicity of operation. The layer method generates Bloch waves from an eigenvalue equation using standard numerical techniques whereas pseudopotential methods involve hunting about the complex-wave vector plane for zeros of a determinant. The pseudopotential method is only feasible at normal incidence when simplifications of symmetry reduce the size of matrices. The layer method on the other hand can handle a general angle of incidence with ease. Typical computing times for copper, at a general angle of incidence, energies around 50 eV, are 30 seconds per energy on the current generation of computers (e.g. an ICL Titan) when 13 beams and 5 phase shifts are included. Times begin to increase rapidly above 100 eV incident energy because of an increasing number of partial waves and Fourier components required to expand the wavefunction.

Other methods of calculating truncate the semi-infinite crystal at a finite number of layers relying on absorption for convergence of reflectivities with increasing number of layers. Reflectivities could be found either by stacking together first two layers, then two pairs, etc., building up the number of layers (the matrix doubling method) or by treating the whole slab as a giant layer and inverting a giant matrix to find all the unknowns (the giant matrix method). The former method is much the faster, giant matrix methods being so slow that generally matrices cannot be made large enough for good convergence.

For situations where a detailed interpretation of spectra is not required the matrix doubling method is to be preferred to Bloch-wave methods because it is faster, and also easier to program. However the renormalised-forward-scattering (RFS) perturbative method described in Chapter V is even faster than the matrix doubling method, and gives the same amount of information. The main value of the matrix doubling method is as an alternative to the R.F.S. scheme at very low energies below 10 eV where weak absorption, sometimes combines with relatively strong back-scattering to prevent the R.F.S. perturbation theory from working.

Distinctions are not so clear cut between the two methods of treating multiple scattering within a plane of ion cores: direct summation and Kambe transformed methods. Transformation of the summation does indeed make for a more convergent series, but the smaller number of terms is partially offset by their complexity and the Kambe method was found to be about 30% faster in calculating planar reflectivities. Against the advantage of time is to be weighed the more complex programming involved and larger core space needed. On balance the direct summation method seems to be best suited to LEED calculations.

The most time consuming part of calculating scattering by a plane was making the summation in $F_{\ell m}$, equation (4.54), which averaged to about 12 seconds in calculations on a copper 100 surface, between 0 and 100 eV incident energy. The next most time consuming part was inversion of the matrix $(1 - X)$ in equation (4.49) which averaged to about 8 seconds. Other steps were much faster, including summation over $\ell'm'$ in equation (4.55).

Computation time can often be spectacularly reduced by applying symmetry reductions of Section H. Factors of 64 have been quoted for reductions of asymptotic times and in practice factors of 10 are easily achieved in configurations of moderate symmetry. Against saving of computing time must be balanced increased effort for the programmer in making symmetry transformations, which often have to be rewritten for configurations of different symmetry.

The computing times given above are for an ICL Titan computer. Transferring programs to an IBM 370/165 speeded up calculations by factors between 10 and 20.

Chapter 5

Perturbative Methods and Related Techniques

VA Introduction

The main thrust in LEED research comes from a desire to use the technique to investigate surface structure. Having gained sufficient understanding of the diffraction process by calculating for surfaces whose structure we can be reasonably sure of, we can go on to investigate systems where the surface structure is not known. The procedure would be to guess a structure and using experience accumulated for simpler systems calculate its reflectivities knowing that if the structure were correct, agreement between theory and experiment would be seen. Evidently more than one calculation will be needed, unless we make a lucky guess, and calculation will be further increased by the greater complexity that diffraction patterns from surface structures have than clean surface diffraction patterns.

The methods described in Chapter IV are ideal for initial investigations of LEED intensities from clean surfaces. They converge under a wide range of parameters, and the Bloch-wave method can be used in principle for any values of the parameters. By eliminating all uncertainty and inaccuracy from the method of solution they enable a fair test to be made of our model. But having verified the accuracy of approximating the surface by a set of spherically symmetric ion-core scatterers embedded in a uniformly absorbing medium, the question occurs of whether we can use understanding we have gained of the diffraction process to invent new methods of solution tailored to take advantage of special properties of the system that our experiences lead us to expect. If this could be done it would bring great advantages in simplicity and increased speed of calculation and extend the power of this method of surface analysis. Certainly structural analysis would be limited to rather simple systems if the Bloch-wave method were to be used, because of computing time involved.

Broadly speaking there are three possibilities for simplifying the theory. It might be that the model we have chosen describes the situation to a much finer degree of accuracy than is needed and simplifications could be made by discarding inessential features. In solving the model it might prove that the important details of the wavefunction can be found by perturbation theory, and we know from experience of band structure calculations that perturbative methods are very fast in execution. Alternatively it may happen that diffracted intensities always contain contributions from complicated processes, but by certain processing of the experimental data the complicated parts of spectra can be eliminated, only parts due to simple, easily calculated processes remaining.

Of course such techniques do not have the generality of other methods. One must have enough experience to know that approximations being made retain their validity at all points of interest or that the perturbation series involved actually converges to the correct answer.

Of the three sorts of simplification, it is elimination of involved details from the model that would appear to give greatest reduction in complications. Most of the time-consuming parts of calculations come about because of complicated angular structure of the ion-core scattering. The more structure there is the more phase shifts are needed, increasing size of matrices involved. McRae (1966) showed at an early stage that isotropic (s-wave only) scattering by ion cores could be treated simply and efficiently, and Duke *et al.* (1970) postulated that it was not important to know details of ion-core scattering. The latter workers introduced the isotropic scattering model which approximated ion cores as s-wave scatterers. The model simplifies calculations very effectively and a large literature of s-wave calculations exists, references to some of which can be found in Tucker and Duke (1971).

Useful as the model was in giving experience of theoretical calculations and investigating qualitative dependence of spectra on various parameters, it had several severe drawbacks which precluded meaningful quantitative comparison with experiment. One limitation is that there is a unitarity limit to the amount of scattering that can take place in the s-wave channel. In equation (2.58) the t matrix, which determines wave amplitudes scattered from an ion-core has a maximum of

$$t_{\max} = -\frac{2\pi i}{\kappa} \tag{4.1}$$

when the $\ell = 0$ phase shift takes the value $\frac{\pi}{2}$. On the other hand the true t-matrix for the ion core scatters much more strongly because it includes more than the s-wave phase shift, reference Fig. 2.13. In the s-wave scattering model scattered intensities are too weak, and multiple scattering is not as strongly developed as it should be.

Holland *et al.* (1971) attempted to remedy this situation by recognising that if the s-wave phase-shift were made complex, scattering could be made arbitrarily strong at the price of sacrificing flux-conservation inside the ion-core. The extra scattering strength comes about by each ion core acting as a radiator of electrons rather than as a passive scatterer.

By means of this extension of the model more multiple scattering structure could be introduced into spectra. Yet still serious discrepancies remained, both in intensities of the curves and energies at which detailed structure occurred. It seemed that when scattering was strong enough to give the pronounced multiple scattering structure observed in spectra, overall intensities of spectra were too high, but when scattering strength of ion cores was reduced to give correct overall intensities of spectra, multiple scattering structure was much too weak relative to the Bragg peaks, especially at higher temperatures.

It is this final observation that led to the abandonment of s-wave scattering models. As we have observed in Chapter II multiple scattering is brought about by strong forward scattering from ion cores. The back scattering that determines reflected intensities is relatively weak. Before any realistic model of LEED can be constructed this basic truth must be built into the formalism, and isotropic scattering cannot do this for us.

It seems from extensive investigations made with various models, published in the past few years, that accurate representation of ion-core scattering with a full complement of phase shifts, and the presence of an absorbing medium are essential ingredients of any model. We might have anticipated as much from discussions put forward in Chapter II. Our hopes for simplification must rest on perturbation schemes or data-simplification techniques.

Perturbation methods are commonly used in complicated quantum mechanical problems to calculate wavefunctions by refining an initial approximation to the wavefunction. Methods differ in the care taken in preparing the initial approximation, and in the number and type of refinements (the order of the perturbation) made. For the method to work well there are two requirements: that the initial approximation be close to the exact solution and that the method of refinement converge rapidly to the exact solution.

For example, we could apply Rayleigh-Schrödinger perturbation theory to calculate Bloch waves. Let us neglect the crystal potential, V, to a zeroth approximation, then the Bloch wave is simply

$$\psi_{\mathbf{k}}^{(0)}(\mathbf{r}) = \Omega^{-\frac{1}{2}} \exp(ik_z z + i\mathbf{k}_{0\parallel}\cdot\mathbf{r}_\parallel), \tag{5.2}$$

$\mathbf{k}_{0\parallel}$ is fixed by the incident conditions; Ω is the volume of the crystal. Applying first order perturbation theory (Dicke and Wittke, 1960) we obtain for the energy

$$\begin{aligned} E^{(1)} &= E^{(0)} + \int V(\mathbf{r}) |\psi_{\mathbf{k}}^{(0)}(\mathbf{r})|^2 d^3\mathbf{r} \\ &= \tfrac{1}{2}(|k_z|^2 + |\mathbf{k}_{0\parallel}|^2) + \Omega^{-1} \int V(\mathbf{r})\, d^3\mathbf{r}, \end{aligned} \tag{5.3}$$

which can be solved for k_z, since the energy is fixed by incident conditions. Next, corrections to the wave function can be found by adding in other waves as in equation (4.97)

$$\psi_{\mathbf{k}}^{(1)} = \psi_{\mathbf{k}}^{(0)} + \Omega^{-\frac{1}{2}} \sum_{\mathbf{G} \neq 0} \psi_{\mathbf{kG}}^{(1)} \exp[i(\mathbf{k} + \mathbf{G})\cdot\mathbf{r}]. \tag{5.4}$$

First order perturbation theory gives for

$$\psi_{\mathbf{kG}}^{(1)} = \frac{\Omega^{-1} \int \exp[-i(\mathbf{k} + \mathbf{G})\cdot\mathbf{r}]\, V \exp[i\mathbf{k}\cdot\mathbf{r}]\, d^3\mathbf{r}}{\frac{1}{2}|\mathbf{k}|^2 - \frac{1}{2}|\mathbf{k} + \mathbf{G}|^2} \tag{5.5}$$

Higher orders of perturbation theory can be applied without undue complication of formulae and the method is rapidly executed.

Unfortunately this most simple application of perturbation theory gives poor results because the initial wavefunction is not at all like the final wavefunction, nor is the method even convergent over successive orders of perturbation. Much of the difficulty stems from trying to calculate the complicated wavefunction in the core region, and can be remedied by adopting a pseudopotential, V_{ps}, in place of the true potential, V, as described in Section IVG.

Strozier and Jones (1971) have applied the pseudopotential perturbation method to LEED. Some of their results for beryllium showed encouraging agreement of structure and intensity but in certain energy ranges and for some angles of incidence agreement was not seen. Figure 5.1 shows some typical results taken from their paper. Even with the help of pseudopotentials Rayleigh-Schrödinger perturbation theory is not sufficiently accurate to give reliable calculations of LEED intensities. Experiments measure small back-reflected components of the wavefunction and calculating *wavefunctions* accurately in perturbation theory is much more difficult than the usual problem encountered of calculating the much less sensitive *energy levels*.

To get a perturbation theory that will work, much more care has to be put into preparing the starting wavefunction. Scattering by individual ion-cores must be treated exactly via the t-matrix and many theories go as far as to calculate scattering by a whole plane of atoms exactly. A variety of techniques exists and Sections B, C and D deal with them in detail.

Of perturbation schemes, one of the simplest is the kinematic theory described in Section IIIA. It treats scattering by each ion core exactly, and

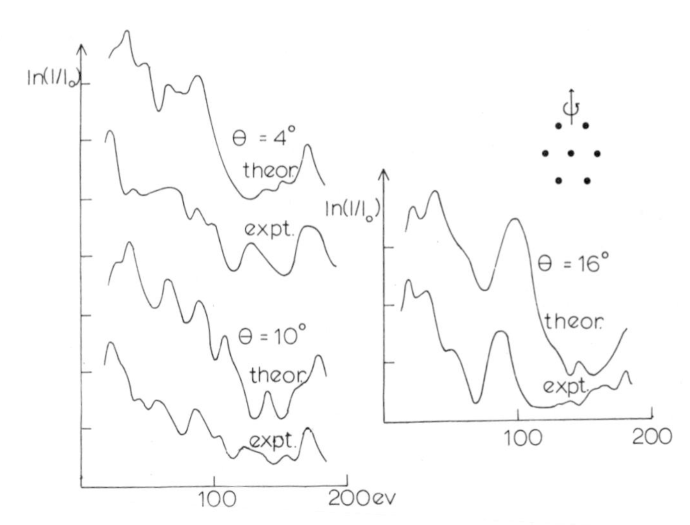

FIGURE 5.1. Comparison of a pseudopotential perturbation theory calculation for beryllium with experiment. The surface is the (1000) and angles refer to rotation away from normal incidence in the manner shown in the inset figure (Strozier and Jones, 1971). The experiments were made by Baker (1970).

takes into account contributions to intensities from waves scattering off only one atom. As we have seen multiple scattering processes are usually important and the theory does not provide an accurate description of spectra. Recently certain methods have been developed (Lagally, Ngoc and Webb, 1971) making the assumption that multiple scattering produces extra structure in a random manner, and go on to average data over a variety of incident conditions retrieving what is assumed to be a kinematic component. Two difficulties remain to be resolved, one is that since multiple scattering produces peaks at least as pronounced as peaks due to the simple kinematic scattering, a large amount of data must be processed to give good statistics on the amount of multiple scattering remaining. The other is that recent theory (Pendry, 1972) only partly justifies the ansatz of random contributions from multiple scattering processes. As theory stands at the moment interpretation of averaged curves involves some complicated calculations.

Nevertheless Lagally, Ngoc and Webb have shown that averaging does considerably simplify structure in curves and if a sufficiently simple theory could be found, there is no doubt that the averaging methods would offer wide scope for interpretation of LEED spectra.

VB Perturbation theory and planar scattering

To treat scattering by a single ion core in a non-perturbative manner is not a complicated matter. We have seen that amplitudes of scattered waves are given by the t-matrix of equation (2.58). It is much harder to take account of all multiple scattering processes in a plane of atoms to find the planar transmission and reflection matrices of Section IVC, and any simplification that can be afforded by perturbation theory would be useful. Here we develop a perturbation expansion for scattering within a plane and test convergence of that expansion for various materials, following closely a paper by Pendry (1971c).

The analysis will tell us more than whether perturbation expansions for single layers converge. It will also tell us whether those schemes converge that do not make a specific separation into inter and intra layer scattering, because if perturbation theory is found not to work for a single layer it cannot be expected to work for the whole assembly of layers that constitutes the crystal.

Let us take a simple example of a plane with only one atom per unit cell, and suppose that a plane wave

$$\begin{aligned}\chi^{(0)} &= \exp(i\mathbf{K}\cdot\mathbf{r}),\\ \tfrac{1}{2}|\mathbf{K}|^2 &= E - V_0,\\ \mathbf{K}_{\|} &= \mathbf{k}_{0\|},\end{aligned} \tag{5.6}$$

is incident on the plane. Each ion core in the plane has a set of spherical waves incident upon it, due to the plane wave. For the jth ion core,

$$\begin{aligned}\chi^{(0)} &= \sum_{\ell m} \exp(i\mathbf{k}_{0\|}\cdot\mathbf{R}_j)\, i^{\ell} 4\pi j_{\ell}(|\mathbf{K}|\,|\mathbf{r}-\mathbf{R}_j|)\\ &\quad\times Y_{\ell m}(\Omega(\mathbf{r}-\mathbf{R}_j))\, Y_{\ell -m}(\Omega(\mathbf{K}))(-1)^m\\ &= \sum_{\ell m} A^{(0)}_{\ell m} j_{\ell}(|\mathbf{K}|\,|\mathbf{r}-\mathbf{R}_j|)\, Y_{\ell m}(\Omega(\mathbf{r}-\mathbf{R}_j))\\ &\quad\times \exp(i\mathbf{k}_{0\|}\cdot\mathbf{R}_j).\end{aligned} \tag{5.7}$$

Each ion core scatters some of the wavefield incident upon it, the wave scattered by the jth atom being

$$\begin{aligned}&\sum_{\ell m} A^{(0)}_{\ell m} \exp(i\mathbf{k}_{0\|}\cdot\mathbf{R}_j)\exp(i\delta_{\ell})\sin(\delta_{\ell})\\ &\quad\times i h^{(1)}_{\ell}(|\mathbf{K}|\,|\mathbf{r}-\mathbf{R}_j|)\, Y_{\ell m}(\Omega(\mathbf{r}-\mathbf{R}_j)).\end{aligned} \tag{5.8}$$

Some of the scattered wave may come to be incident upon a second atom, k, giving for flux incident on the kth atom, that has been scattered once by the jth atom:

$$\begin{aligned}\chi^{(1)} &= \sum_{j\neq k}\sum_{\ell' m'} A^{(0)}_{\ell' m'} \exp(i\mathbf{k}_{0\|}\cdot\mathbf{R}_j)\exp(i\delta_{\ell'})\sin(\delta_{\ell'})\\ &\quad\times i h^{(1)}_{\ell'}(|\mathbf{K}|\,|\mathbf{r}-\mathbf{R}_j|)\, Y_{\ell' m'}(\Omega(\mathbf{r}-\mathbf{R}_j)),\\ &= \sum_{j\neq k}\sum_{\ell' m'} A^{(0)}_{\ell' m'} \exp(i\mathbf{k}_{0\|}\cdot\mathbf{R}_j)\exp(i\delta_{\ell'})\sin(\delta_{\ell'})\end{aligned}$$

$$\times \sum_{\ell m} G_{\ell' m'; \ell m}(\mathbf{R}_k - \mathbf{R}_j) j_\ell(|\mathbf{K}|\,|\mathbf{r} - \mathbf{R}_k|)\, Y_{\ell m}(\Omega(\mathbf{r} - \mathbf{R}_k)). \quad (5.9)$$

Matrix G, originally defined in equation (4.39), has been used to re-express expansions about the jth ion core as expansions about the kth ion core. The once-scattered flux can be expressed more concisely:

$$\chi^{(1)} = \sum_{\ell m} A^{(1)}_{\ell m} \exp(i\mathbf{k}_{0\|} \cdot \mathbf{R}_k) j_\ell(|\mathbf{K}|\,|\mathbf{r} - \mathbf{R}_k|)\, Y_{\ell m}(\Omega(\mathbf{r} - \mathbf{R}_k)), \quad (5.10)$$

$$A^{(1)}_{\ell m} = \sum_{\ell' m'} A^{(0)}_{\ell' m'} X_{\ell' m', \ell m}, \quad (5.11)$$

$$X_{\ell' m', \ell m} = \exp(i\delta_{\ell'}) \sin(\delta_{\ell'}) \sum_{j \neq k} \exp[i\mathbf{k}_{0\|} \cdot (\mathbf{R}_j - \mathbf{R}_k)]\, G_{\ell' m', \ell m}(\mathbf{R}_k - \mathbf{R}_j) \quad (4.12)$$

Once-scattered flux can scatter a second time giving yet another contribution to flux incident on each ion core

$$A^{(2)}_{\ell m} = \sum_{\ell' m'} A^{(1)}_{\ell' m'} X_{\ell' m', \ell m} = \sum_{\ell' m'} A^{(0)}_{\ell' m'} (X^2)_{\ell' m', \ell m}, \quad (5.13)$$

and the total flux incident on each ion core is given by

$$A_{\ell m} = \sum_{n=0}^{\infty} A^{(n)}_{\ell m} = \sum_{\ell' m'} A^{(0)}_{\ell' m'} \sum_{n=0}^{\infty} (X^n)_{\ell' m', \ell m}, \quad (5.14)$$

$$= \sum_{\ell' m'} A^{(0)}_{\ell' m'} (1 - X)^{-1}_{\ell' m', \ell m}. \quad (5.15)$$

In (5.14) we have a perturbation expansion for scattering within a plane. Summing the series we retrieve our original expression (4.44) for multiple scattering. In mathematical terms the series is simply a power series expansion of $(1 - X)^{-1}$; in physical terms the nth order of perturbation theory includes all processes wherein an electron scatters consecutively from n-atoms within the plane.

By substituting our nth order approximation for $(1 - X)^{-1}$ into (4.49) we determine forward and backward scattering factors for the plane, correct to nth order perturbations. In so doing we save time by replacing a matrix inversion with matrix multiplications which are simpler and faster. However, calculation of matrix X itself remains unchanged.

There is no guarantee that the series we have derived is a convergent one. We can best illustrate difficulties that might arise by taking the case when only $\ell = 0$ phase shifts are considered. Then X is simplified from being a matrix to a number because of the single partial wave included. The expansion

$$(1 - X_{00})^{-1} = 1 + X_{00} + X_{00}^2 + \ldots \quad (5.16)$$

has well known convergence properties. If

$$|X_{00}| < 1 \quad (5.17)$$

successive terms grow smaller and smaller and convergence is achieved. If

$$|X_{00}| > 1 \tag{5.18}$$

successive terms grow in amplitude and the series diverges.

The criterion for convergence of a matrix series is more involved. Let Q be the matrix that diagonalises X:

$$[QXQ^{-1}]_{\ell'm',\ell m} = X^D_{\ell m}\delta_{\ell\ell'}\delta_{mm'}. \tag{5.19}$$

The diagonal elements $X^D_{\ell m}$ are eigenvalues of X. Multiplying the expansion

$$(1 - X)^{-1} = 1 + X + X^2 + \ldots \tag{5.20}$$

from the left and from the right by $Q^{-1}Q$ and inserting $Q^{-1}Q$ between each pair of matrices on the right-hand side, changes neither the sum, nor each term in the sum, but allows us to rewrite the equation using the diagonal matrix (5.19)

$$(1 - X)^{-1} = Q^{-1}[1 + QXQ^{-1} + (QXQ^{-1})^2 + \ldots]Q. \tag{5.21}$$

Let us consider separately each diagonal element of the series enclosed in square brackets.

$$(1 - X^D_{\ell m})^{-1} = 1 + X^D_{\ell m} + (X^D_{\ell m})^2 + \ldots \tag{5.22}$$

It is evident that the matrix power series involves several power series in the numbers $X^D_{\ell m}$, one for each diagonal element. For the matrix series as a whole to converge, each numerical series must converge. If $X^D_{\max}$ is the eigenvalue of X of largest modulus, the condition is

$$|X^D_{\max}| < 1 \tag{5.23}$$

and the smaller $X^D_{\max}$ is the more rapidly the series will converge.

We have chosen three materials, aluminium, nickel, and xenon, to represent a wide variety of properties. In nickel the unit cell is small, leading to strong elastic scattering per unit volume of crystal, whilst V_{0i} representing inelastic scattering has been estimated to be -3 eV (Pendry, 1969c). Xenon has a very large unit cell, eight times as large as that of nickel and the scattering strength of ion cores is diluted by this large volume. V_{0i} for xenon is known to be about -3.5 eV (Ignatjevs *et al.*, 1971). Xenon is an extreme example of a weak scattering solid; more typical of weak scattering situations is aluminium which has a unit cell only half as large again as nickel, but for which electron-gas theory predicts a large value of V_{0i}, about -5 eV (Lundqvist, 1969).

Eigenvalues of matrix X were calculated for close packed planes parallel to a (111) surface in these materials, and the largest eigenvalue plotted in Fig. 5.2.

Matrix X describes multiple scattering corrections and the stronger scattering in the plane is, and the smaller inelastic damping of multiple scattered waves, the larger its eigenvalues will be. As might be expected nickel has the largest eigenvalue followed by aluminium, then xenon. Eigenvalues will be particularly large at energies where waves reaching an atom have been scattered in phase. Also the more closely packed the plane in question the more atoms can contribute to multiple scattering and the larger eigenvalues will be. We have deliberately chosen the (111) plane as a critical test of the expansion.

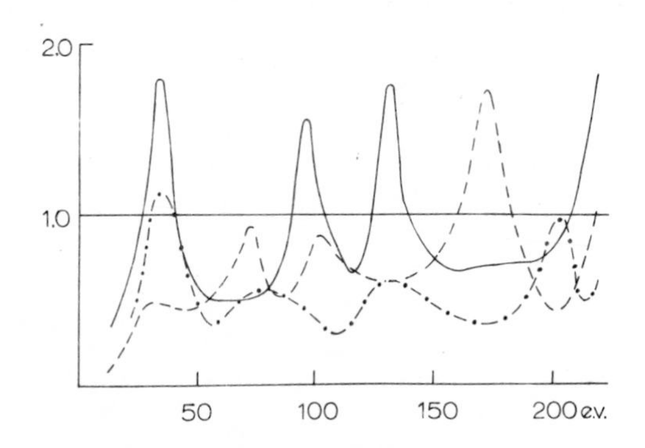

FIGURE 5.2. Modulus of the largest eigenvalue of matrix X for nickel (full curve), aluminium (broken curve) and xenon (chain curve).

Unfortunately in nickel, typical of transition and noble metals, the largest eigenvalue tells us that perturbation expansion (5.14) does not usually converge well and that there are some energies where the expansion does not converge at all. Matters are more hopeful for aluminium, and for xenon it will usually be the case that the expansion converges.

Evidently careful tests must be made for any material to which the perturbation expansion is to be applied.

It will be noted that there is no sign of increasing energy's leading to more convergent expansions, nor does inclusion of corrections for finite temperatures improve convergence much. Once flux is scattered into the plane, strong multiple scattering effects nearly always occur and can only be avoided by arranging that only a small amount of flux ever scatters into the plane. For example, at normal incidence an electron must first scatter through 90° to scatter subsequently within a plane and by working at high energies, say greater than 200 eV, and elevated temperatures the t-matrix for scattering through 90° can be made very small. Scattering within the layer is never excited to an appreciable extent, and we can as well neglect matrix X altogether.

VC Perturbation schemes for interplanar scattering

Between each pair of planes in the surface, sets of forward and backward travelling waves exist,

$$\sum_{\mathbf{g}} a_{j\mathbf{g}}^{+} \exp(i\mathbf{K}_{\mathbf{g}}^{+} \cdot \mathbf{r}) + a_{j\mathbf{g}}^{-} \exp(i\mathbf{K}_{\mathbf{g}}^{-} \cdot \mathbf{r}), \tag{5.24}$$

between the jth and $(j+1)$th layers. $K_{\mathbf{g}}^{\pm}$ take their usual meaning defined in (3.63), and we have assumed that the waves have been referred to an origin half way between planes. The workings of perturbation schemes can most simply be illustrated by treating the case of one plane per layer, and all layers having the same spacing $\mathbf{c}$ except for the first 'layer' which consists not of a set of ion cores but of the step in potential from the constant value between ion cores inside the crystal to the value in the vacuum outside. We take the barrier to be situated $\frac{1}{2}\mathbf{c}$ away from the second plane.

In Section IIIC we discussed how waves were reflected and transmitted by planes, and derived equations governing equilibrium of beam amplitudes between successive planes. Equations (3.67) can be rewritten

$$a_j^{+} = P^{+}(I + M^{++})P^{+}a_{j-1}^{+} + P^{+}M^{+-}P^{-}a_j^{-}, \tag{5.25a}$$

$$a_{j-1}^{-} = P^{-}M^{-+}P^{+}a_{j-1}^{+} + P^{-}(I + M^{--})P^{-}a_j^{-}, \tag{5.25b}$$

where we have suppressed subscripts $\mathbf{g}$ on matrices and vectors for simplicity. $P^{\pm}$ are defined to be

$$P_{\mathbf{gg}'}^{+} = \delta_{\mathbf{gg}'} \exp(i\mathbf{K}_{\mathbf{g}}^{+} \cdot \tfrac{1}{2}\mathbf{c}), \tag{5.26a}$$

$$P_{\mathbf{gg}'}^{-} = \delta_{\mathbf{gg}'} \exp(-i\mathbf{K}_{\mathbf{g}}^{-} \cdot \tfrac{1}{2}\mathbf{c}), \tag{5.26b}$$

and $\mathbf{c}$ is the spacing between planes.

Equations (5.25) describe scattering by planes of ion cores. In addition there is a surface barrier in potential at the surface of the crystal which we shall think of as the first layer. Thus reflected amplitudes a_0^{-} are related to incident amplitudes by

$$a_1^{+} = Q^{\mathrm{I}}(1)a_0^{+} + Q^{\mathrm{II}}(1)a_1^{-}, \tag{5.27a}$$

$$a_0^{+} = Q^{\mathrm{III}}(1)a_1^{+} + Q^{\mathrm{IV}}(1)a_1^{-}. \tag{5.27b}$$

The Q-matrices defining reflection and transmission by the barrier can be found by numerical integration through the barrier, but more usually the barrier is approximated as being very smooth so that it has small reflectivity

$$Q^{\mathrm{II}} \simeq 0 \simeq Q^{\mathrm{III}} \tag{5.28}$$

and unit transmissivity,

$$Q^{\mathrm{I}} \simeq I \simeq Q^{\mathrm{IV}}. \tag{5.29}$$

Compared with planes of ion cores the barrier is a weak scatterer and does not play a crucial role in determining whether a perturbation expansion converges.

Equations (5.25) can be given a simple pictorial interpretation which is most useful in deriving perturbation expansions. (5.25a) tells us that forward travelling waves half way between the jth and $(j + 1)$th layers (a_j^+) are compounded of waves that have originated between the previous pair of layers (a_{j-1}^+), propagated to the jth layer (factor of P^+) been transmitted or forward scattered there (factor of $(I + M^{++})$) then propagated to a point halfway between the jth and $(j + 1)$th layers (factor of P^+), and waves that originated as back travelling waves between the jth and $(j + 1)$th layers (a_j^-) propagated to the jth layer (factor of P^-) were reflected there (factor of M^{+-}) and finally returned to the mid point (factor of P^+).

The first perturbation scheme we discuss treats scattering by each plane exactly, not making use of the perturbation expansion for planar scattering discussed in the last section. We shall call this 'planar perturbation theory' because only scattering by planes is assumed weak.

•One starting point for a perturbation expansion is to assume that in equations (5.25) all scattering by ion cores is small and in equations (5.27) that reflection matrices for the barrier are small. Then to zeroth order

$$a_j^{+(0)} = (P^+)^{2j-2}Q^{\mathrm{I}}(1)a_0^+, \tag{5.30a}$$

$$a_j^{-(0)} = 0. \tag{5.30b}$$

Substituting expressions (5.30) into (5.25) and keeping terms up to first order gives expressions for $a_0^{-(1)}$ which can be interpreted as follows: the incident wave (a_0^+) is transmitted through the barrier

$$Q^{\mathrm{I}}(1)a_0^+,$$

propagates from halfway between the first and second layers to the jth layer

$$(P^+)^{2j-3}Q^{\mathrm{I}}(1)a_0^+,$$

is reflected there

$$M^{-+}(P^+)^{2j-3}Q^{\mathrm{I}}(1)a_0^+,$$

propagates back to halfway between the first and second layers

$$(P^-)^{2j-3}M^{-+}(P^+)^{2j-3}Q^{\mathrm{I}}(1)a_0^+,$$

and finally is transmitted through the first layer, the surface barrier

$$a_0^{-(1)} = \sum_{j=2}^{\infty} Q^{\mathrm{IV}}(1)(P^-)^{2j-3}M^{-+}(P^+)^{2j-3}Q^{\mathrm{I}}(1)a_0^+ + Q^{\mathrm{III}}(1)a_0^- \tag{5.31}$$

In addition there is a contribution from reflection from the surface barrier.

It should now be clear that we can make a more formal statement of the pictorial method of writing down contributions of processes to planar perturbation theory. The following are a set of rules for calculating

(i) Equations (5.25) generate all possible scattering paths for the electron. For nth order contributions to perturbation theory draw all paths that scatter from n planes, whether in the forward, or backward direction.

(ii) With each half-interplanar spacing a path crosses in the forward direction, associate a factor of P^+, and in the backward direction a factor of P^-.

(iii) With each scattering by a plane associate a factor of M^{++}, M^{+-}, M^{-+} or M^{--}, according to the directions of incident and scattered beams or in the case of the surface barrier, factors of $Q(1)^{\text{I,IV}}$ for transmission in the forward or backward directions, and $Q(1)^{\text{II,III}}$ for backwards to forwards, or forwards to backwards reflection.

(iv) Since the factors are matrices, the order in which they are multiplied matters: factors for events early in the path are written to the right, later in the path to the left.

(v) Sum over all nth order contributions.

To illustrate operation of the rules we draw second order contributions to a_0^- in Fig. 5.3. Using rules (i) to (v) we deduce

$$\begin{aligned} a_0^{-(2)} &= \sum_{j=2}^{\infty} \sum_{k=j+1}^{\infty} Q^{\text{IV}}(1)(P^-)^{2k-3} M^{-+}(P^+)^{2(k-j)} M^{++} \\ &\times (P^+)^{2j-3} Q^{\text{I}}(1) a_0^+ + \sum_{j=k+1}^{\infty} \sum_{k=2}^{\infty} Q^{\text{IV}}(1)(P^-)^{2k-3} M^{--} \\ &\times (P^-)^{2j-2k} M^{-+}(P^+)^{2j-3} Q^{\text{I}}(1) a_0^+ . \end{aligned} \tag{5.32}$$

McRae (1968) first suggested approximating a_0^- by $a_0^{-(1)} + a_0^{-(2)}$.

If, as we have assumed, we are dealing with a material in which all scattering

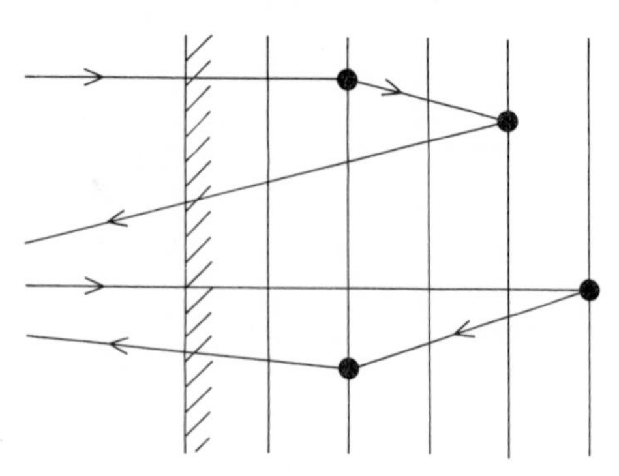

FIGURE 5.3. Two second order contributions to back-scattering. Interactions with planes of ion cores have been calculated exactly and are shown as large black dots.

processes are weak, it is logical to expand not only inter layer scattering in a perturbation series but also to calculate intra layer scattering by the perturbation series discussed in the last section, giving for nth order contributions to scattering within the layer from equation (4.49)

$$M_{\mathbf{g}'\mathbf{g}}^{\pm\pm(n)} = \frac{8\pi^2 i}{|\mathbf{K}_{\mathbf{g}}^{\pm}|\, A\, |K_{\mathbf{g}'z}^{\pm}|} \sum_{\substack{\ell m \\ \ell' m'}} [i^{\ell}(-1)^m Y_{\ell -m}(\Omega(\mathbf{K}_{\mathbf{g}}^{\pm}))] \times X_{\ell m, \ell' m'}^{n-1} [i^{-\ell'} Y_{\ell' m'}(\Omega(\mathbf{K}_{\mathbf{g}'}^{\pm}))] \exp(i\delta_{\ell'}) \sin(\delta_{\ell'}), \quad (5.33)$$

in the case that the layer contains only one atom per unit cell.

The second perturbation scheme for inter layer scattering makes a more restrictive assumption than planar perturbation theory: that scattering by ion cores in the plane is weak, we shall call it 'ion-core perturbation theory'. Kinematic theory corresponds to first order ion-core perturbation theory.

Now, to distinguish between orders of scattering within a layer rules (*i*) and (*iii*) must be modified to read for ion-core perturbation theory

(*i'*) Equations (5.25) generate all possible scattering paths for the electron. For nth order contributions to a perturbation theory in which intra layer scattering is included in the perturbation scheme, draw all paths that scatter r times at one plane, s times at another . . ., whether in the forward or backward direction, such that $r + s + \ldots = n$.

(*iii'*) With r consecutive scatterings within a plane associate a factor of $M^{++(r)}$, $M^{+-(r)}$, $M^{-+(r)}$, or $M^{--(r)}$ according to the directions of incident and scattered beams, or in the case of the surface barrier factors of $Q(1)^{\text{I,IV}}$ for transmission in the forward or backward directions, and $Q(1)^{\text{II,III}}$ for backward to forward, or forward to backward reflection.

Figure 5.4 shows some diagrams of second order events, with intra layer scattering treated in perturbation theory.

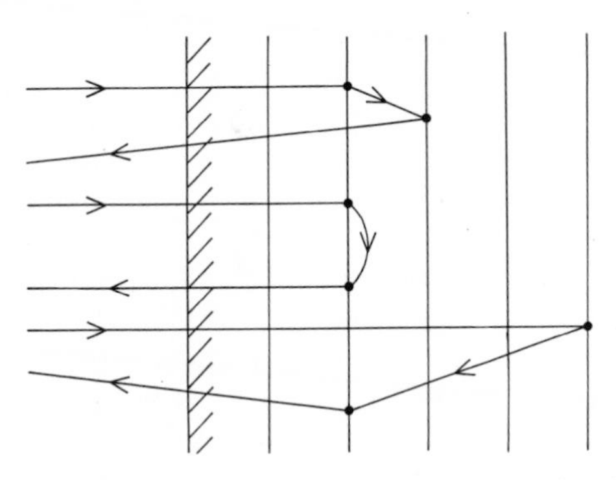

FIGURE 5.4. Second order contributions to back-scattering. Each order of scattering within a plane is represented by a small dot.

Tait, Tong and Rhodin (1972) have proposed such a scheme and carried it to third order. Summations over planes were performed analytically, for example by observing in a process that scatters backwards off plane j and subsequently forwards off plane k, the summation over j and k has the form

$$\sum_{j=3}^{\infty}\sum_{k=2}^{j-1}(P_{\mathbf{gg}}^{-})^{2k-3}(P_{\mathbf{g'g'}}^{-})^{2j-2k}(P_{\mathbf{g''g''}}^{+})^{2j-3}$$

$$= \sum_{k=2}^{\infty}\sum_{j=k+1}^{\infty}\exp\left[-i\mathbf{K}_{\mathbf{g}}^{-}\cdot\tfrac{1}{2}(2k-3)\mathbf{c} - i\mathbf{K}_{\mathbf{g}'}^{-}\cdot\tfrac{1}{2}(2j-2k)\mathbf{c} + i\mathbf{K}_{\mathbf{g}''}^{+}\cdot\tfrac{1}{2}(2j-3)\mathbf{c}\right]$$

$$= \frac{\sum_{k=2}^{\infty}\exp\left[i(\mathbf{K}_{\mathbf{g}''}^{+} - \mathbf{K}_{\mathbf{g}}^{-})\cdot k\mathbf{c} + i(-\tfrac{1}{2}\mathbf{K}_{\mathbf{g}''}^{+} + \tfrac{3}{2}\mathbf{K}_{\mathbf{g}}^{-} - \mathbf{K}_{\mathbf{g}'}^{-})\cdot\mathbf{c}\right]}{1 - \exp\left[i(\mathbf{K}_{\mathbf{g}''}^{+} - \mathbf{K}_{\mathbf{g}'}^{-})\cdot\mathbf{c}\right]}$$

$$= \frac{\exp\left[i(\tfrac{3}{2}\mathbf{K}_{\mathbf{g}''}^{+} - \tfrac{1}{2}\mathbf{K}_{\mathbf{g}}^{-} - \mathbf{K}_{\mathbf{g}'}^{-})\cdot\mathbf{c}\right]}{\left(1 - \exp\left[i(\mathbf{K}_{\mathbf{g}''}^{+} - \mathbf{K}_{\mathbf{g}'}^{-})\cdot\mathbf{c}\right]\right)\left(1 - \exp\left[i(\mathbf{K}_{\mathbf{g}''}^{+} - \mathbf{K}_{\mathbf{g}}^{-})\cdot\mathbf{c}\right]\right)} \quad (5.34)$$

Tracing through all the steps in ion-core perturbation theory shows that at no stage is any matrix inverted, only multiplication of matrices times vectors need be involved, scaling in computing times as the square of the matrices' dimensions.

Tong and Rhodin's third order perturbation scheme is very fast in execution, once matrix X has been calculated. In fact calculation of X dominates computing times taking an average of about 10 seconds per point in the 0–100 eV range. On the other hand the method has limited applicability because of poor convergence of the perturbation expansion for intra layer scattering, as we showed in Section B. Nevertheless from Fig. 5.2 one might expect convergence for calculations on aluminium and in Fig. 5.5 we reproduce Tong and Rhodin's calculation for the (100) surface, using their third order scheme. Corrections for temperature effects have been made following the treatment of Chapter VI.

Figure 5.2 leads us to expect that for nickel and similar metals good convergence of intra-layer perturbation expansions cannot be expected. In calculations for such materials the original planar perturbation scheme of McRae must be followed, scattering by each plane being calculated exactly. The resulting inversion of $(1 - X)$ adds a little extra to times, perhaps another 60%. Calculations have been made for copper (Pendry 1971c) with McRae's scheme and compared not with experiment but with Bloch wave calculations of intensities so that discrepancies can be attributed directly to failure of the perturbation expansion. Figure 5.6 shows the comparison.

It is clear from Fig. 5.6 that the strong scattering nature of the transition

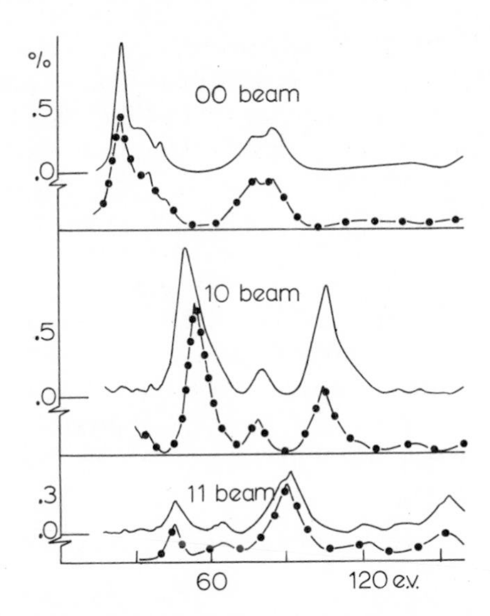

FIGURE 5.5. Comparison between calculations where both inter and intra layer scattering has been calculated in third order perturbation theory (solid lines) and experimental results of Jona (1970) (chain curves) for the 00, 10 and 11 beams from a (100) surface of aluminium. All curves are for normal incidence except the 00 beam for which the crystal has been rotated about the (010) axis by 25°.

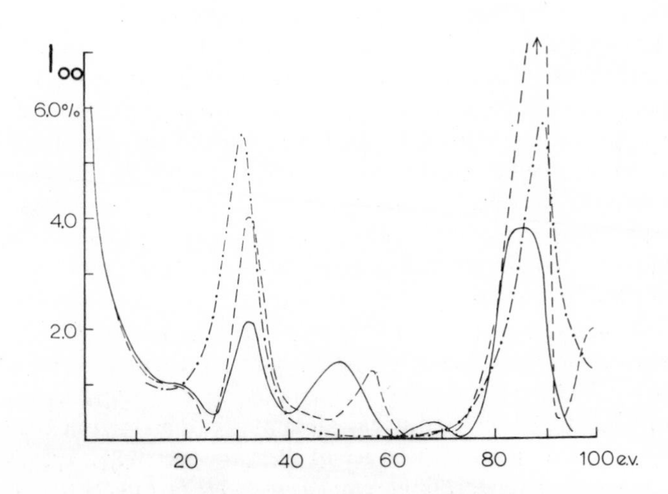

FIGURE 5.6. Comparison of calculations for a copper (100) surface at normal incidence where intra layer scattering is treated exactly and interlayer scattering is calculated to first order (chain curve) and second order (dash curve), with the results of an accurate Bloch wave calculation (full curve).

and noble metals which prevented intra layer perturbation methods from converging in nickel, also interferes with good convergence of inter-layer planar perturbation theory in copper, (and presumably in nickel too). The second order results seem hardly closer to the true solution than first order calculations, and it is clear that many orders of perturbation theory would have to be applied to get accurate answers for transition and noble metals.

In making tests of perturbation expansions it is most important to use realistic values of parameters in the theory, otherwise spurious conclusions can easily be drawn. Such would be the case if instead of using accurate ion core scattering factors, only the *s*-wave component were taken. Neglect of the important higher phase shifts would make scattering factors much weaker than in fact they are, leading to the impression that perturbation expansions are more convergent than is generally the case.

VD Renormalised forward scattering perturbation theory

It is particularly unfortunate that perturbation schemes we have considered so far, fail for transition and noble metals because these are the materials on which LEED experiments are most commonly performed. Failure of a particular perturbation expansion need not spell the end of all perturbative approaches to the problem. Sometimes a detailed examination reveals that the divergence is due only to part of the scattering processes and that this strong scattering component can be treated exactly in a simple manner. The perturbation expansion is replaced with a new one in which all strong scattering events are replaced by an exact calculation of their contribution, or in more technical language the expansion is *renormalised.*

We quote as a simple example of renormalisation the case of an optical diffraction grating immersed in a transparent liquid. One approach would be to consider each element of volume of liquid as a separate scatterer and try to solve the very complicated multiple scattering equations. A much simpler approach recognises that we know the effect of multiple scattering within the liquid to be very simple: light travels just as in vacuo except that its wave vector is larger. Now the whole problem can be solved as simply as in vacuo once the wave vector has been renormalised.

Is renormalisation of the inter layer planar perturbation theory possible? Equations (5.25) reveal that two sorts of scattering take place: in the forward and backward directions, and it has already been observed in Section IIA that forward scattering is the larger of the two. We shall see in the next chapter that thermal vibrations emphasise even further the discrepancy between forward and backward scattering.

If we could somehow solve exactly the forward scattering part of the perturbation series, not only would we have simplified the series but perturbations would involve only weak back-scattering events. Suppose we want to know how waves of amplitudes a_j^+ travel from between the jth and $(j+1)$th layers to between the kth and $(k+1)$th layers taking account of all forward scattering events on the way. Figure 5.7 shows some of the processes that can contribute.

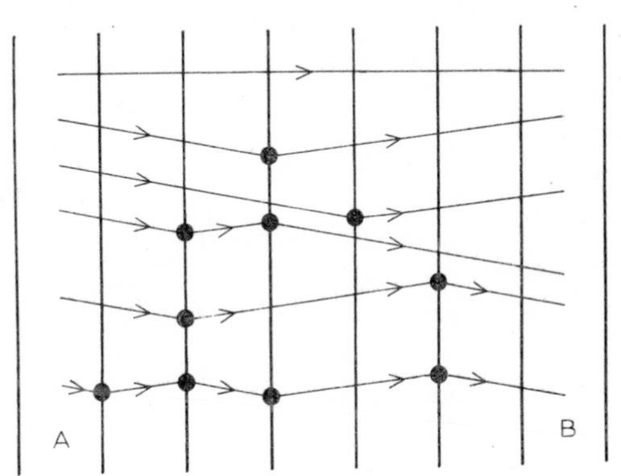

FIGURE 5.7. Some corrections to propagation from A to B arising from forward scattering events.

From rules (i)–(v) of Section C contributions are as follows:

(i) Propagation with no scattering,

$$(P^+)^{2k-2j} \cdot a_j^+ \tag{5.35}$$

(ii) Propagation to the ℓth plane, forward scattering, propagation to between the kth and $(k+1)$th planes,

$$\sum_{j<\ell\leqslant k} (P^+)^{2k-2\ell+1} M^{++} (P^+)^{2\ell-2j-1} a_j^+,$$

(iii) with two intermediate scatterings,

$$\sum_{j<\ell<m\leqslant k} (P^+)^{2k-2m-1} M^{++} (P^+)^{2m-2\ell} M^{++} (P^+)^{2\ell-2j-1} a_j^+,$$

⋮

$(i+k-j)$ Propagation with $k-j$ scatterings

$$P^+ M^{++} (P^+)^2 M^{++} \ldots (P^+)^2 M^{++} P^+ a_j^+.$$

Fortunately all these terms can be summed exactly to give

$$[P_{RFS}^+]^{k-j} a_j^+ = [P^+(I + M^{++})P^+]^{k-j} a_j^+, \tag{5.36}$$

as can be verified by inspection. By simply replacing $(P^+)^2$ with P_{RFS}^+ we take into account all possible forward scattering events. P_{RFS}^+ is the renormalised forward scattering propagator. A similar quantity describes propagation out of the crystal

$$P^{-}_{RFS} = P^{-}(I + M^{--})P^{-} \tag{5.37}$$

Now let us develop a renormalised perturbation theory. To zeroth order in back scattering the incident waves, a_0^+, are transmitted through the surface barrier to give

$$a_1^{+(0)} = Q^{\mathrm{I}}(1)a_0^+,$$

subsequently we use the renormalised forward scattering formalism to calculate

$$a_j^{+(0)} = (P^{+}_{RFS})^{j-1}Q^{\mathrm{I}}(1)a_0^+, \tag{5.38}$$

correct to all orders of forward scattering. Even orders contribute only to forward travelling waves, odd orders to backward travelling waves.

Because of absorption the wavefield dies away exponentially into the crystal and after, say, $j_{\max}$ layers, can be neglected. Next the first order terms in back reflection can be found. The zeroth order forward travelling wave between the $j_{\max}$th and $(j_{\max} + 1)$th layers, propagates to the $(j_{\max} + 1)$th layer, reflects there and propagates back to the half-way point giving

$$a^{-(1)}_{j\max} = P^{-}M^{-+}P^{+}a^{+(0)}_{j\max}. \tag{5.39}$$

In subsequent spaces closer to the surface

$$a_j^{-(1)} = P^{-}M^{-+}P^{+}a_j^{+(0)} + P^{-}_{RFS}a^{-(1)}_{j+1}, \tag{5.40}$$

the additional term coming from transmission of the first order backward travelling wave in the previous space. Thus we can determine $a_1^{-(1)}$ and hence

$$a_0^{-(1)} = Q^{\mathrm{IV}}(1)a_1^{-(1)} + Q^{\mathrm{III}}(1)a_0^+ \tag{5.41}$$

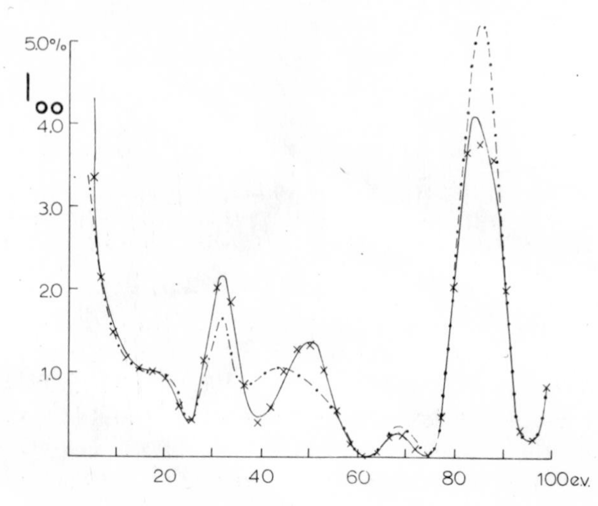

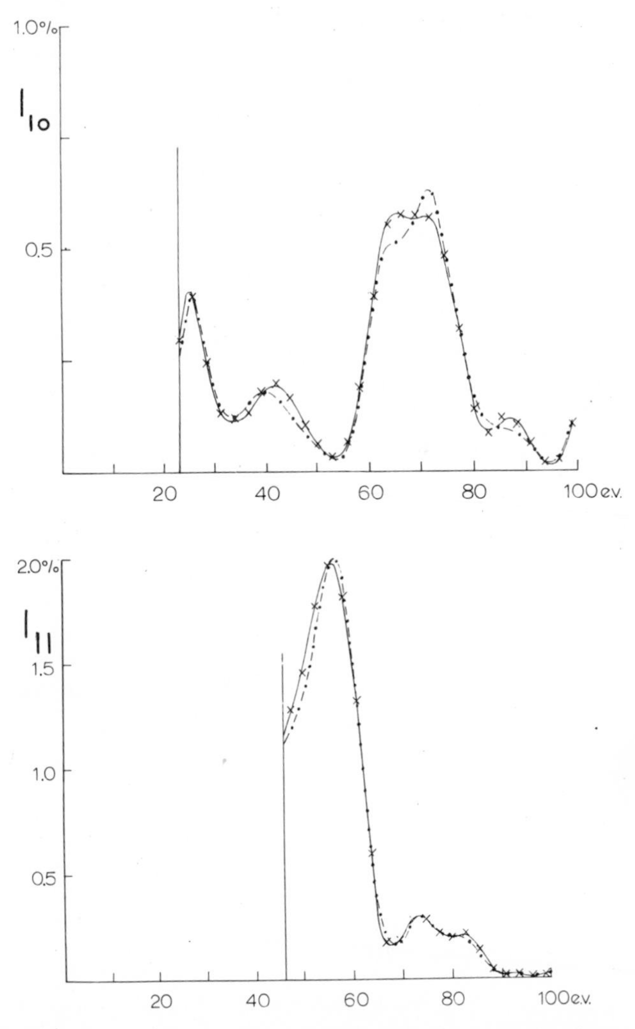

FIGURE 5.8. Comparison of renormalised forward scattering perturbation theory calculations for one pass (chain curve) and for two passes (crosses) with Bloch wave calculations (full curve) for normal incidence on a copper (100) surface. Thirteen beams were used in the calculation.

The first order *RFS* scheme is rapidly executed once transmission and reflection matrices of the layers have been found. Figure 5.8 taken from Pendry (1971c), where the *RFS* scheme was originally described, shows comparisons of first order *RFS* perturbation theory with results of a Bloch wave calculation.

Once the scattering factors of the layer were known, the first order *RFS* calculations took a further 2 seconds on an ICL machine, as opposed to 40 seconds for the Bloch wave calculations. Thirteen beams were included in the calculations.

Agreement with the accurate Bloch wave calculations is good, especially for non-specular beams. They emerge at a more shallow angle from the surface and see effectively a smaller number of layers. Hence the multiple scattering problem is less critical.

The scheme can be extended to higher order with simplicity. The incident wave remains unchanged, therefore the second order contribution to it is zero.

$$a_0^{+(2)} = 0. \tag{5.42}$$

Between the first and second layers the second order contribution is given by reflection of the first order wavefield at the barrier

$$a_1^{+(2)} = Q^{\mathrm{II}}(1)a_1^{-(1)}. \tag{5.43}$$

Between subsequent pairs of layers the second order contribution is composed partly of first order waves reflected at a layer, and partly of second order waves propagating from the previous space.

$$a_j^{+(2)} = P^+M^{+-}P^-a_j^{-(1)} + P_{RFS}^+a_{j-1}^{+(2)}. \tag{5.44}$$

Calculation of second order terms is carried into the crystal until the wavefield is negligible, then calculation of third order terms begins

$$a_{j\max}^{-(3)} = P^-M^{-+}P^+a_{j\max}^{+(2)}, \tag{5.45}$$

subsequently

$$a_j^{-(3)} = P^-M^{-+}P^+a_j^{+(2)} + P_{RFS}^-a_{j+1}^{-(3)} \tag{5.46}$$

until

$$a_0^{-(3)} = Q^{\mathrm{IV}}(1)a_1^{-(3)}. \tag{5.47}$$

The calculation sweeps into the crystal evaluating even order contributions to forward travelling waves, and out again evaluating odd order contributions to backward travelling waves. Each complete process increases the order of perturbation theory by two and we shall refer to it as a 'pass'.

Figure 5.8 also shows intensities when corrections from the second pass have been made. Agreement with the Bloch wave calculation is very close: much closer than can ever be expected between theory and experiment. We have succeeded in our aim of finding a convergent perturbation expansion and reduced the inter-layer scattering to so trivial a computational problem that the most time consuming part of the calculation is now finding of the transmission and reflection coefficients.

In our simple analogy for renormalisation of perturbation theory expansions, the effect of renormalising for the presence of liquid could be understood simply in terms of scattering in vacuo plus a change of wave vector. Modifications brought about to first order electron scattering theory by renormalisation of forward scattering processes can also be given a simple interpretation. In un-renormalised theory an incident wave amplitude $a_{0\mathbf{g}}^{+}$ propagates through each successive layer by a factor of

$$(P^{+})_{\mathbf{gg}}^{2} = \exp(i\mathbf{K}_{\mathbf{g}}^{+}\cdot\mathbf{c}). \tag{5.48}$$

After being reflected into beam $\mathbf{g}'$ it propagates outwards through each successive layer by a factor of

$$(P^{-})_{\mathbf{g}'\mathbf{g}'}^{2} = \exp(-i\mathbf{K}_{\mathbf{g}'}^{-}\cdot\mathbf{c}). \tag{5.49}$$

Thus contributions to total scattered amplitudes by two successive layers differ in phase by a factor

$$\exp\left[i(\mathbf{K}_{\mathbf{g}}^{+} - \mathbf{K}_{\mathbf{g}'}^{-})\cdot\mathbf{c}\right]. \tag{5.50}$$

Bragg peaks are seen when

$$(\mathbf{K}_{\mathbf{g}}^{+} - \mathbf{K}_{\mathbf{g}'}^{-})\cdot\mathbf{c} = 2\pi \times \text{integer}. \tag{5.51}$$

In the *RFS* scheme waves propagate through successive layers by matrix multiplication by P_{RFS}^{+}, rather than by a simple phase factor. The simple picture is retrieved only when the incident wave amplitudes happen to constitute an eigenvector of P_{RFS}^{+}:

$$P_{RFS}^{+}b(\mathbf{k}_{j}^{+}) = \exp(i\mathbf{k}_{j}^{+}\cdot\mathbf{c})b(\mathbf{k}_{j}^{+}). \tag{5.52}$$

In the more general case we must decompose a_{0}^{+} in terms of eigenvectors of P_{RFS}^{+},

$$a_{0}^{+} = \sum_{j} B(\mathbf{k}_{j}^{+})b(\mathbf{k}_{j}^{+}). \tag{5.53}$$

In the renormalised forward scattering scheme the incident plane wave varying in phase by a constant amount from layer to layer is replaced by a linear combination of sets of waves. Each set changes phase from layer to layer in its own way according to (5.52). What renormalisation has done is not only change the wave-vector of the incident beam, as was the case in our optical analogy, but introduce a variety of wave-vectors into the situation. Waves travelling out of the crystal can be similarly decomposed in terms of eigenvectors of P_{RFS}^{-},

$$P_{RFS}^{-}b(\mathbf{k}_{j}^{-}) = \exp(-i\mathbf{k}_{j}^{-}\cdot\mathbf{c})b(\mathbf{k}_{j}^{-}). \tag{5.54}$$

Reflection peaks are seen when any pair of forward and backward travelling wave-vectors satisfy the Bragg condition for constructive interference

$$(\mathbf{k}_j^+ - \mathbf{k}_i^-)\cdot\mathbf{c} = 2\pi \times \text{integer}. \tag{5.55}$$

In this way the *RFS* scheme can generate complicated structure even to first order in back scattering.

Reference to Section IVB reveals that eigenvalues $P_{RFS}^{\pm}$ are the normal modes (Bloch waves) calculated in the approximation that back scattering is neglected.

Successful as the *RFS* scheme is, it can break down under extreme values of parameters. There are two assumptions important to the theory: one is that in (5.38) the zeroth order wavefield does grow smaller deeper inside the crystal. The other is that successive passes do give convergent results.

First consider convergence of the zeroth order wavefield with depth. The zeroth order wavefield only goes to zero at large depths if all eigenvalues of P_{RFS}^{+} are of modulus less than unity, because then whatever incident vector we apply successively higher powers of P_{RFS}^{+} to, in equation (5.38), the result must always grow smaller. In the presence of absorption and providing that back-scattering is weak compared with forward scattering the eigenvalues of P_{RFS}^{+} are always of modulus less than unity, but at very low energies below 10 eV where absorption is often weak, and back scattering comparable with forward scattering, it sometimes happens that an eigenvalue grows too large, and the method fails.

The second possible cause of failure, that increasing the number of passes made in and out of the crystal may not give successively smaller contributions, occurs if back-scattering is too strong relative to absorption. An electron can become almost trapped between pairs of layers and failure of the perturbation series results. Again this type of breakdown can be expected to be confined to the region below 10 *eV*.

At energies above 10 eV the *RFS* perturbation method has proved reliable, accurate, and fast. Its generality of application is second only to that of the matrix doubling method and it is our preferred method for fast reliable calculations on transition and noble metals.

VE Kinematic theory and the averaging postulate

Perturbation methods speed up calculations considerably, to the extent that for most simple surfaces it is easier to make a calculation than to do an experiment. Even so methods of sufficient sophistication to give an accurate account of spectra involve more than trivial calculation. They could not be worked through by hand, nor even on a small desk calculator. Only one perturbation scheme, the kinematic (single scattering) theory, is simple enough to give rise to trivial calculations and that scheme we know is not capable of accounting for spectra from most surfaces.

Lagally, Ngoc, and Webb (1971) reasoned that experimental curves, multiple scattering in nature as they are, contain *contributions* from kinematic processes. If somehow the theoretically simple kinematic component could be extracted, it may be that an experimentalist would find it preferable to make the more sophisticated and detailed experiments needed to do this, whatever they might be, rather than use more sophisticated theories to interpret the simple data.

There are in existence several averaging schemes more or less derivative from that of Lagally *et al.* The most important of these other schemes is the energy-averaging method of Tucker and Duke (1972), which is concerned specifically with the overlayer problem and is most appropriately considered in Chapter VII. The scheme of Lagally *et al.* has more general applicability and we shall concern ourselves with it in some detail.

Beyond the elementary observation that kinematic Bragg peaks occur near the Bragg condition, there is no way of distinguishing between single and multiple scattering structure in an experimental intensity/energy curve. Even those peaks near the Bragg condition have multiple scattering contributions to their intensity. Some more sophisticated way of extracting the kinematic component must be found.

The kinematic amplitude of beam $\mathbf{g}$ reflected by the crystal is given by equation (3.28)

$$R\delta_{\mathbf{g}0} + \frac{T^+(\mathbf{K}_0^+)T^-(\mathbf{K}_\mathbf{g}^-)M(\mathbf{K}_\mathbf{g}^-, \mathbf{K}_0^+)}{1 - \exp\left[i(\mathbf{K}_0^+ - \mathbf{K}_\mathbf{g}^-)\cdot\mathbf{c}\right]}. \tag{5.56}$$

We have assumed unit amplitude of incident wave. R is the reflection coefficient of the surface barrier and T^-, T^+ the transmission coefficients for back and forward travelling waves incident on the barrier. M is the kinematic expression for the reflection coefficient of a layer of ion cores defined by equation (3.23). M is proportional to the ion-core scattering factor for that angle of scattering, $\theta_\mathbf{g}$,

$$M(\mathbf{K}_\mathbf{g}^-, \mathbf{K}_0^+) = -\frac{it(\theta_\mathbf{g})}{A|K_{\mathbf{g}z}^-|}. \tag{5.57}$$

$\mathbf{K}_\mathbf{g}^\pm$ are the wave vectors with which various beams propagate inside the crystal. The denominator in (5.56) determines where peaks are to occur. Other terms in the equation, being generally much less rapidly varying functions of diffraction conditions, merely serve to modify the height of the Bragg peak, as we saw in Section IIIA. Peak positions are determined by

$$(\mathbf{K}_0^+ - \mathbf{K}_\mathbf{g}^-)\cdot\mathbf{c} = 2\pi \times \text{integer}$$

a condition that depends only on the difference between $\mathbf{K}_0^+$ and $\mathbf{K}_\mathbf{g}^-$. If we define

$$\mathbf{S} = \mathbf{K}_0^+ - \mathbf{K}_\mathbf{g}^- \tag{5.58}$$

and plot intensities not as functions of energy, but of $\mathbf{S}\cdot\mathbf{c}$, then changing incident conditions does not change kinematic peak positions in this plot, only intensities of the peaks. On the other hand positions of peaks originating from the much more complicated multiple scattering processes are in general dependent on both $\mathbf{K}_0^+$ and $\mathbf{K}_\mathbf{g}^-$ separately, and in a plot of $I(\mathbf{S}\cdot\mathbf{c})$ their positions will change according to incident conditions under which the curve is taken. Figure 5.9 shows how a variety of configurations of incident wave vectors can keep $\mathbf{S}$ constant.

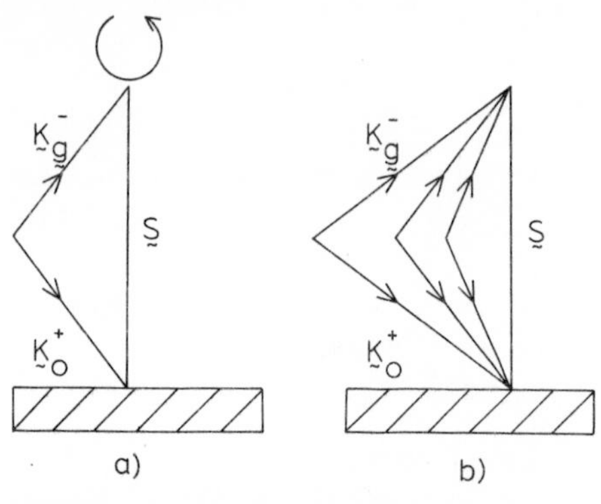

FIGURE 5.9. A variety of incident conditions can keep vector **S** constant, for example when scattering is to the specular beam (a) rotation of the crystal about the surface normal, (b) varying the angle of incidence.

If the postulates are made that

(i) Multiple scattering peaks in a plot of $I(\mathbf{S}\cdot\mathbf{c})$ are as likely to occur for one value of $\mathbf{S}\cdot\mathbf{c}$ as another when all possible incident conditions illustrated in Fig. 5.8 are considered;

(ii) Multiple scattering makes no systematic change in the simple kinematic Bragg peaks, e.g. that they are as likely to be increased in intensities as decreased;

then in an averaging of curves $I(\mathbf{S}\cdot\mathbf{c})$ for all possible incident conditions, $< I(\mathbf{S}\cdot\mathbf{c}) >$, multiple scattering peaks average to a constant background and modifications to kinematic peaks average to zero. The only remaining problem is to perform enough experiments to give good statistical elimination of multiple scattering contributions.

Conditions (i) and (ii) are postulates, and we shall try to derive them from theoretical considerations in the next section. In the meantime we shall describe experimental evidence brought forward by Lagally *et al.* in support of (i) and (ii). They carried out the averaging procedure on the (111) surface of nickel and their results are presented in Fig. 5.9 together with an average over the same incident conditions of a kinematic calculation of intensities (Lagally *et al.*, 1972). In the kinematic calculation the constant potential between muffin tin spheres was adjusted to give the best fit of peak positions and widths in the averaged curves,

$$V_{0r} = -18\,eV, \qquad V_{0i} = -9\,eV, \tag{5.59}$$

and the ion-core scattering factors were taken from a first principles calculation by Fink (unpublished) corrected for thermal effects but scaled to give agreement of intensities at

$$\mathbf{S}\cdot\mathbf{c}/2\pi = 5.$$

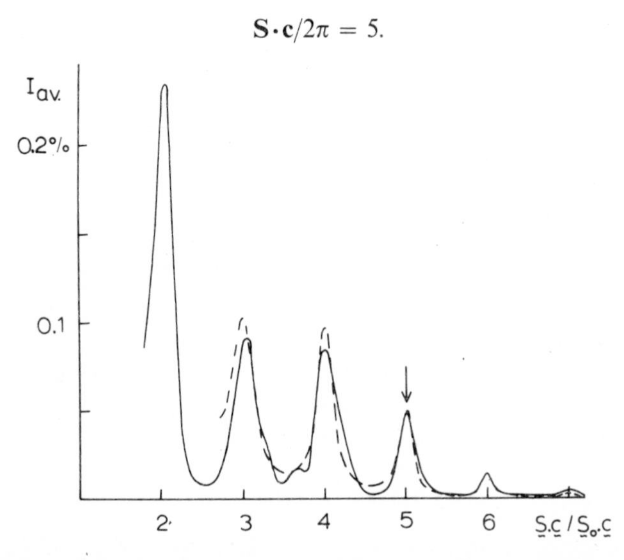

FIGURE 5.10. Experimental $I_{00}(\mathbf{S}\cdot\mathbf{c})$ curve for a nickel (111) surface, averaged over a variety of incident conditions at a temperature of 423°K. Comparison is made with the same average of a kinematic calculation using parameters as described in the text, and corrected for thermal effects.

Excellent agreement is seen in Fig. 5.10 between theory and experiment. Because of the adjustable parameters used in calculating the kinematic average, the test confirms only that averaged data has a kinematic *form*. In particular it is to be noted that the large value of -9 eV used for V_{0i} is larger by a factor of three than that estimated by Pendry (1969c) by fitting multiple scattering calculations to un-averaged data. This point will be important when we develop a theory of averaged data in the next section.

As an indication of quality of the statistical averaging away of multiple scattering contributions Lagally *et al.* averaged their data over two separate ranges of incident conditions and in Fig. 5.11 we can see that in each case

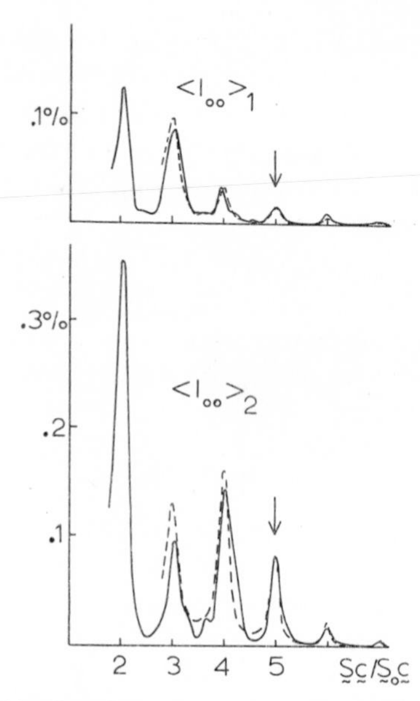

FIGURE 5.11. Experimental $I_{00}(\mathbf{S}\cdot\mathbf{c})$ for a nickel (111) surface averaged over two different ranges of incident conditions, at a temperature of 423°K, and compared with the corresponding averages of a kinematic calculation.

good agreement with the kinematic average is achieved, using the same parameters. They observed further that averaging over rotations about the normal was less effective in reducing data to kinematic form than varying the scattering angle.

VF Theory of averaged LEED data

Before attempting to make a theoretical justification of the postulate that data averaged at constant $\mathbf{S}\cdot\mathbf{c}$ retains only kinematic contributions we should investigate whether the postulate is consistent with the general properties of LEED spectra. Our analysis will follow closely a paper by Pendry (1972). The key to the matter is that multiple scattering should produce no systematic effects in the curves, because after averaging, systematic effects will leave residual structure in intensities plotted against $\mathbf{S}\cdot\mathbf{c}$. There are three important quantities that must be free of systematic multiple scattering effects: intensities of peaks near the Bragg condition, values of $\mathbf{S}\cdot\mathbf{c}$ at which peaks occur, and widths of peaks. These three quantities are directly related in kinematic theory respectively to the ion core t-matrix, the real part of the constant potential between ion-cores V_{0r}, and the imaginary part of the constant potential, V_{0i}. We shall examine each of these quantities in turn.

Firstly modifications to intensities of peaks in reflectivity from inclusion of multiple scattering processes: several systematic effects can be detected, all tending to reduce intensities of peaks. When multiple scattering is strong electrons that multiply scatter follow a more tortuous path in the crystal than in the single scattering case. In consequence of the longer path length in the crystal intensities are reduced. We have already discussed the effect in Chapter III and Fig. 3.18 shows how multiple scattering increases attenuation suffered by an electron inside a crystal.

Another influence at work is 'beam sharing'. Suppose that kinematic processes are producing a strong peak in the 00 beam. As the 00 beam passes through layers on the way out of the crystal, strong ion-core scattering will remove flux from the 00 beam producing non-kinematic peaks in other beams. The scattered flux is effectively lost because on averaging it contributes only to the uniform background. Yet another influence related to the last one occurs when flux is scattered out of the incident beam into other beams not necessarily in Bragg reflecting conditions, reducing flux available in the incident beam for kinematic processes.

Three systematic effects work to reduce multiple scattering reflectivities therefore we must expect averaged multiple scattering, i.e. experimental curves, to be much lower in intensity than averages of kinematic calculations. This conclusion is not incompatible with the findings of Lagally *et al.* because of their adjustment of overall intensities by a scaling factor.

Secondly there is the possibility of systematic shifts in peak positions which can be interpreted as ion cores acting as either net attractive or net repulsive scatterers. At low energies the effective potential of the ion cores, the pseudopotential, is weak because of cancellation in the core region as we discussed in Section IIC, but at higher energies, above say 100–200 eV, the electron begins to penetrate the strongly negative core region and its average energy in the crystal is accordingly lowered. Therefore ion-cores can cause a systematic shift of structure to lower energies. The higher the incident energy the greater the shift to be expected.

Thirdly there are the peak widths. Equation (2.11) deduced from the uncertainty principle states that peaks can never be narrower than $2|V_{0i}|$, whether formed by multiple scattering processes or otherwise. Further, we can verify directly from (5.56) that kinematic peaks have the limiting width: no peaks are more narrow than those formed by kinematic processes. Thus the requirement on multiple scattering modifications to kinematic peaks are very stringent. If they push peaks too far away from Bragg conditions, even if the average displacement is zero, the averaged peak width will be broadened and be too wide to compare with narrow kinematic peaks. In fact displacements of peaks away from the Bragg condition of the order of 5 to 10 eV are not uncommon and we are forced to conclude that multiple scattering causes averaged peaks to be systematically broadened in comparison with kinematic peaks. We recall that the value of V_{0i} needed to fit the averaged data of Fig. 5.9 was three times as big as the generally accepted value.

The presence of these systematic effects which in the case of intensities and peak widths appear to be strong, implies that averaged data cannot be kinematic. On the other hand Fig. 5.9 convincingly demonstrates that the averaged data have kinematic form. We conclude that any theory we derive must show averaged data to be *quasi-kinematic*, the intensities being given by averaging (5.6) with modified values substituted for the three parameters, t corresponding to ion core scattering, V_{0r} the shift in zero of energy between ion cores, and V_{0i} the absorption. Hopefully the theory will give a simple prescription for these modified parameters.

Some of the scattering processes must be examined in detail and checked to find which of them do in fact average to a constant contribution. For simplicity we shall leave off the surface barrier,

$$R = 0, T(\mathbf{K}_{\mathbf{g}}^{+}) = T(\mathbf{K}_{\mathbf{g}}^{-}) = 1. \tag{5.60}$$

We can represent kinematic events in a diagram shown in Fig. 5.12.

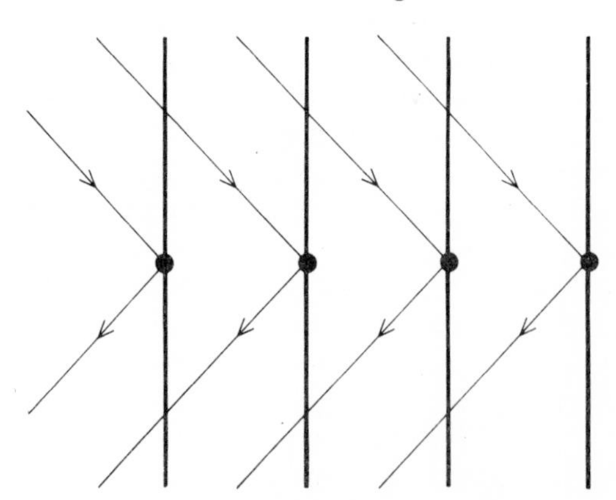

FIGURE 5.12. Contributions to kinematic amplitudes.

Some second order multiple scattering events are shown in Fig. 5.13 and in Section C we saw how to write down contributions to the scattered amplitudes:

$$\sum_{j=2}^{\infty} \exp(-\tfrac{1}{2}i\mathbf{K}_{\mathbf{g}}^{-}\cdot\mathbf{c})\, M(\mathbf{K}_{\mathbf{g}}^{-}, \mathbf{K}_{\mathbf{g}'}^{-}) \exp[-i(j-1)\mathbf{K}_{\mathbf{g}'}^{-}\cdot\mathbf{c}]$$
$$\times\; M(\mathbf{K}_{\mathbf{g}'}^{-}, \mathbf{K}_{0}^{+}) \exp[i(j-\tfrac{1}{2})\mathbf{K}_{0}^{+}\cdot\mathbf{c}],$$

where we have used the single scattering expression for scattering by a layer.

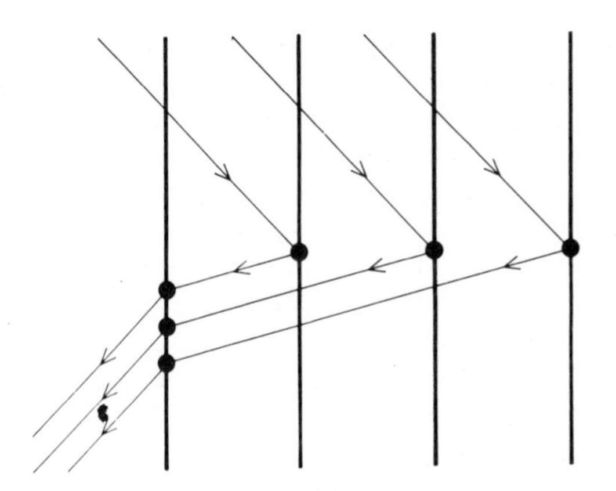

FIGURE 5.13. Second order multiple scattering contributions.

Processes (5.61) sum to give a peak when

$$(\mathbf{K}_{\mathbf{g}}^{+} - \mathbf{K}_{\mathbf{g}'}^{-})\cdot\mathbf{c} = \mathbf{S}'\cdot\mathbf{c} = 2\pi \times \text{integer}, \tag{5.62}$$

in contrast with the kinematic condition. When angles of incidence are changed keeping $\mathbf{S}\cdot\mathbf{c}$ constant, $\mathbf{S}'\cdot\mathbf{c}$ will not be constant and secondary peaks associated with (5.58) will change their positions averaging to a constant or at least a slowly varying background, if incident conditions are varied over a wide enough range. Interference terms between kinematic processes and second order contributions of Fig. 5.13 average to zero because the phase of (5.61) varies with incident conditions.

The sort of multiple scattering considered above does average away to a constant background in accordance with the postulate, but if there were any

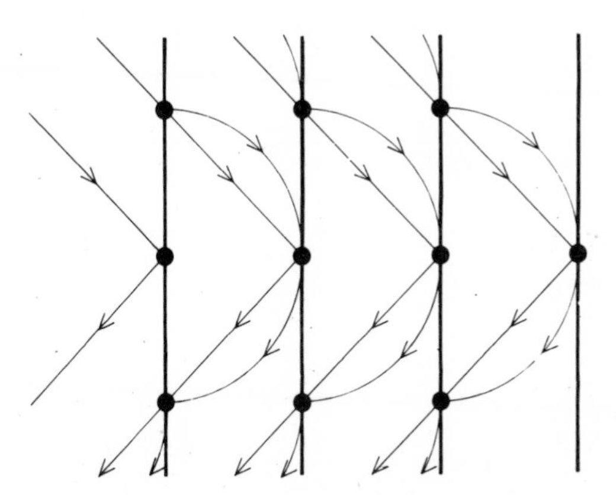

FIGURE 5.14. Zero angle scattering contributions to quasi-kinematic intensities.

higher order diagrams that also gave peaks at the kinematic condition, the postulate would be spoilt. There are diagrams that do just this. In Fig. 5.14 we show a set of processes in which the electron is scattered by the ion cores in the same direction as it was going originally, but with a change of phase. In these diagrams the electron follows the same path through the crystal as in the kinematic case and all the phase factors coming about because of propagation, are the same as the kinematic ones. Therefore peaks due to these processes are determined by $\mathbf{S}\cdot\mathbf{c}$ as in the kinematic case and do not average to a constant.

Thus in making our theory of averaged data we must sum all zero angle scattering contributions. What appears to be a complicated problem can in fact be solved quite easily. For the moment we are interested only in the zero angle scattering factor of the ion cores,

$$t(E, \theta = 0) = -2\pi\kappa^{-1} \sum_{\ell} (2\ell + 1) \sin(\delta_\ell) \exp(i\delta_\ell) \times P_\ell(\cos(\theta = 0)). \tag{5.63}$$

We observe that potentials that scatter only through zero angles must be constant inside the unit cell, because otherwise they would diffract electrons through finite angles. The scattering by a constant potential, V_f, per unit cell of volume Ω is

$$\Omega\, V_f. \tag{5.64}$$

Therefore by equations (5.63) and (5.64) we deduce that correcting for zero angle scattering corresponds to introducing a constant potential

$$V_f(E) = t(E, \theta = 0)/\Omega. \tag{5.65}$$

Thus all the complicated diagrams of Fig. 5.14 are equivalent to the kinematic diagrams of Fig. 5.12, provided we modify wave-vectors of beams because of the additional effective potential needed to account for zero angle scattering

$$\tilde{\mathbf{K}}_{\mathbf{g}}^{\pm} = [(\mathbf{k}_{0\|} + \mathbf{g})_x, (\mathbf{k}_{0\|} + \mathbf{g})_y, \pm\tilde{\kappa}_{\mathbf{g}}], \tag{5.66}$$

$$\tilde{\kappa}_{\mathbf{g}} = (2E - 2V_0 - 2V_f - |\mathbf{k}_{0\|} + \mathbf{g}|^2)^{\frac{1}{2}}. \tag{5.67}$$

Inclusion of zero angle scattering goes some way towards correcting the deficiencies we have observed. The averaged data still have kinematic form, but inclusion of V_f in (5.67) lowers the zero of energy because of the real part of $t(\theta = 0)$, meeting the second objection of systematic shifts in peak energies. The imaginary part of V_f has the same effect as increasing absorption, broadening peaks to

$$\Delta E_{ave.} = 2|V_{0i} + V_{fi}|, \tag{5.68}$$

meeting objection three and decreasing peak intensities, going some way towards meeting objection one.

If ion-core scattering factors are known it is easy to calculate V_f. Figure 5.15 shows V_f for nickel.

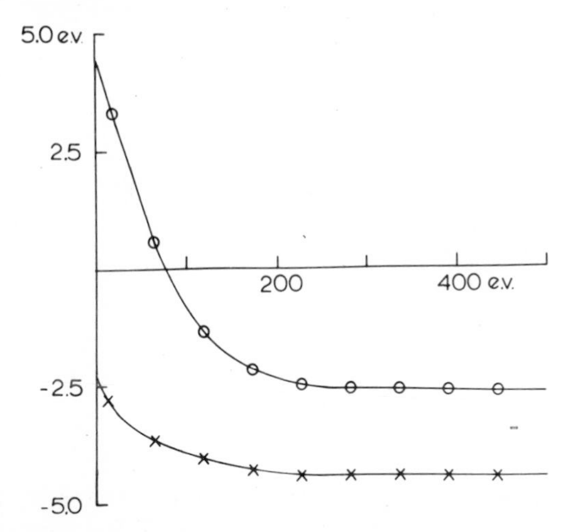

FIGURE 5.15. Corrections to the potential in nickel to take account of zero angle scattering; —o— real part, —×— imaginary part.

We must compare the absorption parameter Lagally *et al.* used to fit their curves not with V_{0i}, but with $V_{0i} + V_{fi}$. From Fig. 5.15 we deduce that typically

$$|V_{0i} + V_{fi}| \simeq |-3-5| = 8 \text{ eV}, \tag{5.59}$$

to be compared with 9 eV from Lagally *et al.* Corrections to the real part of V_0 are relatively smaller. Taking the value of

$$V_{0r} = -18 \text{ eV}, \tag{5.70}$$

found by Andersson and Pendry (1972) in the 0–100 eV range, gives

$$V_{0r} + V_{fr} = -18 - 2.5 = -20.5 \text{ eV}. \tag{5.71}$$

again in good agreement with the fitted value of -18 eV.

In Fig. 5.16 we show how much different curves calculated with quasi-kinematic parameters corrected for zero angle scattering are from those calculated with the straightforward kinematic parameters. The new parameters have the expected effects of lowering intensities and broadening peaks.

It was fortunate that zero angle scattering diagrams could be so simply corrected for, and certainly they seem to account for peak widths and positions satisfactorily. No comparison between averaged experimental and quasi-kinematic intensities has yet been made, and we must ask whether there

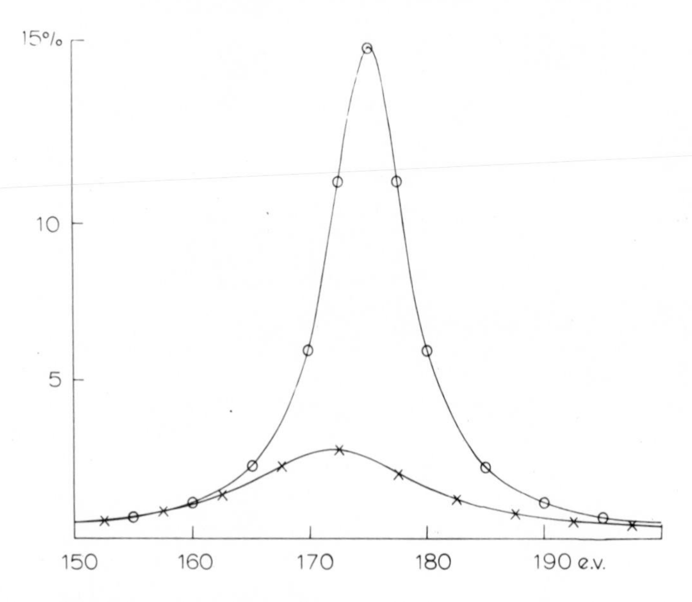

FIGURE 5.16. Calculations of the 00 beam for normal incidence on a nickel (001) surface at $T = 0°K$: —⊖— kinematic theory, —×— quasi-kinematic theory using parameters corrected for zero angle scattering.

are even more multiple scattering diagrams that can affect intensities by modifying the scattering factor. That would leave peak widths and positions in agreement with experiment. Unfortunately several such processes can be picked out. One is shown in Fig. 5.17. Its contribution can be found to be

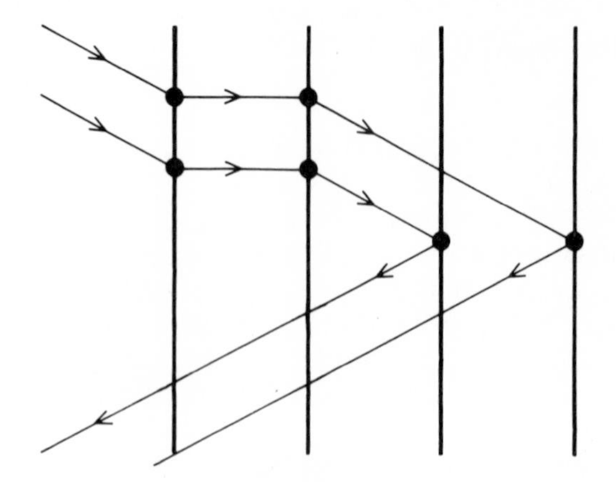

FIGURE 5.17. Contributions to quasi-kinematic intensities that change the effective scattering factor.

$$\sum_{j=3}^{\infty} \exp\left[\tfrac{1}{2}i\mathbf{K}_0^+ \cdot \mathbf{c} + i\mathbf{K}_{\mathbf{g}'}^+ \cdot \mathbf{c} + (j-2)i\mathbf{K}_0^+ \cdot \mathbf{c} - (j-\tfrac{1}{2})i\mathbf{K}_{\mathbf{g}}^- \cdot \mathbf{c}\right]$$
$$\times\ M(\mathbf{K}_{\mathbf{g}}^-, \mathbf{K}_0^+)M(\mathbf{K}_0^+, \mathbf{K}_{\mathbf{g}'}^+)M(\mathbf{K}_{\mathbf{g}'}^+, \mathbf{K}_0^+), \quad (5.72)$$

which can be rearranged in the form

$$\sum_{j=1}^{\infty} \tilde{M}(\mathbf{K}_{\mathbf{g}}^-, \mathbf{K}_0^+)\exp\left[(j-\tfrac{1}{2})i\mathbf{K}_0^+ \cdot \mathbf{c} - (j-\tfrac{1}{2})i\mathbf{K}_{\mathbf{g}}^- \cdot \mathbf{c}\right], \quad (5.73)$$

$$\tilde{M}(\mathbf{K}_{\mathbf{g}}^-, \mathbf{K}_0^+) = M(\mathbf{K}_{\mathbf{g}}^-, \mathbf{K}_0^+)M(\mathbf{K}_0^+, \mathbf{K}_{\mathbf{g}'}^+)M(\mathbf{K}_{\mathbf{g}'}^+, \mathbf{K}_0^+)$$
$$\times \exp(-2i\mathbf{K}_{\mathbf{g}}^- \cdot \mathbf{c} + i\mathbf{K}_{\mathbf{g}'}^+ \cdot \mathbf{c} + i\mathbf{K}_0^+ \cdot \mathbf{c}) \quad (5.74)$$

(5.73) has the same form as the kinematic expression and averages to give peaks at the kinematic condition, therefore $\tilde{M}$ must be added to the scattering factor M when calculating the total effective scattering remaining after averaging. It is not hard to think of other diagrams that change the effective scattering factor, including corrections for multiple scattering within the layer.

As yet no simple way of making these corrections has been found beyond direct calculation of the processes. It could be argued that the corrections will not be large because of the decaying exponential factor in (5.74), but this must be verified by direct calculation and comparison with experiment.

At the time of writing the averaging method is at an encouraging stage but has not yet been explored sufficiently thoroughly to be able to get a theory of averaged data that is complete, and has the required simplicity of form.

Chapter 6

Temperature Effects

VIA The experimental situation

When the temperature of a crystal is raised, two changes become apparent in the diffraction pattern: firstly intensities of individual diffraction spots decrease, and secondly the intensity of electron flux collected between spots grows stronger. Typical examples showing reduction in beam intensities are seen in Fig. 6.1. The effect is one of general reduction in intensities: greater at high energies than low, but otherwise without drastic interference with relative structure of the curve.

It is the thermally induced vibrations of atoms that are responsible. First discussions of the effect assumed electrons to scatter only weakly from ion-

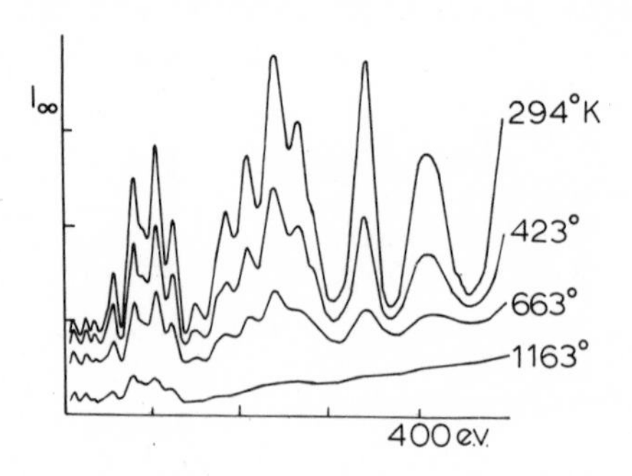

FIGURE 6.1. (00) reflection from a niobium (100) surface, 8° from normal incidence (Tabor and Wilson, 1970). Curves for various temperatures are shown displaced relative to one another. Intensities have not been corrected for variation of incident current with energy, nor for accidental inclusion of thermal diffuse background intensity, at high energies.

cores so that kinematic theory could be used to describe diffraction. Even though this assumption is incorrect for most materials, as we have seen, we shall adopt it for the time being as forming a simple framework within which effects can be described.

It will be recalled that the kinematic formula for reflected amplitudes is,

$$A(\mathbf{k}', \mathbf{k}) \propto \sum_j \exp(-i\mathbf{k}' \cdot \mathbf{P}_j)\, t(\mathbf{k}'\mathbf{k}) \exp(i\mathbf{k} \cdot \mathbf{P}_j). \tag{6.1}$$

$\mathbf{k}$ is the wave vector of the incident wave, $\mathbf{k}'$ that of the reflected wave. $\mathbf{P}_j$ is the position of the jth atom and we consider the case of only one atom per unit cell, for simplicity. For the same reason we have omitted the surface barrier. When the lattice is heated atoms are set in motion, vibrating about a mean position $\mathbf{P}_j$, if we neglect thermal expansion. It is a good approximation to assume that diffracting electrons move so quickly compared with thermal motions of atoms, typically: faster by a factor of 10^4, that atoms can be thought of as having displacements unchanged during the process of diffraction. This is the Born-Oppenheimer approximation and corresponds to neglecting energy loss to phonons. Typical energy losses would be $\ll 0.1$ eV and few LEED experiments can resolve energies on this scale. The reflected intensity follows from (6.1),

$$I(\mathbf{k}', \mathbf{k}) = \left|\frac{k'_z}{k_z}\right| |A(k', k)|^2 \propto \sum_{u \cdot v} |t(\mathbf{k}'\mathbf{k})|^2 \left|\frac{k'_z}{k_z}\right|$$

$$\times\, exp[i(\mathbf{k} - \mathbf{k}') \cdot (\mathbf{P}_u - \mathbf{P}_v)] \exp[i(\mathbf{k} - \mathbf{k}') \cdot (\Delta\mathbf{P}_u - \Delta\mathbf{P}_v)]. \tag{6.2}$$

$\Delta\mathbf{P}_v$ represents a displacement from the mean atomic position, due to thermal motion.

Although atoms are effectively stationary on the time scale of the diffraction process, their motion is still sufficiently rapid for the detector, Faraday cup or spot-photometer, to register intensities averaged over displacements $\Delta\mathbf{P}_v$.

$$\langle I \rangle_T \propto \sum_{uV} |t(\mathbf{k}'\mathbf{k})|^2 \left|\frac{k'_z}{k_z}\right| \exp[i(\mathbf{k} - \mathbf{k}') \cdot (\mathbf{P}_u - \mathbf{P}_V)]$$

$$\times \langle \exp[i(\mathbf{k} - \mathbf{k}') \cdot (\Delta\mathbf{P}_u - \Delta\mathbf{P}_v)] \rangle_T. \tag{6.3}$$

If displacements of the uth atom are uncorrelated with those of the Vth atom,

$$\langle \exp[i(\mathbf{k} - \mathbf{k}') \cdot (\Delta\mathbf{P}_u - \Delta\mathbf{P}_v)] \rangle_T =$$

$$\langle \exp[i(\mathbf{k} - \mathbf{k}') \cdot \Delta\mathbf{P}_u] \rangle_T \langle \exp[-i(\mathbf{k} - \mathbf{k}') \cdot \Delta\mathbf{P}_v] \rangle_T. \tag{6.4}$$

It has been proved by Glauber (1955) that

$$\langle \exp[i(\mathbf{k} - \mathbf{k}') \cdot \Delta\mathbf{P}_u] \rangle_T = \exp[-\tfrac{1}{2}\langle [(\mathbf{k} - \mathbf{k}') \cdot \Delta\mathbf{P}_u]^2 \rangle_T]. \tag{6.5}$$

It is usual to associate thermal fluctuation effects with the scattering factor

$$t(\mathbf{k}'\mathbf{k})\exp(-M_u), \tag{6.6}$$

$$M_u = \tfrac{1}{2}\langle[(\mathbf{k} - \mathbf{k}')\cdot\Delta\mathbf{P}_u]^2\rangle_T. \tag{6.7}$$

$\exp(-2M)$ is the *Debye-Waller factor.* Given a knowledge of the lattice dynamics of the crystal, M_u can be calculated. For a simple crystal with only one atom per unit cell, for atoms not in the immediate surface region, M is approximated by the formula

$$M = \frac{3|\mathbf{k} - \mathbf{k}|^2 T}{2mk_B \Theta^2}, \tag{6.8}$$

where m is the mass of the atom in electron mass units, k_B is Boltzmann's constant in Hartrees/°K, T is the temperature in °K. Θ has the dimensions of temperature and is called the Debye temperature. It is a measure of the compressibility of the solid; the higher Θ the harder the solid.

Modification (6.6) to the scattering factor implies that at finite temperatures all the beams are less intense than at zero temperature. Yet merely by displacing an atom, no intrinsic change has been made in the scattering factor. The true interpretation of (6.6) is that atoms scatter just as much flux, but because the lattice is no longer perfect, only part of the flux is used in a coherent manner to form beams. The rest is scattered incoherently and goes to form the thermal diffuse background intensity that exists between beams at finite temperatures. McKinney, Jones and Webb (1967) have made a careful study of thermal diffuse scattering (TDS) in silver. Fig. 6.2 shows some of their measurements. The important results they arrived at were that the TDS has a slowly varying nature away from the discrete beams. Around each beam the TDS increases, being inversely proportional to the angular

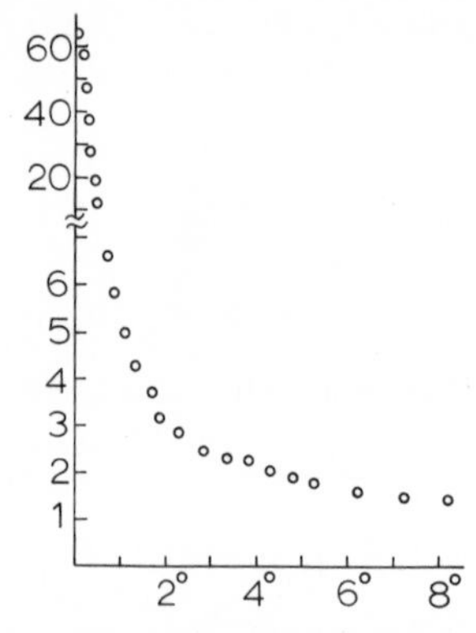

FIGURE 6.2. Variation of thermal diffuse intensity measured as a function of angular deviation from specular reflection for a (111) surface of silver at 92 eV incident energy, temperature 295°K.

distance from that beam. The ratio of the total TDS to that in the discrete beam is approximately $\exp(2M)$. A consequence is that experiments must be corrected for inclusion of TDS in measurements of the discrete beams when thermal effects are large.

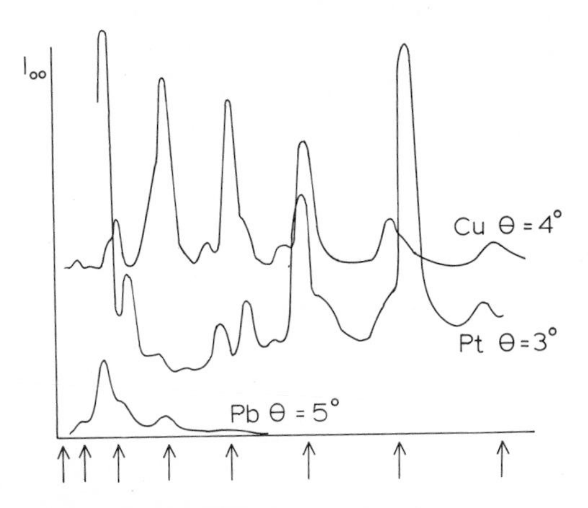

FIGURE 6.3. Intensities of 00 beams reflected from (111) surfaces of platinum, copper and lead measured at room temperature. Because of its very low Debye-Waller factor, lead has a spectrum that cannot be easily observed at energies above 150 eV at room temperature. Each of the curves has been scaled in energy so that kinematic Bragg peaks would fall in the same place. The highest energy is about 700 eV for the copper spectrum.

Returning to intensities of discrete beams: we can appreciate qualitatively what is happening to the curves in Fig. 6.1. As the temperature rises, the Debye Waller factor, $\exp(-2M)$, decreases. Increasing energy also decreases $\exp(-2M)$ because the change in momentum, $|\mathbf{k} - \mathbf{k}'|$, required to reflect a beam becomes larger. It is the Debye Waller factor that limits observation of LEED spectra to the range of energies less than 1000 eV, by weakening beams until they cannot be distinguished from the thermal diffuse background. In especially soft materials the upper limit of incident energy at room temperature is even smaller, for example, in Fig. 6.3 spectra published by Goodman, Farrell and Somorjai (1968) for copper ($\Theta = 343°$ K), platinum ($\Theta = 229°$ K), and lead ($\Theta = 95°$ K).

Two drastic assumptions have gone into derivation of equations (6.6), (6.7): that diffraction processes are kinematic, and that all atoms involved vibrate to the same extent as a bulk atom. We should examine to what extent experimental results conform to our simple theory. A large amount of work has been performed on the temperature dependence of LEED peaks (Barnes, Lagally and Webb, 1968; Reid, 1972; amongst others). The usual procedure is to plot $\ell n(I(E))$ versus T, adjusting the energy at each temperature so that $I(E)$ is always at a peak value (in this way lattice expansion is corrected for) and being careful to subtract any thermal diffuse scattering accidentally included.

Despite dubious assumptions in the theory, plots of $\ell n(I)/T$ are usually quite linear as Fig. 6.4 shows, but gradients give values of Ⓗ that are not

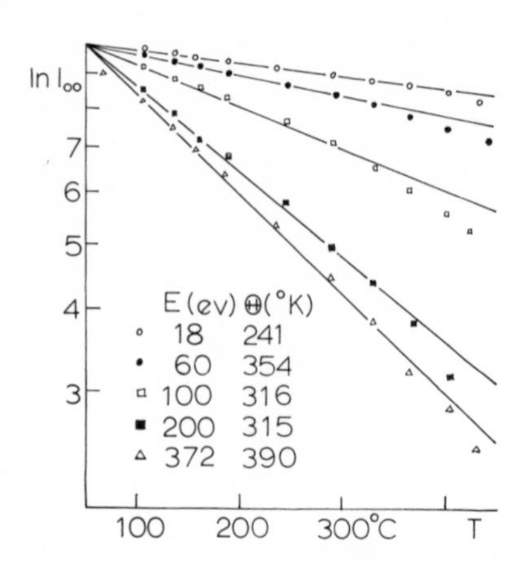

FIGURE 6.4. $\ell n(I)/T$ for a molybdenum 100 surface, measured at 6° from normal incidence by Wilson (1971). The incident beam lies along the [11] azimuth.

constant from one peak to another. It is surprising even that the plots are linear, in view of the fact that some of the peaks on which measurements were made, should not even exist according to the kinematic theory used!

More detailed variation of Ⓗ for molybdenum is shown in Fig. 6.5. Several interesting facts emerge from this figure. At high energies Ⓗ approximates to the bulk value obtained by X-ray scattering, at lower energies, behaviour is more erratic, but with an overall tendency to lower temperatures. There are two possible causes of departure from the bulk value of Ⓗ: multiple scattering nature of diffraction, and enhanced vibration of surface atoms. At high energies, penetration of the crystal increases, reducing influence of enhanced surface vibration. Also multiple scattering at high energies, though always strong, is increasingly confined to a narrow cone about the incident direction because of the large peak in forward scattering amplitudes. Large angle scattering where most of the temperature dependence comes about is weak and simple kinematic theory holds for these events.

At lower energies more complicated behaviour takes over because electrons stay relatively close to the surface, and multiply scatter through larger angles than is possible at high energies. Suppose we decompose the wavefield inside the crystal into Bloch-waves as we did in Chapter III. One of these

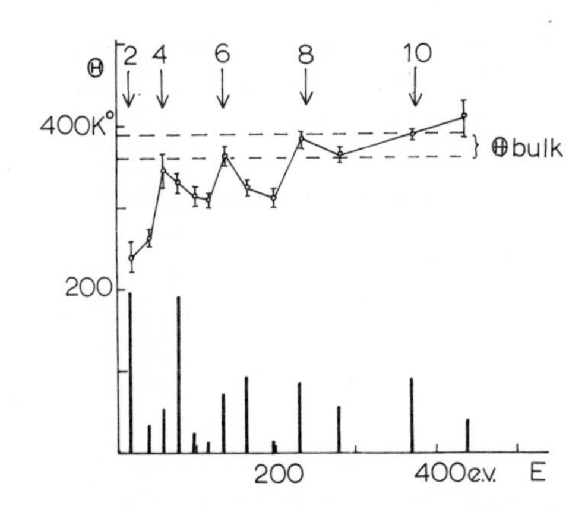

FIGURE 6.5. Variation of Θ for molybdenum as a function of energy. Relative intensities at room temperature of peaks used to determine Θ are shown by solid bars. Vertical lines numbered with even n denote positions where kinematic Bragg peaks are expected (Wilson, 1971).

Bloch waves might involve much multiple scattering, electrons being scattered from many ion cores, and because of the tortuous path they pursue the depth to which they penetrate is restricted. Many scatterings implies that reflected intensities due to this Bloch wave will be much affected by thermal motion, and restriction in penetration makes the electron sample the enhanced vibrations of the surface region. Hence at lower energies peaks are

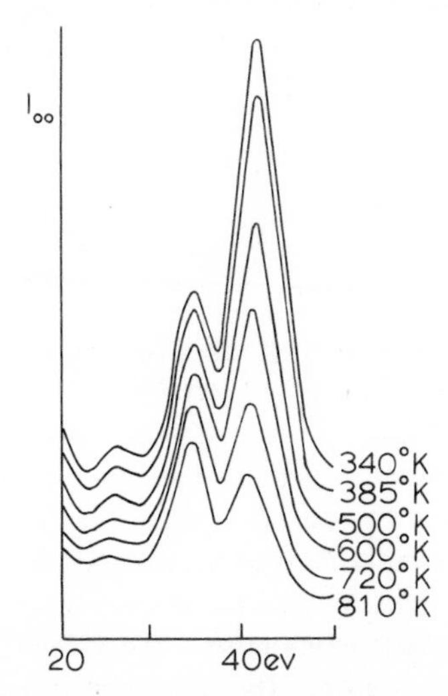

FIGURE 6.6. The 00 beam from a copper (100) surface at 11° from normal incidence in a [110] azimuth. $\ell n(I)/T$ plots yield Debye temperatures of 311°K for the 34 eV peak and 202°K for the 42 eV peak. Absolute scales of intensity have been adjusted for each temperature.

more strongly attenuated by temperature, leading to much lower effective Debye temperatures. We can even go further and say that those Bloch waves in which multiple scattering is especially strong, that is those Bloch waves producing peaks far away from the kinematic condition, will be particularly susceptible to temperature. Reference to Fig. 6.5 shows that lower Debye temperatures are generally found for peaks far from the kinematic condition.

A striking example of anomalous temperature dependence of intensities has been published by Reid (1971). Figure 6.6 shows his results for copper, in which two peaks were observed to vary with intensity. Despite the small energy separation, one peak is much more sensitive to temperature than the other.

From the point of view of using LEED to measure enhanced vibration at surfaces, it is unfortunate that multiple scattering has pronounced influence on temperature effects in the same range of energies as is sensitive to surface vibrations. Yet by selecting peaks close to the kinematic condition we can hope at least to minimise multiple scattering, and if we are to use kinematic theory these are certainly the peaks we must concentrate upon. Using this philosophy Jones, McKinney and Webb (1966) have investigated surface dynamics of silver. By selecting peaks near kinematic conditions they obtained effective Debye temperatures shown in Fig. 6.7. The characteristic

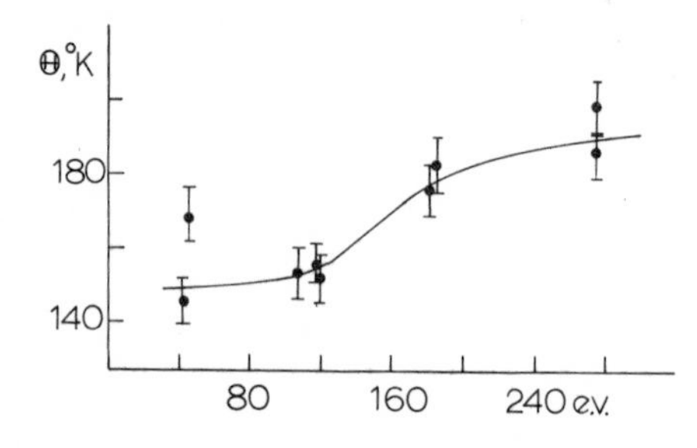

FIGURE 6.7. Measured effective Debye temperature versus energy for a silver (111) surface, 00 beam, 22° from normal incidence (azimuth is unspecified).

fall in Debye temperature is seen at low energies, and by interpreting the low energy limit as dominated by the top layer of atoms they deduced that mean square displacements normal to the surface for the top layer are 1.8 times the bulk value, in good agreement with theoretical predictions of a factor of 2 by Maradudin and Melnagailis (1964). However there were certain disagreements with the theory in that the experiments of Jones, McKinney, and Webb interpreted according to kinematic theory implied isotropic vibration of surface atoms, whereas theoretical calculations deduce that vibrations parallel to are 35% smaller than vibrations normal to the surface, in the top layer of atoms.

VIB Lattice vibrations

This section is concerned to give a brief account of thermal vibrations. Details can be had by referring to Born and Huang (1954), Cochran (1963), Maradudin and Melnagailis (1964), Clark, Herman and Wallis (1965). The two latter papers address themselves specifically to vibrations at surfaces. A clear account of thermal vibrations with special relevance to scattering of electromagnetic radiation can be found in Sherwood (1972).

To discover how a system vibrates at finite temperature it is necessary to find the normal modes. Then statistical mechanics tells us that with each normal mode there is associated kinetic energy of $\frac{1}{2}k_B T$, in the classical approximation. Knowing to what extent each normal mode is excited, the total vibration of an atom can be found by summing over contributions from each normal mode.

Normal vibrational modes associated with the bulk of a solid are waves: sound waves in fact, at the low frequency end of the spectrum. The displacements $\Delta\mathbf{P}_{jm}$ of the mth atom in the jth unit cell due to the nth normal mode, wave-vector $\mathbf{q}$, can be written

$$\Delta\mathbf{P}_{jm}(t) = \mathbf{a}_{\mathbf{q}n}\cos\left[\mathbf{q}\cdot(\mathbf{P}_j + \mathbf{p}_m) + \phi_{\mathbf{q}mn}\right] \times \cos(\omega_{\mathbf{q}n}t + \psi_{\mathbf{q}n}), \tag{6.9}$$

where $\mathbf{P}_j$ and $\mathbf{p}_m$ are the positions of the jth unit cell, and the mth atom in the jth cell relative to the origin of that cell respectively, in the absence of vibrations. $\phi_{\mathbf{q}mn}$ is determined by boundary conditions at the surface and $\psi_{\mathbf{q}n}$ is determined by the history of the crystal; it is assumed to vary randomly from one mode to the next.

For example, in a simple-cubic crystal with only one atom per unit cell and $\mathbf{q}$ parallel to a symmetry direction, there would be three different types of normal mode indexed by $n = 1, 2, 3$ corresponding to one longitudinal sound wave vibrating parallel to $\mathbf{q}$, and two transverse waves vibrating perpendicular to $\mathbf{q}$. With a more complicated unit cell more different types of modes occur: three for every atom in the unit cell.

Values of wave vector $\mathbf{q}$ for which modes exist depend on boundary conditions at the surface, but in the bulk the important quantity is the number of modes per unit volume of $\mathbf{q}$-space, which is independent of the boundary conditions. The density of modes can easily be shown to be $\Omega/(2\pi)^3$ by arranging that (6.9) satisfy some simple condition at the faces of a big cube, volume Ω, and finding the allowed values of $\mathbf{q}$.

The points $\mathbf{P}_j$ describe the Bravais lattice of the crystal and can be expressed as

$$\mathbf{P}_j = N_a\mathbf{a} + N_b\mathbf{b} + N_c\mathbf{c}, \tag{6.10}$$

where N_a, N_b, and N_c are integers, implying that if $\mathbf{q}$ is increased by

$$2\pi n_c \frac{\mathbf{a} \times \mathbf{b}}{(\mathbf{a} \times \mathbf{b})\cdot\mathbf{c}} + 2\pi n_b \frac{\mathbf{a} \times \mathbf{c}}{(\mathbf{a} \times \mathbf{b})\cdot\mathbf{c}} + 2\pi n_a \frac{\mathbf{b} \times \mathbf{c}}{(\mathbf{a} \times \mathbf{b})\cdot\mathbf{c}} \tag{6.11}$$

the values of the displacements given by (6.9) are unchanged. Therefore all possible modes are completely described by values of $\mathbf{q}$ lying inside a region of $\mathbf{q}$-space called the 'first Brillouin zone', the boundaries of which are defined by the planes

$$\frac{\mathbf{q}\cdot(\mathbf{a} \times \mathbf{c})}{(\mathbf{a} \times \mathbf{b})\cdot\mathbf{c}} = \pm\pi, \frac{\mathbf{q}\cdot(\mathbf{b} \times \mathbf{c})}{(\mathbf{a} \times \mathbf{b})\cdot\mathbf{c}} = \pm\pi, \frac{\mathbf{q}\cdot(\mathbf{b} \times \mathbf{c})}{(\mathbf{a} \times \mathbf{b})\cdot\mathbf{c}} = \pm\pi. \tag{6.12}$$

Taking into account the density of allowed points there is a total of

$$\Omega/((\mathbf{a} \times \mathbf{b})\cdot\mathbf{c}) \tag{6.13}$$

allowed values of $\mathbf{q}$ for each mode, a number that is simply the total number of unit cells in the crystal. Thus there are three normal modes for every atom in the crystal, as there should be from consideration of degrees of freedom.

Frequencies of vibration are important because they enter into expressions for the kinetic energy of a mode, and so determine amplitudes of excitation at a given temperature. They must be determined either by calculation or from neutron diffraction experiments. This applies also to directions along which atoms vibrate, unless the crystal is one of high symmetry when the directions are sometimes determined by the symmetry.

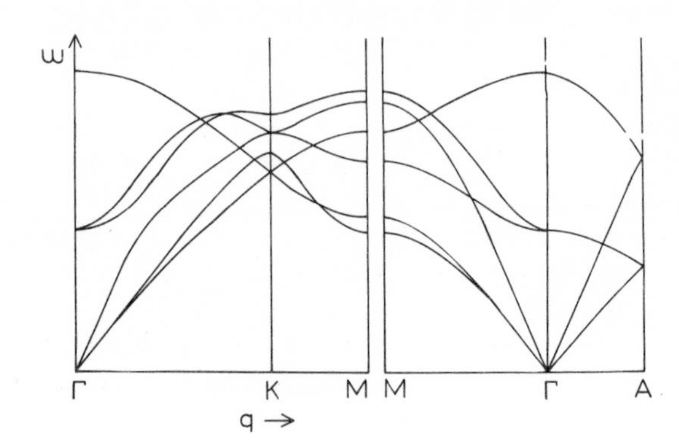

FIGURE 6.8. Phonon spectrum for magnesium calculated for $\mathbf{q}$ lying along various symmetry directions.

Calculations are long and involved, and can only be carried out with accuracy for simple metals; the main source of information is experiment. To illustrate the nature of dispersion curves, as $\omega(\mathbf{q})$ plots are called, we reproduce in Fig. 6.8 some calculations by Shaw and Pynn (1969) for magnesium. For comparison experimental determinations of the dispersion curves are shown in Fig. 6.9, reproduced from Shaw and Pynn, compounded of date by

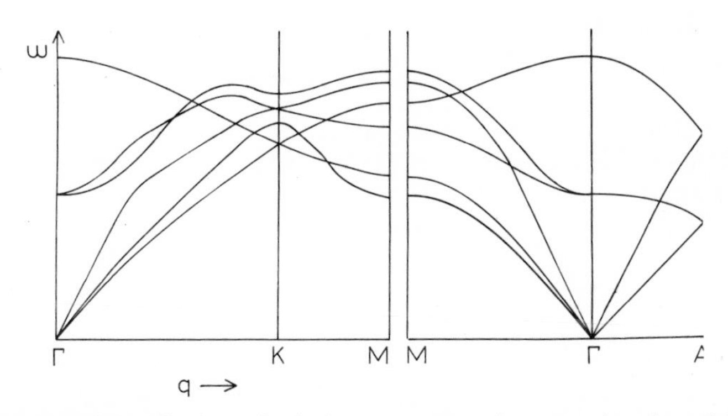

FIGURE 6.9. Experimentally determined phonon spectrum for magnesium for $\mathbf{q}$ lying along various symmetry directions.

Squires (1966), Pynn and Squires (1968) and Iyengar *et al.* (1965). Magnesium has an hexagonal close-packed structure with two atoms per unit cell, hence six modes are to be expected.

It is clear that to take account of details of crystal dynamics in calculating temperature effects on LEED spectra, is a complicated process. In other diffraction situations where temperature is of importance, e.g. for X-rays, James (1962), or high energy electrons, Hirsch *et al.* (1965), it has been found that crude approximations to the crystal dynamics give satisfactory accounts of diffraction at finite temperatures.

There are several approximations. The simplest of all is called the Einstein approximation and replaces $\omega_{\mathbf{q}n}$ by a constant,

$$\omega_{\mathbf{q}n} = \omega_E, \tag{6.14}$$

the Einstein frequency, a constant for all modes. Each atom is thereby assumed to vibrate independently of its neighbours, correlation of motion being neglected. ω_E can be determined either directly by requiring that it give the correct temperature dependence, or alternatively by adjusting it to fit specific heat data.

A more realistic model is due to Debye. He noted that low frequency behaviour of normal modes can be determined by macroscopic investigations of sound propagation in crystals.

$$\omega_{\mathbf{q}n} = c_n|\mathbf{q}|, \tag{6.15}$$

where c_n is the velocity of sound appropriate to the nth mode. The assumption is made that (6.15) is true for all allowed values of $\mathbf{q}$.

By way of example, consider a simple cubic crystal lattice with one atom per unit cell, in which

$$c_1 = c_2 = c_3$$

The normal modes consist of two transverse modes and one longitudinal mode. Statistical mechanics gives for the mean kinetic energy of a mode defined by (6.9)

$$\left[\frac{1}{2} + \frac{1}{[\exp(\hbar\omega_{\mathbf{q}n}/k_B T) - 1]}\right]\hbar\omega_{\mathbf{q}n} = \frac{\Omega}{b^3} m \frac{\omega_{\mathbf{q}n}^2 |\mathbf{a}_{\mathbf{q}n}|^2}{4} \tag{6.16}$$

Ω/b^3 is the number of atoms in the crystal, m is the mass of an atom. $|\mathbf{a}_{\mathbf{q}n}|$ is given by

$$|a_{\mathbf{q}n}| = \left[\frac{4b^3\hbar\omega_{\mathbf{q}n}}{m\omega_{\mathbf{q}n}^2\Omega}\left[\frac{1}{2} + \frac{1}{(\exp(\hbar\omega_{\mathbf{q}n}/k_B T) - 1)}\right]\right]^{\frac{1}{2}} \tag{6.17}$$

The direction of $\mathbf{a}_{\mathbf{q}n}$ is parallel to $\mathbf{q}$ for longitudinal and perpendicular for transverse, in our example. At low temperatures (6.17) shows that higher frequency modes are not strongly excited, and for the important part of the spectrum (6.15) is a good approximation.

What we require for our LEED calculations is the mean square displacement along $(\mathbf{k} - \mathbf{k}')$ cf. equation (6.7). Knowing amplitudes from (6.17) we can determine the contribution of each normal mode to

$$\Delta\mathbf{P}_j \cdot (\mathbf{k} - \mathbf{k}') \tag{6.18}$$

Because of the arbitrary phase, $\psi_{\mathbf{q}n}$, of each normal mode, when (6.18) is squared, cross terms between different normal modes will average to zero,

$$\begin{aligned} \langle[\Delta\mathbf{P}_j \cdot (\mathbf{k} - \mathbf{k}')]^2\rangle_T &= \sum_{\mathbf{q}n} |\mathbf{a}_{\mathbf{q}n} \cdot (\mathbf{k} - \mathbf{k}')|^2 \\ &\times \cos^2[\mathbf{q} \cdot \mathbf{P}_j + \phi_{\mathbf{q}n}] \cos^2[\omega_{\mathbf{q}n} t + \psi_{\mathbf{q}n}] \end{aligned} \tag{6.19}$$

Phase of the time-cosine function varies randomly from one $\psi_{\mathbf{q}n}$ to another, and that of the space-cosine function varies rapidly provided we are well away from the surface. Further, in a solid of macroscopic dimensions, allowed values of $\mathbf{q}$ are closely spaced and it is a good approximation to replace the rapidly varying cosine functions by their average values.

$$\langle[\Delta\mathbf{P}_j\cdot(\mathbf{k}-\mathbf{k}')]^2\rangle_T = \sum_{\mathbf{q}} \frac{|k-k'|^2 4b^3\hbar\omega_{\mathbf{q}}}{4m\omega_{\mathbf{q}}^2\Omega}\left[\frac{1}{[\exp(\hbar\omega_{\mathbf{q}}/kT)-1]}+\frac{1}{2}\right]$$

$$= |\mathbf{k}-\mathbf{k}'|^2 \int_{-\frac{\pi}{b}}^{+\frac{\pi}{b}}\int_{-\frac{\pi}{b}}^{+\frac{\pi}{b}}\int_{-\frac{\pi}{b}}^{+\frac{\pi}{b}} dq_x dq_y dq_z \frac{\Omega}{(2\pi)^3}\frac{b^3\hbar\omega_{\mathbf{q}}}{m\omega_{\mathbf{q}}^2\Omega}$$

$$\times\left[\frac{1}{[\exp(\hbar\omega_{\mathbf{q}}/kT)-1]}+\frac{1}{2}\right] \tag{6.20}$$

In the first step we have used the assumption that all three types of modes have the same frequency $\omega_{\mathbf{q}}$ to sum over n, and in the second step the summation over finely spaced values of $\mathbf{q}$ has been replaced by an integral.

Since high-$\mathbf{q}$ modes are inaccurately approximated in any case, it is usual to replace the cube of integration by a sphere of the same volume, making errors in only large-$\mathbf{q}$ terms

$$\langle[\Delta\mathbf{P}_j\cdot(\mathbf{k}-\mathbf{k}')]^2\rangle_T = |\mathbf{k}-\mathbf{k}'|^2$$

$$\times\frac{4\pi\hbar b^3}{8m\pi^3}\int_0^Q \frac{q^2dq}{cq}\left[\frac{1}{\exp(\hbar cq/k_BT)-1}+\frac{1}{2}\right], \tag{6.21}$$

$$Q = \frac{2\pi}{b}\left(\frac{3}{4\pi}\right)^{\frac{1}{3}}$$

Equation (6.21) can be re-expressed as

$$\langle[\Delta\mathbf{P}_j\cdot(\mathbf{k}-\mathbf{k}')]^2\rangle_T = \frac{3\hbar^2|\mathbf{k}-\mathbf{k}'|^2}{mk_B\Theta}\left[\frac{T^2}{\Theta^2}\int_0^{\frac{\Theta}{T}}\frac{xdx}{\exp(x)-1}+\frac{1}{4}\right], \tag{6.22}$$

$$\Theta = \frac{\hbar c}{k_B}\cdot\frac{2\pi}{b}\left(\frac{3}{4\pi}\right)^{\frac{1}{3}}$$

In (6.22) the second term inside square brackets is the contribution of zero point motion to displacements of atoms and is temperature independent. At room temperature it is the first term that dominates.

There are two cases in which approximations to (6.23) can be found. At low temperatures the limit on the integral can be taken as infinity, and using the identity

$$\int_0^\infty \frac{xdx}{\exp(x)-1} = 1.642, \tag{6.24}$$

(6.22) is approximated by

$$\langle[\Delta\mathbf{P}_j\cdot(\mathbf{k}-\mathbf{k}')]^2\rangle_T \simeq \frac{3\hbar^2|\mathbf{k}-\mathbf{k}'|^2}{mk_B\Theta}\left[1.642\frac{T^2}{\Theta^2}+\frac{1}{4}\right]$$

The high temperature limit, Θ/T small, is more appropriate to most LEED experiments:

$$\langle[\Delta \mathbf{P}_j \cdot (\mathbf{k} - \mathbf{k}')]^2\rangle_T \simeq \frac{3\hbar^2 T|\mathbf{k} - \mathbf{k}'|^2}{m k_B \Theta^2} \tag{6.26}$$

For intermediate temperatures, a table of integrals occurring in (6.22) can be found in the International Crystallographic Tables (Ed. Lonsdale, 1959).

The theory of mean square displacements outlined above works well for X-ray experiments and also for high energy electron microscopy.

When information about mean square displacements in the bulk is required, it is probably most simply obtained from the Debye temperature measured in X-ray experiments, to be found by reference to the International Crystallographic Tables. A short list of Debye temperatures for some common materials is reproduced in Table 6.1.

substance	Θ°K	substance	Θ°K
Ag	225	Mg	406
Aℓ	418	Mo	425
Au	165	Nb	252
Be	1160	Ni	456
C (diamond)	2000	Pb	95
Cr	402	Pd	275
Cu	343	Pt	229
Fe	467	Si	658
Ge	366	Zn	308
Hg	60–90	W	379

TABLE 6.1. Debye temperatures for some common substances.

Within the framework of kinematic theory, to calculate intensity of coherent scattering only the mean square displacements of atoms are needed, but in multiple scattering situations correlations between displacements on neighbouring atoms can enter into the theory. Correlation between displacements of the sth and tth atoms is measured by

$$C_{st} = \frac{\langle \Delta \mathbf{P}_s \cdot (\mathbf{k} - \mathbf{k}') \times \Delta \mathbf{P}_t \cdot (\mathbf{k} - \mathbf{k}')\rangle_T}{(\langle[\Delta \mathbf{P}_s \cdot (\mathbf{k} - \mathbf{k}')]^2\rangle_T \langle[\Delta \mathbf{P}_t \cdot (\mathbf{k} - \mathbf{k}')]^2\rangle_T)^{\frac{1}{2}}} \tag{6.27}$$

Evidently if atoms s and t are perfectly correlated, C_{st} is unity, but if they are uncorrelated the numerator will average to zero.

The Debye model can be applied to calculate

$$\langle \Delta \mathbf{P}_s \cdot (\mathbf{k} - \mathbf{k}') \Delta \mathbf{P}_t \cdot (\mathbf{k} - \mathbf{k}')\rangle_T = \langle[\sum_{\mathbf{q}n} (\mathbf{k} - \mathbf{k}') \cdot \mathbf{a}_{\mathbf{q}n} \cos(\mathbf{q} \cdot \mathbf{P}_s + \phi_{\mathbf{q}n})$$

$$\cos(\omega_{\mathbf{q}n}t + \psi_{\mathbf{q}n})] \times [\sum_{\mathbf{q}'n'} (\mathbf{k} - \mathbf{k}')\cdot\mathbf{a}_{\mathbf{q}'n'} \cos(\mathbf{q}'\cdot\mathbf{P}_t + \phi_{\mathbf{q}'n'})$$

$$\cos(\omega_{\mathbf{q}'n'}t + \psi_{\mathbf{q}'n'})]\rangle_T \tag{6.28}$$

Proceeding as in the previous calculation we can show that averaging over rapidly varying functions of $\mathbf{q}$, cross terms between different modes vanish, leaving

$$\langle \Delta\mathbf{P}_s\cdot(\mathbf{k} - \mathbf{k}')\Delta\mathbf{P}_t\cdot(\mathbf{k} - \mathbf{k}')\rangle_T = \tfrac{1}{4}|\mathbf{k} - \mathbf{k}'|^2 \sum_{\mathbf{q}} |\mathbf{a}_{\mathbf{q}}|^2 \cos[\mathbf{q}\cdot(\mathbf{P}_t - \mathbf{P}_s)] \simeq$$

$$\tfrac{1}{4}|\mathbf{k} - \mathbf{k}'|^2 \int_0^{2\pi} d\phi \int_0^{\pi} \sin(\theta)d\theta \int_0^{Q} q^2 dq\Omega(2\pi)^{-3} \cos[\mathbf{q}\cdot(\mathbf{P}_t - \mathbf{P}_s)]$$

$$\times \left[\frac{1}{\exp(\hbar\omega_{\mathbf{q}}/k_BT) - 1} + \frac{1}{2}\right] \frac{4\hbar b^3}{m\omega_{\mathbf{q}n}\Omega}$$

$$= \frac{\hbar b^3 |\mathbf{k} - \mathbf{k}'|^2}{2\pi^2 mc|\mathbf{P}_t - \mathbf{P}_s|} \int_0^{Q} \left[\frac{1}{\exp(\hbar cq/k_BT) - 1} + \frac{1}{2}\right] \sin(q|\mathbf{P}_t - \mathbf{P}_s|)dq \tag{6.29}$$

At high temperature a simplification can be made to give

$$C_{st} = \frac{\hbar b^3|\mathbf{k} - \mathbf{k}'|^2}{2\pi^2 mc|\mathbf{P}_t - \mathbf{P}_s|}\cdot\frac{\hbar Q}{k_B\Theta}\cdot\frac{k_BT}{\hbar}\int_0^{Q|\mathbf{P}_t - \mathbf{P}_s|}\frac{\sin(x)}{x}dx \Bigg/ \left[\frac{3\hbar^2 T|\mathbf{k} - \mathbf{k}'|^2}{mk_B\Theta^2}\right]$$

$$= \frac{2}{3}\left(\frac{3}{4\pi}\right)^{\frac{2}{3}} \frac{b}{|\mathbf{P}_t - \mathbf{P}_s|}\int_0^X \frac{\sin(x)}{x}dx, \tag{6.30}$$

$$X = 2\pi\left(\frac{3}{4\pi}\right)^{\frac{1}{3}}\frac{|\mathbf{P}_t - \mathbf{P}_s|}{b}. \tag{6.31}$$

In the Debye model, the correlation coefficient at high temperatures is independent of Θ and the temperature. Table 6.2 shows C_{st} evaluated for various ratios $|\mathbf{P}_t - \mathbf{P}_s|/b$.

Correlation between nearest neighbours is strong but thereafter drops to a small value, followed by further slow decrease as $|\mathbf{P}_t - \mathbf{P}_s|^{-1}$. The long range correlations of motion are due to the low-q phonons and the short range ones to the high-q phonons. There are many more high than low q phonons and this is why only short range correlations are strong.

In the above we have been concerned with vibrations in the bulk of a crystal. Usually vibrations become typical of the bulk more than two layers deep in the crystal, the first few layers having rather different vibrational properties. First of all mechanical forces between layers change, secondly the specific boundary condition that bulk modes obey at the surface is of importance, and thirdly additional modes localised near the surface need to be considered.

$\frac{\lvert \mathbf{P}_t - \mathbf{P}_s \rvert}{b}$	C_{st}
0	1.00
1	0.46
2	0.20
3	0.13
4	0.11
5	0.08
6	0.07
7	0.06

TABLE 6.2. Correlation between displacements of atoms separated by $(\mathbf{P}_t - \mathbf{P}_s)$ evaluated in the high temperature limit of the Debye model.

The qualitative effects are obvious enough: vibration in the top layers will be greater than in the bulk. Detailed knowledge is available from a few LEED experiments and only then based on a kinematic interpretation. Neutron diffraction studies so successful in determining bulk vibrational properties are not sensitive to surface conditions.

Theoretical treatments have been made (Allen and Dewette, 1969; Allen, Dewette and Rahman, 1969; Clark, Herman and Wallis, 1965; Maradudin and Melnagailis, 1964) which assume an inter-atomic force law and solve equations of motion to find the normal modes of a finite slab of crystal. Thus the surface is built into the calculation and for a sufficiently thick slab, surface vibrations are typical of those at the surface of a semi-infinite slab. For crystals of inert gases, inter-atomic force laws are known accurately, but for other materials are fitted to experimental bulk phonon spectra.

Results show enhanced amplitudes of vibration in surface layers, of the order of 50% greater than in the bulk. Also the asymmetry introduced by the

layer number	(100)		(111)		(110)	
	//	⊥	//	⊥	//	⊥
1	1.56	2.05	1.33	2.05	1.61	2.01
2	1.15	1.29	1.11	1.29	1.19	1.67
3	1.07	1.13	1.06	1.14	1.10	1.24
4	1.04	1.07	1.04	1.09	1.06	1.15
5	1.02	1.04	1.03	1.06	1.03	1.08

TABLE 6.3. Enhanced mean square vibrational amplitudes parallel and perpendicular to the surface, for various faces of nickel (f.c.c.), normalised to the bulk value. The semi-infinite crystal is approximated by a slab 20 layers thick.

surface leads to different amplitudes for vibrations perpendicular and parallel to the surface. In Table 6.3 the results of Clark, Herman and Wallis are reproduced.

As yet a sufficient bulk of experimental and theoretical LEED work on temperature effects has not been amassed to be able to say with what accuracy such calculations can describe the situation.

VIC Theoretical treatment

In interpreting experiments made on temperature effects, a simple analysis was made in Section A making use of kinematic theory. The resulting formulae were analogous to those used in X-ray diffraction. But as we have seen in earlier chapters, kinematic theory does not suffice to describe even zero temperature intensity/energy spectra, and it is not surprising that many examples have been found of the simple temperature dependence given by kinematic theory being violated. What is surprising is that in many instances the simple law is nearly obeyed. These facts suggest an approach to finite temperature scattering in which an expansion is used having the kinematic expression as its first term.

We shall follow Holland's (1971) derivation of the result. Another derivation of the same result, that makes the same assumptions is to be found in papers by Duke and Laramore (1970), and Laramore and Duke (1970).

If a plane wave

$$\exp(i\mathbf{k}\cdot\mathbf{r}), \quad \tfrac{1}{2}|\mathbf{k}|^2 = E - V_0$$

is incident on an array of atoms at positions $\mathbf{P}_j$, the amplitude scattered once by the atoms is from equation (2.56)

$$\phi_s^{(1)} = \sum_j \sum_{\ell m} 4\pi i^{\ell+1} \sin(\delta_\ell(j)) \exp(i\delta_\ell(j))\, Y_{\ell m}(\Omega(\mathbf{k})) \times \exp(i\mathbf{k}\cdot\mathbf{P}_j)(-1)^m Y_{\ell -m}(\Omega(\mathbf{r}-\mathbf{P}_j))h_\ell^{(1)}(|\mathbf{k}|\,|\mathbf{r}-\mathbf{P}_j|). \tag{6.32}$$

A spherical wave representation is not suitable for our purposes. Instead we use the identity,

$$h_\ell^{(1)}(|\mathbf{k}|\,|\mathbf{r}|)\, Y_{\ell -m}(\Omega(\mathbf{r})) = \frac{i^{1-\ell}}{2\pi^2|\mathbf{k}|}\int \frac{\exp(i\mathbf{k}'\cdot\mathbf{r})}{2E - 2V_0 - |\mathbf{k}'|^2 + i\varepsilon} Y_{\ell -m}(\Omega(\mathbf{k}'))d^3\mathbf{k}', \quad \varepsilon = +0, \tag{6.33}$$

to express

$$\phi_s^{(1)} = \sum_j \frac{\exp(i\mathbf{k}\cdot\mathbf{P}_j)}{4\pi^3}\int t_j(\mathbf{k}',\mathbf{k})\frac{\exp[i\mathbf{k}'\cdot(\mathbf{r}-\mathbf{P}_j)]}{2E - 2V_0 - |\mathbf{k}'|^2 + i\varepsilon}d^3\mathbf{k}', \tag{6.34}$$

$$t_j(\mathbf{k}',\mathbf{k}) = -2\pi|\mathbf{k}|^{-1}\sum_{\ell}(2\ell+1)\sin(\delta_\ell(j))\exp(i\delta_\ell(j))P_\ell(\cos\theta_{\mathbf{k}'\mathbf{k}}). \quad (6.35)$$

The multiple scattering problem raises its head at this point because these once scattered waves can be scattered a second time to waves represented by

$$\phi_s^{(2)} = \sum_{n\neq j}\sum_j \iint d^3\mathbf{k}''d^3\mathbf{k}'(16\pi^6)^{-1}\exp(i\mathbf{k}\cdot\mathbf{P}_j)t_j(\mathbf{k}'\mathbf{k})$$
$$\times \frac{\exp[i\mathbf{k}'\cdot(\mathbf{P}_n-\mathbf{P}_j)]}{2E-2V_0-|\mathbf{k}'|^2+i\varepsilon}t_n(\mathbf{k}'',\mathbf{k}')\frac{\exp[i\mathbf{k}''\cdot(\mathbf{r}-\mathbf{P}_n)]}{2E-2V_0-|\mathbf{k}''|^2+i\varepsilon}. \quad (6.36)$$

The total amplitude of plane waves scattered is

$$\phi_s^{(t)} = \sum_u \int d^3\mathbf{k}^{(1)}\exp(i\mathbf{k}^{(1)}\cdot\mathbf{r})G(\mathbf{k}^{(1)})\exp[i(\mathbf{k}-\mathbf{k}^{(1)})\cdot\mathbf{P}_u]t_u(\mathbf{k}^{(1)}\mathbf{k}),$$
$$+\sum_{v\neq u}\sum_u\iint d^3\mathbf{k}^{(2)}d^3\mathbf{k}^{(1)}\exp(i\mathbf{k}^{(2)}\cdot\mathbf{r})G(\mathbf{k}^{(2)})\exp[i(\mathbf{k}^{(1)}-\mathbf{k}^{(2)})\cdot\mathbf{P}_v]t_v(\mathbf{k}^{(2)};\mathbf{k}^{(1)})$$
$$\times G(\mathbf{k}^{(1)})\exp[i(\mathbf{k}-\mathbf{k}^{(1)})\mathbf{P}_u]t_u(\mathbf{k}^{(1)};\mathbf{k}) +$$
$$+\sum_{\omega\neq v}\sum_{v\neq u}\sum_u\iiint d^3\mathbf{k}^{(3)}d^3\mathbf{k}^{(2)}d^3\mathbf{k}^{(1)}\times\ldots, \quad (6.37)$$

$$G(\mathbf{k}) = \frac{1}{4\pi^3}\frac{1}{2E-2V_0-|\mathbf{k}|^2+i\varepsilon}. \quad (6.38)$$

At finite temperatures coordinates of atoms do not generally lie on lattice points $\mathbf{P}_n$ but differ from them by $\Delta\mathbf{P}_n$ as discussed previously. It is evident from (6.37) that replacement of zero temperature atomic positions by their finite temperature counterparts is equivalent to replacing the factors t_n by

$$t_n(\mathbf{k}^{(1)};\mathbf{k})\exp[i(\mathbf{k}-\mathbf{k}^{(1)})\Delta\mathbf{P}_n]. \quad (6.39)$$

Therefore the general term becomes

$$\sum_{x\neq w}\sum_{w\neq v}\cdots\sum_u\iint\ldots\int d^3\mathbf{k}^{(n)}d^3\mathbf{k}^{(n-1)}\ldots d^3\mathbf{k}^{(1)}$$
$$\times\left[\begin{array}{c}\text{factor}\\ \text{independent}\\ \text{of temperature}\end{array}\right]\times t_x(\mathbf{k}^{(n)},\mathbf{k}^{(n-1)})$$
$$\times\exp[i(\mathbf{k}^{(n-1)}-\mathbf{k}^{(n)})\cdot(\mathbf{P}_x+\Delta\mathbf{P}_x)]t_w(\mathbf{k}^{(n-1)};\mathbf{k}^{(n-2)})\exp[i(\mathbf{k}^{(n-2)}-\mathbf{k}^{(n-1)})\cdot$$
$$(\mathbf{P}_w+\Delta\mathbf{P}_w)]\times\ldots\times t_u(\mathbf{k}^{(1)},\mathbf{k})\exp[i(\mathbf{k}-\mathbf{k}^{(1)})\cdot(\mathbf{P}_u+\Delta\mathbf{P}_u)] \quad (6.40)$$

The point of so arranging the series is that we are trying to take account of temperature by defining a temperature dependent scattering factor for each atom.

Experiments observe the finite temperature average of $|\phi_s^{(t)}|^2$ at some point remote from the crystal, an expression that involves cross-products of terms such as (6.40), i.e. we have to average expressions which are products of terms like

$$t_x(\mathbf{k}^{(n)}; \mathbf{k}^{(n-1)}) \exp\left[i(\mathbf{k}^{(n-1)} - \mathbf{k}^{(n)}) \cdot \Delta\mathbf{P}_x\right]. \tag{6.41}$$

If all the variables $\Delta\mathbf{P}_x$, $\Delta\mathbf{P}_w \ldots$ are uncorrelated, statistical theory gives the result that averaging the product of terms like (6.41) gives the same result as taking the product of averages. The average of (6.41) is

$$\langle t_x \exp\left[i(\mathbf{k}^{(n-1)} - \mathbf{k}^{(n)}) \cdot \Delta\mathbf{P}_x\right]\rangle_T = \exp(-M_x) t_x(\mathbf{k}^{(n)}; \mathbf{k}^{(n-1)}) \tag{6.42}$$

which is to be compared with the kinematic treatment in (6.6). Finite temperature multiple scattering theory in this approximation becomes a simple extension of the kinematic theory of temperature effects: all scattering factors used in the zero temperature theory are replaced at finite temperatures by (6.42).

To what extent can neglecting correlation in taking averages be justified? Returning to the cross-products involved in $|\phi_s^{(t)}|^2$: if $\Delta\mathbf{P}_u$ occurs in the $\phi_s^{(t)}$ half of the product, and $\Delta\mathbf{P}_v$ in the $\phi_s^{(t)*}$ half, then u and v and refer to any pair of atoms in the crystal. We have already estimated in Table 6.2 how atomic displacements correlate over distance. It is evident that only a very small fraction of possible pairs of atoms will happen to be close enough to be highly correlated.

If $\Delta\mathbf{P}_u$ and $\Delta\mathbf{P}_v$ both come from terms in $\phi_s^{(t)}$, or both from $\phi_s^{(t)*}$, the situation changes. To see why, it is convenient to represent different contributions to $\phi_s^{(t)}$ as paths in the crystal. A term in which the electron scatters from atom i to atom j to ... to atom p, and out of the crystal is represented by lines joining these atoms: Fig. 6.10. Only finite path-lengths contribute because of absorp-

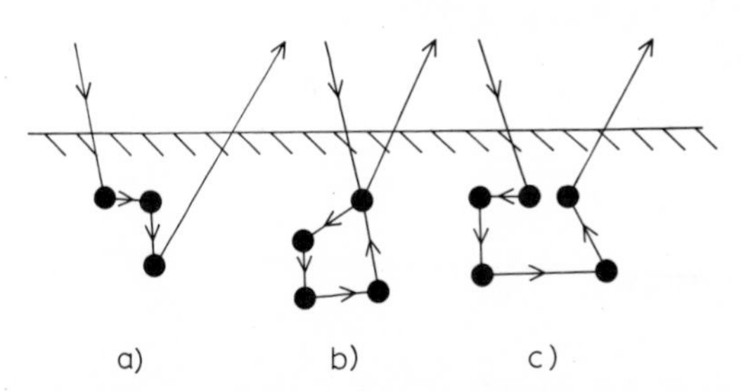

FIGURE 6.10. Possible scattering paths in a crystal: (a) a relatively open path with little correlation between scattering centres, (b) a closed path where correlation plays an important role, (c) an almost closed path where correlation might be important.

tive processes. Therefore the important contributions to $\phi_s^{(t)}$ all have $\Delta\mathbf{P}_u$ and $\Delta\mathbf{P}_v$ within a finite distance of one another; the shorter the path length the greater probability there is of $\Delta\mathbf{P}_u$ and $\Delta\mathbf{P}_v$ being correlated. In particular if the path passes through the same atom twice, there will be 100% correlation between $\Delta\mathbf{P}_u$ and itself. Paths with loops in them, or even paths that double back to come within a nearest-neighbour distance of a previously encountered atom, will be inaccurately represented by any theory neglecting correlations of atomic motions.

No practical formalism has yet been constructed to take account of correlated motions in a strong multiple scattering situation, therefore it is important to know whether the uncorrelated theory is accurate.

Several arguments lead to the conclusion that neglect of correlation is not so serious as may be imagined. The fraction of the total number of paths that double back on themselves sufficiently to involve correlation is small, because the *amplitude*-extinction distance is typically 10–20Å and a large number of paths of this length can be constructed of which the loops form only a small fraction. There is another reason why the loops should not contribute strongly. It has been observed that atomic scattering in the forward direction is much stronger than in the back direction, a statement that is increasingly true at higher energies. Therefore most of the multiple scattering, the cause of the trouble in our theory, takes place in the forward direction, i.e. with small $|\mathbf{k}^{(n-1)} - \mathbf{k}^{(n)}|$. Back scattering events are rare occurrences. Correlation changes mainly the forward scattering Debye Waller factor. When the temperature is such as to leave the forward scattering Debye Waller factor virtually unity, errors in this factor are not important because the effect is small in any case.

The effective temperature dependent scattering factor is given by equation (6.42). In this representation t appears as a matrix element between plane-waves, but in our multiple scattering formulation of Chapter IV, ion core scattering was treated in terms of partial waves. For spherically symmetric ion cores we had relationship (2.58)

$$t(\mathbf{k}',\mathbf{k}) = -2\pi(\kappa)^{-1}\sum_{\ell}(2\ell + 1)\sin(\delta_\ell)\exp(i\delta_\ell)P_\ell(\cos\theta_{\mathbf{k}'\mathbf{k}})$$
$$\kappa^2 = 2E - 2V_0$$

which can be rewritten

$$t(\mathbf{k}',\mathbf{k}) = 8\pi^2\sum_{\ell m}\sum_{\ell' m'} t(\ell' m',\ell m)\, Y_{\ell' -m'}(\Omega(\mathbf{k}'))\, Y_{\ell m}(\Omega(\mathbf{k}))(-1)^{m'} \tag{6.43}$$

$$t(\ell' m',\ell m) = -\delta_{\ell\ell'}\delta_{mm'}\kappa^{-1}\exp(i\delta_\ell)\sin(\delta_\ell) \tag{6.44}$$

(The delta function with two subscripts, $\delta_{\ell\ell'}$, which by common convention is zero unless the subscripts are equal, otherwise unity, should not be confused with δ_ℓ, the phase shift.) $t(\ell'\text{m}',\ell m)$ is a matrix connecting ℓm to $\ell' m'$, and takes the diagonal form (6.44) in consequence of the spherically symmetric potential. At finite temperatures we define

$$t(T,\mathbf{k}',\mathbf{k}) = t(0,\mathbf{k}',\mathbf{k})\exp(-\tfrac{1}{2}\langle[(\mathbf{k}-\mathbf{k}')\cdot\Delta\mathbf{P}]^2\rangle_T)$$
$$= 8\pi^2 \sum_{\ell m}\sum_{\ell' m'} t(T,\ell'm',\ell m)\, Y_{\ell'-m'}(\Omega(\mathbf{k}'))\, Y_{\ell m}(\Omega(\mathbf{k}))(-1)^{m'} \tag{6.45}$$

Multiplying by $Y_{\ell' m'}$ and $Y_{\ell -m}(-1)^m$, and integrating over angles we deduce that

$$t(T,\ell'm',\ell m) = (8\pi^2)^{-1}\int\int t(T,\mathbf{k}',\mathbf{k})\, Y_{\ell -m}(\Omega(k)) \times Y_{\ell' m'}(\Omega(k'))(-1)^m d\Omega(\mathbf{k})d\Omega(\mathbf{k}') \tag{6.46}$$

since $\int Y_{\ell' m'}(\Omega)\, Y_{\ell -m}(\Omega)(-1)^m d\Omega = \delta_{\ell\ell'}\delta_{mm'}$

It is not hard to trace the generalisation of non-spherical scattering through Section IVC to deduce in place of (4.49)

$$M^{\pm\pm}_{\mathbf{g}'\mathbf{g}} = \frac{8\pi^2 i}{|\mathbf{K}^{\pm}_{\mathbf{g}}|A|K^{\pm}_{\mathbf{g}'z}|}\sum_{\substack{\ell' m' k'\\ \ell m k}} \exp(i\mathbf{K}^{\pm}_{\mathbf{g}}\mathbf{r}_k - i\mathbf{K}^{\pm}_{\mathbf{g}'}\mathbf{r}_{k'})$$
$$\times [i^{\ell}(-1)^m Y_{\ell -m}(\Omega(\mathbf{K}^{\pm}_{\mathbf{g}}))](1 - X(T))^{-1}_{\ell mk,\ell' m' k'}$$
$$\times \sum_{\ell'' m''}(-\kappa t_{k'}(T,\ell'' m'',\ell' m'))[i^{-\ell''} Y_{\ell'' m''}(\Omega(\mathbf{K}^{\pm}_{\mathbf{g}'}))], \tag{6.47}$$

with a temperature dependent definition of X, in place of (4.46).

$$X_{\ell mk,\ell'' m'' s} = \sum_{LM} \kappa t_k(T,LM,\ell m) \sum_{\ell' m'} 4\pi(\kappa)^{-1}(-1)^{\frac{1}{2}(\ell''-L-\ell')}$$
$$(-1)^{m''} B^L(\ell' m',\ell'' m'')[\kappa i(4\pi)^{-\frac{1}{2}}\delta_{m'0}\delta_{\ell' 0} + D_{\ell' m'}(ks)]. \tag{6.48}$$

As it stands (6.46) is not a convenient expression for evaluation of t in the spherical wave representation. It can be made more tractable by substituting

$$\exp(-M_x) = \exp(-\tfrac{1}{2}\langle[(\mathbf{k}'-\mathbf{k})\cdot\Delta\mathbf{P}_x]^2\rangle_T)$$
$$= \sum_{\substack{LM\\ L'M'}} W_{LM,L'M'}\, Y_{L'M'}(\Omega(\mathbf{k}'))\, Y_{L-M}(\Omega(\mathbf{k}))(-1)^M \tag{6.49}$$

into equation (6.42). Now (6.46) becomes

$$t(T,\ell'm',\ell m) = \sum_{\substack{LM \\ L'M'}} \sum_{\substack{\ell''m'' \\ \ell'''m'''}} W_{LM,L'M'} t(0,\ell'''m''',\ell''m'')(-1)^{M+m+m'''}$$
$$\times \int Y_{L-M}(\Omega)\, Y_{\ell''m''}(\Omega)\, Y_{\ell-m}(\Omega) d\Omega$$
$$\times \int Y_{L'M'}(\Omega')\, Y_{\ell'''-m'''}(\Omega')\, Y_{\ell'm'}(\Omega') d\Omega'$$
$$= \sum_{\substack{LM \\ L'M'}} \sum_{\substack{\ell''m'' \\ \ell'''m'''}} W_{LM,L'M'} t(0,\ell'''m''',\ell''m'')(-1)^{M+m'+m'''}$$
$$\times B^{L}(\ell m,\ell''m'') B^{L'}(\ell'm',\ell'''m''') \tag{6.50}$$

B has been defined in equation (4.40), and can be derived from an analytic expression to be found in the appendix.

In the special case that vibrations are isotropic we regain a temperature dependent scattering factor that is diagonal in a spherical-wave representation, and can be expressed in terms of phase shifts, just as the zero temperature scattering factor could, except that the new phase shifts will be temperature dependent. We write

$$\exp(-M_x) = \exp(-\alpha|\mathbf{k} - \mathbf{k}'|^2) \tag{6.51}$$

where α might be approximated by

$$\alpha \simeq \frac{3\hbar^2 T}{2mk_B \Theta^2}, \tag{6.52}$$

from equation (6.26). The exponential can be expanded as follows,

$$\exp(-\alpha|\mathbf{k} - \mathbf{k}'|^2) = \exp[-\alpha(|\mathbf{k}|^2 + |\mathbf{k}'|^2)] \times \exp[2\alpha|\mathbf{k}|\,|\mathbf{k}'|\cos\theta_{\mathbf{kk}'}]$$
$$= \exp[-2\alpha\kappa^2] \sum_{\ell m} 4\pi i^{\ell} j_{\ell}(-i2\alpha\kappa^2) \times Y_{\ell m}(\Omega(\mathbf{k}))\, Y_{\ell-m}(\Omega(\mathbf{k}'))(-1)^m \tag{6.53}$$

since

$$|\mathbf{k}|^2 = |\mathbf{k}'|^2 = \kappa^2$$

and from (6.49) we identify

$$W_{LM,L'M'} = \delta_{LL'}\,\delta_{MM'} \exp[-2\alpha\kappa^2] 4\pi i^L j_L(-i2\alpha\kappa^2) \tag{6.54}$$

Therefore on substituting (6.54) and (6.44) into (6.50) and making use of various properties of spherical harmonics:

$$t(T,\ell'm',\ell m) = \sum_{L\ell''} 4\pi \exp(-2\alpha\kappa^2)\, \delta_{\ell\ell'}\, \delta_{mm'}$$
$$\times\, i^L j_L(-2i\alpha\kappa^2)(-\kappa)^{-1} \exp(i\delta_{\ell''}) \sin(\delta_{\ell''})$$
$$\times (2L+1)(2\ell''+1)(8\pi)^{-1} \int P_L(\cos\theta) P_{\ell''}(\cos\theta) P_{\ell}(\cos\theta) \sin(\theta) d\theta$$
$$= -\,\delta_{\ell\ell'}\,\delta_{mm'}\,\kappa^{-1} \exp(i\delta_{\ell}(T)) \sin(\delta_{\ell}(T)) \tag{6.55}$$

which can be manipulated to give;

$$\exp(i\delta_\ell(T))\sin(\delta_\ell(T)) = \sum_{L\ell''} i^L \exp(-2\alpha\kappa^2) j_L(-2i\alpha\kappa^2)$$

$$\exp(i\delta_{\ell''})\sin(\delta_{\ell''})\left[\frac{4\pi(2L+1)(2\ell''+1)}{(2\ell+1)}\right]^{\frac{1}{2}} B^{\ell''}(L0,\ell 0). \qquad (6.56)$$

A computer program for finding the temperature dependent phase shifts is provided in the appendix.

$\delta_\ell(T)$ calculated in this way is not real. The imaginary component is always positive implying that less flux is scattered than is incident on the ion core. This is a result of our averaging procedure in which we ignore all flux scattered incoherently, therefore sacrificing flux conservation in the scattering process.

When thermal vibrations are isotropic, and temperature corrections can be summarised by a new set of phase shifts, it is a simple matter to modify zero-temperature calculations by substituting $\delta_\ell(T)$.

VID Relationship of theory with experiment

The prescription for taking account of temperature in a multiple scattering picture—replacement of t by the te^{-M}—is the same as for simple kinematic theory given in Section A of this chapter, but changes produced in multiple scattering intensities by this substitution are more complicated. In kinematic theory the Debye Waller factor merely modifies all intensities by an appropriate e^{-2M}, but in multiple scattering theory changes in the magnitude of t affect details of the process itself.

For example, at a particular energy reflected intensity in a given beam may be low because of destructive interference between two second order scatterings. One process might involve two t-matrices $t(\theta_1)$, $t(\theta_2)$; the competing process, $t(\theta_3)$ and $t(\theta_4)$, at 0°K. At finite temperature each t matrix takes a different Debye-Waller factor and the change in total intensity is not simple because of changing interference between the two beams. In extreme cases it could happen that exact cancellation occurs at 0°K. At higher temperatures it would not, hence an *increase* of intensity with temperature could conceivably come about. Although possible such a case is improbable. More usual is the situation in which an intensity peak comes about through strong multiple scattering, the final intensity having many factors of 't' in it. At finite temperatures many Debye Waller factors occur instead of only one, attenuating the peak strongly and giving the appearance of an anomalously low Debye temperature if a kinematic interpretation is attempted. Such effects can easily be confused with Debye temperatures enhanced by increased amplitudes of surface atom vibrations, especially as multiple scattering through large angles tends to be stronger at lower energies, between 20 and 100 eV, where the shorter extinction distance makes for increased contributions from the surface.

Some calculations have been made to illustrate these points. Figure 6.11–6.13 compare theoretical 0°K curves for a copper (001) surface with experiments by Andersson (1969). Phase shifts were calculated accurately in the Hartree-Fock approximation. V_0 was fixed to have the same value we used in Chapter III

$$V_0 = (-15 - 4i)\,\text{eV}, \tag{6.57}$$

and reflectivity of the surface barrier was neglected. The RFS perturbation technique was used, further speeded up by taking advantage of symmetry at normal incidence. Sufficient beams and phase shifts were included to calculate

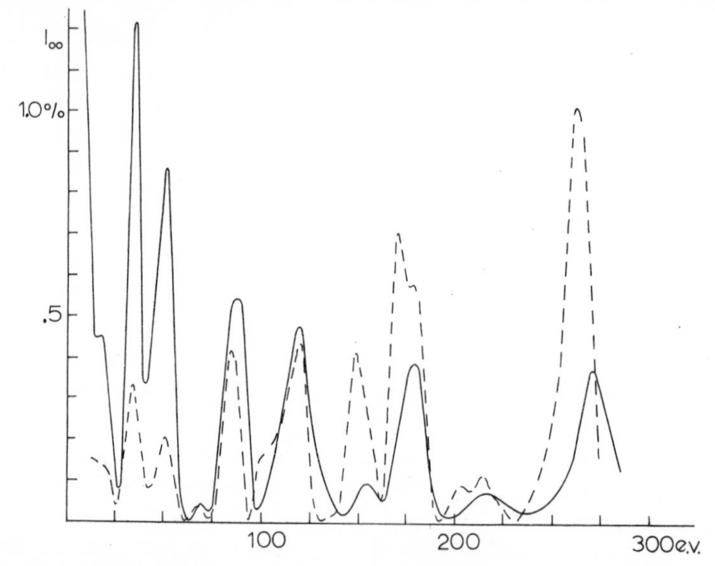

FIGURE 6.11. Intensity of the 00 beam from a copper (001) surface. ——— experiment at 3° from normal incidence, $T = 300$°K; --- theory at normal incidence, $T = 0$°K. Theoretical intensities have been divided by 7.5.

at higher energies where temperature effects are most pronounced. Below 100 eV, 21 beams and 6 phase shifts were used: more than adequate for convergence. Above 100 eV, 37 beams and 7 phase shifts were used. Repetition of the calculations at selected energies using 69 beams and 10 phase shifts revealed only small changes in intensities. As many passes of the RFS scheme were made as were required to give beams accurate to 1 part in 100. Thus a model that has all the essential features of electron scattering is being solved accurately. This is important if meaningful conclusions are to be drawn from calculations.

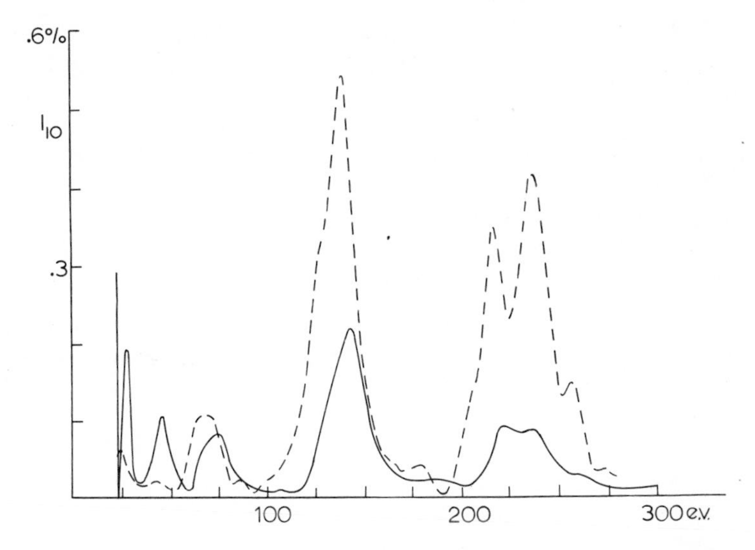

FIGURE 6.12. Intensity of the 10 beam from a copper (001) surface. ——— experiment, $T =$ 300°K; --- theory, $T = 0$°K; both taken at normal incidence. Theoretical intensities have been divided by 6.

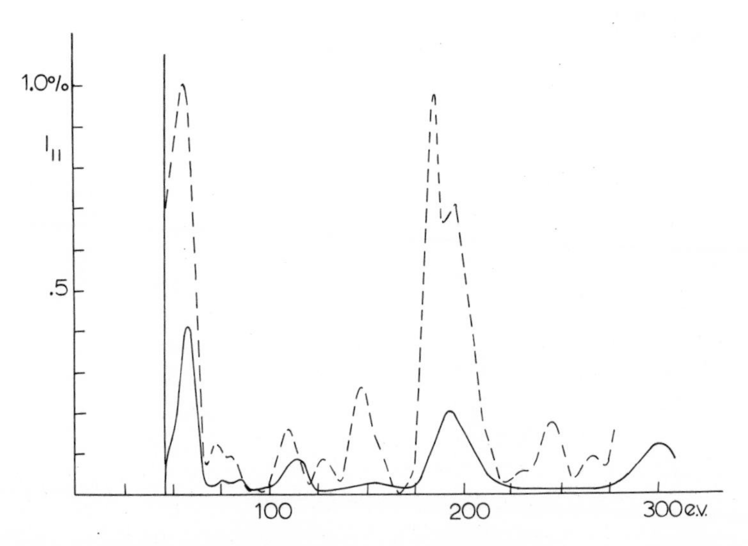

FIGURE 6.13. Intensity of the 11 beam from a copper (001) surface. ——— experiment, $T =$ 300°K; --- theory, $T = 0$°K; both taken at normal incidence. Theoretical intensities have been divided by 2.

The most obvious disagreement is a discrepancy in intensity increasing markedly with energy. Evidently large temperature corrections are needed. On the other hand, despite gross disagreement of absolute intensity almost all structure is reproduced in relative intensity, and positions of peaks are accurately accounted for. This fact will be used in Chapter VII to justify making use of relative structure, and peak positions to analyse spectra for which the thermal parameters are not available. A few differences in the 00 beam, principally the low experimental intensity of the peak at 80 eV, are attributable to the 3° difference in angles of incidence for that beam.

The first corrections to be made for temperature assume that every atom vibrates isotropically with the same amplitude given by the Debye temperature

$$\Theta = 343°\text{K}$$

At 300°K and 600°K the t-matrix is modified as shown in Fig. 6.14. To make calculations at finite temperatures the temperature dependent t-matrix is decomposed into phase shifts

$$t(\theta, T) = t(\theta, T = 0) \exp(-M)$$

$$= -2\pi\kappa^{-1} \sum_{\ell=0}^{\infty} (2\ell + 1) P_\ell(\cos\theta) \exp(i\delta_\ell) \sin(\delta_\ell) \qquad (6.58)$$

as described in Section C. There is a danger that should be pointed out: the same phase shifts in (6.58) must combine for $\theta = 0$ to give a strong peak in forward scattering, and for $\theta = 180°$ to give the much weakened back scattering. Evidently for back scattering there must be strong cancellation between

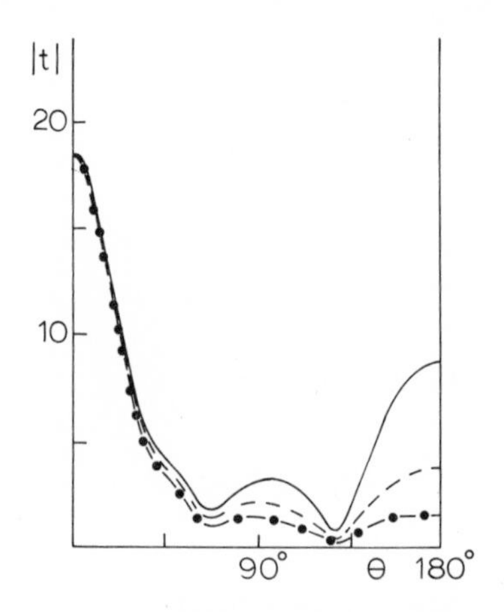

FIGURE 6.14. $|t(\theta)|$ calculated for copper at 250 eV. ——— $T = 0°$K, --- $T = 300°$K, -·-·- $T = 600°$K.

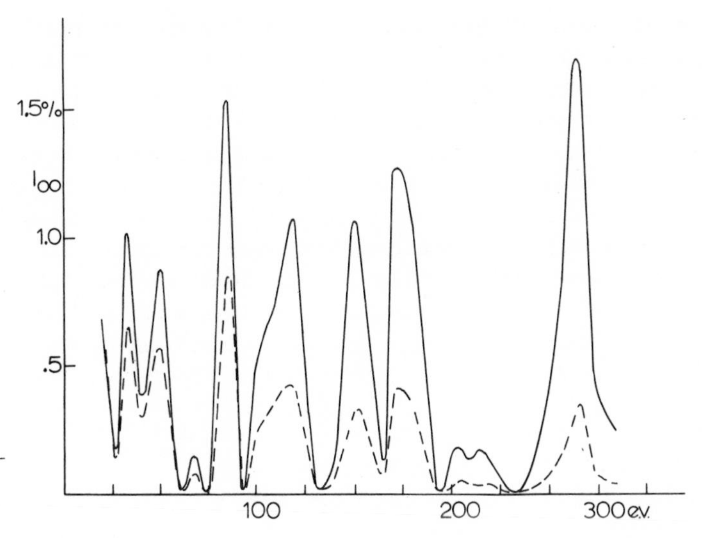

FIGURE 6.15. Intensity of the 00 beam from a copper (001) surface for normal incidence calculated at 300° K, ———, and 600° K, – – –.

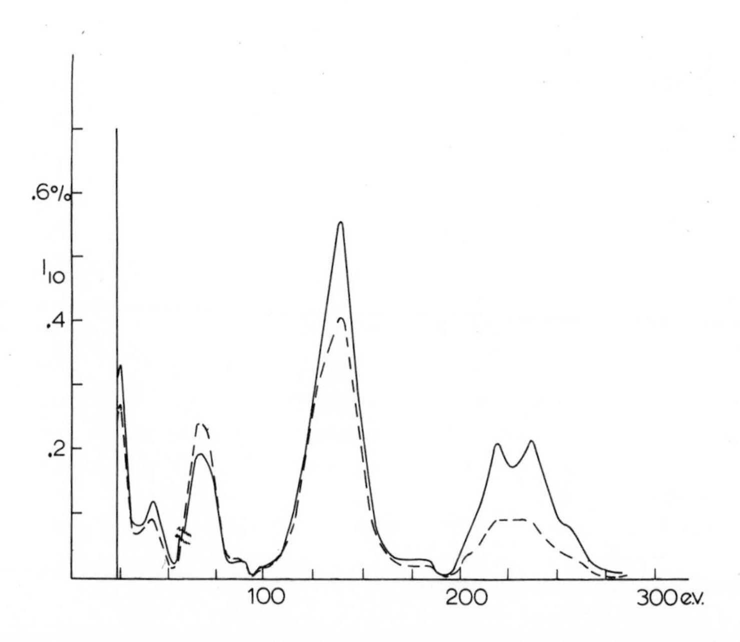

FIGURE 6.16. Intensity of the 10 beam from a copper (001) surface for normal incidence calculated at 300°K, ———, and 600°K, – – – . Above 60 eV the 300°K curve has been multiplied by $\frac{1}{2}$.

contributions from different phase shifts. The higher the temperature and energy the more this is true. Evidently requirements on accuracy of $\delta_\ell(T)$ are more stringent than on $\delta_\ell(T = 0)$, and care must be taken especially over truncation of the number of phase shifts included in the calculation. At 250 eV 7 phase shifts were used, sufficient to reduce errors in the scattering, expressed in terms of equivalent variations in Debye temperature, of from 5 to 10°K.

Finite temperature intensity/energy spectra were calculated by substituting the complex finite temperature phase shifts into the RFS scheme. Calculations are even faster at high temperatures than at 0°K, because reduced back scattering makes for better convergence of the scheme. Comparison of intensities in Figs 6.15–6.17 with experimental intensities in Figs 6.11–13 shows that

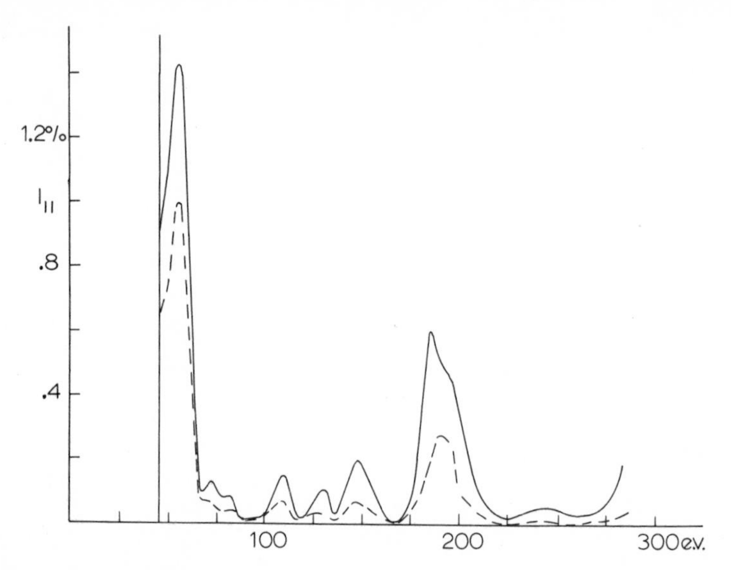

FIGURE 6.17. Intensity of the 11 beam from a copper (001) surface for normal incidence calculated at 300°K, ———, and 600°K, - - - -.

thermal vibrations have produced changes in the right direction. Intensities calculated at 0°K disagree with experiments at 300°K by factors between 10 and 20 at higher energies, whereas for calculations at 300°K disagreement is reduced to factors of 3–4.

We must next investigate how calculated intensities change with temperature. Every atom has been given a Debye temperature of 343°K, but do intensities follow the kinematic law,

$$I_{\mathbf{k}'\mathbf{k}}(T) = I_{\mathbf{k}'\mathbf{k}}(T=0)\exp\left(-\frac{3|\mathbf{k}-\mathbf{k}'|^2 T}{mk_B\Theta^2}\right), \tag{6.59}$$

or does multiple scattering completely change the picture? Energies near peak positions in various beams were chosen, and for the 300°K and 600°K calculations, changes in intensities were interpreted in terms of an effective Debye temperature,

$$\frac{I(T)}{I(T=0)} = \exp\left(-\frac{3|\mathbf{k}-\mathbf{k}'|^2 T}{mk_B \Theta_{\text{eff}}^2}\right) \tag{6.60}$$

Wave-vectors of beams were calculated in the region of constant potential between muffin tins. The results are given in Figs 6.18 to 6.20.

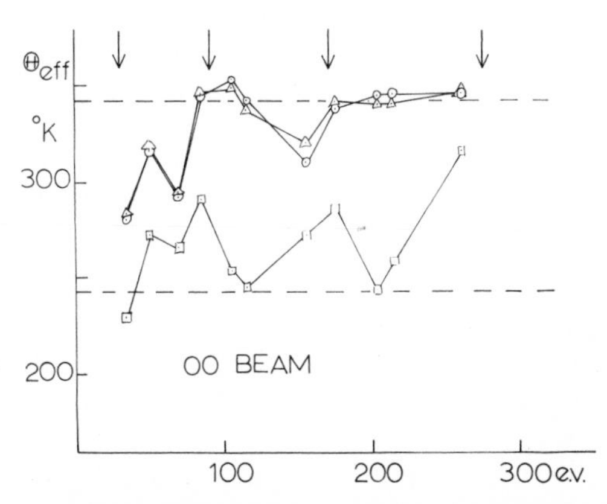

FIGURE 6.18. Effective Debye temperatures of various peaks in the 00 beam from a copper (001) surface at normal incidence, calculated when all atoms have a true Debye temperature of 343°K, ⊙ $T = 300$°K, △ $T = 600$°K, and when the surface is given a different Debye temperature, $343/\sqrt{2}$°k, ⊡ $T = 300$°K. Arrows indicate energies where kinematic Bragg peaks are expected.

For a given peak, the 300°K and 600°K calculations yield almost identical values of Θ_{eff}. Variation of the ℓnI with temperature is linear to a high degree of accuracy. Even though Θ_{eff} is well defined, its numerical value differs from the true Debye temperature, generally taking a lower value, but in a manner varying from one peak to another. As pointed out earlier the reason is that multiple scatterings accumulate Debye Waller factors. Intensity scattered from **k** to **k**″ to **k**‴ to **k**′ would be reduced by

$$\exp\left[-\frac{3T}{mk_B\Theta^2}\left(|k-k''|^2+|k''-k'''|^2+|k'''-k'|^2\right)\right], \tag{6.61}$$

a factor that will usually (though not always) be smaller than the single scattering value

$$\exp\left[-\frac{3T}{mk_B\Theta^2}|\mathbf{k}-\mathbf{k}'|^2\right],$$

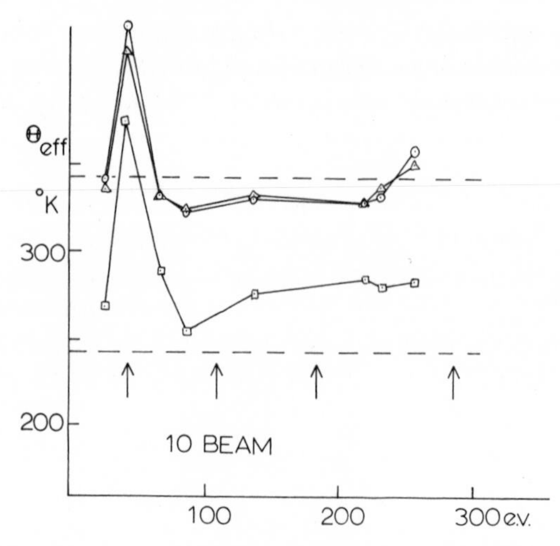

FIGURE 6.19. Effective Debye temperatures calculated as in Fig. 6.18, but for peaks in the 10 beam.

but still leaves ℓnI a linear function of temperature. Variations are largest at low energies because multiple scattering at high energies is confined to a narrow cone of angles within which Debye Waller factors are much nearer unity than for back scattering.

A lesson to be drawn from these calculations is the danger of making purely kinematic interpretation of temperature dependence. In our calculation all atoms have the same Ⓗ, i.e. mean square thermal vibrations, yet if kinematic theory were applied to interpret Fig. 6.18 in terms of thermal vibrations,

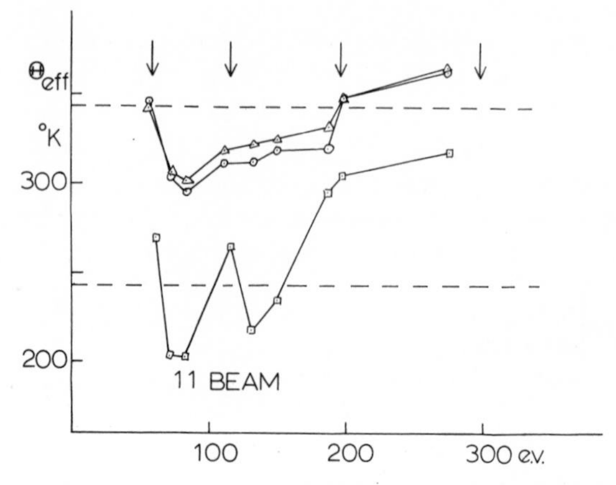

FIGURE 6.20. Effective Debye temperatures calculated as in Fig. 6.18, but for peaks in the 11 beam.

spurious conclusions might be drawn. The dip in effective Debye temperature at low energies where incident electrons penetrate only a few layers, would erroneously imply enhanced vibrations of atoms at the surface. Enhanced vibrations of the top layer of atoms do indeed depress Θ_{eff} at low energies, but they are only part of the story.

Table 6.3 shows that the top layer of copper atoms vibrates with up to twice the mean square amplitude of the bulk. Yet another calculation was made giving atoms in the first layer twice the mean square amplitude of the bulk ($\Theta_s = 343/\sqrt{2}$°K), but assuming that vibration is still isotropic, thus overestimating attenuation of certain diffraction events. Results are shown for $T = 300$°K in Figs 6.21–6.23. Effective Debye temperatures were calculated

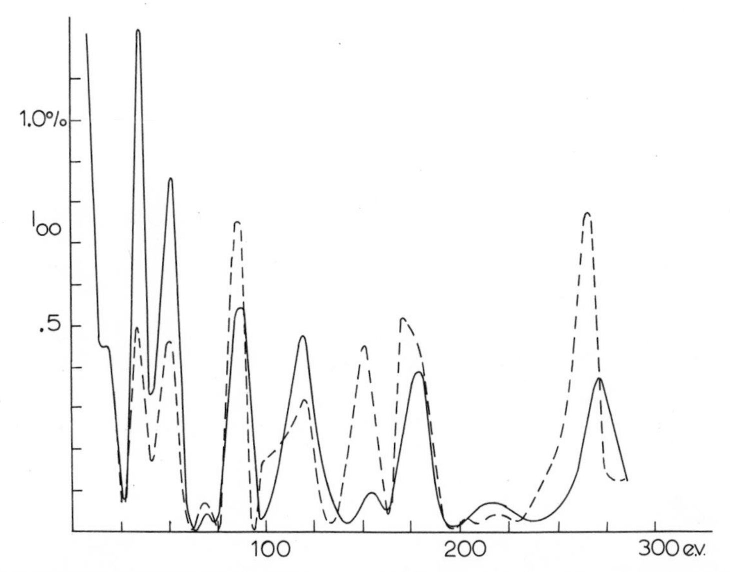

FIGURE 6.21. Intensity of the 00 beam from a copper (001) surface. ——— experiment at 3° from normal incidence, $T = 300$°K; - - - theory at normal incidence, $T = 300$°K with enhanced vibration of the top atomic layer. Theoretical intensities have been multiplied by 0.75.

using (6.60) and plotted in Fig. 6.18–6.20. Not surprisingly Θ_{eff} falls between the bulk and surface values, with a tendency to the bulk value at higher energies. The sort of variations of Θ_{eff} noted by Wilson (1971) (see Fig. 6.5) are evident in the 00 and 11 beams. Electrons reflecting near the kinematic Bragg condition have penetrated further into the crystal because there is less multiple scattering to impede them and consequently their Θ_{eff} is more typical of the bulk.

Figures 6.18 to 6.20 make clear how hard it is to infer vibrational properties

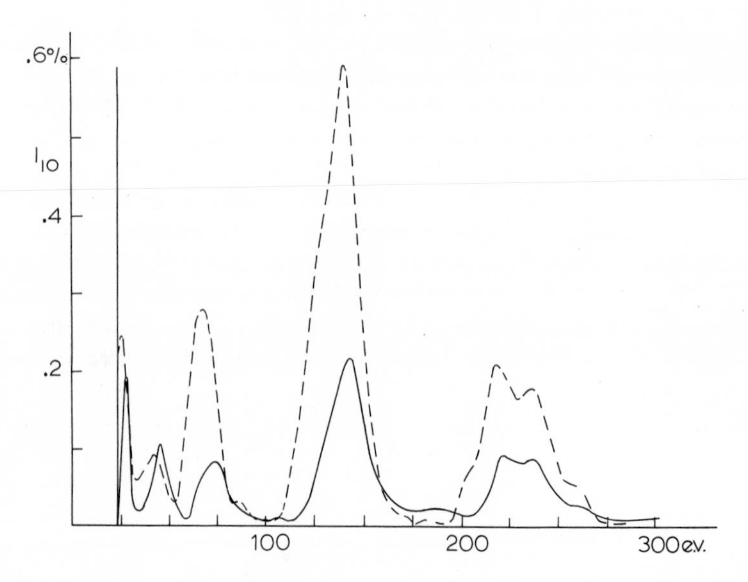

FIGURE 6.22. Intensity of the 10 beam from a copper (001) surface. ——— experiment, $T = 300°K$; --- theory, $T = 300°K$ with enhanced vibration of the top layer. Both curves are taken at normal incidence.

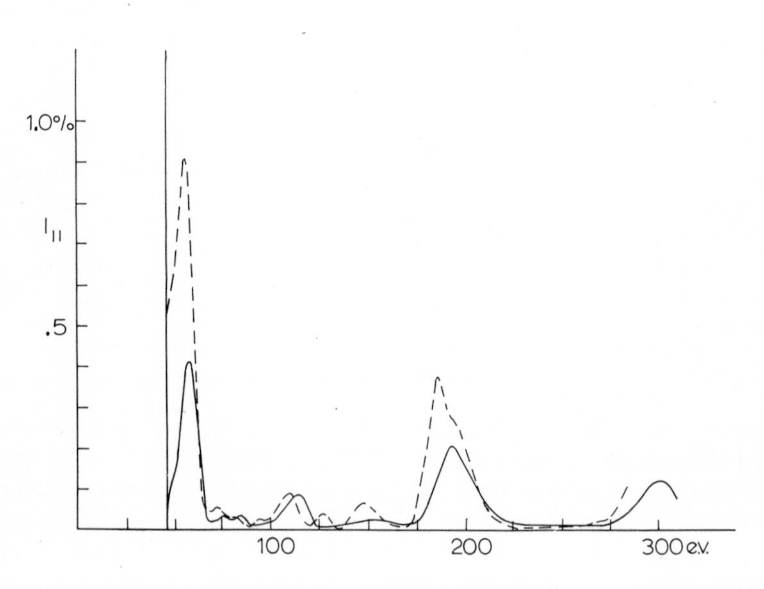

FIGURE 6.23. Intensity of the 11 beam from a copper (001) surface. ——— experiment, $T = 300°K$; ---theory, $T = 300°K$ with enhanced vibration of the top atomic layer. Both curves were taken at normal incidence.

of a crystal from a knowledge of Θ_{eff} alone. For really accurate estimates analysis must proceed via multiple scattering calculations.

When vibrations of atoms in the top layer are enhanced comparison with experiment is improved but there are still discrepancies of the order of $\times 2$ or 3 at high energies. Possible causes of the disagreement fall into two classes: inaccuracies in the non-thermal parameters such as the phase shifts or absorption parameters, and inaccuracies in the thermal properties such as bulk or surface Debye temperatures. It is possible that V_{0i}, the absorption parameter, increases somewhat at higher energies, but Ing (1972) calculates that increases of the order 50 or 100% needed to explain the discrepancies do not occur in copper. The phase shifts are certainly too accurately known to be responsible.

If the error lies with thermal properties, to bring our calculations into agreement with experiment would require Θ_{eff} to lie in the range 220–260°K whereas even curves calculated with enhanced surface vibration have Θ_{eff} in the range 250–290°K. That peaks do in fact have Θ_{eff} in the lower range has been confirmed by Reid (1972). His experimental results for the copper (001) face gave values of Θ_{eff} between 220 and 260°K. Whereas these measurements were not at normal incidence they were sufficiently close to give the order of Θ_{eff}.

Evidently a complete analysis would go on to adjust Θ for bulk and surface atoms until agreement is seen with intensities and Θ_{eff}. It is certainly permissible to adjust the bulk Debye temperature by up to 40°K because, as pointed out in the International Crystallographic Tables (Ed. Lonsdale, 1959), different determinations of Θ for copper vary between 304° and 343°K. However, even with the unrefined thermal parameters, discrepancies in intensities at high energies are reduced from factors of 10–20, by an order of magnitude to factors of 2–3. Intensities in the range 0–100 eV where corrections are small, are for the main part well accounted for.

Another question that may be asked of the theory is, to what extent do temperature effects change penetration into the crystal of coherently scattered electrons? In Fig. 6.24 has been plotted

$$\sum_{\mathbf{g}}(|a_{i\mathbf{g}}^{+}|^2 + |a_{i\mathbf{g}}^{-}|^2)/\sum_{\mathbf{g}}(|a_{0\mathbf{g}}^{+}|^2 + |a_{0\mathbf{g}}^{-}|^2), \tag{6.62}$$

where $a_{i\mathbf{g}}^{\pm}$ are the amplitudes of forward (+) and backward (−) travelling waves between the ith and $(i + 1)$th layers. Compared with the 0°K situation, at room temperature penetration is considerably impeded.

The finite temperature effective t-matrix shows the cause of this effect

$$t(T, \theta_{\mathbf{k}'\mathbf{k}}) = t(0, \theta_{\mathbf{k}'\mathbf{k}}) \exp\left(-\frac{3|\mathbf{k}' - \mathbf{k}|^2 T}{2mk_B\Theta^2}\right) \tag{6.63}$$

The important point is that $t(T, \theta)$ for forward scattering, $\theta = 0$, is unchanged from $t(0, 0)$. In particular, its imaginary part is unchanged,

$$\simeq -2\pi\kappa^{-1} \sum_{\ell} (2\ell + 1) \sin^2 \delta_\ell$$

and gives the total flux scattered, which does not change. Although total flux scattered is unchanged at finite temperatures, not all of it reappears in the coherent beams; some is lost to the incoherent thermal diffuse scattering. Thus at finite temperatures a new mechanism exists for loss of coherent flux.

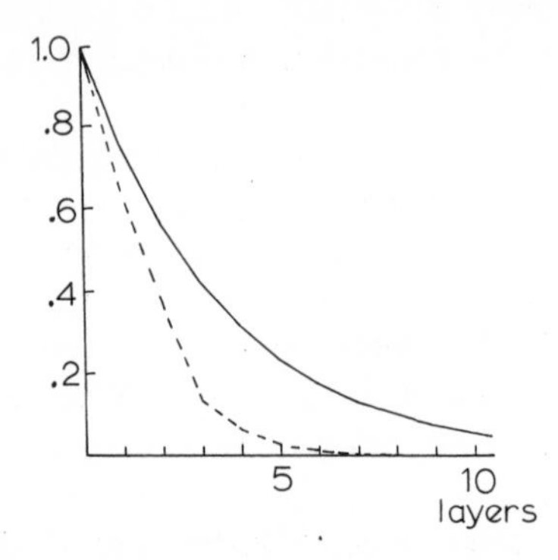

FIGURE 6.24. Plot of expression (6.61), a measure of the amount of flux between layers i and $i + 1$, against layer number, for $T = 0°K$, ———, and $T = 300°K$ --- at 150 eV in copper.

Unlike that due to plasmons, it does not produce absorption that is uniform inside the crystal. An electron whose wavefunction is concentrated near a thermally vibrating ion core is more likely to be incoherently scattered than one staying away from ion cores.

An electron's coherent lifetime governs widths of peaks that can be formed,

$$\Delta E \geqslant \frac{1}{\tau}$$

as we deduced in Chapter I, and it might be anticipated that temperature will affect peak widths through its influence on lifetimes. Experimental results are often in conflict on this matter. Complications arise because many peaks measured in LEED experiments are in fact conglomerates of several simple Bragg peaks. It can happen that the outline of the conglomerate changes because of different Debye Waller factors associated with component peaks. Thus an impression of strong temperature dependence of widths can be created. Conversely the width of a conglomerate may be determined primarily by the separation of partially overlapping components. Thus any change in width of individual components is masked by the overlap. These effects probably account for the conflict of experimental evidence.

The complications caused by non-uniform thermal absorption go away in the limit of infinite temperature. Because of the Debye Waller factor all scattering except that in the forward direction is zero.

$$t(T = 0, \theta = 0) = -2\pi\kappa^{-1}\sum_{\ell}(2\ell + 1)\sin(\delta_\ell)\exp(i\delta_\ell). \tag{6.64}$$

In Chapter V, Section F, we saw that it is easy to write down a potential that produces only forward scattering. It is a uniform potential taking the value,

$$V_f = -2\pi(\kappa\Omega)^{-1}\sum_{\ell}(2\ell + 1)\sin(\delta_\ell)\exp(i\delta_\ell), \tag{6.65}$$

where Ω is the volume of the unit cell. At $T = \infty$ the ion cores are smeared into a uniformly absorbing background. Together with the uniform potential present at $T = 0$, an electron experiences a total of

$$V_f + V_0 \tag{6.66}$$

at $T = \infty$. Each contribution has an imaginary component, that of V_0 is around -4 eV and that of V_f around -5 eV in copper.

Diffracted intensities go to zero as the infinite temperature limit is reached, but it might be supposed that stopping just short of the limit a simple theory could be found. Unfortunately the rapid change of scattering strength with angle at high temperatures prevents any separation in magnitude between orders of scattering, and a simple kinetic theory does not result in general. What we can say is that in consequence of zero angle scattering remaining strong and contributing to the uniform potential as in (6.66) peaks, in the high temperature limit are subject to an additional broadening

$$\Delta E \geqslant |2(V_{0i} + V_{fi})| \simeq 18 \text{ eV} \tag{6.67}$$

Chapter 7

Applications of LEED to Surface Structure Analysis

VIIA Surface structures and LEED experiments

In constructing and confirming a theory of LEED we have used experiments made on surfaces which are as simple and as well determined as possible, in order to eliminate uncertainty in our work. Yet the goal we have been working towards is use of experiment and theory of LEED as a technique for determination of structure. Now that the theory has been set on a firm basis by showing that it accounts for diffraction from well-understood surfaces, we can turn our attention to more complicated systems.

An electron diffracting from a surface collects information about the potential. There are many details of the potential we may wish to discover, but by far the most gross factors are the positions and types of ion cores near the surface. In speaking of surface structures we shall mainly be concerned with how coordinates of ion cores change diffraction, and how we can deduce the coordinates from LEED.

There are two ways in which surfaces can deviate from the ideal situation in which a plane divides space into perfect bulk crystal on one side, and vacuum on the other. Either the last few layers of bulk atoms can rearrange themselves, or foreign atoms can adhere to the surface forming a new layer.

Consider the rearrangements in a clean surface. We have already seen how the top layers are often less strongly bound, and their spacings dilate by of the order of 5%, but more drastic changes are possible. The most violent is called faceting. High index surfaces often have much larger surface energies than low index ones, and in some cases the difference may be so large that it is energetically favourable for a flat high index surface to rearrange by migration of atoms into a corrugated arrangement of interlocking low index surfaces. Movement of a large number of atoms is needed to produce the rearrangements which commonly take place on a scale greater than 100Å, and high temperatures are usually required to raise mobilities sufficiently, but sometimes faceting occurs at lower temperatures as is the case for silver (110) surfaces (Morabito, Steiger, and Somorjai, 1969).

Less drastic rearrangements, on a microscopic scale are also met with. In many semiconductors a surface disrupts the bonding mechanism and surface atoms rearrange themselves to a new structure which has a larger unit cell than the simple surface, but whose periodicity is related to that of the bulk. A list of surface structures found on clean surfaces of some semiconductors and metals is given in Table 7.1. The notation $(n \times m)$ tells us that the **a** axis of the larger unit cell is n times the length of that of the simple cell, and the **b** axis is m times greater.

Metals seem much less prone to surface instability than insulators. The structures are often temperature dependent, and by heating the surfaces through a range of temperatures phase transformations can be induced between different structures. For example the silicon (111) surface shows a

material	surface structure
Au (100)	5 × 1
Au (110)	2 × 1
Bi $(1\bar{1}20)$	2 × 10
CdS (0001)	2 × 2
GaAs (111)	2 × 2
GaSb (111)	2 × 2
Ge (100)	4 × 4
Ge (111)	2 × 8
Ge (110)	2 × 1
InSb (100)	2 × 2
InSb (111)	2 × 2
Ir (100)	5 × 1
Pd (100)	C(2 × 2)
Pt (100)	5 × 1
Sb $(1\bar{1}20)$	6 × 3
Si (100)	4 × 4
Si (111)	7 × 7
Te (0001)	2 × 1

TABLE 7.1. Some surface structures found on clean materials (taken from Somorjai and Farrell (1971).

LEED pattern consistent with no rearrangement having taken place below 700°C. In the range 700°–800°C a complicated pattern is seen due to the 7 × 7 structure, whilst above 800°C yet another structure appears: 2 × 2 (Schlier and Farnsworth, 1959; Lander and Morrison, 1962). So far none of the detailed geometry of these structures has been unravelled despite the importance they might have in determining electrical properties of surfaces.

Even more prolific variety of structure is found in systems of foreign atoms adsorbed on surfaces. Atoms are preferably introduced from the gaseous phase by raising the partial pressure of those atoms in the system, except in the case of materials that are not easily vapourised where an ion beam can be used instead. The amount of adsorbate on a surface is difficult to estimate accurately especially when deposited from the gaseous phase. Some assumptions about sticking probabilities combined with elementary kinetic theory of gases are often employed, but more accurate determinations await development of a tool for quantitative surface analysis such as Auger spectra may eventually provide.

A distinction is made between physisorption and chemisorption, based on binding energies. Physisorption involves energies of less than about 10 k cal./mole. An example would be adsorption of xenon on an iridium (100) surface (Ignatjevs, Jones and Rhodin, 1972). The low binding energies require that a cooling system be employed in physisorption experiments.

Chemisorption systems are of greater importance: because of their higher binding energies they exist over a wide range of temperatures, and constitute at least the initial stage of all gas-solid chemical reactions amongst which are many oxidation and catalytic processes.

Chemical forces binding atoms to surfaces are as yet little understood. The complication of lying halfway between the theoretical extremes of bulk solid and isolated molecule, combined with the small amount of data available on inter-atomic distances has discouraged many workers. There have been some enterprising beginnings by Bennett, McCarroll and Messmer (1971), Cyrot-Lackmann (1970), Grimley (1971), and Newns (1969). Occasionally a picture of chemical bonds is useful in predicting surface structures but more usually the surface environment is so radical a departure from a molecular environment that in so far as can be judged from the data, bonding considerations are not reliable indications of what is going to happen for a given adsorbate and substrate. Of greater importance are more elementary considerations such as the size of atoms involved, which used in conjunction with a knowledge of the unit cell obtained from the diffraction pattern, can sometimes dictate what the structure must be by considerations of packing atoms into the cell.

Electrostatic considerations can also be powerful. If it is thought that an adsorbate is in an ionised state then its geometry is subject to the restriction that the dipole moment of the structure must be equal to the work function change. Since work function changes are always small, $\simeq \pm 2\,\text{eV}$, these arguments usually imply that at high coverages, ionic species must be paired within a plane parallel to the surface to keep the dipole moment small.

The large energy available in chemisorption systems rivals the binding energy of the substrate itself, and sometimes the substrate plays a more active role in the surface structure. It can be reconstructed by adsorbing atoms to form a structure in which substrate atoms are included. Such possibilities provide obvious complications when it comes to elucidating the structure by comparison of calculations with experiment.

An enormously wide variety of ordered surface structures exists and in Table 7.2 we give a few of the more common examples taken from Somorjai and Farrell (1971).

We must now turn our attention to how LEED responds to surface structures. Generally the first indication that an adsorbate is being deposited is a fading of the spot pattern and increase in background intensity on the fluorescent screen. Atoms deposit in a random manner giving initially a disordered structure. If the surface has been well prepared and cleaned prior to adsorption, annealing of the adsorbate at an elevated temperature usually leads to an ordered structure. The background intensity diminishes; spots become sharper; and frequently additional spots appear. Only rarely is it the case that adsorbing atoms find it energetically favourable to occupy positions that preserve the original unit cell of the surface. More usually a structure with a larger unit cell is formed, but still correlated with the lattice of the bulk. The number of spots in a diffraction pattern is increased in the ratio of areas of the adsorbate to clean surface unit cells: a grating with twice the spacing gives twice as many orders of diffraction.

If sodium is deposited on a nickel (001) surface, the large sodium atom

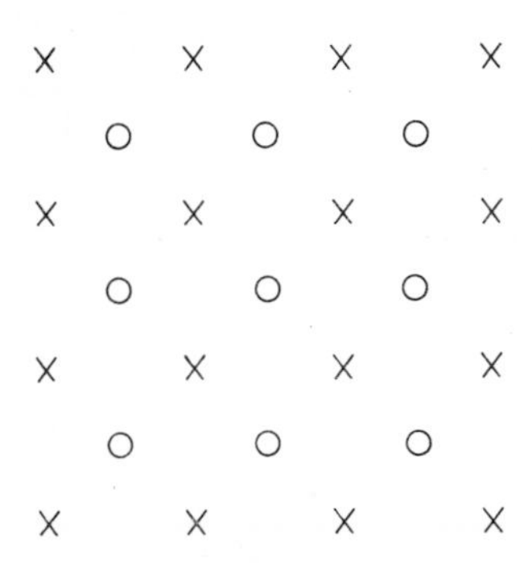

FIGURE 7.1. Diffraction pattern from a nickel (001) surface with sodium atoms adsorbed, taken after annealing. x spots present when the surface was clean, o additional spots introduced by the overlayer.

Surface	Adsorbed gas	Surface structure
Face-centred cubic structures		
Ni(111)	O_2	(2 × 2)–O
		($\sqrt{3} \times \sqrt{3}$)R30°–O
	CO	(2 × 2)–CO
	H_2	(1 × 1)–H
		Disordered
	N_2	Not adsorbed
	CO_2	(2 × 2)–CO_2
		(2 × $\sqrt{3}$)–CO_2
Ni(100)	O_2	(2 × 2)–O
		c(2 × 2)–O
	CO	c(2 × 2)–O
	H_2	Disordered
	N_2	Not adsorbed
	C	
Ni(110)	O_2	(2 × 1)–O
		(3 × 1)–O
		(5 × 2)–O
		(5 × 1)–O
	CO	(1 × 1)–CO
	H_2, D_2	(1 × 2)–H
	H_2O	(2 × 1)–H_2O
	C	c(2 × 2)–C
Ni(210)	I_2	Facet to Ni(540)
Pt(111)	O_2	(2 × 2)–O
	CO	c(4 × 2)–CO
Cu(111)	O_2	(11 × 5)R5°–O
Cu(100)	O_2	c(2 × 2)–O
		(1 × 1)–O
		(2 × 1)–O
	N_2	(1 × 1)–N
Cu(110)	O_2	(2 × 1)–O
		c(6 × 2)–O
Cu(035)	(1 × 1)–O	
Cu(014)	O_2	(1 × 1)–O
Al(100)	O_2	Disordered

Surface	Adsorbed gas	Surface structure
Body-centred cubic structures		
W(110)	O_2	(2 × 1)–O
		c(14 × 7)–O
		c(21 × 7)–O
		c(48 × 16)–O
		c(2 × 2)–O
		(2 × 2)–O
		(1 × 1)–O
	CO	Disordered
		c(9 × 5)–CO
W(111)	O_2	To (211) facets
W(211)	O_2	(2 × 1)
		(4 × 3)
		(1 × 2)
		(1 × n)–O, n = 1,2,3,4
	CO	c(6 × 4)–CO
		(2 × 1)–CO
		c(4 × 2)–CO
W(100)	O_2	(4 × 1)–O
		(2 × 1)–O
	CO	c(2 × 2)–CO
	N_2	c(2 × 2)–N
	H	c(2 × 2)–H
		(2 × 5)–H
		(4 × 1)–H
Diamond structures		
Si(111)– (1 × 1)	O_2	(1 × 1)
(7 × 7)		Disordered
	H_2S	(2 × 2)–S
	H_2	Not adsorbed
	H_2Se	(2 × 2)–Se
Si(100)	O_2	(1 × 1)
Ge(111)	O_2	(1 × 1)
		Disordered
	I	(1 × 1)
Ge(100)	O_2	(1 × 1)
		Disordered
	I_2	(3 × 3)
Ge(110)	O_2	(1 × 1)
		Disordered

TABLE 7.2. Some surface structures formed by adsorbates.

cannot fit within the small nickel unit cell and a diffraction pattern given in Fig. 7.1 appears. It is not hard to show that the pattern is consistent with sodium atoms being arranged as shown in Fig. 7.2.

Figure 7.1 shows a relatively straightforward diffraction pattern. Another more complicated example is given in Fig. 7.3 which is observed when potassium is adsorbed on a nickel (001) surface. The additional spots almost imply that the large potassium atoms form a structure whose unit cell is sixteen times larger than the original clean surface cell (Fig. 7.4). The key to interpreting this pattern is that systematic absences are seen of spots at $(\frac{1}{4}\frac{1}{4})$, $(\frac{1}{4}\frac{3}{4})$ etc., for which there is no explanation if the structure of Fig. 7.4 is correct. The true structure is more complicated.

Potassium in fact forms a rectangular unit cell, such as is shown in Fig. 7.5. The spots in Fig. 7.3 are caused by there being present two types of structures, *A* and *B* which are equivalent and differ in that the unit cells are rotated by

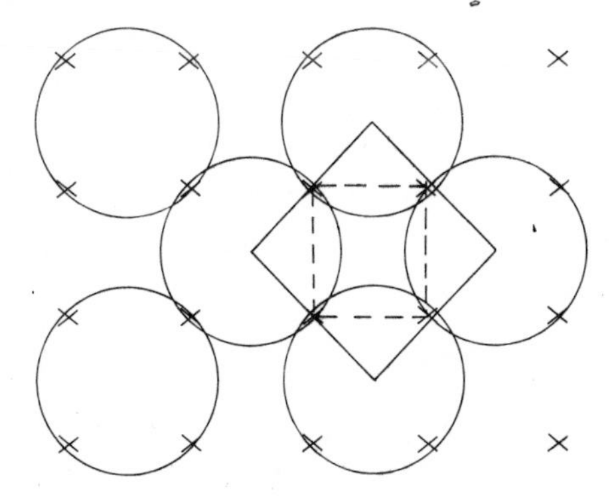

FIGURE 7.2. A surface structure consistent with the diffraction pattern of Fig. 7.1 x nickel atoms in the top nickel layer, o sodium atoms. ——— unit cell of the surface plus adsorbate, - - - original unit cell. Sodium atoms are drawn with radii appropriate to bulk sodium.

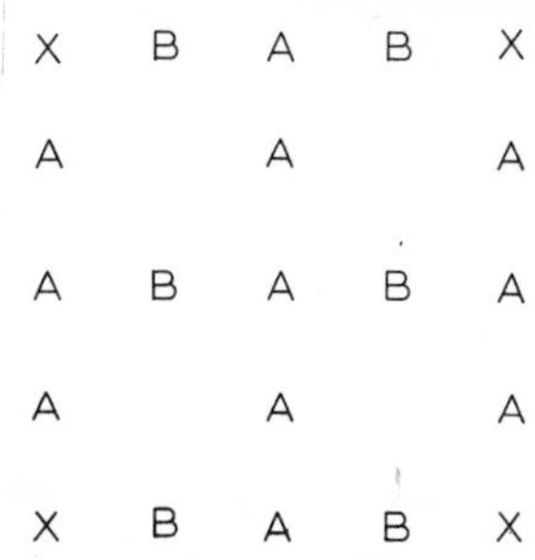

X	B	A	B	X
A		A		A
A	B	A	B	A
A		A		A
X	B	A	B	X

FIGURE 7.3. Diffraction pattern from a nickel (001) surface with potassium atoms adsorbed, taken after annealing. x spots present when the surface was clean, A and B: additional spots introduced by the overlayer.

90° relative to one another. There are assumed to be islands of each type on the surface. Spots labelled A are caused by islands of type A, introduction of type B islands adds the spots labelled B. The systematic absences are accounted for.

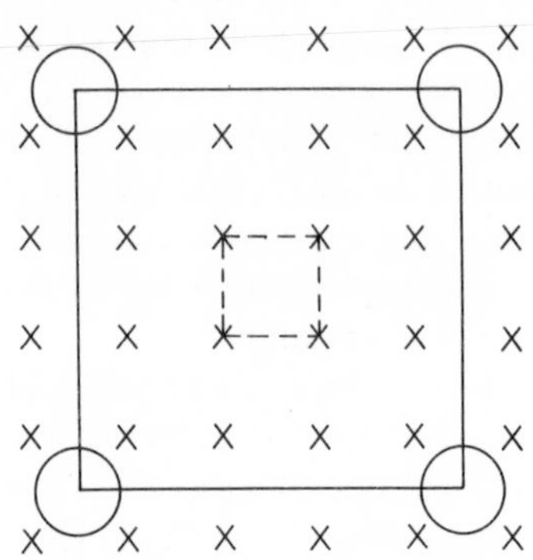

FIGURE 7.4. A surface structure for potassium on a nickel (001) surface, consistent with the diffraction pattern of Fig. 7.3 except that the systematic absence of spots at $(\frac{1}{4}\frac{1}{4})$ etc. is not explained.

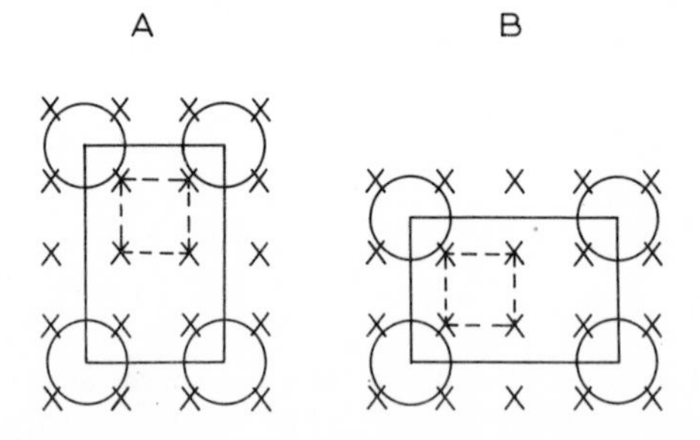

FIGURE 7.5. When an adsorbate forms a structure with a unit cell less symmetrical than that of the clean surface, equivalent domains coexist on the surface. Here we show a possible structure for potassium on a nickel (001) surface. x nickel atoms in the top layer of the bulk, o potassium atoms.

The existence of equivalent but different islands, or domains as they are often called, is a complication in the analysis of diffraction patterns for overlayers. The full consequences of their existence will be dealt with when a more formal description of surface structures is made in Section B.

It has been carefully stressed that structures suggested as interpreting diffraction patterns are not unique. In themselves the patterns can only ever yield information about the unit cell of the surface. In Fig. 7.2 the layer of sodium atoms could be translated laterally and vertically, placing the sodium atoms anywhere within the larger unit cell, and still the same spots would appear in the pattern. Only size and shape of unit cell are implied by the pattern. To measure the disposition of contents of the cell we must turn to intensities.

Suppose we wish to know the separation between the adsorbed layer and the bulk. Beams reflected from the overlayer and from the bulk differ in phase, partly because of the different reflection coefficients but also because of the path difference between the two beams (Fig. 7.6). Intensity and phase of the total will both be sensitive to the path difference, and the intensity, if not the phase, is experimentally accessible.

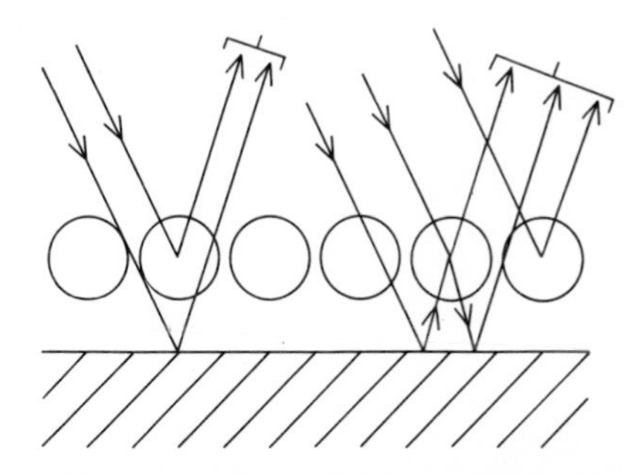

FIGURE 7.6. The distance between overlayer and bulk introduces phase differences between reflections. In this way intensities convey information about geometry.

Farnsworth has made a wide study of adsorbate systems. Some of his recent work with Zehner (Zehner and Farnsworth, 1972) will be used to illustrate influence of adsorbates on intensities. The rhenium $(10\bar{1}0)$ surface will adsorb carbon monoxide, taking one molecule for every atom in the topmost layer of rhenium. The structure of the adsorbate has the same unit cell as the clean surface and hence there are no additional beams to provide evidence of adsorbtion. However, diffracted intensities can be expected to change. Figure 7.7 compares plots of intensity/energy spectra.

There can be no doubt that spectra show sensitivity to the adsorbed monolayer. In some cases curves are completely transformed by adsorption: new peaks occur and some old ones disappear. Where the old peaks can be identified in the new curves, they are frequently changed in relative intensity and shifted in position. From the gross nature of the many changes it is clear that much information resides in spectra, promising that here indeed is a technique that can unravel detailed surface geometry. There was initially a substantial barrier to extraction of structural information contained in spectra, due to complexity of the diffraction process. We have seen in preceding chapters that recent advances enable a good account to be given of diffraction at a clean surface, and there is no reason why diffraction in the presence of an adsorbate should make any changes in principle.

In the rest of this chapter the relatively simple step will be accomplished,

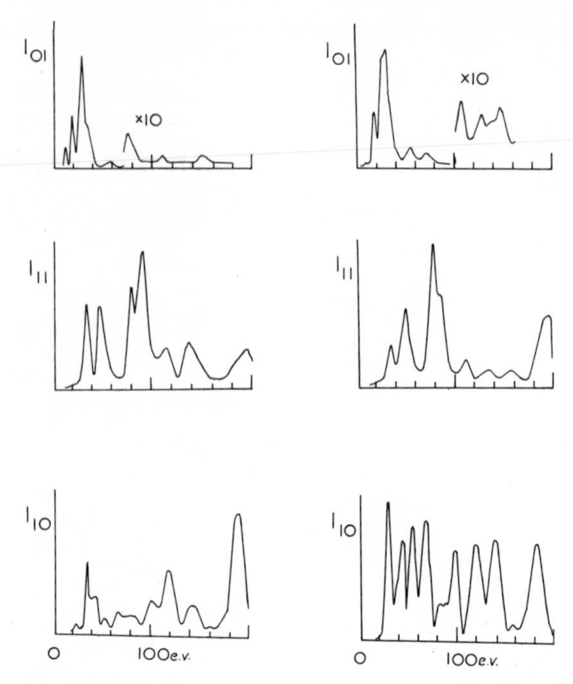

FIGURE 7.7. Intensity/energy plots of various beams taken at normal incidence on a $(10\bar{1}0)$ rhenium surface. The graphs on the left are for a clean surface, those on the right for a surface with a (1×1) overlayer of carbon monoxide.

of extending theory for clean surfaces without superstructures, to the more complicated general surface problem.

VIIB Formal description of surface structure

Suppose that we are dealing with a crystal whose surface is described by terminating the bulk crystal at a plane. The surface unit cell would have sides given by vectors of the bulk unit cell

$$\mathbf{a}_B, \quad \mathbf{b}_B \tag{7.1}$$

lying in the plane of the surface, as we described in Chapter I. In this chapter we must admit the more general case where layers of atoms at the surface do not have the same periodicity as those in the bulk. Great complexities can be imagined, but to illustrate principles it suffices to consider a situation where the topmost layer of atoms has a structure different from the bulk. This layer may be either the top layer of the crystal itself rearranged to a new structure, or a layer of adsorbed molecules. We shall assume that it has periodicity described by a unit cell with sides

$$\mathbf{a}_A, \quad \mathbf{b}_A \tag{7.2}$$

It is possible to express these vectors in terms of those for the bulk:

$$\mathbf{a}_A = G_{11}\mathbf{a}_B + G_{12}\mathbf{b}_B \tag{7.3a}$$

$$\mathbf{b}_A = G_{21}\mathbf{a}_B + G_{22}\mathbf{b}_B \tag{7.3b}$$

It is easy to show that the ratio

$$\frac{\text{area of adsorbate cell}}{\text{area of bulk cell}} = \det G. \tag{7.4}$$

We wish to consider symmetry of the complete system of clean surface plus adsorbate. There are three situations that occur:
(i) Take the lattice with the larger unit cell (in practice the unit cell of the adsorbate is usually largest); if every point in this lattice is also a point in the finer lattice, the lattices are *simply related.*
(ii) If the lattices are not simply related, it still may be true that a third lattice can be defined with unit cell sides

$$\mathbf{a}_C, \quad \mathbf{b}_C, \tag{7.5}$$

such that every point in lattice C is a point in both the A and B lattices. The points of lattice C are coincidence sites and the two original lattices are *rationally related.*
(iii) If neither (i) or (ii) is true there is no coincidence lattice and A and B are *irrationally related.*

In case (ii) the complete system has translational symmetry described by a unit cell of sides given by (7.5). Case (i) is a special case of (ii) in which the unit cell of the complete system is equal to the larger of unit cells A and B. In case (iii) the system has no two-dimensional translational symmetry.

Again relationships can be written down in terms of the bulk cell

$$\mathbf{a}_C = P_{11}\mathbf{a}_B + P_{12}\mathbf{b}_B, \tag{7.6a}$$

$$\mathbf{b}_C = P_{21}\mathbf{a}_B + P_{22}\mathbf{b}_B, \tag{7.6b}$$

or of the adsorbate cell,

$$\mathbf{a}_C = Q_{11}\mathbf{a}_A + Q_{12}\mathbf{b}_A, \tag{7.7a}$$

$$\mathbf{b}_C = Q_{21}\mathbf{a}_A + Q_{22}\mathbf{b}_A. \tag{7.7b}$$

Since lattice C by definition coincides with points in both A and B, elements of P and Q must be integers.

A more common but less generally applicable way of describing the total symmetry is by the expression

$$\left(\frac{|\mathbf{a}_C|}{|\mathbf{a}_B|}, \frac{|\mathbf{b}_C|}{|\mathbf{b}_B|}\right)\phi. \tag{7.8}$$

It has been assumed that unit cells B and C have the same included angle. ϕ is the angle of rotation of cell C relative to B; if ϕ is zero it is usually omitted. This is the notation used in Table 7.1 and 7.2. Sometimes it is convenient to use a non-primitive centred unit cell for the coincidence lattice. In that case the notation (7.8) is preferred with a 'c'. An example is given in Fig. 7.8.

An observation that will be important in the theoretical treatment of diffraction from these structures is that unit cell C contains an integral number of unit cells of B (and of A), see Fig. 7.9. The number of cells of B is det P, and of A, det Q. Each point of lattice B within a unit cell of C has a displacement relative to the origin of that unit cell of C.

$$m\mathbf{a}_B + n\mathbf{b}_B \tag{7.9}$$

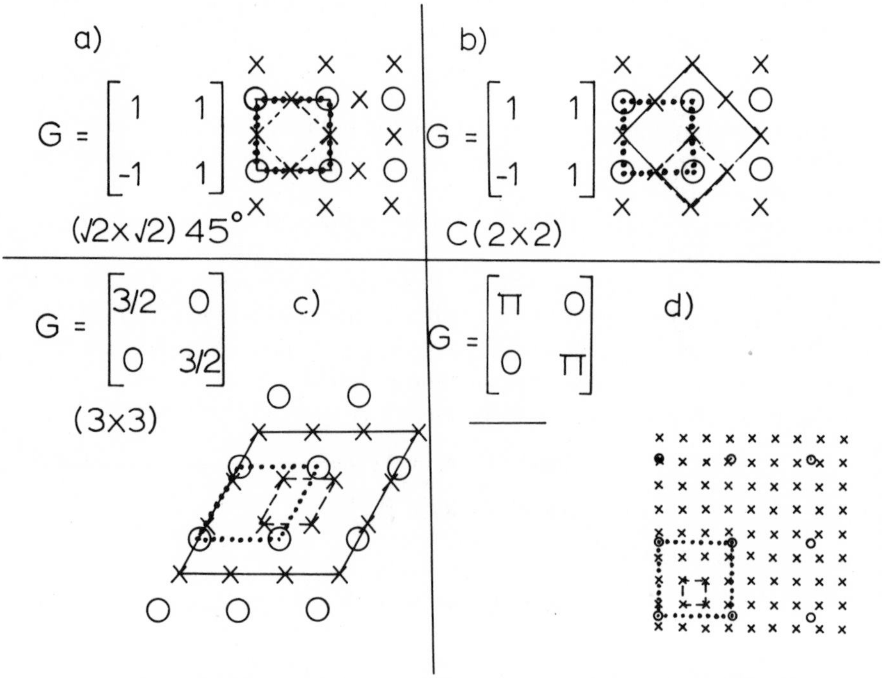

FIGURE 7.8. Examples of some lattices – – – unit cell of B, · · · · unit cell of A, ——— unit cell of C. (a) an example of simply related structures, (b) the same structure but a non-primitive unit cell is used for C, (c) rationally related structures, (d) irrationally related structures.

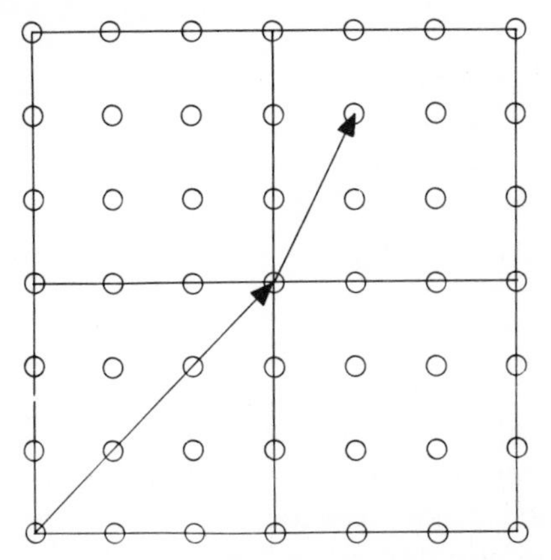

FIGURE 7.9. Any point in lattice B or in lattice A, can be expressed in terms of the unit cell of lattice C in which it is located, plus its displacement relative to the origin of that cell.

where m and n are integers. Only a finite number of pairs (m, n) are needed to account for all points. The position of any point in lattice B, $\mathbf{R}^{(B)}$, can be expressed in terms of the position of the cell of C in which it finds itself, $\mathbf{R}^{(C)}$, plus its position relative to the origin of that cell, $\mathbf{R}^{(0)}$.

$$\mathbf{R}^{(B)} = \mathbf{R}^{(C)} + \mathbf{R}^{(0)} = M\mathbf{a}_C + N\mathbf{b}_C + m\mathbf{a}_B + n\mathbf{b}_B. \tag{7.10}$$

M and N can take any integer value between $\pm\infty$. Lattice B has been broken up into a finite number of sub lattices each characterised by a different $\mathbf{R}^{(0)}$. When speaking in terms of sub-lattices it will be convenient to label $\mathbf{R}^{(B)}$ as follows

$$R_j^{(B)}(s) = \mathbf{R}_j^{(C)} + \mathbf{R}_s^{(0)}. \tag{7.11}$$

A LEED experiment measures the reciprocal rather than the real lattice. Equations (1.28) show how to construct the reciprocal from the real lattice, and can be rewritten as follows

$$\mathbf{A}_B = \frac{2\pi}{\mathbf{a}_B \cdot \tilde{\mathbf{b}}_B} \tilde{\mathbf{b}}_B, \tag{7.11a}$$

$$\mathbf{B}_B = -\frac{2\pi}{\mathbf{a}_B \cdot \tilde{\mathbf{b}}_B} \tilde{\mathbf{a}}_B, \tag{7.11b}$$

where for example the vector

$$\tilde{\mathbf{b}} = (\tilde{b}_x, \tilde{b}_y) = (b_y, -b_x). \tag{7.12}$$

By equations (7.6) the real coincidence lattice is related to the real lattice of the bulk crystal, implying the following relationships between the corresponding reciprocal lattices.

$$\mathbf{A}_C = \frac{2\pi}{\mathbf{a}_B \cdot \tilde{\mathbf{b}}_B \det P}(P_{21}\tilde{\mathbf{a}}_B + P_{22}\tilde{\mathbf{b}}_B) = \tilde{P}_{11}\mathbf{A}_B + \tilde{P}_{12}\mathbf{B}_B, \tag{7.13a}$$

$$\mathbf{B}_C = -\frac{2\pi}{\mathbf{a}_B \cdot \tilde{\mathbf{b}}_B \det P}(P_{11}\tilde{\mathbf{a}}_B + P_{12}\tilde{\mathbf{b}}_B)$$

$$= \tilde{P}_{21}\mathbf{A}_B + \tilde{P}_{22}\mathbf{B}_B, \tag{7.13b}$$

$$\tilde{\mathrm{P}} = \frac{1}{\det P}\begin{bmatrix} P_{22} & -P_{21} \\ -P_{12} & P_{11} \end{bmatrix}. \tag{7.14}$$

In reciprocal space the roles of lattices B and C are reversed. In real space, lattice C had a larger unit cell and was a sub-lattice of B; P, relating unit cells in equation (7.6) had integer elements. In reciprocal space, by taking the inverse of (7.14) reciprocal unit cell B can be related to reciprocal unit cell C by equations (7.15)

$$\mathbf{A}_B = P_{11}\mathbf{A}_C + P_{21}\mathbf{B}_C, \tag{7.15a}$$

$$\mathbf{B}_B = P_{12}\mathbf{A}_C + P_{22}\mathbf{B}_C. \tag{7.15b}$$

Clearly P_{11} etc. are integers, and we conclude that reciprocal lattice B is a sub-lattice of reciprocal lattice C, and further that the ratio of unit cell areas is the inverse of the ratio of real space unit cell areas. Thus a point in reciprocal lattice C can by analogy with (7.10) be written

$$\mathbf{g}^{(C)} = M\mathbf{A}_B + N\mathbf{B}_B + m\mathbf{A}_C + n\mathbf{B}_C = \mathbf{g}^{(B)} + \mathbf{g}^{(0)} \tag{7.16}$$

where $\mathbf{g}^{(B)}$ is a vector of reciprocal lattice B. $\mathbf{g}^{(0)}$ is the position of lattice point $\mathbf{g}_j^{(C)}$ relative to the origin of the unit cell of reciprocal lattice B in which it finds itself. Reciprocal lattice C can be divided into a finite number of sub-lattices, each characterised by a value of $\mathbf{g}^{(0)}$. When working in terms of sub-lattices the following labelling will be used

$$\mathbf{g}_j^{(C)}(s) = \mathbf{g}_j^{(B)} + \mathbf{g}_s^{(0)} \tag{7.17}$$

In the case that bulk and adsorbate structures are rationally related, the surface has two-dimensional translational symmetry, and the picture we have adopted throughout the book holds good, of incident electrons being diffracted by the surface into a discrete set of beams determined by the reciprocal lattice. Since the surface in the presence of an adsorbate has real lattice structure C, the diffraction pattern is determined by vectors of reciprocal lattice C, $\mathbf{g}^{(C)}$.

It might be supposed that a much more gross conceptual change is involved in passing to systems in which bulk and adsorbate have irrationally related structures. Then the surface as a whole has no two-dimensional translational symmetry and diffraction of discrete beams no longer follows from group theoretical considerations. Figure 7.10 shows irrationally related cells *A* and *B* for copper adsorbed onto a tungsten (110) surface (Taylor, 1966). An understanding of patterns formed on the screen is most easily had by treating adsorbate and bulk separately.

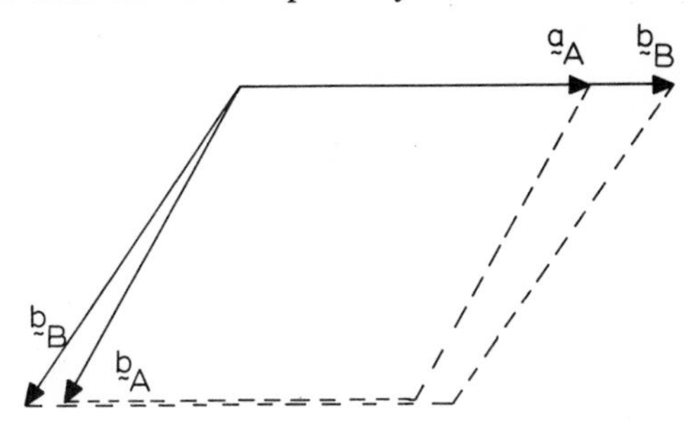

FIGURE 7.10. Unit cells of bulk and adsorbate structures for copper adsorbed on a tungsten (110) surface.

As electrons pass through the adsorbate they are diffracted into a set of beams characteristic of reciprocal lattice *A*. Those back scattered form discrete spots on the screen, each spot labelled by the amount of momentum parallel to the surface it has acquired $\mathbf{g}^{(A)}$. Those forward scattered reflect from the bulk, each beam changing its parallel momentum by units characteristic of reciprocal lattice *B*, $\mathbf{g}^{(B)}$. The wavefield now consists of beams with momenta of the form

$$\mathbf{g}^{(A)} + \mathbf{g}^{(B)} \tag{7.18}$$

Subsequent transmission through the adsorbate introduces no new momenta, and beams striking the screen have parallel momenta given by (7.18). There is a mathematical theorem which states that if lattices *A* and *B* are irrationally related, any vector can be approximated to any desired degree of accuracy by (7.18). In principle it is possible to diffract the beam to any part of the screen by a judicious combination of $\mathbf{g}^{(A)}$ and $\mathbf{g}^{(B)}$. Of course to reach some points on the screen, beams with extremely large values of $\mathbf{g}^{(A)}$ and $\mathbf{g}^{(B)}$ have to be excited at some intermediate stage. Large values of parallel momentum imply that a beam decays rapidly between layers and is not important in the diffraction process. The only points on the screen that receive beams of appreciable strength are those corresponding to relatively small values of $\mathbf{g}^{(A)}$ and $\mathbf{g}^{(B)}$ in expression (7.18). Thus a discrete set of spots is observed even in the case of irrationally related lattices. The number of beams it is possible to produce with strong amplitude is increased if there is appreciable scattering back and forth between overlayer and bulk, and also by increasing the energy which increases the number of momenta $\mathbf{g}^{(A)}$ and $\mathbf{g}^{(B)}$ that can be excited.

Another problem we are now equipped to handle is that of antiphase domains. In Section A we hinted briefly that an adsorbate with structure rationally related to the bulk could occupy a number of equivalent sites. Consider an adsorbate/bulk system. The adsorbate is bonded to atoms in the layer below. If the adsorbate is lifted from the surface and translated parallel to the surface a distance

$$n\mathbf{a}_B + m\mathbf{b}_B \tag{7.19}$$

it sees below it an environment equivalent to the one it came from because (7.19) is a translational symmetry of the bulk. The adsorbate can be set down on a new site.

If (7.19) is also a translational symmetry of the adsorbate the adsorbate atoms occupy the same sites as they did previous to the translation, but with a different atom on a given site. But (7.19) need not be a translational symmetry of the adsorbate, and if not the new structure occupies sites not previously occupied and a different but equivalent structure has been generated. The idea is best explained by the pictorial example in Fig. 7.11.

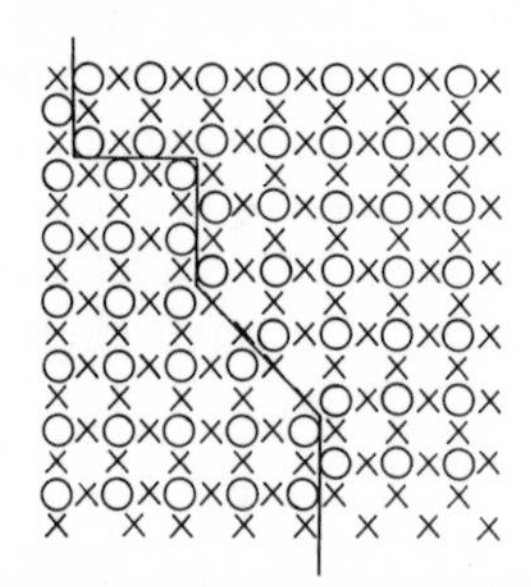

FIGURE 7.11. On this bulk lattice (x) two domains of adsorbate (o) are possible. Adsorbate domains are related to one another by a displacement that is a symmetry of the bulk lattice, but not of the adsorbate lattice.

The argument applies not only for translational symmetries of the bulk lattice, but for all types of symmetry. Since all the different domains generated in this way are completely equivalent with respect to chemistry, crystallography, and therefore probability of formation, we must always expect statistically equal numbers of each type of domain.

An example of domains related by a rotational symmetry of the bulk has already been encountered in Section A. Their effect on the diffraction pattern was that two sets of spots were seen, from an equal mixture of each type of domain, with the result that the diffraction pattern had rotational symmetry of the bulk, rather than of any separate domain.

The new case we consider here of domains related by translational symmetry does not lead to any additional diffracted beams. The diffraction pattern is affected in a more subtle way. Suppose each domain is large enough so that diffraction processes within a single domain are essentially those taking place in an infinite surface. Each domain diffracts a set of beams described by the change in parallel momentum, (7.16)

$$\mathbf{g}_j^{(C)}(s) = \mathbf{g}_j^{(B)} + \mathbf{g}_s^{(0)}$$

Some of the beams, those with

$$\mathbf{g}_s^{(0)} = 0$$

were present before the adsorbate was introduced. Diffraction from each domain differs because positions of its atoms have been shifted by a translational symmetry of the bulk that is not a symmetry of the adsorbate. From (7.10) the displacement can be written

$$\mathbf{R}_t^{(B)}(u) = \mathbf{R}_t^{(C)} + \mathbf{R}_u^{(0)}$$

The displacement being a symmetry of the bulk, we could as well generate the domain by translating not the adsorbate alone but adsorbate plus that piece of bulk on which it rests, through the same distance. With this observation it is easy to see that the effect of the translation on diffraction of the (js)th beam, is to change its phase an amount

$$\exp[i\mathbf{g}_j^{(C)}(s)\cdot\mathbf{R}_t^{(B)}(u)] = \exp[i\mathbf{g}_s^{(0)}\cdot\mathbf{R}_u^{(0)}] \tag{7.20}$$

We have made use of the result that the scalar product of a real lattice vector with a vector from the corresponding reciprocal lattice is a multiple of 2π.

When adding amplitudes diffracted by all domains we must sum the phase factor over all displacements, $\mathbf{R}_u^{(0)}$, of domains. For those beams that were present before the adsorbate was introduced, (7.20) ensures that the phase factor is always unity, and for those beams amplitudes add in phase. For all other beams the sum of (7.20) over equivalent domains is zero, a result that can be proven by observing that $\mathbf{R}_u^{(0)}$ is a vector in lattice B, which is simply related to the larger lattice C,

$$N_2\mathbf{R}_u^{(0)} = \mathbf{R}^{(C)}. \tag{7.21}$$

$\mathbf{g}_s^{(0)}$ is a reciprocal vector of lattice C:

$$\mathbf{R}^{(C)}\cdot\mathbf{g}_s^{(0)} = 2\pi N_1, \tag{7.22}$$

$$\mathbf{R}^{(0)}\cdot\mathbf{g}_s^{(0)} = 2\pi N_1/N_2. \tag{7.23}$$

If there are domains with displacements $\mathbf{R}_u^{(0)}$, there are sure to be a statistically equal number with displacements

$$2\mathbf{R}_u^{(0)}, 3R_u^{(0)}, \ldots, N_2\mathbf{R}_u^{(0)}$$

In consequence of (7.21)

$$(N_2 + 1)\mathbf{R}_u^{(0)}$$

generates the same sort of domain as $\mathbf{R}_u^{(0)}$. Therefore summing over this set of domains

$$\sum_{n=1}^{N_2} \exp(n2\pi i N_1/N_2)$$

$$= \frac{\exp(2\pi i N_1/N_2) - \exp([N_2 + 1]2\pi i N_1/N_2)}{1 - \exp(2\pi i N_1/N_2)} = 0, \tag{7.24}$$

because N_1/N_2 is neither integer nor zero provided (7.19) does not hold.

The result is that the presence of domains makes it impossible to observe the extra beams introduced by the adsorbate, except when the domains grow so large that there are not enough of them within the area illuminated by the beam to provide good statistical equality of the number of domains of each sort: a result that is often observed to be violated in experiments! The reason is that we have omitted from the phase factor in equation (7.20) an additional term due to the incident beam not being a perfect plane wave. Its phase wanders away from the mean value across the surface, and we saw in Chapter I that regions separated by more than about 500Å are illuminated by waves of randomly related phase. Therefore we must only add amplitudes for domains within the coherence area; outside, intensities are to be added. These facts modify the result to be that if there are many domains within a *coherence area*, amplitudes of extra beams will average to zero. If domains are more than a few hundred Å across, intensities for each domain add, and extra spots reappear on the screen. At intermediate sizes of domains, weakening of intensities of extra beams occurs accompanied by spot broadening associated with an effective reduction in the area of coherent diffraction. When domains do not occur with a random distribution, more complicated results ensue in the intermediate regime: spots can be split or streaked.

VIIC Scattering processes in the presence of overlayers

Before discussing theories of diffraction from systems of adsorbates one must pause to consider what generalisations are needed to our model of the scattering potential. It will be recalled that there were three ingredients: ion-core scattering which was assumed spherically symmetric and represented by phase shifts, and real and imaginary parts of the potential V_0 outside muffin tins containing the ion-cores. V_0 was assumed constant within the solid but fell to zero beyond the last layer of atoms in a way that is not known accurately, but which is not critical to accurate calculations.

Ion-core scattering was found to be the most important factor in calculations for a clean surface. It is at least equally so for an adsorbate system. We have seen that diffraction measures distances between bulk and overlayer by comparing phases of waves reflected from each. Clearly large errors in phases of ion-core scattering factors will change apparent distances. Suppose an error π is made in the phase at 100 eV (3.7 Hartrees). This would be equivalent to an error in distance, Δd, of

$$\Delta d \simeq \frac{\pi}{(\sqrt{2E - 2V_0})_r} \simeq 1.2 \text{ a.u.} = 0.6\text{Å},$$

and worse than that, it would so interfere with our calculations of scattering within the layer as to produce even more serious errors.

Fortunately work on clean surfaces shows that ion-cores are not easily deformed, and whatever the environment they are placed in their scattering properties remain much the same. We are going to assume that ion-cores of the bulk material in the surface region have the same phase shifts as those in the bulk, i.e. that they are not distorted by the surface environment. The phase shifts of adsorbate ion-cores are to be calculated by drawing around them non-overlapping muffin tins, just as we would have done for a bulk atom. In the absence of a precise knowledge of geometry it is sufficiently accurate to guess what radius they will take, from crystallography of compounds of the adsorbate atom, preferably compounds with the bulk atom. Having found the muffin tin in which the ion-core is confined, the phase shifts can be calculated from the atomic wave functions in the same way as was described in Chapter II.

Adjustment of V_0, the potential between the muffin tins, is more complicated. Away from the surface region, in the bulk, the absorptive part, iV_{0i}, takes the same value as for the clean surface situation, but the real part, V_{0r}, may be changed because the adsorbate can change the dipole moment of the surface. Corrections are easily made: raising V_0 raises the whole band structure including the Fermi level, and will be reflected in a work function change, $\Delta\phi$, which can be measured experimentally. Thus in the bulk V_0 has only to be corrected for the surface dipole layer,

$$\Delta V_0 = -\Delta\phi$$

In close-packed materials with one atom per unit cell V_0 takes a constant value between muffin tins in the bulk. At a clean surface it remains constant through the last layer of atoms and then falls to the vacuum zero. In the presence of an adsorbate, transition to the vacuum will naturally be more gradual. Between bulk and adsorbate V_0 changes towards a value typical of the adsorbate, with a further change to the vacuum zero on the far side of the overlayer.

Changes in V_{0i} between adsorbate and bulk, to a first approximation alter only reflected intensity but changes in V_{0r} have a direct influence on determination of distances. The phase change of a wave across the gap between overlayer and bulk is of the order of

$$(\sqrt{2E - 2V_0})_r d,$$

therefore a change in V_{0r} could as well be interpreted as a change in d of

$$\Delta d = -\frac{\Delta V_{0r} d}{[2E - 2V_0]_r}. \tag{7.25}$$

Lang (1971) has made calculations for a model of an adsorbate system (Fig. 7.12) in which the bulk is represented as an electron gas of density appropriate to the material under consideration, neutralised by a uniform fixed background of positive charge. The adsorbate is represented by a slab of electron gas of density appropriate to the adsorbate, again neutralised

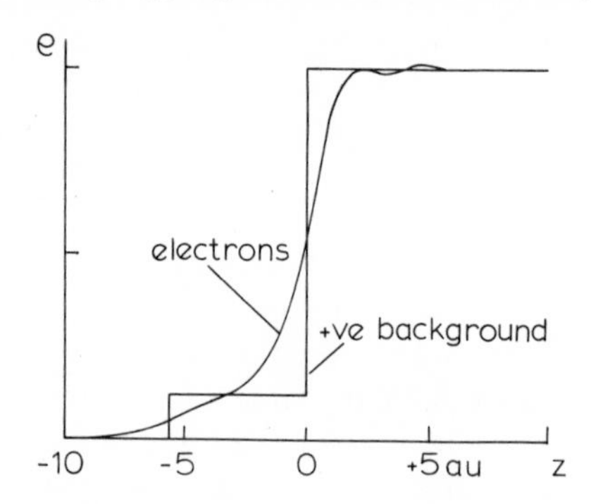

FIGURE 7.12. Density of electrons and positive background charge calculated by Lang for his model of the sodium on nickel system.

by a fixed positive background. The electron density is allowed to readjust until consistent with the electrostatic and exchange potentials generated. Figure 7.13 shows the variation of Lang's potential (which we can equate with V_{0r}) through the surface region. Densities have been chosen to be typical of a monolayer of sodium adsorbed on a nickel surface. The sodium atoms should be imagined positioned at $z \simeq -3.0$ a.u.

For the sort of variation that Lang calculates it is not a bad approximation to take V_0 as retaining the bulk value right through the region between the last bulk layer and the overlayer, only beginning to fall on the far side of the overlayer. Using arguments given above we can show that even for sodium on nickel, which is a bad case because of the large discrepancy between electron densities in sodium and nickel and the large spacing of the overlayer, the errors in d introduced by this approximation are of the order of ± 0.1Å at 20 eV incident energy. This number is not larger than limitations on accuracy imposed from other sources.

Finally there is the question of correction for thermal vibrations. These change effective ion-core scattering by a Debye-Waller factor. Here there are problems in making corrections because even for clean surfaces of known configuration calculations of thermal motion are not well advanced. The

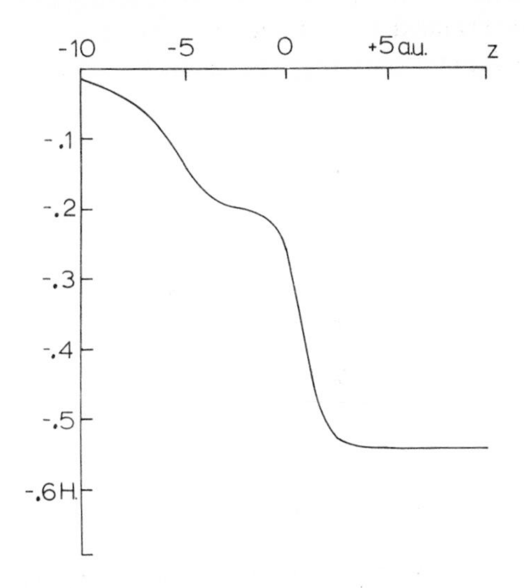

FIGURE 7.13. Potential acting on an electron for the model shown in Fig. 7.12. I am indebted to Dr. Lang for private communication of this figure.

problems for overlayers where even the geometry of atoms is yet to be determined, are considerable. On the other hand thermal vibrations are known to change mainly the intensities of peaks, and since we are confined to lower incident energies for sensitivity to the overlayer, we can hope that just as for clean surfaces, temperature corrections will not be essential to correlation of theory and experiment. This will be especially true of systems where the adsorbate is strongly bound.

For more weakly bonded adsorbates there are two alternatives. The most reliable is to make experimental corrections for thermal motion, either by lowering the temperature or by investigating temperature dependence of structure so that it can be extrapolated to 0°K. This approach has the advantage that when the static positions of atoms in the overlayer have been found, attention can be turned to determining what their thermal vibrations must be to reproduce correct variation with temperature. The other alternative is to introduce a crude model for thermal vibrations, possibly with parameters that can be fitted to give the proper intensities. However, only in extreme cases is ignorance of thermal vibrations a serious impediment to dynamic structural analysis.

VIID Averaging methods and determination of structure using kinematic theory

Our model of the scattering potential is shown in Fig. 7.14. The overlayer sits a distance $\mathbf{c}_2$ away from the last layer of bulk atoms and this is one of the unknowns to be determined. For simplicity we shall assume that the overlayer leaves structure of the bulk unchanged but alters V_0 so that it remains constant

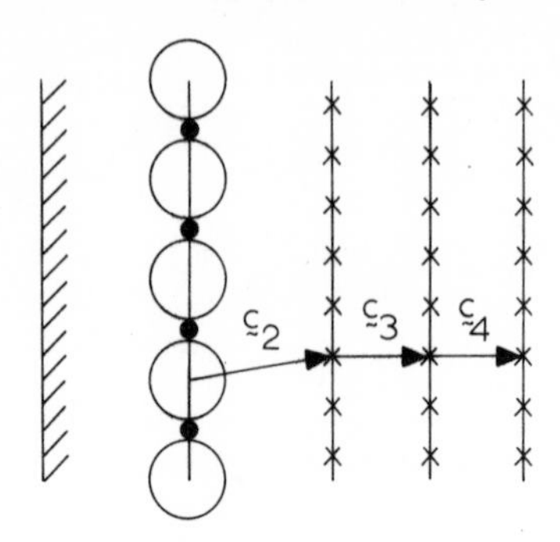

FIGURE 7.14. Model of an overlayer system showing the surface barrier outside the last layer of atoms and positions of ion-cores immersed in a constant potential.

through the overlayer, the surface barrier being moved to just outside the overlayer. This is probably a good approximation in many systems.

As in Section IIIA, assume that a wave

$$B \exp(i\mathbf{k}_0 \cdot \mathbf{r}), \tag{7.26}$$
$$\tfrac{1}{2}|\mathbf{k}_0|^2 = E,$$

is incident on the surface. Part of the wave is reflected from the surface barrier

$$R.B. \exp(i\mathbf{k}_{0\|} \cdot \mathbf{r}_\| - ik_{0z}z), \tag{7.27}$$

but (7.27) can be taken as being very weak at all but the lowest energies and we shall neglect it. There is transmitted into the crystal

$$T^+ B \exp(i\mathbf{k} \cdot \mathbf{r}), \tag{7.20}$$
$$\tfrac{1}{2}|\mathbf{k}|^2 = E - V_0, \qquad \mathbf{k}_\| = \mathbf{k}_{0\|}.$$

The position of the sth atom in the jth unit cell of the overlayer is

$$\mathbf{R}_j^{(A)} + \mathbf{u}_s^{(A)} \tag{7.29}$$

and hence this atom has a wavefield amplitude

$$T^+ B \exp[i\mathbf{k} \cdot (\mathbf{R}^{(A)} + \mathbf{u}_s^{(A)})] \tag{7.30}$$

incident upon it. In the kinematic approximation each atom in the overlayer scatters waves only weakly, in which case the sum over atoms in the overlayer gives scattered waves totalling

$$\sum_{\mathbf{k}'} M^{(A)}(\mathbf{k}',\mathbf{k})\, T^{+}(\mathbf{k})\, B \exp(i\mathbf{k}'\cdot\mathbf{r}), \tag{7.31}$$

$$M^{(A)}(\mathbf{k}',\mathbf{k}) = S^{(A)}(\mathbf{k}'_{\parallel} - \mathbf{k}_{\parallel}) \frac{2\pi i}{A|\mathbf{k}|\,|k'_z|} \sum_{\ell s} (2\ell + 1) \sin\left(\delta_{\ell}^{(A)}(s)\right)$$
$$\times \exp\left(i\delta_{\ell}^{(A)}(s)\right) P_{\ell}\left(\frac{\mathbf{k}\cdot\mathbf{k}'}{|\mathbf{k}|\,|\mathbf{k}'|}\right) \exp\left[i(\mathbf{k} - \mathbf{k}')\cdot\mathbf{u}_s^{(A)}\right]. \tag{7.32}$$

Details of the derivation of (7.31) are available in Section IIIA. $S^{(A)}(\mathbf{k}'_{\parallel} - \mathbf{k}_{\parallel})$ is a function that is zero, unless $\mathbf{k}'_{\parallel} - \mathbf{k}_{\parallel}$ is a reciprocal lattice vector of the overlayer, when it is unity. Transmission through the surface barrier gives

$$\phi_s^{(A)} = \sum_{\mathbf{k}'} M^{(A)}(\mathbf{k}',\mathbf{k})\, T^{+}(\mathbf{k})\, T^{-}(\mathbf{k}')\, B \exp(i\mathbf{k}'_0\cdot\mathbf{r}).$$

Proceeding in a similar way for the bulk layers gives an analogous expression except that each layer is now displaced further into the crystal. From the first layer of the bulk there is scattered and transmitted through the surface barrier

$$\sum_{\mathbf{k}'} \exp\left[i(\mathbf{k} - \mathbf{k}')\cdot\mathbf{c}_2\right] M^{B}(\mathbf{k}',\mathbf{k})\, T^{+}(\mathbf{k}) \times T^{-}(\mathbf{k}')\, B \exp(i\mathbf{k}'_0\cdot\mathbf{r}) \tag{7.33}$$

It is easy to see that in the kinematic approximation all bulk layers being displaced from the surface barrier by an extra distance $\mathbf{c}_2$, contribute to diffracted amplitudes in exactly the same way they did when the surface was clean except for a phase factor

$$\exp\left[i(\mathbf{k} - \mathbf{k}')\cdot\mathbf{c}_2\right].$$

Suppose that in the kinematic approximation the clean surface scatters

$$\sum_{\mathbf{k}'} R(\mathbf{k}',\mathbf{k})\, B \exp(i\mathbf{k}'_0\cdot\mathbf{r}),$$

then in the presence of an overlayer each of these waves is phase shifted

$$\phi_s^{(B)} = \sum_{\mathbf{k}'} \exp\left[i(\mathbf{k} - \mathbf{k}')\cdot\mathbf{c}_2\right] R(\mathbf{k}',\mathbf{k})\, B \exp(i\mathbf{k}'_0\cdot\mathbf{r}). \tag{7.34}$$

The total wavefield is

$$\phi_s = \phi_s^{(A)} + \phi_s^{(B)}. \tag{7.35}$$

There are two sorts of beams in ϕ_s. There are those beams that were present before the overlayer was introduced. The overlayer will in general produce additional contributions to these beams and interfere with the contributions from the bulk. In this way information is to be had about relative orientation of overlayer and bulk.

The second sort of beam in ϕ_s is introduced by the presence of the overlayer, and only the overlayer contributes to these beams in the kinematic approximation. This affords a most useful factorisation of the problem. Locations of extra beams tell us about size and shape of the overlayer's unit cell. Concentrating on the intensities of additional beams we can dismiss the bulk from our considerations whilst we use equation (7.32) to interpret scattered intensities which are determined by the scattering factor of the overlayer unit cell. (7.32) shows that coordinates of ion-cores within this cell enter into the scattering via a phase factor

$$\exp[i(\mathbf{k} - \mathbf{k}')\cdot\mathbf{u}_s]$$

Therefore intensities of the extra beams can be used to determine $\mathbf{u}_s$.

Having elucidated structure of the overlayer from additional beams, it remains to find its position relative to the bulk, $\mathbf{c}_2$. This information is contained in interference between $\phi_s^{(A)}$ and $\phi_s^{(B)}$. The kinematic bulk reflectivity is small except at Bragg conditions implying that the information is to be had by interpreting modifications made by the overlayer to Bragg peaks. The sort of changes taking place can be appreciated by considering the specularly reflected beam as a function of incident energy. Assume for the moment that overlayer scattering does not change appreciably with energy and that each Bragg peak produced by the clean surface has the same amplitude and phase. Then interference between $\phi_s^{(B)}$ and $\phi_s^{(A)}$ changes from one Bragg peak to another in a manner determined by the phase factor

$$\exp[i(\mathbf{k} - \mathbf{k}')\cdot\mathbf{c}_2] \tag{7.36}$$

in equation (7.34) resulting in interference that differs from one peak to another in a manner determined by $\mathbf{c}_2$. Variation with energy of scattering factors for layers complicates this simple picture, but can be calculated and factor (7.36) inferred from the intensities.

In addition to intensity modulation there will be changes as energy is increased through the width of a Bragg peak. If it tends towards constructive interference the peak will move to higher energies.

The kinematic theory with its elegant factorisation into the extra beams of structure within the overlayer, combined with simplicity of mathematical formulae, would be ideal for interpretation of data except for one difficulty. The theory is totally inadequate for describing real diffraction conditions! The experimental data must be treated to extract the kinematic component before any of the theory above can be used.

A scheme for reducing complicated dynamical curves to a simpler form has already been described in Chapter V: that due to Lagally, Ngoc and Webb

(1972). It will be recalled that averaging data at constant scattering vector reduced processes that did not peak at the kinematic condition, to a constant background. The remaining curve had kinematic form but could only be interpreted with a quasi-kinematic theory in which the three scattering parameters: the ion-core scattering factor, V_{0r}, and V_{0i} all took on modified values. Theory of averaging is incomplete for diffraction from clean surfaces in that no simple prescription is known for finding the effective scattering factor. Similar difficulties arise for adsorbate systems and in addition the prescription for making modifications to V_0 which was simple in the clean surface case, is no longer well defined in the vicinity of an overlayer. It will be appreciated from (7.31), (7.32) and (7.34) that an erroneous effective scattering factor, especially one in which the phase is erroneous, prevents accurate determination of the phase factors

$$\exp[i\mathbf{k} - \mathbf{k}')\cdot\mathbf{c}_2], \qquad \exp[i(\mathbf{k} - \mathbf{k}')\cdot\mathbf{u}_s]$$

and hence of $\mathbf{c}_2$ and $\mathbf{u}_s$.

As we argued in Chapter V it is possible that under certain favourable circumstances the effective scattering factors will not be very much different from the unmodified ones appearing in the original kinematic expression. Then an analysis of kinematic data can proceed in a manner outlined above. Work is proceeding on an investigation of this matter, but as yet no definite results have been obtained and it remains to be seen whether the averaging technique will prove an effective tool for structural analysis.

A less ambitious scheme of averaging has been proposed by Tucker and Duke (1970, 1972). They concentrate their attention on extra beams introduced by the overlayer. In the kinematic approximation (7.31) and (7.32) tell us that the kinematic contribution to the extra beams varies with energy in a manner determined by variations in the scattering of a single unit cell of the overlayer. There are no Bragg-type peaks and the structure is relatively slowly varying. In the experimental curves multiple scattering enables the bulk to contribute to the extra beams by diffracting a second time electrons forward scattered by the overlayer. The multiple scattering contributions vary much more rapidly with energy than the kinematic contribution.

If the multiple scattering contribution is less than the kinematic part the rapid variation means that the kinematic part will now be enhanced, now diminished by interference over a relatively short scale of energies. Averaging over a range of energies should remove the multiple scattering part leaving a number given by kinematic contributions. If on the other hand the multiple scattering contribution is larger than the kinematic part the argument can be reversed, as the kinematic part now adds, now subtracts from the multiple scattering contribution and the averaged intensity is typical of the multiple scattering component; the kinematic part is eliminated. In favour of the former case we have that an electron must scatter from the overlayer at least once to appear in one of the extra beams, therefore the strength of scattering of the overlayer always determines the scale of intensity. On the other hand multiple scattering, even that within a layer, as shown in Section VB, is of the same order as kinematic scattering and we might anticipate that trouble will sometimes occur. Favourable situations would occur for very sparse overlayers on weak-scattering substrates. The method is at its most reliable handling systems for which kinematic theory only just fails to work for the unaveraged data.

If $I_{\mathbf{k}'}(E)$ is the intensity of the $\mathbf{k}'$ beam, then Tucker and Duke define

$$\bar{I}_{\mathbf{k}'}(E_0, \Delta) = \int_{E_0 - \frac{1}{2}\Delta}^{E_0 + \frac{1}{2}\Delta} I_{\mathbf{k}'}(E)\,dE, \tag{7.37a}$$

$$\bar{I}_{0\mathbf{k}'}(E_0, \Delta) = \int_{E_0 - \frac{1}{2}\Delta}^{E_0 + \frac{1}{2}\Delta} I_{0\mathbf{k}'}(E)\,dE, \tag{7.37b}$$

where $\bar{I}_{\mathbf{k}'}$ and $\bar{I}_{0\mathbf{k}'}$ are the averaged experimental and kinematic intensities. Tucker and Duke then postulate

$$\bar{I}_{\mathbf{k}'} = \bar{I}_{0\mathbf{k}'} \tag{7.38}$$

arguing along the lines of the reasoning above. They test the postulate by calculating $\bar{I}_{\mathbf{k}'}$ and $\bar{I}_{0\mathbf{k}'}$ for a simplified model in which only the s-wave phase shifts, δ_0, are included. Results are given in Fig. 7.15.

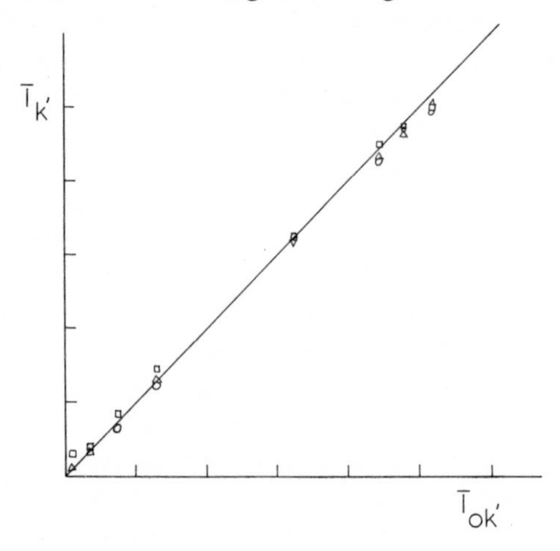

FIGURE 7.15. Comparison of $\bar{I}_{\mathbf{k}'}$ and $\bar{I}_{0\mathbf{k}'}$ for overlayer systems on surfaces with the same geometry as aluminium (100), □, (110), ○, and (111), △. The overlayer in each case has two atoms per unit cell at (0, 0) and ($\frac{1}{8}$, $\frac{1}{3}$) in units of the cell side and is situated a distance above the top bulk layer equal to the spacing between first and second bulk layers. Only s-wave phase shifts were considered and δ_0 was set equal to $\pi/2$ for all ion cores. $V_{0r} = 0$, $V_{0i} = -5$ eV.

Because only the s-wave component of the full complement of phase shifts has been evaluated, Fig. 7.15 corresponds to a weak scattering situation. Scattering is smaller than that encountered in, say, nickel or copper by typically $\times 1/5$. Nevertheless Fig. 7.15 does provide encouraging evidence that as we predicted postulate (7.38) is obeyed in the weak scattering case.

VIIE Extension of dynamical methods to overlayers

The overlayer fulfils three functions in the diffraction process: to act as a filter presenting the 'clean surface' underneath with a more complex incident beam and mixing together outgoing waves; to back-scatter electrons directly from itself; and to let electrons multiply scatter between itself and the bulk. All three processes contribute to some extent but the first is the most important.

As regards direct reflection of the incident beam from the overlayer, there is no difficulty. Once we know or have postulated a structure for the overlayer and calculated phase shifts for the ion-cores, the methods of Chapter IV can be applied to calculate $M^{\pm}_{\mathbf{g}_1(A)}{}^{\pm}_{\mathbf{g}_1(A)}(2)$ for the overlayer. Difficulties arise when stacking together layers with different two-dimensional lattices. So as to approach the problem gradually we shall consider cases of ascending complexity.

In the case that the overlayer has the same unit cell as bulk layers we can take over the formalism developed in section IIIC for reflectivities of clean surfaces in which the top layer is different from the bulk, but has the same unit cell. It will be recalled that amplitudes of forward and backward travelling waves between the nth and $(n+1)$th layers were $a_n^{\pm}$ referred to an origin half way between the layers shown in Fig. 3.14. The method of solution there was first to strip away as many layers as were atypical of the bulk leaving an ideal surface. Here we shall remove only the surface barrier and the overlayer assuming that the first layer of bulk atoms is a typical layer. This is usually a good approximation. We have now to relate amplitudes of waves reflected from the ideal surface, denoted by vector a_2^-, to the incident waves a_2^+. This can be done either by matching to a sum of Bloch waves (Section IIIC) or by using the matrix doubling method (Section IVF).

$$a_2^- = R(2)a_2^+ \tag{7.39}$$

Now the overlayer can be put back into place again. $M^{\pm\pm}(2)$ describe transmission and reflection of waves with amplitudes referred to the centre of the layer as origin. A simple change of origin must be made to work in terms of amplitudes half way between layers

$$Q^{\mathrm{I}}_{\mathbf{gg}'}(2) = \big(I + M^{++}(2)\big)_{\mathbf{gg}'}\exp(i\mathbf{K}_{\mathbf{g}}^{+}\cdot\tfrac{1}{2}\mathbf{c}_2 + i\mathbf{K}_{\mathbf{g}'}^{+}\cdot\tfrac{1}{2}\mathbf{c}_1), \tag{7.40a}$$

$$Q^{\mathrm{II}}_{\mathbf{gg}'}(2) = M^{+-}_{\mathbf{gg}'}(2)\exp(i\mathbf{K}_{\mathbf{g}}^{+}\cdot\tfrac{1}{2}\mathbf{c}_2 - i\mathbf{K}_{\mathbf{g}'}^{-}\cdot\tfrac{1}{2}\mathbf{c}_2), \tag{7.40b}$$

$$Q^{\mathrm{III}}_{\mathbf{gg}'}(2) = M^{-+}_{\mathbf{gg}'}(2)\exp(-i\mathbf{K}^{-}_{\mathbf{g}}\cdot\tfrac{1}{2}\mathbf{c}_1 + i\mathbf{K}^{+}_{\mathbf{g}'}\cdot\tfrac{1}{2}\mathbf{c}_1), \tag{7.40c}$$

$$Q^{\mathrm{IV}}_{\mathbf{gg}'}(2) = (I + M^{--}(2))_{\mathbf{gg}'}\exp(-i\mathbf{K}^{-}_{\mathbf{g}}\cdot\tfrac{1}{2}\mathbf{c}_1 - i\mathbf{K}^{-}_{\mathbf{g}'}\cdot\tfrac{1}{2}\mathbf{c}_2), \tag{7.40d}$$

so that we can write

$$a_2^+ = Q^{\mathrm{I}}(2)a_1^+ + Q^{\mathrm{II}}(2)a_2^-, \tag{7.41a}$$

$$a_1^- = Q^{\mathrm{III}}(2)a_1^+ + Q^{\mathrm{IV}}(2)a_2^-. \tag{7.41b}$$

Equations (7.40) and (7.41) compare directly with equations (3.67). (7.41) and (7.39) can be solved for a_1^- in terms of a_1^+

$$a_1^- = R(1)a_1^+, \tag{7.42}$$

$$R(1) = [Q^{\mathrm{III}}(2) + Q^{\mathrm{IV}}(2)[1 - R(2)Q^{\mathrm{II}}(1)]^{-1}R(2)Q^{\mathrm{I}}(1). \tag{7.43}$$

Knowing how the surface barrier transmits and reflects electrons, i.e. $Q(1)$, the reflectivity of the whole system can be found, $R(0)$.

The above amounts merely to a reminder of what we have already written down for clean surfaces. Next we complicate the problem by giving the overlayer a structure different from that of the bulk layers, but still keeping a rational relationship between the structures. Instead of waves between layers having parallel components of momentum related by reciprocal lattice vectors of the bulk

$$\mathbf{k}_{0\parallel} + \mathbf{g}^{(B)}, \tag{7.44}$$

as we had before, the overlayer filters in new beams. The enlarged set of beams has parallel components of momentum differing by reciprocal lattice vectors of the coincidence lattice,

$$\mathbf{k}_{0\parallel} + \mathbf{g}^{(C)}. \tag{7.45}$$

Consider the bulk layers first. The problem is that we only know how to calculate reflection and transmission coefficients of layers for incident waves whose parallel components of momentum have the form of (7.44).

We can use result (7.16) of Section B to re-express (7.45) as

$$\mathbf{k}_{0\parallel} + \mathbf{g}_j^{(C)}(s) = (\mathbf{k}_{0\parallel} + \mathbf{g}_s^{(0)}) + \mathbf{g}_j^{(B)}. \tag{7.46}$$

The interpretation of (7.46) is that an overlayer can be thought of as introducing several different incident wave vectors,

$$\mathbf{k}_{0\parallel} + \mathbf{g}_s^{(0)}.$$

For a coincidence lattice, as has been pointed out, there is a finite number of different values of $\mathbf{g}_s^{(0)}$. It is as if instead of one LEED gun directed at the surface, there were now several (but of a very high quality, so that their beams are in phase). Our course of action must be to make a separate calculation in the bulk for each set of beams characterised by a given $(\mathbf{k}_{0\|} + \mathbf{g}_s^{(0)})$. The bulk layers have too much symmetry to couple together beams with different $(\mathbf{k}_{0\|} + \mathbf{g}_s^{(0)})$ and each set is completely independent of any other set, in the bulk.

Amplitudes of beams between layers, formerly written as single vectors $a_n^{\pm}$ can now be divided into sets of beams, each corresponding to a different $\mathbf{g}_s^{(0)}$ in equation (7.46). $a_n^{+}(s)$ is a vector containing amplitudes of forward travelling beams with parallel components of momentum

$$(\mathbf{k}_{0\|} + \mathbf{g}_s^{(0)} + \mathbf{g}_j^{(B)}).$$

As previously reflectivity of the bulk can be determined but in this case each set of beams has its own reflectivity,

$$a_2^{-}(s) = R_s(2)\, a_2^{+}(s). \tag{7.47a}$$

A more formal way of saying that components of $a_2^{\pm}$ corresponding to different values of $(\mathbf{k}_{0\|} + \mathbf{g}_s^{(0)})$ behave independently is that, if (7.47) is rewritten in the old notation using the complete vectors a_2^{+}, the matrix linking them factorises. Equation (7.47b) shows what is meant by this.

$$\begin{bmatrix} a_2^{-}(s=1) \\ a_2^{-}(s=2) \\ \cdot \\ \cdot \end{bmatrix} = \begin{bmatrix} R_{s=1}(2), & 0 & ,\ldots \\ 0 & , R_{s=2}(2), & \ldots \\ \cdot & \cdot & \\ \cdot & \cdot & \end{bmatrix} \begin{bmatrix} a_2^{+}(s=1) \\ a_2^{+}(s=2) \\ \cdot \\ \cdot \end{bmatrix} \tag{7.47b}$$

Division of the problem in equation (7.47) into smaller matrices, each of about the same size as would be encountered in the clean surface diffraction problem, means that calculation of reflectivities of the bulk in presence of an adsorbate involves times proportional to the number of different values, n, of $(\mathbf{k}_{0\|} + \mathbf{g}_s^{(0)})$. If the matrix in (7.47b) had not factorised we would have been dealing with a single matrix n times as large involving computing times scaling as n^3.

So much for relationships involving only layers of bulk atoms. If the relationship between bulk and adsorbate lattices is simple the reciprocal vectors $\mathbf{g}_j^{(C)}$ coincide with those of the adsorbate (condition (i) of Section B). Beams incident on the overlayer differ in parallel momentum by reciprocal vectors of the overlayer lattice,

$$\mathbf{g}^{(B)} = \mathbf{g}^{(C)}$$

and there are no complications. If the relationship between bulk and adsorbate lattices is only rational (condition (ii)) we must decompose

$$\mathbf{k}_{0\|} + \mathbf{g}_j^{(C)}(t) = \mathbf{k}_{0\|} + \mathbf{g}_j^{(A)} + \mathbf{G}_t^{(0)} \tag{7.48}$$

in the same way as for the bulk lattice. There will be a different number of vectors $\mathbf{G}_t^{(0)}$ needed than for the bulk, usually less. Again, it appears to the lattice that more than one electron gun is being used. For each set of beams, $a_1^{\pm}(t), a_2^{\pm}(t)$, reflection and transmission matrices must be calculated and from them the Q-matrices. Working in terms of the complete amplitude vectors $a_1^{\pm}, a_2^{\pm}$ again leads to factorisation of matrices relating them, into as many diagonal blocks as there are values of $(\mathbf{k}_{0\|} + \mathbf{G}_t^{(0)})$. This number will be different from the number of values of $(\mathbf{k}_{0\|} + \mathbf{g}_s^{(0)})$.

$$\begin{bmatrix} a_2^+(t=1) \\ a_2^+(t=2) \\ \cdot \\ \cdot \end{bmatrix} = \begin{bmatrix} Q^{\mathrm{I}}_{t=1}, 0 & , \cdot\cdot \\ 0 & , Q^{\mathrm{I}}_{t=2}, \cdot\cdot \\ \cdot & \cdot \\ \cdot & \cdot \end{bmatrix} \begin{bmatrix} a_1^+(t=1) \\ a_1^+(t=2) \\ \\ \end{bmatrix} + \begin{bmatrix} Q^{\mathrm{II}}_{t=1}, 0 & \cdot\cdot \\ 0 & , Q^{\mathrm{II}}_{t=2} \cdot\cdot \\ \cdot & \cdot \\ \cdot & \cdot \end{bmatrix} \begin{bmatrix} a_2^-(t=1) \\ a_2^-(t=2) \\ \cdot \\ \cdot \end{bmatrix} \tag{7.49a}$$

$$\begin{bmatrix} a_1^-(t=1) \\ a_1^-(t=2) \\ \cdot \\ \cdot \end{bmatrix} = \begin{bmatrix} Q^{\mathrm{III}}_{t=1}, 0 & , \cdot\cdot \\ 0 & , Q^{\mathrm{III}}_{t=2}, \cdot\cdot \\ \cdot & \cdot \\ \cdot & \cdot \end{bmatrix} \begin{bmatrix} a_1^+(t=1) \\ a_1^+(t=2) \\ \cdot \\ \cdot \end{bmatrix} + \begin{bmatrix} Q^{\mathrm{II}}_{t=1}, 0 & \cdot\cdot \\ 0 & , Q^{\mathrm{II}}_{t=2} \cdot\cdot \\ \cdot & \cdot \\ \cdot & \cdot \end{bmatrix} \begin{bmatrix} a_2^-(t=1) \\ a_2^-(t=2) \\ \cdot \\ \cdot \end{bmatrix} \tag{7.49b}$$

Equations (7.47) and (7.49) can be solved for $a_1^{\pm}$. The matrix connecting $a_1^{\pm}$

$$a_1^- = R(1)a_1^+$$

never factorises.

Lastly the surface barrier can be put in place in the same way shown in Section IIIC. There are no complications in this part of the problem. The Bloch wave method has been used as an example of how calculations proceed. Generalisation to other methods can be inferred; the key is that as far as bulk layers are concerned, wave amplitudes must be divided into groups, each with a different $(\mathbf{k}_{0\|} + \mathbf{g}_s^{(0)})$. Each group is completely independent of the other groups as far as bulk layers are concerned. For the overlayer, wave amplitudes divide into a different set of groups with different values of $(\mathbf{k}_{0\|} + \mathbf{G}_t^{(0)})$ and in scattering from the overlayer it is these groups that are independent. The surface barrier in the approximation we use at the moment reflects and transmits without changing $\mathbf{k}_{0\|} + \mathbf{g}_j^{(C)}$. Therefore no problems arise in generalising the surface barrier scattering to the overlayer case.

Thus all the methods we have proposed for the clean surface can be generalised to the adsorbate case when overlayer and substrate are rationally related. When they are irrationally related calculations become fearsome. Only perturbation methods are applicable. Let us trace through the steps in an *RFS* perturbation calculation. A single beam, parallel momentum $\mathbf{k}_{0\|}$, is transmitted by the surface barrier and impinges upon the overlayer. In the *RFS* scheme the transmitted wave is calculated. It will consist of many beams each defined by parallel momentum $(\mathbf{k}_{0\|} + \mathbf{g}_j^{(A)})$. Because of the irrational relationship between bulk and overlayer, no subdivision into groups of these beams is possible. Each behaves as a separate independent beam incident on the bulk. For every value of $(\mathbf{k}_{0\|} + \mathbf{g}_j^{(A)})$ separate transmission and reflection coefficients of bulk layers must be calculated. Each original beam is scattered into many other beams by the bulk, with parallel momenta

$$(\mathbf{k}_{0\|} + \mathbf{g}_j^{(A)} + \mathbf{g}_s^{(B)}).$$

Eventually when the path of each separate beam has been traced through the bulk layers, there emerge from the bulk sets of beams. Each single incident beam with momentum $(\mathbf{k}_{0\|} + \mathbf{g}_j^{(A)})$ gives rise to its own set, none of the members of which need coincide with any other set. To calculate transmission of the wavefield through the adsorbate, each group of parallel momenta with a different $(\mathbf{k}_{0\|} + \mathbf{g}_s^{(B)})$ must be treated separately. So the *RFS* perturbation approach can be extended to irrational structures, but after studying the method carefully it will be seen that at each pass the number of beams needed in the calculation increases, and becomes infinite after an infinite number of passes. Calculations are only possible for a small number of passes.

VIIF Efficient calculation of intra-layer scattering

In the overlayer case some of the quantities that we evaluated only once in the clean surface problem, must be calculated several times for slightly different values of parameters. For example, transmission and reflection matrices must be evaluated in the bulk for each of the different parallel components of momentum, $(\mathbf{k}_{0\|} + \mathbf{g}_p^{(0)})$. Many details of the calculation are the same in each case, and it is possible to rearrange the calculation more economically to save repetition.

One of the most time-consuming steps in the *RFS* perturbation calculation, is evaluation of lattice sums (4.45) which take account of multiple scattering within a layer. For the bulk layers (4.45) becomes after substituting $(\mathbf{k}_{0\|} + \mathbf{g}_p^{(0)})$ as the effective parallel momentum,

$$D_{\ell m}(ks) = -\kappa i(4\pi)^{-\frac{1}{2}}\delta_{m0}\delta_{\ell 0} - i\kappa(-1)^{\ell+m} \times \sum_j{}' h_\ell^{(1)}(\kappa|\mathbf{R}_{0s}^{(B)} - \mathbf{R}_{jk}^{(B)}|)\, Y_{\ell -m}(\Omega(\mathbf{R}_{0s}^{(B)} - \mathbf{R}_{jk}^{(B)})) \times \exp[i(\mathbf{k}_{0\|} + \mathbf{g}_p^{(0)})\cdot\mathbf{R}_j^{(B)}]. \quad (7.50)$$

Evidently there is much repetition because only the last factor changes when calculations are made for another $(\mathbf{k}_{0\|} + \mathbf{g}_p^{(0)})$. Much more efficient calculations can be performed.

The vector $\mathbf{R}_{jk}^{(B)}$, the position of the kth atom of the jth unit cell, can be divided into the position of the kth atom in the 0th unit cell, $\mathbf{R}_{0k}^{(B)}$, plus a vector of the bulk lattice giving the position of the jth unit cell, which expressed in current terminology gives

$$\mathbf{R}_{jk}^{(B)} = \mathbf{R}_{0k}^{(B)} + \mathbf{R}_u^{(B)}(t) = \mathbf{R}_{0k}^{(B)} + \mathbf{R}_u^{(C)} + \mathbf{R}_t^{(0)} \quad (7.51)$$

Subscript j has been replaced by the pair (ut) which give more detailed information about which unit cell of the coincidence lattice the point occurs in. We can rewrite

$$\exp[i(\mathbf{k}_{0\|} + \mathbf{g}_p^{(0)})\cdot\mathbf{R}_u^{(B)}(t)] = \exp[i(\mathbf{k}_{0\|} + \mathbf{g}_p^{(0)})\cdot(\mathbf{R}_u^{(C)} + \mathbf{R}_t^{(0)})] = \exp[i\mathbf{k}_{0\|}\cdot(\mathbf{R}_u^{(C)} + \mathbf{R}_t^{(0)})]\exp[i\mathbf{g}_p^{(0)}\cdot\mathbf{R}_t^{(0)}], \quad (7.52)$$

since $\mathbf{g}_p^{(0)}$ is a reciprocal vector of the coincidence lattice. A more compact expression can be given for D by defining

$$F_{\ell m}(kst) = i\kappa(-1)^{\ell+m}\sum_u{}' h_\ell^{(1)}(\kappa|\mathbf{R}_{0s}^{(B)} - \mathbf{R}_{0k}^{(B)} - \mathbf{R}_t^{(0)} - \mathbf{R}_u^{(C)}|) \times Y_{\ell -m}(\Omega(\mathbf{R}_{0s}^{(B)} - \mathbf{R}_{0k}^{(B)} - \mathbf{R}_t^{(0)} - \mathbf{R}_u^{(C)}))\exp[i\mathbf{k}_{0\|}\cdot(\mathbf{R}_u^{(C)} + \mathbf{R}_t^{(0)})] \quad (7.53)$$

The summation is over a sublattice, not the whole bulk lattice. D is constructed from F by summing contributions from different sublattices, being careful to insert the proper factors.

$$D_{\ell m}(ks) = -\kappa i(4\pi)^{-\frac{1}{2}}\delta_{m0}\delta_{\ell 0} - \sum_t F_{\ell m}(kst)\exp[i\mathbf{g}_p^{(0)}\cdot\mathbf{R}_t^{(0)}] \quad (7.54)$$

Evidently function F does not depend on which of the several parallel components of momentum $(\mathbf{k}_{0\|} + \mathbf{g}_p^{(0)})$, the sum is being made for. $\mathbf{g}_p^{(0)}$ enters only in the sum over sublattices. Much the greater part of work is done in the first summation which involves a hundred points, $\mathbf{R}_u^{(C)}$, or more. The second summation is over the number of sublattices and is a trivial computation compared with the first summation. By parcelling into separate summations over sublattices the bulk of the labour has been done for all groups of beams. Evaluation of D by the old method takes 10 seconds per call. Thus for a (2×2) adsorbate structure with four sublattices, 40 seconds would have been needed to calculate for each different $(\mathbf{k}_{0\|} + \mathbf{g}_p^{(0)})$ but with the new technique, only 10 seconds plus overheads of about 1 second, are needed. Times quoted are for an ICL Titan.

The same simplification can be made in calculating for the overlayer. In simple systems it is often possible in the overlayer calculation to make use of the quantities F calculated for the bulk.

Once the quantities D have been calculated, the rest of the calculation for intra-layer scattering proceeds as before, for each $(\mathbf{k}_{0\parallel} + \mathbf{g}_p^{(0)})$ separately.

VIIG Dynamical methods and their sensitivity

Dynamical calculations are capable of interpreting experimental LEED intensities directly without any averaging of data to remove troublesome components. Structural determinations can proceed by postulating a surface structure, calculating, and comparing directly with experimental intensities. The method is economical in experimentation but the price paid is that calculations are more complicated than in the kinematic case. Dynamical methods have the advantage at the time of writing that they are the only methods not based on only partially established hypotheses and the only methods by which complete, reliable structural determinations have as yet been made.

Before describing some structural determinations we must complete our picture of how experimental, i.e. dynamical, diffraction measurements react to an overlayer. We have shown in detail how a change in unit cell of the surface changes the diffraction pattern, but how intensities react, for example, to change of relative positions of overlayer and bulk, is not something accessible via experiments. Atoms sit on the surface in fixed positions and it is not possible in an experiment to vary their positions continuously. We must proceed with the investigation through dynamical calculations. At the same time dynamical theory will enable us to explain and interpret the changes.

In order to maximise our insight the Bloch wave method of calculation will be employed in this section. It is not the fastest way of proceeding (the *RFS* perturbation scheme is much better in this respect) but the band structure it generates affords much more information than intensities alone.

The system we shall use is one of the simplest that occurs: the $(\sqrt{2} \times \sqrt{2})$ 45° sodium structure on a nickel (001) surface. The sodium on nickel system has been the subject of a number of experiments (Gerlach and Rhodin, 1970; Andersson and Kasemo, 1972) and it was for the $(\sqrt{2} \times \sqrt{2})$ 45° structure that the first complete structural determination was made (Andersson and Pendry, 1972, 1973).

The diffraction pattern is consistent with an arrangement of atoms shown in Fig. 7.2. Because the bulk and overlayer structures are simply related, the coincidence lattice is the same as the overlayer lattice. Figure 7.16 shows how

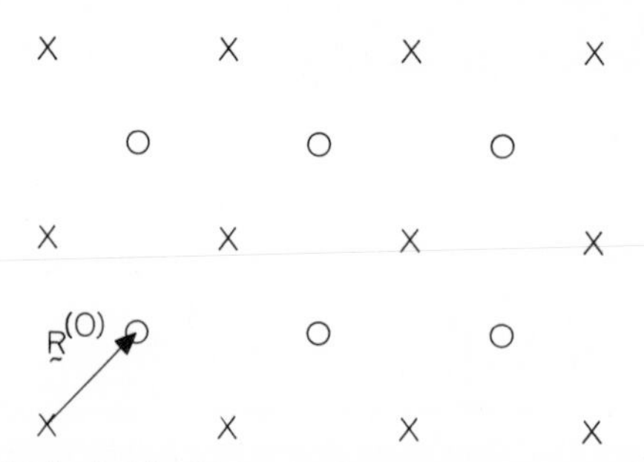

FIGURE 7.16. The bulk lattice in Fig. 7.2 can be described in terms of two coincidence lattices displaced from one another by $\mathbf{R}^{(0)}$.

to divide the bulk lattice into sublattices. There are two for this example.

The structure can form two sorts of antiphase domains, and the assumption will be made that domains are larger than the coherence area of the incident beam so that intensities rather than amplitudes of beams from each domain add. Phase shifts of sodium and nickel ion cores were constructed as prescribed and gave t-matrices that are plotted in Fig. 7.17. The different amplitudes and

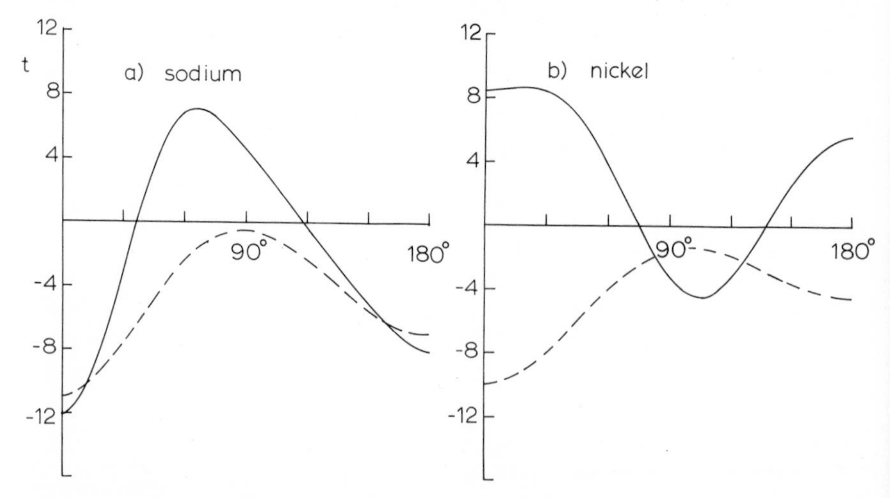

FIGURE 7.17. $t(\theta)$ at 20 eV. (a) for sodium, (b) for nickel. ——— real part, - - - - - - imaginary part.

phases of these scattering factors serve to emphasise the importance of calculating them accurately even in kinematic situations: an error in phase implies an error in apparent position of an atom.

The constant potential between atoms was given the values

$$V_0 = -16.3 - i1.0\,\text{eV}, \quad E < 15\,\text{eV},$$
$$= -16.3 - i4.0\,\text{eV}, \quad E > 15\,\text{eV}. \tag{7.55}$$

The surface barrier was given the form

$$\frac{V_0}{1 + \exp(-2z)} \tag{7.56}$$

with the centre of the barrier placed on the vacuum side of the sodium layer in a plane that just skirts muffin tins of the sodium ion-cores. The scale on which V_0 goes to zero, is of the order of an electronic screening length in nickel. The surface barrier was not a critical ingredient. Section H will describe how the numbers in (7.55) were arrived at; here our only concern is to have realistic models of all important influences on diffraction.

We have stressed, and will stress further in Section H, that peak positions are liable to be much more accurately determined than intensities. The least well known quantities are those influencing intensities rather than energies. When scrutinising curves for their sensitivity to geometry particular attention must be paid to energies of features. From clean surface calculations we saw that energies could be reproduced to within ± 1 or 2 eV, which will in turn fix how accurately geometry can be determined, once it is known how sensitive curves are to ion-core positions.

Beams observed in the presence of an overlayer can be described by two bulk reciprocal lattices displaced by $\mathbf{g}^{(0)}$ relative to one another (Fig. 7.18).

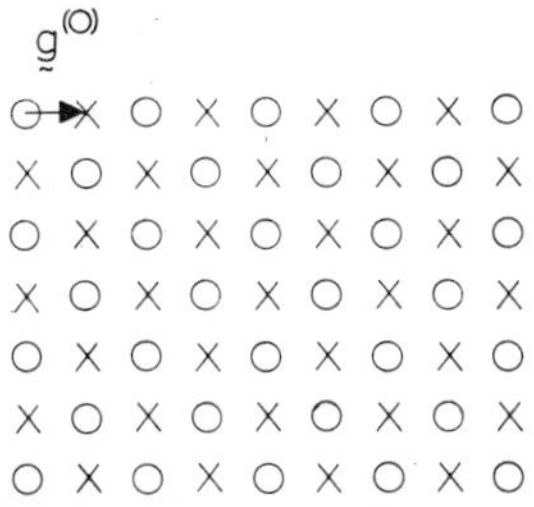

$g^{(O)}$

O→X O X O X O X O
X O X O X O X O X
O X O X O X O X O
X O X O X O X O X
O X O X O X O X O
X O X O X O X O X
O X O X O X O X O

FIGURE 7.18. The reciprocal lattice of a $(\sqrt{2} \times \sqrt{2})$ 45° sodium on nickel (001) structure can be decomposed into two bulk reciprocal lattices with relative displacements $\mathbf{g}^{(0)}$.

In determining the response of the bulk to beams transmitted by the overlayer, two sets of Bloch waves must be calculated, one for a set of beams with parallel momentum

$$\mathbf{k}_{\|} = \mathbf{k}_{0\|}$$

and the other with

$$\mathbf{k}_{\|} = \mathbf{k}_{0\|} + \mathbf{g}^{(0)},$$

modulo a reciprocal vector of the bulk lattice. Two sets of band structure are excited simultaneously, and these have been calculated for the case of normal incidence

$$\mathbf{k}_{0\parallel} = 0$$

and plotted in Fig. 7.19 and 7.20. In conventional terminology the two sections are Γ to X and X to W cuts.

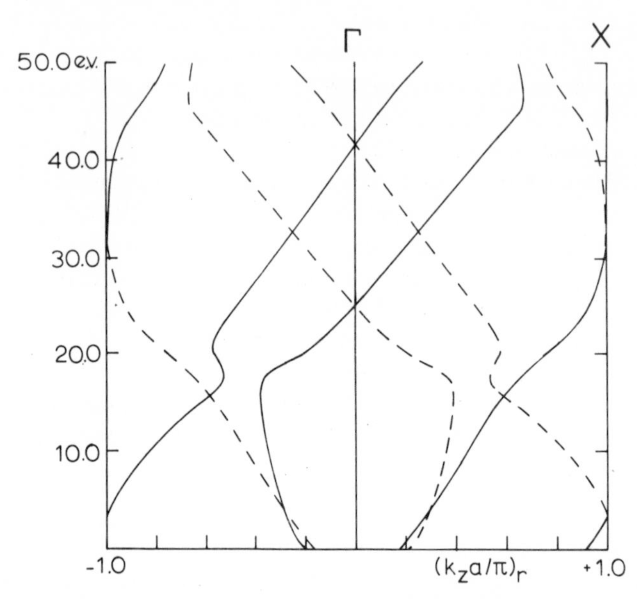

FIGURE 7.19. Band structure for $\mathbf{k}_{\parallel} = 0$ on the (001) surface of nickel. These bands are relevant to both the clean surface and $(\sqrt{2} \times \sqrt{2})45°$ overlayer problems. ——— bands excited by incident electrons. - - - bands with sources at $+\infty$. Only the real part of k_z is plotted.

Following the arguments of Chapter III peaks are expected at places where a forward travelling Bloch wave is nearly degenerate with a backward travelling one.

First we investigate changes in reflectivities as the overlayer is displaced vertically. Figures 7.21–23 show calculations with sodium atoms sitting in lateral positions given by Fig. 7.2, for various values of the vertical displacement, c_{2z}, between centres of the nickel and sodium layers.

At all but the lowest energies the overlayer contributes relatively weakly to back-scattering. Its main function is to change the wave incident on the bulk and to mix together waves reflected from the bulk. Peaks are caused mainly by the bulk whereas the overlayer acts so as to modify this basic structure of peaks by changing amplitudes of excitation of Bloch waves. It changes the intensity seen in given peaks, and also peak positions by, for example, exciting the Bloch wave responsible for a peak with increasing

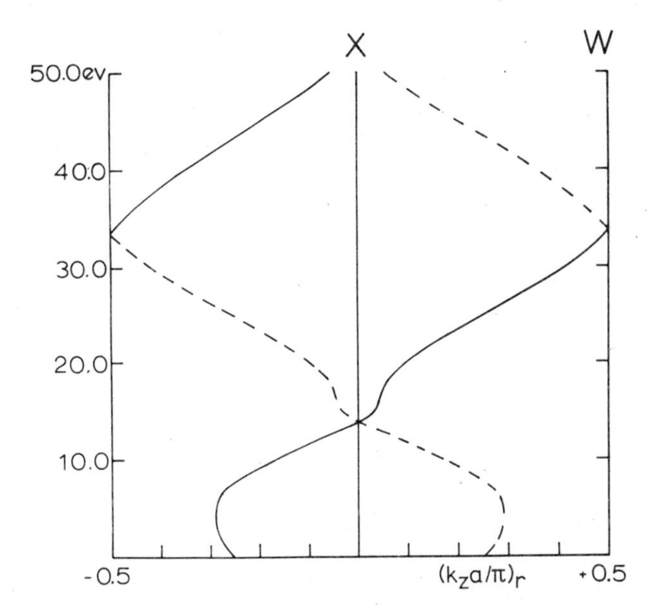

FIGURE 7.20. Band structure for $\mathbf{k}_{\parallel} = \mathbf{k}_{0\parallel} + \mathbf{g}^{(0)}$ on the (001) surface of nickel. This is the other section of band structure brought into play by the $(\sqrt{2} \times \sqrt{2})45°$ overlayer. ——— bands excited by incident electrons. - - - bands with sources at $+\infty$. Only the real part of k_z is plotted.

amplitude across the energy of the peak. Thus the peak would appear shifted to higher energies.

At low energies, especially less than 10 eV, back-scattering by the overlayer is important. The ion cores are able to provide the smaller momenta needed at low energies to turn the electrons around, and at the same time absorption is smaller. Reflections from the vacuum side of the overlayer interfere with reflections produced by other mechanisms, and reflections at the inside surface of the overlayer change the wavefield incident on the bulk. In extreme cases in which both overlayer and bulk reflect strongly, electrons can be multiply reflected between bulk and overlayer in the manner of light in a Fabry-Perôt interferometer.

Structure in Figs 7.21–23 can be interpreted as follows. In the 00 beam at low energies near $E = 0$, is a large peak that is very broad. It is caused by the band at $E = 3$ eV in Fig. 7.19 satisfying a Bragg reflection condition with

$$\text{real part } (k_z c_z/\pi) = 1.0$$

The same peak is present in clean surface spectra but in the presence of an overlayer the broad peak is crossed by a narrow interference fringe caused by a Fabry-Perôt type resonance. It is the only example in these curves of such an effect. The next piece of structure in the 00 beam is the peak at 15–20 eV

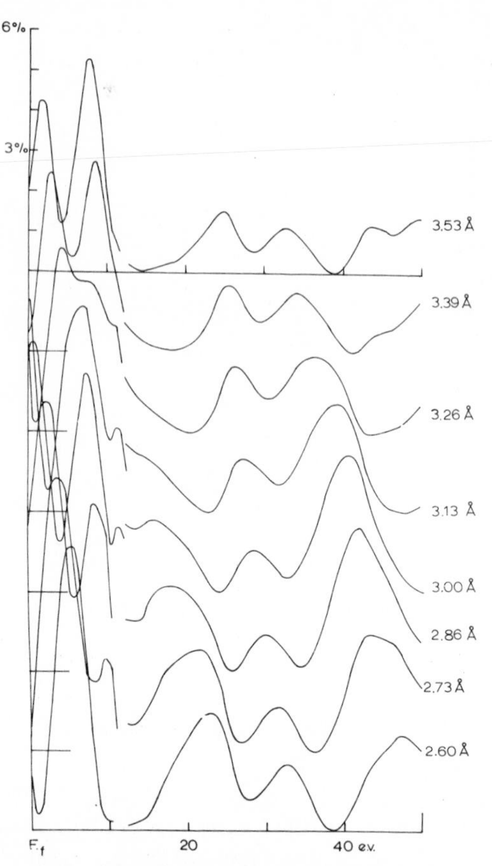

FIGURE 7.21. $I(E)$ for the 00 beam taken at normal incidence on a nickel (001) surface covered by a $(\sqrt{2} \times \sqrt{2})45^\circ$ sodium overlayer, calculated for various values of the vertical spacing between the last nickel layer and overlayer. The sodium atoms were placed above the hollows in the last nickel layer. In this figure the zero of energy is measured from the Fermi level.

caused by the band in Fig. 7.20. As can be seen from distortions in this band in this energy range, the Bragg reflection is a strong one. The same reflection is responsible for a peak in the $\frac{1}{2}\frac{1}{2}$ beam in the same energy range at 24, 33, and 42 eV. There is a triplet of Bragg reflections in Fig. 7.19 causing three peaks seen in the 00 beam for $c_{2z} = 3.53\text{Å}$, and also in the 10 beam, for example, at $c_{2z} = 3.00\text{Å}$. In the $\frac{1}{2}\frac{1}{2}$ beam the large peak at 33 eV appears to be caused by the Bragg reflection at 33 eV in Fig. 7.20.

Evidently not all possible Bragg reflections are excited at a given spacing and the absence of certain peaks provides information as useful as the presence of others.

Of crucial importance is sensitivity of spectra to c_{2z}. Both intensities and positions of peaks change. The positions, on which we shall depend most heavily for structural analysis, vary at different rates, but a typical value is 10–20 eV per Å displacement in c_{2z}. Bearing in mind that calculations are expected to be accurate to within 1–2 eV, a sensitivity of

$$\Delta c_{2z} = \pm 0.1\text{Å} \tag{7.57}$$

is implied. This estimate is of the same order as inaccuracies introduced by approximating V_0 as a constant through the overlayer.

There will be a self-consistency check on (7.57) in that, since more than one peak will be used in a structural determination, the variation in agreement with experimental position will be an indication of the accuracy we have achieved.

Next changes with lateral displacement of the overlayer are investigated.

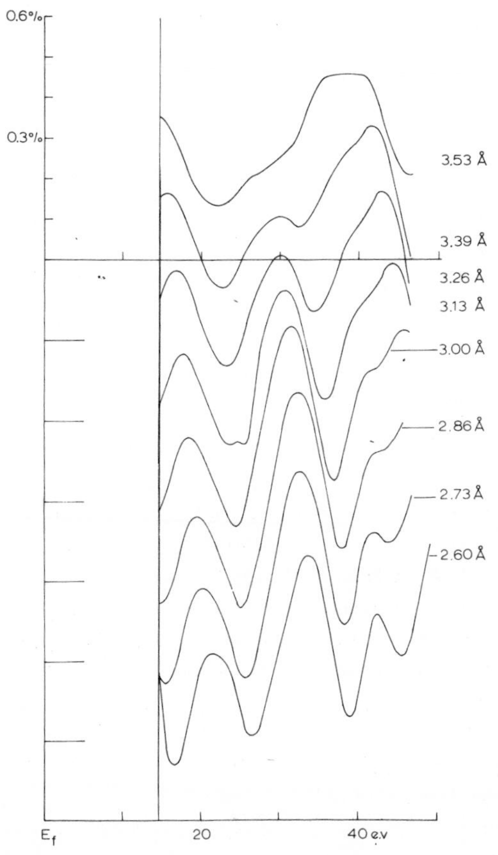

FIGURE 7.22. The $\frac{1}{2}\frac{1}{2}$ beam for the same conditions as Fig. 7.21. This beam is a new one introduced by the overlayer.

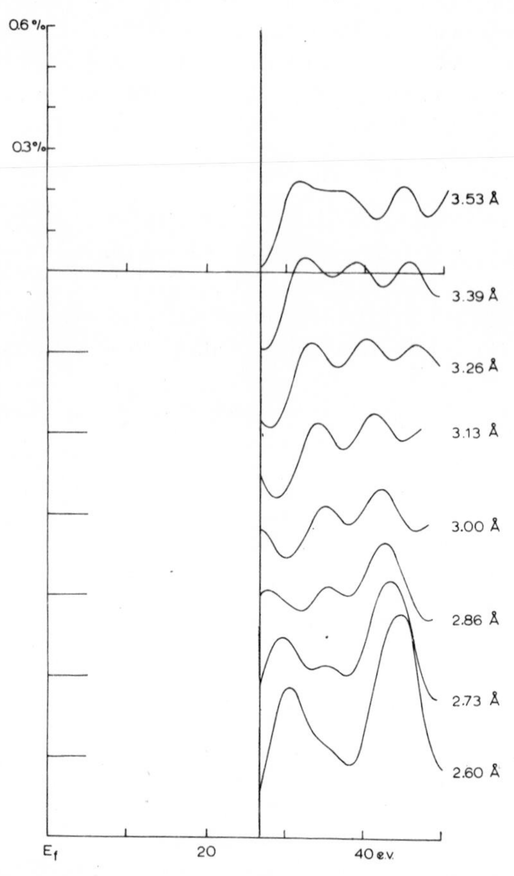

FIGURE 7.23. The 10 beams for the same conditions as Fig. 7.21.

Figure 7.24 compares curves calculated for two lateral positions of the overlayer: sodium atoms over hollows, and directly above nickel atoms in the layer below. The first point to note is that the 00 beam despite its sensitivity to vertical displacements, is not at all sensitive even to a considerable lateral displacement. However, the other beams do show sensitivity and it should be possible to distinguish between the two sites by comparison of non-specular beams with experiment.

Calculations shown for the sodium/nickel system included 25 beams: 13 beams originally present and 12 beams introduced by the overlayer. This number was more than adequate for convergence in the range 0–40 eV. The fourfold symmetry of the structure was used to simplify calculations, which were completed rapidly. Computing times averaged to around 5 seconds per point, which is well inside the range of feasibility. It is estimated that without

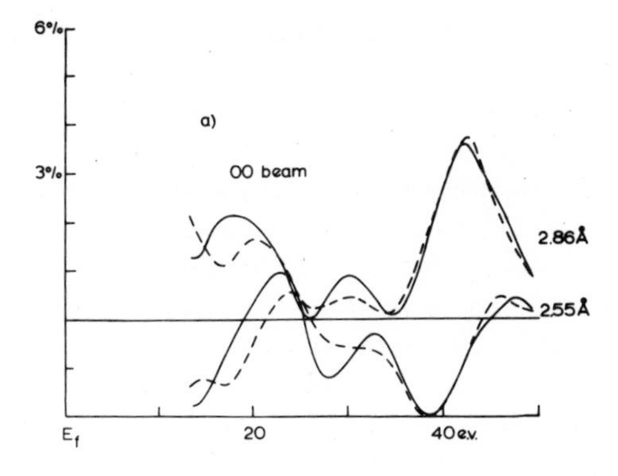

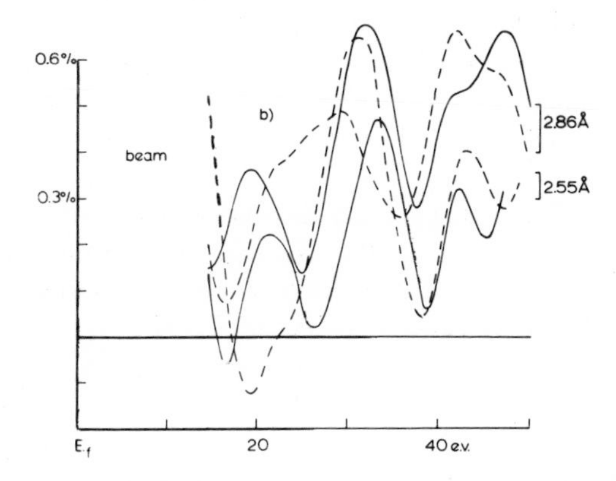

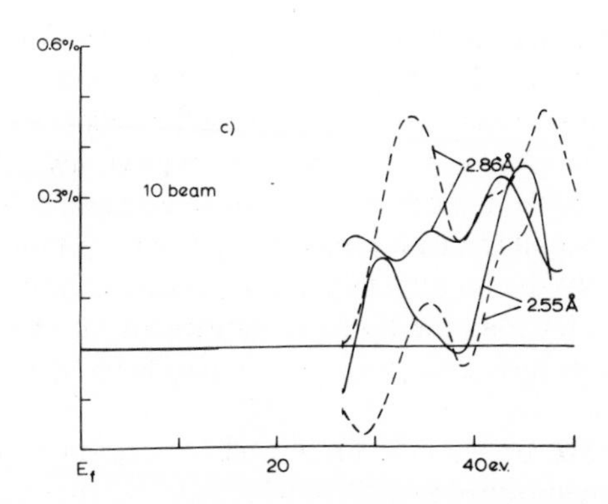

FIGURE 7.24. $I(E)$ for various beams taken at normal incidence on a nickel (001) surface covered with a $(\sqrt{2} \times \sqrt{2})45°$ overlayer of sodium, for two lateral positions of the overlayer. ——— sodium atoms above the hollows. --- sodium atoms above nickel atoms. Two different vertical spacings are used.

increasing times unduly, similar calculations could be made for structures as complicated as (2×4), such as potassium forms on a nickel (001) surface.

It is fortunate that certain practical considerations such as requirements that the energy range we use be sensitive to surface structure and involve weakest possible temperature corrections, force us to work in the lower range of energies, 0–100 eV, where calculations are fastest. Even the multiple scattering nature of experiments which complicates the theory, does have some helpful consequences. If scattering were kinematic and only kinematic Bragg peaks present, then in the range 0–40 eV there would be only three peaks in the curves of Figs 7.21–23. Considerably more information has been introduced by multiple scattering events. Also we might expect multiple scattering structure in spectra to be more sensitive to surface geometry because waves scattered twice by an atom whose position we are trying to find will be twice as sensitive. An example is furnished by the interference fringes that cross the peak near $E = 0$ in the 00 beam, caused by multiple scattering between bulk and overlayer. It is well known that in a Fabry-Perôt interferometer increasing reflectivity of the mirrors increases resolution of the instrument.

Advantages of the much simpler theoretical nature of kinematic scattering are partly cancelled by reduction in sensitivity and information content of the curves. In a kinematic analysis a much wider range of energy and/or variety of angles of incidence must be employed.

VIIH Some determinations of surface structures

The most appealing candidates for structural determination are clean surfaces of simple metals. In Chapter II it was shown that even in such simple examples the surface is not given by truncating the bulk at a plane. Theoretical calculations imply dilations of the top layer of atoms away from the bulk. It would be a useful check on the theory if this displacement could be determined by LEED. Unfortunately displacements expected are typically $\simeq 0.1$Å and could only be resolved with difficulty at the present stage of refinement. That this is the case is shown by the excellent agreement with LEED experiments that several authors have obtained for calculations assuming no dilation of the top layer.

Calculations have been made for nickel (Andersson and Pendry, 1973) with phase shifts determined from first principles, V_{0r} obtained from the work-function and bandwidth, and V_{0i} fitted to peak widths as prescribed in Chapter II. V_{0r} was found to be

$$V_{0r} = -18 \text{ eV} \tag{7.58}$$

by this means. Peak widths were reproduced best with

$$\begin{aligned} V_{0i} &= -1\,\text{eV}, \quad E < 15\,\text{eV}, \\ &= -4\,\text{eV}, \quad E > 15\,\text{eV}. \end{aligned} \tag{7.59}$$

V_{0i} is not an important parameter in structural analysis. It changes peak widths and intensities but has a weak influence on peak positions on which we rely most heavily for structural information.

Figure 7.25 shows the 00 beam calculated for normal incidence on a nickel (001) surface in the same way as the curves for copper in Fig. 3.17. A second calculation is also shown in which the surface layer is dilated by 0.13Å, or

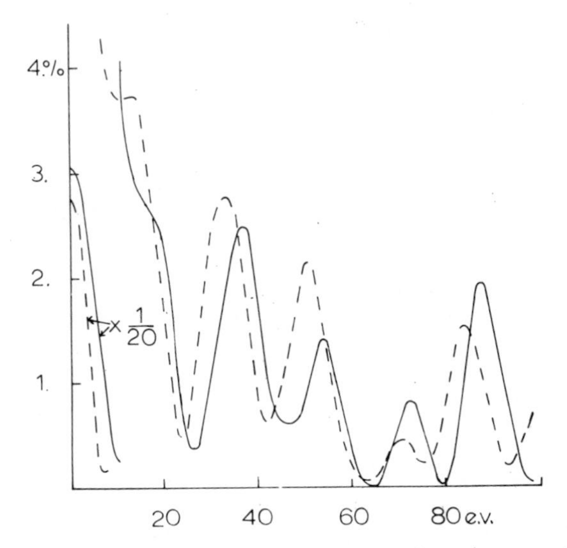

FIGURE 7.25. Intensity of the 00 beam calculated for normal incidence on a clean nickel (001) surface. ——— for the undilated top layer, --- for the top layer dilated by 0.13Å.

about as much as theoretical estimates. These calculations were made neglecting reflectivity of the surface barrier, which does not change the conclusions.

Curves show the same sensitivity of peak positions as did sodium overlayer calculations; the 0.13Å displacement moves peaks by about 2 eV. A disappointing aspect of the curves is that all peaks are displaced to lower energies by the same amount. Unlike the case of a sodium overlayer, a 1–2 eV displacement in curves could be interpreted either as a dilation or as an error in V_{0r}.

This is a general feature of dilations in the top layer spacing and can be crudely interpreted as follows. Dilation of the top layer influences calculations in first order approximation by increasing average spacing between layers seen by the beams, and constructive interferences are forced to lower energies. The higher the energy of a peak the more it is shifted by a change in average spacing. At the same time penetration increases with increasing energy, reducing the average increase in spacing and the net result is the approximately constant shift in energies.

In this case resolution turns upon the accuracy with which V_{0r} is fixed: about ± 1–2 eV. Figure 7.26 shows experimental results of Andersson and Kasemo (1971). If a higher value of -17 eV is chosen for V_{0r} a dilation of 0.13Å is implied. The lower value, -19 eV, implies no dilation at all and theoretical curves using this value are also shown in Fig. 7.26. We conclude that the top layer is dilated by no more than 0.13Å.

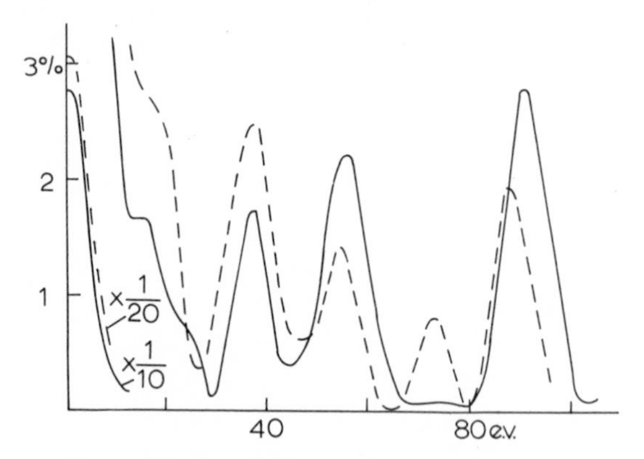

FIGURE 7.26. Comparison of 00 beams for normal incidence on a nickel (001) surface. ——— experiment, --- theory for an undilated top layer, and $V_{0r} = -19$ eV.

Every structural determination for an overlayer system must be preceded by a thorough analysis of the clean surface. Apart from providing the assurance that diffraction by bulk layers is accounted for, the value of V_{0i} is fixed so that no arbitrary parameters other than the geometry to be determined are needed in the overlayer case.

When a $(\sqrt{2} \times \sqrt{2})$ 45° sodium structure is formed on the nickel (001) surface, the work function decreases by 2.5 eV implying formation of a dipole layer in the surface that increases V_{0r} to

$$V_{0r} = -19.0 + 2.5 = -16.5 \text{ eV}. \tag{7.60}$$

The value -19.0 eV is chosen for the clean surface consistent with the assumption of an undilated top layer of nickel atoms. The nickel phase shifts and V_{0i} remain unaffected by the dipole layer.

It remains to compare theoretical curves for the $(\sqrt{2} \times \sqrt{2})$ 45° sodium structure with experiment. Figure 2.27 taken from Andersson and Pendry (1972) presents experimental results together with two sets of theoretical curves: those giving best agreement with experiment, $c_{2z} = 2.87$Å and those for a spacing between overlayer and substrate deduced from atomic radii, $c_{2z} = 2.55$Å. Radii were calculated from bulk sodium and nickel lattice constants. Both theoretical curves are for the lateral positioning of the overlayer shown in Fig. 7.2.

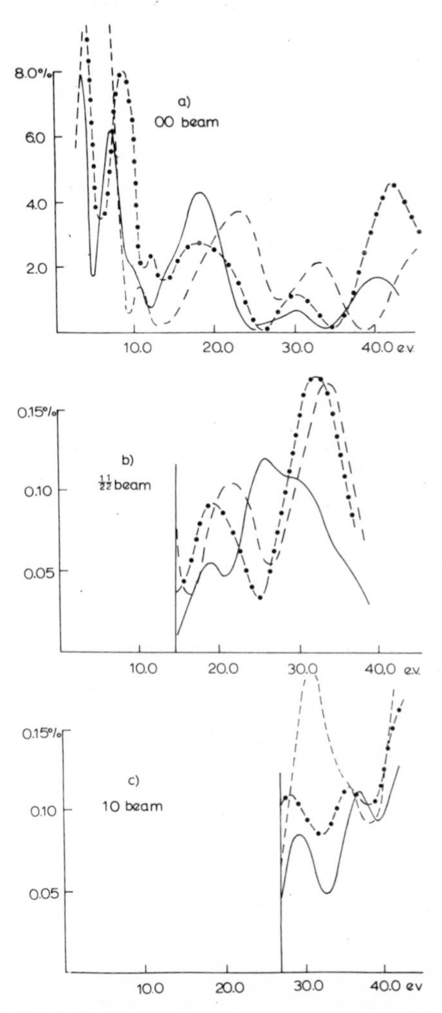

FIGURE 7.27. LEED spectra taken at normal incidence from a $(\sqrt{2} \times \sqrt{2})45°$ sodium structure on a nickel (001) surface. ——— experiment, --- theory for $c_{2z} = 2.55$Å, -·-·- theory for $c_{2z} = 2.87$Å. In this figure the zero of energy is measured from the Fermi level.

As suspected, intensities are not to be brought into excellent agreement with experiment, but all prominent features of the experimental curves are reproduced in the 2.87Å calculation, except for a peak at 25 eV in the $\frac{1}{2}\frac{1}{2}$ beam. In the 00 beam there are seven prominent features, all sensitive to c_{2z}: splitting of the low energy peak, peaks at 18, 30 and 40 eV, and three minima at 11, 26 and 34 eV. All are correctly aligned at 2.87Å usually within 1 eV, always within 2 eV. In the $\frac{1}{2}\frac{1}{2}$ beam there are three experimental peaks at 19, 25, and 30 eV and a minimum at 21 eV of these the peaks at 19 and 30 eV are correctly reproduced but the one at 25 eV is not, also resulting in the minimum not being correctly placed. The missing peak is almost certainly a resonance (Section IIIE) associated with the emergence of the 10 beam. Failure of the calculation for this feature reflects sensitivity of resonance effects to details of the variation in V_0 near the surface. In the 10 beam the four prominent features: two peaks at 29 and 37 eV, two minima at 33 and 39 eV, are all reproduced to within 1 eV.

Further comparisons with experiment of calculations for other lateral positions lead to the conclusion that placing sodium atoms above hollows in the top nickel layer with

$$c_{2z} = 2.87 \pm 0.1\text{Å}$$

gives best agreement with experiment.

The ability to align structure to within 1–2 eV confirms that this is indeed the order of accuracy achieved in calculations. Thermal vibrations which were omitted from the calculations are responsible for most of the poor agreement with intensities, but even for the weakly bound sodium overlayer, it seems possible to determine structure to good accuracy by relying first on

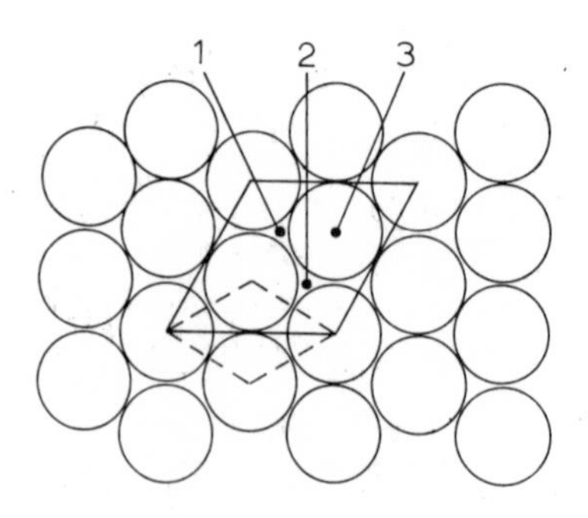

FIGURE 7.28. The silver (111) surface showing silver atoms, ○, the unit cell of the clean surface ---, and of the surface in the presence of an overlayer of iodine, ———. Iodine atoms can sit anywhere inside the unit cell: for example on sites 1, 2 or 3. Note that sites 1 and 2 are not equivalent because they differ in their orientation relative to deeper layers of silver. Below position 2 there is an atom in the second silver layer; there is none in position 1.

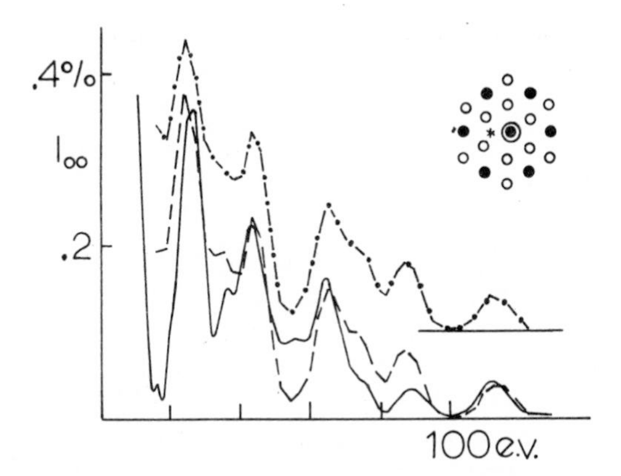

FIGURE 7.29. Intensity oʲ the 00 beam from a $(\sqrt{3} \times \sqrt{3})30°$ iodine structure on silver (111), taken at 8° from normal incidence. ——— experiment, --- theory for position 1, -·--·- theory for position 2. In each case $c_{2z} = 2.25$Å. The theoretical curves have been divided by 1.45 to facilitate comparison with experiment. The insert shows the diffracted beams: ● those present in the clean surface pattern, 0 extra beams, * position of incident beam. The 00 beam is ringed.

reproduction of qualitative features in the curves, and for greater accuracy, on the energies of these features.

Although the technique is sensitive enough to distinguish between the true geometry and that implied by atomic radii, the distances are not greatly different, implying an increase in effective radius of sodium of 14%.

Another system investigated by Forstmann Berndt and Büttner (1973) using the same technique, is that of iodine on a silver (111) surface. Their analysis began with a careful interpretation of spectra for clean silver (111), which they were able to calculate with very good accuracy (Forstmann and Berndt, 1973). The parameters derived for clean silver

$$V_{0r} = -10.5 \text{ eV},$$

$$V_{0i} = -4 \text{ eV},$$

modified for the 0.5 eV work function change on depositing iodine, were carried through to the overlayer calculation. Some corrections were made for thermal vibrations by assuming that overlayer curves had the same temperature dependence as curves for clean silver, and multiplying by the appropriate factor.

Diffraction patterns imply a $(\sqrt{3} \times \sqrt{3})$ 30° structure. Figure 7.28 shows the geometry of the close packed silver (111) surface. Forstmann *et al.* found the same insensitivity of the 00 beam to lateral displacements of the overlayer observed by Andersson and Pendry for the sodium/nickel system, and made use of it in two ways. First the vertical height of the overlayer was determined using the 00 beam only. Then before trying to distinguish between lateral

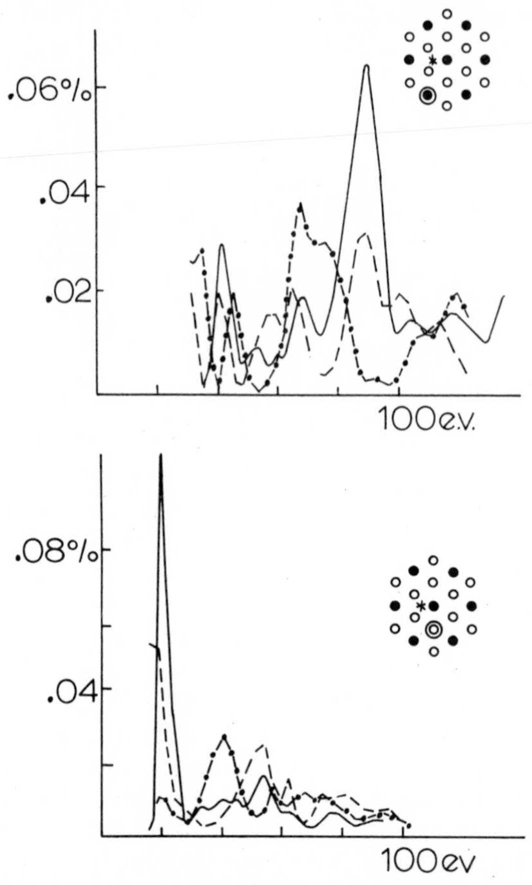

FIGURE 7.30. Comparison with experiment of non-specular beams under the same conditions described in Fig. 7.29. In each case the insert shows incident beam position, *, and the beam measured is ringed. ——— experiment, --- theory for position 1, – · – · – theory for position 2. In each case $c_{2z} = 2.28$Å.

positions of the overlayer the phase shifts for iodine were adjusted by taking a linear combination of phase shifts for I and for I^- to give best agreement with the 00 beam. The phase shifts for I were found to be best. Results obtained are shown in Fig. 7.29. The agreement with experiment is evident, as is insensitivity to lateral displacement.

Non-specular beams are shown in Fig. 7.30 for positions 1 and 2, vertical displacement fixed at 2.28Å. There is clear evidence that position 1 is correct. Iodine sits in those hollows between atoms in the first layer of silver such that there is no atom in the second silver layer directly below the iodine. Accuracy estimated for this determination was ±0.1Å in agreement with previous estimates. The geometry was consistent with an Ag-*I* bond length of 2.8Å found in the silver iodide crystal.

It would appear that the dynamical method is well established at least for the simple structures.

Chapter 8

Appendices

Appendix A—Definitions of Special Functions

Legendre polynomials and associated Legendre functions

Differential equation

$$\left[(1 - z^2)\frac{d^2}{dz^2} - 2z\frac{d}{dz} + \ell(\ell + 1) - \frac{m^2}{1 - z^2}\right]P_\ell^{|m|} = 0$$

$$\ell = 0, 1, 2 \ldots, |m| = 0, 1, 2 \ldots, 0 \leqslant |m| \leqslant \ell, \tag{A.1}$$

$P_\ell = P_\ell^0$ is the Legendre polynomial, $P_\ell^{|m|}$ is an associated Legendre polynomial.

Recurrence relations

$$(2\ell + 1)zP_\ell^{|m|} = (\ell + 1 - |m|)P_{\ell+1}^{|m|} + (\ell + |m|)P_{\ell-1}^{|m|} \tag{A.2}$$

Special values

$$P_\ell^\ell(z) = \frac{(2\ell)!}{2^\ell \ell!}(1 - z^2)^{\frac{1}{2}\ell} \tag{A.3}$$

$$P_\ell^{\ell-1}(z) = \frac{(2\ell)!}{2^\ell \ell!}z(1 - z^2)^{\frac{1}{2}\ell - \frac{1}{2}} \tag{A.4}$$

$$P_\ell(1) = 1, \quad P_\ell(-1) = (-1)^\ell \tag{A.5}$$

$$P_\ell^{|m|}(1) = P_\ell^{|m|}(-1) = 0 \quad \text{if } m \neq 0 \tag{A.6}$$

$$P_\ell^{|m|}(0) = (-1)^p(2p + 2|m|)!/[2^\ell p!(p + |m|)!], \quad \text{if} \quad \ell - |m| = 2p,$$

$$= 0 \quad \text{if} \quad \ell - |m| = 2p + 1. \tag{A.7}$$

Explicit expressions

$$P_0 = 1, \quad P_1 = z, \quad P_2 = \tfrac{1}{2}(3z^2 - 1) \tag{A.8}$$

Orthonormality relations

$$\int_{-1}^{+1} P_\ell^{|m|}(z) P_{\ell'}^{|m|}(z) dz = \frac{2}{2\ell + 1} \frac{(\ell + |m|)!}{(\ell - |m|)!} \delta_{\ell\ell'} \tag{A.9}$$

Spherical harmonics

Differential equations

$$-\left[\frac{1}{\sin(\theta)} \frac{\partial}{\partial\theta} \sin(\theta) \frac{\partial}{\partial\theta} + \frac{1}{\sin^2(\theta)} \frac{\partial^2}{\partial\phi^2}\right] Y_{\ell m}(\theta\phi) = \ell(\ell + 1) Y_{\ell m}(\theta\phi) \tag{A.10}$$

$$\frac{\partial}{\partial\phi} Y_{\ell m}(\theta\phi) = im Y_{\ell m}(\theta\phi) \tag{A.11}$$

Relation to associated Legendre functions

$$Y_{\ell|m|}(\theta\phi) = (-1)^{|m|} \left[\frac{2\ell + 1}{4\pi} \frac{(\ell - |m|)!}{(\ell + |m|)!}\right]^{\frac{1}{2}} P_\ell^{|m|}(\cos\theta) \exp(i|m|\phi) \tag{A.12}$$

$$Y_{\ell -|m|}(\theta\phi) = \left[\frac{2\ell + 1}{4\pi} \frac{(\ell - |m|)!}{(\ell + |m|)!}\right]^{\frac{1}{2}} P_\ell^{|m|}(\cos\theta) \exp(-i|m|\phi) \tag{A.13}$$

Recurrence relations

$$\cos(\theta) Y_{\ell m}(\theta\phi) = \left[\frac{(\ell + 1 + m)(\ell + 1 - m)}{(2\ell + 1) \quad (2\ell + 3)}\right]^{\frac{1}{2}} Y_{\ell+1,m}(\theta\phi) + \left[\frac{(\ell + m)\ (\ell - m)}{(2\ell + 1)(2\ell - 1)}\right]^{\frac{1}{2}} Y_{\ell-1,m}(\theta\phi) \tag{A.14}$$

Special values

$$Y_{\ell 0}(\theta\phi) = \left(\frac{2\ell + 1}{4\pi}\right)^{\frac{1}{2}} P_\ell(\cos\theta) \tag{A.15}$$

$$Y_{\ell\ell} = (-1)^\ell \left[\frac{(2\ell + 1)(2\ell)!}{4\pi 2^{2\ell}(\ell!)^2}\right]^{\frac{1}{2}} \sin^\ell(\theta) \exp(i\ell\phi) \tag{A.16}$$

$$Y_{\ell,\ell-1} = (-1)^{\ell-1} \left[\frac{(2\ell + 1)2\ell(2\ell)!}{4\pi 2^{2\ell}(\ell!)^2}\right]^{\frac{1}{2}} \sin^{\ell-1}(\theta) \cos(\theta) \exp[i(\ell - 1)\phi] \tag{A.17}$$

Explicit expressions

$$Y_{00} = \left(\frac{1}{4\pi}\right)^{\frac{1}{2}}, \quad Y_{10} = \left(\frac{3}{4\pi}\right)^{\frac{1}{2}} \cos(\theta)$$

$$Y_{11} = -\left(\frac{3}{8\pi}\right)^{\frac{1}{2}} \sin(\theta) \exp(i\phi)$$

$$Y_{1-1} = \left(\frac{3}{8\pi}\right)^{\frac{1}{2}} \sin(\theta) \exp(-i\phi) \tag{A.18}$$

Orthonormality relations

$$\int_0^{\pi} \sin(\theta) d\theta \int_0^{2\pi} d\phi \, Y_{\ell' m'}(\theta\phi) \, Y_{\ell -m}(\theta\phi)(-1)^m = \delta_{\ell\ell'} \delta_{mm'} \tag{A.19}$$

Other relations

$$Y_{\ell m}(\pi - \theta, \phi + \pi) = (-1)^{\ell} \, Y_{\ell m}(\theta\phi) \tag{A.20}$$

$$Y^*_{\ell m}(\theta, \phi) = (-1)^m \, Y_{\ell - m}(\theta\phi), \quad \theta \text{ and } \phi \text{ both real} \tag{A.21}$$

$$\sum_{\ell=0}^{\infty} \sum_{m=-\ell}^{+\ell} (-1)^m \mathrm{Y}_{\ell - m}(\theta\phi) \mathrm{Y}_{\ell m}(\theta'\phi') = \frac{\delta(\theta - \theta')\delta(\phi - \phi')}{\sin(\theta)} = \delta(\Omega - \Omega') \tag{A.22}$$

Warning
Complex conjugation is not an analytic operation. Only those formulae given above that do not make complex conjugation of $Y_{\ell m}$ can be analytically continued to complex values of the arguments. This difficulty can always be circumvented by substituting from above

$$Y^*_{\ell m}(\theta\phi) = (-1)^m \, Y_{\ell - m}(\theta\phi)$$

Spherical Bessel and Hankel functions

j_ℓ and n_ℓ are spherical Bessel functions of the first and second kinds. $h_\ell^{(1)}$ and $h_\ell^{(2)}$ are Hankel functions of the first and second kinds. They are related by

$$h_\ell^{(1)} = j_\ell + i n_\ell \tag{A.23}$$

$$h_\ell^{(2)} = j_\ell - i n_\ell \tag{A.24}$$

Differential equation

$$\left[\frac{d^2}{dz^2} + \frac{2}{z}\frac{d}{dz} + 1 - \frac{\ell(\ell + 1)}{z^2}\right] f_\ell = 0 \tag{A.25}$$

$$f_\ell = j_\ell, n_\ell, h_\ell^{(1)} \quad \text{or} \quad h_\ell^{(2)}$$

Recurrence relations

$$f_{\ell-1}(z) + f_{\ell+1}(z) = \frac{2\ell + 1}{z} f_\ell(z) \tag{A.26}$$

Warning: in the case of $j_\ell(z)$ the recurrence relations for increasing ℓ are unstable at small z, because $j_\ell(z)$ becomes successively smaller as $z^\ell/(2\ell + 1)!!$.

Series expansions

$$j_\ell(z) = \frac{z^\ell}{(2\ell + 1)!!}\left[1 - \frac{\frac{1}{2}z^2}{1!(2\ell + 3)} + \frac{(\frac{1}{2}z^2)^2}{2!(2\ell + 3)(2\ell + 5)} - \dots\right] \tag{A.27}$$

$$n_\ell(z) = -\frac{(2\ell + 1)!!}{(2\ell + 1)z^{\ell+1}}\left[1 - \frac{\frac{1}{2}z^2}{1!(1 - 2\ell)} + \frac{(\frac{1}{2}z^2)^2}{2!(1 - 2\ell)(3 - 2\ell)} - \dots\right] \tag{A.28}$$

$$\text{where} \quad (2\ell + 1)!! = 1.3.5\dots(2\ell + 1).$$

Explicit expressions

$$j_0 = z^{-1}\sin(z), j_1 = z^{-2}\sin(z) - z^{-1}\cos(z) \tag{A.29}$$

$$n_0 = -z^{-1}\cos(z), n_1 = -z^{-2}\cos(z) - z^{-1}\sin(z) \tag{A.30}$$

$$h_0^{(1)} = -iz^{-1}\exp(iz), h_1^{(1)} = -iz^{-2}\exp(iz) - z^{-1}\exp(iz) \tag{A.31}$$

$$h_0^{(2)} = iz^{-1}\exp(-iz), h_1^{(2)} = iz^{-2}\exp(-iz) - z^{-1}\exp(-iz) \tag{A.32}$$

Asymptotic forms as $z \to \infty$

$$j_\ell \sim z^{-1}\sin(z - \tfrac{1}{2}\ell\pi), n_\ell \sim -z^{-1}\cos(z - \tfrac{1}{2}\ell\pi) \tag{A.33}$$

$$\begin{aligned} h_\ell^{(1)} &\sim -iz^{-1}\exp(iz - \tfrac{1}{2}i\ell\pi)\left[1 + i\frac{\ell(\ell + 1)}{2z}\right] \\ h_\ell^{(2)} &\sim iz^{-1}\exp(-iz + \tfrac{1}{2}i\ell\pi)\left[1 - i\frac{\ell(\ell + 1)}{2z}\right] \end{aligned} \tag{A.34}$$

Some expansions

Addition theorem

$$P_\ell(\cos\alpha) = \frac{4\pi}{2\ell + 1}\sum_{m=-\ell}^{+\ell}(-1)^m Y_{\ell-m}(\theta\phi)Y_{\ell m}(\theta'\phi') \tag{A.35}$$

where α is the angle between directions $(\theta\phi)$, $(\theta'\phi')$.

Expansions of Green's functions

$$\frac{1}{|\mathbf{r}_1 - \mathbf{r}_2|} = \sum_{\ell=0}^{\infty} \frac{\mathbf{r}_<^{\ell}}{\mathbf{r}_>^{\ell+1}} P_\ell\left(\frac{\mathbf{r}_1 \cdot \mathbf{r}_2}{|\mathbf{r}_1|\,|\mathbf{r}_2|}\right) \tag{A.36}$$

$$\frac{\exp(ik|\mathbf{r}_1 - \mathbf{r}_2|)}{|\mathbf{r}_1 - \mathbf{r}_2|} = ik \sum_{\ell=0}^{\infty} \sum_{m=-\ell}^{+\ell} 4\pi j_\ell(kr_<) h_\ell^{(1)}(kr_>)$$

$$(-1)^m Y_{\ell-m}[\Omega(\mathbf{r}_1)]\, Y_{\ell m}[\Omega(\mathbf{r}_2)] \tag{A.37}$$

$$\frac{-1}{4\pi} \frac{\exp(ik|\mathbf{r}_1 - \mathbf{r}_2|)}{|\mathbf{r}_1 - \mathbf{r}_2|} = \frac{1}{(2\pi)^3} \int_{-\infty}^{\infty}\int_{-\infty}^{\infty}\int_{-\infty}^{\infty} dK_x dK_y dK_z$$

$$\times \frac{\exp[i\mathbf{K}\cdot(\mathbf{r}_1 - \mathbf{r}_2)]}{|\mathbf{k}|^2 - |\mathbf{K}|^2 + i\delta} \tag{A.38}$$

$$\delta = +0.$$

also to be deduced from the formulae above

$$h_\ell^{(1)}(|\mathbf{k}|\,|\mathbf{r}|)\, Y_{\ell-m}[\Omega(\mathbf{r})] = \frac{i^{1-\ell}}{2\pi^2|\mathbf{k}|} \int Y_{\ell-m}[\Omega(K)] \frac{\exp(i\mathbf{K}\cdot\mathbf{r})}{|\mathbf{k}|^2 - |\mathbf{K}|^2 + i\delta} d^3\mathbf{K} \tag{A.39}$$

Expansion of a plane wave

$$\exp(i\mathbf{k}\cdot\mathbf{r}) = \sum_{\ell=0}^{\infty} (2\ell + 1) i^\ell j_\ell(kr) P_\ell\left(\frac{\mathbf{k}\cdot\mathbf{r}}{|\mathbf{k}|\,|\mathbf{r}|}\right)$$

$$= \sum_{\ell=0}^{\infty} \sum_{m=-\ell}^{+\ell} 4\pi i^\ell j_\ell(kr)(-1)^m \times Y_{\ell-m}[\Omega(\mathbf{r})]\, Y_{\ell m}[\Omega(\mathbf{k})] \tag{A.40}$$

Spherical waves and changes of origin

Spherical waves about point **R**, can also be expressed as spherical waves about **R**′:

$$h_\ell^{(1)}(\kappa|\mathbf{r} - \mathbf{R}|)\, Y_{\ell m}[\Omega(\mathbf{r} - \mathbf{R})] = \sum_{\ell'' m''} G_{\ell m, \ell'' m''} j_{\ell''}(\kappa|\mathbf{r} - \mathbf{R}'|)\, Y_{\ell'' m''}[\Omega(\mathbf{r} - \mathbf{R}')] \tag{A.41}$$

$$G_{\ell m, \ell'' m''} = \sum_{\ell' m'} 4\pi i^{\ell - \ell' - \ell''} (-1)^{m' + m''} h_{\ell'}^{(1)}(\kappa|\mathbf{R}' - \mathbf{R}|)$$

$$\times Y_{\ell' -m'}[\Omega(\mathbf{R}' - \mathbf{R})] \int Y_{\ell m} Y_{\ell' m'} Y_{\ell'' -m''} d\Omega \tag{A.42}$$

and by complex conjugation

$$h_\ell^{(2)}(\kappa|\mathbf{r} - \mathbf{R}|)\, Y_{\ell m}[\Omega(\mathbf{r} - \mathbf{R})]$$

$$= \sum_{\ell'' m''} G'_{\ell m, \ell'' m''} j_{\ell''}(\kappa|\mathbf{r} - \mathbf{R}'|)\, Y_{\ell'' m''}[\Omega(\mathbf{r} - \mathbf{R}')] \tag{A.43}$$

$$G'_{\ell m,\ell''m''} = \sum_{\ell'm'} 4\pi i^{\ell-\ell'-\ell''}(-1)^{m'+m''} h_{\ell'}^{(2)}(\kappa|\mathbf{R}'-\mathbf{R}|)$$
$$\times\ Y_{\ell'-m'}[\Omega(\mathbf{R}'-\mathbf{R})] \int Y_{\ell m} Y_{\ell'm'} Y_{\ell''-m''}\, d\Omega \qquad \text{(A.44)}$$

the expansions are valid provided that

$$|\mathbf{r}-\mathbf{R}'| < |\mathbf{R}-\mathbf{R}'|$$

Using equations (A.23) and (A.24) expansions involving j_ℓ and n_ℓ can be found.

Appendix B—Evaluation of the Integral for Kinematic Scattering

$$\int \chi_s(\mathbf{r}) \exp(-i\mathbf{k}''\cdot\mathbf{r}) d^2\mathbf{r}_{||}$$
$$= \sum_\ell (2\ell+1) \sin[\delta_\ell(s)] \exp[i\delta_\ell(s)]\, i^{\ell+1}$$
$$\int h_\ell^{(1)}(|\mathbf{k}|\,|\mathbf{r}|)\, P_\ell\left(\frac{\mathbf{k}\cdot\mathbf{r}}{|\mathbf{k}|\,|\mathbf{r}|}\right) \exp(-i\mathbf{k}''\cdot\mathbf{r})\, d^2\mathbf{r}_{||} \qquad \text{(A.45)}$$

Two substitutions from Appendix A are made, firstly:

$$P_\ell\left(\frac{\mathbf{k}\cdot\mathbf{r}}{|\mathbf{k}|\,|\mathbf{r}|}\right) = \frac{4\pi}{2\ell+1} \sum_m (-1)^m Y_{\ell m}[\Omega(\mathbf{k})]\, Y_{\ell-m}[\Omega(\mathbf{r})], \qquad \text{(A.35)}$$

secondly the result

$$h_\ell^{(1)}(|\mathbf{k}|\,|\mathbf{r}|)\, Y_{\ell-m}[\Omega(\mathbf{r})]$$
$$= \frac{i^{1-\ell}}{2\pi^2|\mathbf{k}|} \int Y_{\ell-m}[\Omega(\mathbf{K})] \frac{\exp(i\mathbf{K}\cdot\mathbf{r})}{|\mathbf{k}|^2-|\mathbf{K}|^2+i\delta}\, d^3\mathbf{K}. \qquad \text{(A.39)}$$

Therefore

$$\int \chi_s(\mathbf{r}) \exp(-i\mathbf{k}''\cdot\mathbf{r})\, d^2\mathbf{r}_{||}$$
$$= -\frac{2}{|\mathbf{k}|\pi} \sum_{\ell m} (-1)^m \sin[\delta_\ell(s)] \exp[i\delta_\ell(s)]\, Y_{\ell m}[\Omega(\mathbf{k})]$$
$$\times \int\int Y_{\ell-m}[\Omega(\mathbf{K})] \frac{\exp[i(\mathbf{K}-\mathbf{k}'')\cdot\mathbf{r}]}{|\mathbf{k}|^2-|\mathbf{K}|^2+i\delta}\, d^2\mathbf{r}_{||}\, d^3\mathbf{K} \qquad \text{(A.46)}$$

The integral over $d^2\mathbf{r}_{||}$ is very small unless

$$(\mathbf{K}_{||} - \mathbf{k}''_{||}) \simeq 0,$$

because otherwise the oscillatory nature of the exponential means that a great deal of cancellation takes place within the integral. Therefore

$$\int\int Y_{\ell-m}[\Omega(\mathbf{K})]\frac{\exp[i(\mathbf{K}-\mathbf{k}'')\cdot\mathbf{r}]}{|\mathbf{k}|^2-|\mathbf{K}|^2+i\delta}d^2\mathbf{r}_\parallel\, d^3\mathbf{K}$$

$$\int d\mathbf{K}_z\frac{[Y_{\ell-m}(\Omega(\mathbf{K}))]\,\mathbf{K}_\parallel=\mathbf{k}''_\parallel}{|\mathbf{k}|^2-\mathbf{K}_z^2-|\mathbf{k}''_\parallel|^2+i\delta}\exp[i(\mathbf{K}_z-\mathbf{k}''_z)z]$$

$$\times\int\int d^2\mathbf{r}_\parallel\, d^2\mathbf{K}_\parallel \exp[i\mathbf{K}_\parallel-\mathbf{k}''_\parallel)\cdot\mathbf{r}_\parallel] \tag{A.47}$$

and substituting

$\mathbf{K}'_\parallel=\mathbf{K}_\parallel-\mathbf{k}''_\parallel$,

$$\int\int d^2\mathbf{r}_\parallel\, d^2\mathbf{K}_\parallel\exp[i(\mathbf{K}_\parallel-\mathbf{k}''_\parallel)\cdot\mathbf{r}_\parallel]$$

$$\underset{L\to\infty}{=}\int d^2\mathbf{K}'_\parallel\int_{-L}^{L}\int_{-L}^{L}\exp[i(K'_x x+K'_y y)]\,dx\,dy$$

$$\underset{L\to\infty}{=}\int_{-\infty}^{\infty}\int_{-\infty}^{\infty}dK'_x\,dK'_y\frac{1}{iK'_x}[\exp(iK'_xL)-\exp(-iK'_xL)]$$

$$\times\frac{1}{iK'_y}[\exp(iK'_yL)-\exp(-iK'_yL)] \tag{A.48}$$

The integral over dK'_x can be written

$$\lim_{\substack{L\to\infty\\ \delta\to 0}}\int_{-\infty}^{\infty}\frac{dK'_x}{i(K'_x-i\delta)}[\exp(iK'_xL)-\exp(-iK'_xL)] \tag{A.49}$$

The contour can be closed in the upper-half plane when integrating $\exp(+iK'_xL)$ and in the lower-half plane for $\exp(-iK'_xL)$, picking up a contribution from the pole in the upper-half plane of 2π. We can now write

$$\int d^2\mathbf{r}_\parallel\int d^2\mathbf{K}_\parallel\exp[i(\mathbf{K}_\parallel-\mathbf{k}''_\parallel)\cdot\mathbf{r}_\parallel]=(2\pi)^2 \tag{A.50}$$

and substituting into (A.47) gives

$$\int\int Y_{\ell-m}[\Omega(\mathbf{K})]\frac{\exp[i(\mathbf{K}-\mathbf{k}'')\cdot\mathbf{r}]}{|\mathbf{k}|^2-|\mathbf{K}|^2+i\delta}d^2\mathbf{r}_\parallel\, d^3\mathbf{K}$$

$$=(2\pi)^2\int_{-\infty}^{\infty}\frac{[Y_{\ell-m}(\Omega(\mathbf{K}))]\,\mathbf{K}_\parallel=\mathbf{k}''_\parallel}{(k''_z)^2-K_z^2+i\delta}\exp[i(K_z-k''_z)z]\,dK_z \tag{A.51}$$

since $|\mathbf{k}|^2=|\mathbf{k}''|^2=2E-2V_0$. The dK_z contour can be closed in the upper- or lower-half plane depending on whether z is positive or negative, picking up a residue

$$(2\pi)^3 i \frac{Y_{\ell-m}[\Omega(\mathbf{k}'')]}{-2|k_z''|} \tag{A.52}$$

substituting (A.52) into (A.46)

$$\int \chi_s(\mathbf{r}) \exp(-i\mathbf{k}''\cdot\mathbf{r})\, d^2\mathbf{r}_{\|}$$
$$= \frac{2\pi^2 i}{|\mathbf{k}|\,|k_z''|} \sum_{\ell} (2\ell + 1) \sin[\delta_\ell(s)] \exp[i\delta_\ell(s)]$$
$$\times P_\ell\left(\frac{\mathbf{k}\cdot\mathbf{k}''}{|\mathbf{k}|\,|\mathbf{k}''|}\right) \tag{A.53}$$

Appendix C—LEED computer programs

1. Calculation of LEED intensities

The following set of subroutines calculates intensities from a clean surface by four different methods. They contain no surface barrier, so that the lower potential between muffin tins, V_{or}, is taken account of by translating the curves to lower energies by hand. All energies in the program are relative to the muffin tin zero. The arrangement of subroutines has been designed so that sophistications can easily be accommodated into the programs with a minimum of change. Programming was in standard ASA fortran IV, and the subroutines have been tested both on an ICL Titan, and an IBM 370/165 under the WATFIV and FORTG compilers on the latter machine.

The first step calculates reflection and transmission matrices for a single layer (MSMAT), returned as

$$TC = I + M^{++} = I + M^{--}$$
$$RC = M^{+-} = M^{-+}$$

It is assumed that within each layer there is only one atom per unit cell, therefore the symmetry above holds. Matrices M were defined in Section IVC; equation (4.49)

$$M_{\mathbf{g}'\mathbf{g}}^{\pm\pm} = \frac{8\pi^2 i}{|\mathbf{K}_{\mathbf{g}}^{\pm}|\, A\, |K_{\mathbf{g}'z}^{\pm}|} \sum_{\substack{\ell' m' \\ \ell m}} [i^\ell (-1)^m Y_{\ell-m}(\Omega(\mathbf{K}_{\mathbf{g}}^{\pm}))]$$
$$\times (1 - X)^{-1}_{\ell m,\ell' m'} [i^{-\ell'} Y_{\ell' m'}(\Omega(\mathbf{K}_{\mathbf{g}'}^{\pm}))] \exp(i\delta_{\ell'}) \sin(\delta_{\ell'}). \tag{A.54}$$

Subscript 'k' has been removed because there is only one atom per unit cell, and wave amplitudes are referred to a centre of an atom. Matrix X, appearing in the program in factorised form, XODD and XEVEN, is evaluated by a separate subroutine, XMAT, from formulae

$$X_{\ell m,\, \ell'' m''} = \sum_{\ell' + m' = even} C^{\ell}(\ell' m', \ell'' m'')\, F_{\ell' m'} \exp(i\delta_{\ell}) \sin(\delta_{\ell}) \qquad (4.55')$$

$$C^{\ell}(\ell' m', \ell'' m'') = 4\pi(-1)^{\frac{1}{2}(\ell - \ell' - \ell'')}(-1)^{m' + m''}\, Y_{\ell' -m'}(\tfrac{1}{2}\pi, 0)$$
$$\times \int Y_{\ell m}(\Omega)\, Y_{\ell' m'}(\Omega)\, Y_{\ell'' -m''}(\Omega)\, d\Omega. \qquad (4.56')$$

$$F_{\ell' m'} = \sum_j{}' \, i h_{\ell'}^{(1)}(\kappa |\mathbf{R}_j|) \exp(-im'\phi(\mathbf{R}_j))$$
$$\times (-1)^{m'} \exp(i\mathbf{k}_{0\parallel} \cdot \mathbf{R}_j). \qquad (4.54')$$

Summation (4.54′) converges only if there is finite absorption i.e.

$$VPI < 0$$

This is always the case for LEED experiments, but if the case

$$VPI = 0$$

is needed for application of the programs to conventional band structure or surface state calculations, we describe an alternative subroutine XMAT. The current subroutine is the preferred one for LEED calculations because of its greater simplicity and flexibility. Time consuming steps in finding RC and TC are calculation of FLM and to a lesser extent inversion of $(1 - X)$. Typical times on the ICL Titan machine were 10 seconds per step. Times for the IBM 370/165 were typically 1/10th or 1/20th of the Titan times.

The assembly of layers into a surface can be done in several ways. Either by renormalised perturbation theory (Section VD), which is very fast and accurate, or by the Bloch wave method (Sections IIIC and IVB), which is slow but provides a lot of information, or by finding the reflectivity of first two, then four, then eight layers until convergence to the infinite crystal reflectivity is achieved (Section IVF). The latter method is intermediate in speed between RFS perturbation theory and the Bloch wave method, but works at low energies below 10 eV where the RFS scheme sometimes fails. Using 13 beams a call of PERTA might use 4 or 5 seconds, of BMMATR + REAMPR 40 seconds, and of CDBL 10–15 seconds on Titan. Again these times are to be divided by 10 or 20 to give IBM 370/165 times.

The subroutines are complete except in the case of the Bloch wave methods where in subroutines REAMP and REAMPR standard eigenvalue/eigenvector packages must be supplied. These are generally available but differ in form from machine to machine. Comments in REAMP and REAMPR make clear how to adapt the program to a given package. On machines with only 7 figures available in single precision, most eigenvector packages have to be run in double precision.

Specimen data and output streams follow the subroutines, and later in the appendix a subroutine for calculating the quantities CLM is provided.

It is envisaged that these will be calculated once for all, and read in from file each time the programs are used.

Basic variables in the main program for LEED intensities

LMAX	maximum value of ℓ for the phase shifts.
CLM	Values of the quantities defined in (4.56′) tabulated in the order required by the program. 809, appropriate to LMAX = 4, are read in.
SPA	characteristic length in the crystal. E.g. in a cubic lattice SPA is the length of unit cell side in atomic units.
SURF	a 3D vector normal to the surface, pointing inwards.
ASM1,ASM2	two orthogonal 3D vectors lying in the plane of the surface, to serve as axes. (By making use of SURF,ASM1,ASM2, it is possible to retain x, y and z axes parallel to the conventional crystallographic axes of the crystal.)
AR1,AR2	two 2D vectors forming the sides of unit cell within a layer, referred to axes ASM1 and ASM2. They are read in as 3D vectors referred to x, y and z axes, *in units of SPA*. Multiplication by SPA and projection onto ASM1 and ASM2 being done in the program.
AS	a 3D vector describing displacement of successive layers deeper into the crystal. The data stream expresses AS in units of SPA, referred to axes along SURF, ASM1, ASM2 in that order.
TV	area of unit surface cell.
IDEG	order of the axis of symmetry normal to the surface passing through the centres of atoms. For a Bravais lattice IDEG $\geqslant$ 2.
N1	defines the number of lattice points ($\simeq$ IDEG*N1**2) to be included in summation (4.54′).
N	number of beams included in the calculation.
RAR1,RAR2	2D vectors forming sides of the unit cell of the layer's reciprocal lattice, referred to axes ASM1, ASM2.
PQF(I,J)	the Ith component of the reciprocal lattice vector corresponding to the Jth beam, expressed in units of RAR1, and RAR2.
PQ(I,J)	the Ith component of the reciprocal lattice vector corresponding to the Jth beam, referred to axes ASM1 and ASM2.
VPI	the imaginary part of the potential in the crystal, V_{0i}, in Hartrees (VPI must always be negative).

E,E1,E2, PHS,PHS1, PHS2	phase shifts at energy E, PHS, are found by linear interpolation between the phase shifts at energies E1 and E2: PHS1, PHS2. Energies are measured in Hartrees (27.2 eV) relative to the muffin-tin zero.
THETA,FI	THETA is the angle between the normal to the surface and the incident beam, FI the angle between the plane containing the normal and the incident beam, and ASM1 (rotations towards ASM2 are positive). Angles are in radians.
RC,TC	reflection and transmission matrices for a single layer referred to an origin at the centre of an atom in the layer.
WV	the reflected amplitudes.
AT	the reflected intensities.

```
COMMENT MAIN PROGRAM FOR CALCULATION OF LEED
C        INTENSITIES BY VARIOUS METHODS
C        DIMENSIONS ARE FIXED FOR N=13,LMAX=4
      DIMENSION AS(3),AT(13),SURF(3),ASM1(3),ASM2(3)
     1PHS(5),PQF(2,13),PQ(2,13),RC(13,13),TC(13,13)
      DIMENSION PHS1(5),PHS2(5),WV(26),AR1(2),AR2(2)
      DIMENSION RAR1(2),RAR2(2),CLM(809),AF(5)
COMMENT COMPILE THE NEXT LINE ONLY IF BMMAT AND
C        REAMP ARE USED
      DIMENSION BM(26,26)
COMMENT COMPILE THE NEXT LINE ONLY IF BMMATR AND
C        REAMPR ARE USED
      DIMENSION CM(13,13),RM(13,13)
      COMPLEX AF,RC,TC,WV,BM,CM,RM
      COMMON E,SPA,THETA,FI,VPI,CLM
      COMMON /X1/TV,PQ,AS
      COMMON /X2/PHS,AR1,AR2
COMMENT EMACH IS THE MACHINE ACCURACY AND HAS BEEN
C        FIXED FOR AN IBM COMPUTER
      EMACH=1.0E-6
      EPSD=10.0*EMACH
      PI=3.14159265
      LMAX=4
      READ (2,43) (CLM(J),J=1,809)
43    FORMAT(5E14.5)
      READ (5,3) SPA
      READ (5,2) SURF,ASM1,ASM2
2     FORMAT(3F7.1)
      READ (5,3) AS
3     FORMAT(3F12.5)
      WRITE (6,4) SPA
4     FORMAT(5H SPA=,1F12.5)
      WRITE (6,1)
1     FORMAT(///,15H SURF ASM1 ASM2)
      WRITE (6,2) SURF,ASM1,ASM2
      A=0.0
      B=A
      C=A
      D=A
      DO 44 J=1,3
      A=A+ASM1(J)**2
      B=B+ASM2(J)**2
      C=C+AS(J)*ASM1(J)*SPA
44    D=D+AS(J)*ASM2(J)*SPA
      AR1(1)=C/SQRT(A)
      AR1(2)=D/SQRT(B)
      C=0.0
      D=C
      WRITE (6,46) AS
46    FORMAT(13H SURFACE VECS,/ ,3F12.5)
      READ (5,3) AS
      DO 45 J=1,3
      C=C+AS(J)*ASM1(J)*SPA
45    D=D+AS(J)*ASM2(J)*SPA
      AR2(1)=C/SQRT(A)
      AR2(2)=D/SQRT(B)
      WRITE (6,3) AS
      READ (5,3) AS
      DO 30 J=1,3
30    AS(J)=AS(J)*SPA
      WRITE (6,22) AS
```

```
22      FORMAT(19H LAYER DISPLACEMENT,/,3F12.5)
        TV=ABS(AR1(1)*AR2(2)-AR1(2)*AR2(1))
        READ (5,5) IDEG,N1,N
5       FORMAT(3I5)
        READ (5,6) (PQF(1,J),PQF(2,J),J=1,N)
6       FORMAT(2F7.1)
        RAR1(1)=AR2(2)*2.0*PI/TV
        RAR1(2)=-AR2(1)*2.0*PI/TV
        RAR2(1)=AR1(2)*2.0*PI/TV
        RAR2(2)=-AR1(1)*2.0*PI/TV
        DO 12 J=1,N
        PQ(1,J)=PQF(1,J)*RAR1(1)+PQF(2,J)*RAR2(1)
12      PQ(2,J)=PQF(1,J)*RAR1(2)+PQF(2,J)*RAR2(2)
        WRITE (6,13) IDEG,N1,N
13      FORMAT(6H IDEG=,1I5,4H N1=,1I5,3H N=,1I5)
        WRITE (6,41)
        DO 34 J=1,N
34      WRITE (6,6) PQF(1,J),PQF(2,J)
41      FORMAT(15H BEAMS INCLUDED,//,14H   PQF1   PQF2)
        READ (5,3) VPI
        WRITE (6,8) VPI
8       FORMAT(5H VPI=,1F12.5)
25      READ (5,14) E1,PHS1,E2,PHS2
14      FORMAT(/,2(1F12.5,/,5F12.5,/))
9       WRITE (6,7) E1,PHS1,E2,PHS2
7       FORMAT(2(4H E1=,1F12.5,/,4H PHS,/,5F12.5,/))
10      READ (5,3) THETA,FI
26      READ (5,3) E
        IF (E) 50,24,24
50      IF (E+1.5) 51,10,10
51      IF (E+2.5) 19,25,25
24      FAC=(E-E1)/(E2-E1)
COMMENT LINEAR INTERPOLATION OF PHASE SHIFTS
        DO 15 L=1,5
15      PHS(L)=PHS1(L)+FAC*(PHS2(L)-PHS1(L))
COMMENT REFLECTION,RC,AND TRANSMISSION,TC,
C          COEFICIENTS FOR A SINGLE LAYER ARE FOUND
        CALL MSMAT(RC,TC,N,LMAX,N1,IDEG,EMACH)
COMMENT COMPILE THE NEXT LINE IF RFS PERTURBATION THEORY
C          IS USED
        CALL PERTA(N,RC,TC,WV,EMACH)
COMMENT COMPILE THE NEXT THREE LINES IF THE BLOCH
C          WAVE METHOD IS USED(NO MIRROR PLANE)
        NN=N+N
        CALL BMMAT(N,NN,RC,RC,TC,TC,BM,EMACH)
        CALL REAMP(N,NN,BM,WV,RC,TC,EMACH)
COMMENT COMPILE THE NEXT TWO LINES IF THE BLOCH
C          WAVE METHOD IS USED,AND EACH LAYER LIES
C          ON A MIRROR PLANE
        CALL BMMATR(N,RC,TC,CM,RM,EMACH)
        CALL REAMPR(N,CM,RM,WV,RC,TC,EMACH)
COMMENT COMPILE THE NEXT LINE IF THE LAYER-
C          STACKING METHOD IS USED
        CALL CDBL(N,RC,TC,WV,EMACH)
COMMENT REFLECTED AMPLITUDES,WV,ARE USED TO CALCULATE
C          REFLECTED INTENSITIES,AT
        CALL RINT2(N,WV,PQF,PQ,AT)
        GO TO 26
19      STOP
        END
C
```

```
C
COMMENT PERTA,GIVEN RC AND TC,RETURNS THE REFLECTED
C        AMPLITUDES IN WV BY RFS PERTURBATION THEORY
C        DIMENSIONS ARE FIXED FOR N=13
      SUBROUTINE PERTA(N,RC,TC,WV,EMACH)
      DIMENSION RC(N,N),TC(N,N),WV(N),AKP(13),AKM(13)
      DIMENSION AS(3),PQ(2,13),ANEW(13,12),AW(13,3),CLM(809)
      COMMON E,SPA,THETA,FI,VPI,CLM
      COMMON /X1/TV,PQ,AS
      COMPLEX EK,EKM,ANEW,RC,TC,AKP,AKM,CZERO,CI,WV,AW
      COMPLEX CMPLX,CSQRT,CEXP,CSIN,CCOS,CLOG
      CRIT=0.003
      CZERO=CMPLX(0.0,0.0)
      CI=CMPLX(0.0,1.0)
COMMENT AKP AND AKM ARE PHASE FACTORS DESCRIBING THE
C        PROPAGATION OF FORWARD AND BACKWARD TRAVELLING
C        WAVES FROM A LAYER TO HALF WAY BETWEEN
C        LAYERS
      AK=SQRT(2.0*E)*SIN(THETA)
      BK2=AK*COS(FI)
      BK3=AK*SIN(FI)
      DO 2 J=1,N
      AK2=BK2+PQ(1,J)
      AK3=BK3+PQ(2,J)
      A=2.0*E-AK2*AK2-AK3*AK3
      EK=CMPLX(A,-2.0*VPI+0.000001)
      EK=CSQRT(EK)*AS(1)
      EKM=-EK+AK2*AS(2)+AK3*AS(3)
      EK=EKM+2.0*EK
      AKP(J)=CEXP(0.5*CI*EK)
2     AKM(J)=CEXP(-0.5*CI*EKM)
COMMENT ANEW(J,I) IS THE CONTRIBUTION OF THE CURRENT
C        ORDER OF RFS PERTURBATION THEORY TO THE
C        AMPLITUDE OF THE J TH BEAM BETWEEN THE (I-1) TH
C        AND I TH LAYERS.
C        IO IS THE NUMBER OF THE PASS BEING EXECUTED.
      DO 5 I=1,N
      WV(I)=CZERO
      DO 5 J=1,12
5     ANEW(I,J)=CZERO
COMMENT INWARD PASS STARTS HERE
      ANEW(1,1)=CMPLX(1.0,0.0)
      IO=0
      IFAIL=0
      ANORM1=1.0E-6
6     ANORM2=0.0
      ALPHI=1.0
      IO=IO+1
COMMENT OLD AMPLITUDES AT THE (I-1) TH LAYER
      I=1
7     BNORM=0.0
      I=I+1
      DO 3 K=1,N
      AW(K,2)=AKP(K)*ANEW(K,I-1)
3     AW(K,3)=AKM(K)*ANEW(K,I)
      DO 8 IG=1,N
      AW(IG,1)=CZERO
COMMENT TRANSMISSION AND REFLECTION BY THE (I-1) TH LAYER
      DO 8 K=1,N
8     AW(IG,1)=AW(IG,1)+TC(IG,K)*AW(K,2)+RC(IG,K)*
     1AW(K,3)
```

```
COMMENT NEW AMPLITUDES BETWEEN THE (I-1) TH AND I TH LAYERS
      DO 10 IG=1,N
      ANEW(IG,I)=AKP(IG)*AW(IG,1)
      AB=CABS(ANEW(IG,I))
10    BNORM=BNORM+AB*AB
COMMENT TESTS AND CHECKS ON THE DEPTH OF PENETRATION INTO THE
C       CRYSTAL.FOR INCREASED PENETRATION INCREASE THE SECOND
C       DIMENSION OF ANEW AND CHANGE IF STATEMENT 11
      IF(IO-1)40,40,41
40    IF(BNORM-CRIT)14,14,11
41    ALPHI=ALPHI*ALPHA
      IF(SQRT(ALPHI*BNORM)-CRIT)14,14,11
11    IF(I-11)7,12,12
12    WRITE (6,13)
13    FORMAT(23H0**FIRST ORDER TOO DEEP,/)
      IFAIL=1
      GOTO 16
14    WRITE (6,15)IO,CRIT
15    FORMAT(7H0ORDER ,I3,5X,7HCRIT = ,F5.3)
16    IMAX=I-1
      WRITE (6,17)IMAX,BNORM
17    FORMAT(8H IMAX = ,I2,5X,8HBNORM = ,E12.4,/,
     1 4X,3HWVR,9X,3HWVI)
      IMAX=IMAX+1
COMMENT IT IS ASSUMED THAT THE BEAM IS NOW SO STRONGLY
C       ATTENUATED THAT IT HAS EFFECTIVELY ZERO AMPLITUDE
      DO 18 IG=1,N
18    ANEW(IG,IMAX+1)=CZERO
COMMENT OUTWARD PASS STARTS HERE
      DO 20 I=1,IMAX
      II=IMAX-I+1
COMMENT OLD AMPLITUDES AT THE II TH LAYER
      DO 9 K=1,N
      AW(K,2)=AKM(K)*ANEW(K,II+1)
9     AW(K,3)=AKP(K)*ANEW(K,II)
COMMENT TRANSMISSION AND REFLECTION BY THE II TH LAYER
      DO 19 IG=1,N
      AW(IG,1)=CZERO
      DO 19 K=1,N
19    AW(IG,1)=AW(IG,1)+TC(IG,K)*AW(K,2)+RC(IG,K)*
     1AW(K,3)
COMMENT NEW AMPLITUDES BETWEEN THE (II-1) TH AND II TH LAYERS
      DO 20 IG=1,N
20    ANEW(IG,II)=AKM(IG)*AW(IG,1)
COMMENT CONTRIBUTION OF THE CURRENT PASS TO AMPLITUDES AT THE
C       SURFACE (ASSUMED HALF WAY BETWEEN TWO LAYERS)
      DO 21 IG=1,N
      AB=CABS(ANEW(IG,1))
      ANORM2=ANORM2+AB*AB
      WV(IG)=WV(IG)+ANEW(IG,1)
21    ANEW(IG,1)=CZERO
COMMENT CHECKS ON CONVERGENCE OVER PASSES
      DO 23 I=1,N
      WRITE (6,22)WV(I)
22    FORMAT(1H ,2E12.4)
23    CONTINUE
      ANORM1=ANORM1+ANORM2
      WRITE (6,24)ANORM1
24    FORMAT(9H ANORM = ,E12.4,/)
      IF(IFAIL-1)25,26,25
25    IF(ANORM2/ANORM1-0.001)26,26,50
```

```
50      IF(IO-1)51,51,52
51      ALPHA=EXP(ALOG(BNORM)/FLOAT(IMAX-1))
        GOTO 6
52      IF (IO-6) 6,53,53
53      WRITE (6,54) IO
54      FORMAT(/,24H ***NO CONVERGENCE AFTER,
       1I4,11H ITERATIONS)
26      RETURN
        END
C
C
COMMENT MSMAT RETURNS THE TRANSMISSION,TC,AND REFLECTION,
C         RC,COEFICIENTS OF A LAYER OF ATOMS,REFERRED TO AN
C         ORIGIN AT ONE OF THE ATOMIC SITES.
C         THE LAYER MUST BE A BRAVAIS LATTICE.
C         DIMENSIONS ARE FIXED FOR N=13,LMAX=4
        SUBROUTINE MSMAT(RC,TC,N,LMAX,N1,IDEG,EMACH)
        DIMENSION RC(N,N),TC(N,N),AF(5),IPO(10),IPE(15),
       1PHS(5),XODD(10,10),XEVEN(15,15),CYLM(13,25),YLM(25),CLM(809)
        DIMENSION AMULT(13),PQ(2,13),AS(3),AR1(2),AR2(2)
        DIMENSION YLO(10),YLE(15)
        COMPLEX AF,RC,TC,CYLM,YLM,AMULT,XA,YA,CI,ST,CF,CT
        COMPLEX XODD,XEVEN,YLO,YLE
        COMPLEX CMPLX,CSQRT,CEXP,CSIN,CCOS,CLOG
        COMMON E,SPA,THETA,FI,VPI,CLM
        COMMON /X1/TV,PQ,AS
        COMMON /X2/PHS,AR1,AR2
COMMENT LMMAX=NUMBER OF VALUES OF L AND M IN FLM
C         NEVEN=NUMBER OF EVEN VALUES OF L+M IN XEVEN
C         NODD=NUMBER OF ODD VALUES OF L+M IN XODD
        LMMAX=(LMAX+1)**2
        PI=3.14159265
        CI=CMPLX(0.0,1.0)
        LIM=LMAX+1
        DO 1 L=1,LIM
1       AF(L)=SIN(PHS(L))*CEXP(CI*PHS(L))
        B=FLOAT(LMAX)
        A=0.5*(B+1.0)*(B+2.0)
        NEVEN=IFIX(A)
        A=0.5*B*(B+1.0)
        NODD=IFIX(A)
        AK=SQRT(2.0*E)
        AK2=AK*SIN(THETA)*COS(FI)
        AK3=AK*SIN(THETA)*SIN(FI)
        YA=CMPLX(2.0*E,-2.0*VPI+0.000001)
        YA=CSQRT(YA)
COMMENT FOR EACH DIFFRACTED WAVE BETWEEN LAYERS,WITH X
C         AND Y WAVE VECTORS BK2 AND BK3,THE Z WAVE VECTOR,
C         XA,IS CALCULATED
        DO 4 JG=1,N
        BK2=PQ(1,JG)+AK2
        BK3=PQ(2,JG)+AK3
        C=BK2*BK2+BK3*BK3
        XA=CMPLX(2.0*E-C,-2.0*VPI+0.000001)
        XA=CSQRT(XA)
COMMENT THE ARGUMENTS THETA AND FI OF THE SPHERICAL
C         HARMONIC ARE CALCULATED FOR THE CURRENT WAVE AS
C         CF=EXP(I*FI),CT=COS(THETA),ST=SIN(THETA)
        B=0.0
        CF=CMPLX(1.0,0.0)
        IF (C-1.0E-7) 3,3,2
```

```
2       B=SQRT(C)
        CF=CMPLX(BK2/B,BK3/B)
3       CT=XA/YA
        ST=B/YA
        AMULT(JG)=-8.0*PI*PI*CI/(AK*TV*XA)
COMMENT ALL THE SPHERICAL HARMONICS UP TO L=LMAX ARE
C         CALCULATED AND SET IN YLM ORDERED THUS:LM=
C         (00),(1-1),(10),(11),(2-2)....,AND STORED IN CYLM
        CALL SPHRM4(YLM,CT,ST,CF,LMAX)
        DO 4 K=1,LMMAX
4       CYLM(JG,K)=YLM(K)
        APQ=PQ(1,1)
        BPQ=PQ(2,1)
COMMENT THE MATRICES XODD AND XEVEN DESCRIBING MULTIPLE
C         SCATTERING WITHIN A LAYER ARE CALCULATED.THEIR
C         SUBSCRIPTS ARE ORDERED THUS:XODD,LM=(10),(2-1),
C         (21),....,XEVEN,LM=(00),(1-1),(11),(2-2),....
        CALL XMAT(XODD,XEVEN,LMAX,N1,IDEG,APQ,BPQ)
        DO 5 J=1,NODD
5       XODD(J,J)=XODD(J,J)-1.0
        DO 15 J=1,NEVEN
15      XEVEN(J,J)=XEVEN(J,J)-1.0
COMMENT GAUSSIAN ELIMINATION PRIOR TO INVERSION OF XODD
C         AND XEVEN,DETAILS STORED IN IPO AND IPE
        CALL ZGE(XODD,IPO,NODD,NODD,EMACH)
        CALL ZGE(XEVEN,IPE,NEVEN,NEVEN,EMACH)
COMMENT YLM(JGP)*AF*(I)**-L IS SET INTO YLO AND YLE WITH
C         PROPER ORDERING OF SUBSCRIPTS
        DO 10 JGP=1,N
        ST=CMPLX(1.0,0.0)
6       L=0
        K=1
        J=1
        JL=1
8       M=-L
        JL=JL+L+L
        YLE(K)=CYLM(JGP,JL+M)*AF(L+1)/ST
        K=K+1
        GO TO 14
C
7       YLE(K)=CYLM(JGP,JL+M)*AF(L+1)/ST
        YLO(J)=CYLM(JGP,JL+M-1)*AF(L+1)/ST
        J=J+1
        K=K+1
C
14      M=M+2
        IF (L-M) 17,7,7
17      L=L+1
        ST=ST*CI
        IF (LMAX-L) 21,8,8
COMMENT YLO AND YLE ARE MULTIPLIED BY INVERSE XODD,XEVEN,
C         AND THE RESULT RETURNED IN YLO AND YLE
21      CALL ZSU(XODD,IPO,YLO,NODD,NODD,EMACH)
        CALL ZSU(XEVEN,IPE,YLE,NEVEN,NEVEN,EMACH)
COMMENT SCALAR PRODUCT TAKEN AND ACCUMULATED IN XA FOR TC
C         AND YA FOR RC
        DO 10 JG=1,N
        XA=CMPLX(0.0,0.0)
        YA=CMPLX(0.0,0.0)
        ST=CMPLX(1.0,0.0)
        L=0
```

```
      JL=1
      K=1
      J=1
9     M=-L
      JL=JL+L+L
      CF=CMPLX(0.0,0.0)
      CT=YLE(K)*CYLM(JG,JL-M)/ST
      K=K+1
      GO TO 22
C
23    CT=YLE(K)*CYLM(JG,JL-M)/ST
      CF=YLO(J)*CYLM(JG,JL-M+1)/ST
      K=K+1
      J=J+1
22    XA=XA+(CT-CF)
      YA=YA+(CT+CF)
C
      M=M+2
      IF (L-M) 24,23,23
24    L=L+1
      ST=ST*CI
      IF (LMAX-L) 28,9,9
C
28    CONTINUE
      YA=YA*AMULT(JGP)
      XA=XA*AMULT(JGP)
COMMENT RC IS THE REFLECTION MATRIX AND TC THE TRANSMISSION
C        MATRIX
      RC(JGP,JG)=YA
      TC(JGP,JG)=XA
      IF(JGP-JG) 10,16,10
16    TC(JG,JG)=TC(JG,JG)+(1.0,0.0)
10    CONTINUE
      RETURN
      END
C
C
COMMENT RINT2,GIVEN THE REFLECTED AMPLITUDES,WV,
C        RETURNS THE REFLECTED INTENSITIES,AT
C        DIMENSIONS ARE FIXED FOR N=13
      SUBROUTINE RINT2(N,WV,PQF,PQ,AT)
      DIMENSION WV(N),PQF(2,13),PQ(2,13),CLM(809),AT(N)
      COMMON E,SPA,THETA,FI,VPI,CLM
      COMPLEX WV
1     AK=SQRT(2.0*E)
      BK2=AK*SIN(THETA)*COS(FI)
      BK3=AK*SIN(THETA)*SIN(FI)
      DO 3 J=1,N
      AK2=PQ(1,J)+BK2
      AK3=PQ(2,J)+BK3
      A=2.0*E-AK2**2-AK3**2
      AT(J)=0.0
COMMENT IT IS ESTABLISHED WHETHER THE J TH BEAM EMERGES
C        FROM THE CRYSTAL
      IF (A) 3,3,2
2     A=SQRT(A)
      B=CABS(WV(J))
COMMENT THE REFLECTED INTENSITY IS PROPORTIONAL TO THE
C        Z COMPONENT OF MOMENTUM OF THE WAVE
      AT(J)=B*B*A/(AK*COS(THETA))
3     CONTINUE
```

```
      WRITE (6,6) E,THETA,FI
6     FORMAT(3H E=,1F10.4,7H THETA=,1F10.4,4H FI=,1F10.4)
      WRITE (6,4)
4     FORMAT(29H     BEAM             INTENSITY)
      WRITE (6,5)  (PQF(1,J),PQF(2,J),AT(J),J=1,N)
5     FORMAT(2F7.1,1E15.5)
      WRITE (6,7)
7     FORMAT(////)
      RETURN
      END
C
C
COMMENT XMAT CALCULATES THE MATRIX DESCRIBING MULTIPLE
C        SCATTERING WITHIN A LAYER,RETURNED AS XODD
C        CORRESPONDING TO ODD L+M,WITH LM=(10),(2-1),(21),
C        ....,AND XEVEN CORRESPONDING TO EVEN L+M WITH
C        LM=(00),(1-1),(11),(2-2)....
C        THE PROGRAM ASSUMES THAT THE LAYER IS A BRAVAIS
C        LATTICE.DIMENSIONS ARE FIXED FOR LMAX=4
      SUBROUTINE XMAT(XODD,XEVEN,LMAX,N1,IDEG,APQ,BPQ)
      DIMENSION CLM(809),R(2),AK(2),RTAB(4),XODD(10,10),
     1AF(5),FLM(9,17),XEVEN(15,15),PHS(5),AR1(2),AR2(2)
      DIMENSION ANC(6),ANS(6),SA(48),SCC(17)
      COMPLEX RTAB,XODD,XEVEN,FLM,AF,HL,Z,KAPPA,CZERO
      COMPLEX CI,CL,ACS,ACC,ELP,EMP,SA,SCC,SD,SE,SC
      COMPLEX CMPLX,CSQRT,CEXP,CSIN,CCOS,CLOG
      COMMON E,SPA,THETA,FI,VPI,CLM
      COMMON /X2/PHS,AR1,AR2
COMMENT AK(1) AND AK(2) ARE THE X AND Y COMPONENTS OF THE
C        MOMENTUM PARALLEL TO THE SURFACE,MODULO A RECIPROCAL
C        LATTICE VECTOR
      PI=3.14159265
      CZERO=CMPLX(0.0,0.0)
      CI=CMPLX(0.0,1.0)
      KAPPA=CMPLX(2.0*E,-2.0*VPI+0.000001)
      KAPPA=CSQRT(KAPPA)
      AG=SQRT(2.0*E)*SIN(THETA)
      AK(1)=AG*COS(FI)+APQ
      AK(2)=AG*SIN(FI)+BPQ
C
      JL=LMAX+1
      DO 65 J=1,JL
65    AF(J)=SIN(PHS(J))*CEXP(CI*PHS(J))
COMMENT NEVEN AND NODD ARE THE DIMENSIONS OF XEVEN AND XODD
      B=FLOAT(LMAX)
      A=0.5*(B+1.0)*(B+2.0)
      NEVEN=IFIX(A)
      A=0.5*B*(B+1.0)
      NODD=IFIX(A)
COMMENT ANC,ANS AND SA ARE PREPARED TO BE USED IN THE SUM
C        OVER SYMMENTRICALLY RELATED SECTORS OF THE LATTICE
      L2MAX=LMAX+LMAX
      LIM=L2MAX*IDEG
      ANG=2.0*PI/FLOAT(IDEG)
      D=1.0
      DO 72 J=1,IDEG
      ANC(J)=COS(D*ANG)
      ANS(J)=SIN(D*ANG)
72    D=D+1.0
      D=1.0
      DO 71 J=1,LIM
```

```
      SA(J)=CEXP(-CI*D*ANG)
71    D=D+1.0
      J2=L2MAX+1
      K2=J2+J2-1
      DO 62 J=1,J2
      DO 62 K=1,K2
62    FLM(J,K)=CZERO
COMMENT THE LATTICE SUM STARTS.THE SUM IS DIVIDED INTO ONE   .
C       OVER A SINGLE SECTOR,THE OTHER (IDEG-1) SECTORS
C       ARE RELATED BY SYMMETRY EXCEPT FOR FACTORS
C       INVOLVING THE DIRECTION OF R
      LI1=N1
      LI2=N1-1
      JUMP=1
      ADD=1.0
      AST=0.0
      AN1=-1.0
76    DO 75 I1=1,LI1
      AN1=AN1+ADD
      AN2=AST
      DO 75 I2=1,LI2
      AN2=AN2+1.0
COMMENT R=THE CURRENT LATTICE VECTOR IN THE SUM
C       AR=MOD(R)
C       RTAB(1)=-EXP(I*FI(R))
      R(1)=AN1*AR1(1)+AN2*AR2(1)
      R(2)=AN1*AR1(2)+AN2*AR2(2)
      AR=SQRT(R(1)*R(1)+R(2)*R(2))
      RTAB(1)=-CMPLX(R(1)/AR,R(2)/AR)
      J2=L2MAX+L2MAX+1
      DO 73 J=1,J2
73    SCC(J)=CZERO
      ABC=1.0
      ABB=0.0
      IF (AG-1.0E-4) 70,70,61
61    ABC=(AK(1)*R(1)+AK(2)*R(2))/(AG*AR)
      ABB=(-AK(2)*R(1)+AK(1)*R(2))/(AG*AR)
70    SC=CI*AG*AR
COMMENT SCC CONTAINS FACTORS IN THE SUMMATION DEPENDENT ON
C       THE DIRECTION OF R.CONTRIBUTIONS FROM SYMMETRICALLY
C       RELATED SECTORS CAN BE GENERATED SIMPLY.THE SUBSCRIPT
C       M IS ORDERED:M=(-L2MAX),(-L2MAX+1)....(+L2MAX)
      DO 74 J=1,IDEG
      AD=ABC*ANC(J)-ABB*ANS(J)
      SD=CEXP(SC*AD)
      SCC(L2MAX+1)=SCC(L2MAX+1)+SD
      MJ=0
      SE=RTAB(1)
      DO 74 M=1,L2MAX
      MJ=MJ+J
      SCC(L2MAX+1+M)=SCC(L2MAX+1+M)+SD*SA(MJ)/SE
      SCC(L2MAX+1-M)=SCC(L2MAX+1-M)+SD*SE/SA(MJ)
74    SE=SE*RTAB(1)
      Z=AR*KAPPA
      ACS=CSIN(Z)
      ACC=CCOS(Z)
COMMENT RTAB(3)=SPHERICAL HANKEL FUNCTION OF THE FIRST KIND,L=0
C       RTAB(4)=SPHERICAL HANKEL FUNCTION OF THE FIRST KIND,L=1
      RTAB(3)=(ACS-CI*ACC)/Z
      RTAB(4)=((ACS/Z-ACC)-CI*(ACC/Z+ACS))/Z
      AL=0.0
```

```
      LIML=L2MAX+1
COMMENT THE SUMMATION OVER FACTORS INDEPENDENT OF THE
C        DIRECTION OF R IS ACCUMULATED IN FLM
      DO 64 JL=1,LIML
      JM=L2MAX+2-JL
      DO 63 KM=1,JL
      FLM(JL,JM)=FLM(JL,JM)+SCC(JM)*RTAB(3)*CI
      JM=JM+2
63    CONTINUE
COMMENT SPHERICAL HANKEL FUNCTIONS FOR HIGHER L ARE
C        GENERATED BY RECURRENCE RELATIONS
      ACS=(2.0*AL+3.0)*RTAB(4)/Z-RTAB(3)
      RTAB(3)=RTAB(4)
      RTAB(4)=ACS
      AL=AL+1.0
64    CONTINUE
75    CONTINUE
COMMENT SPECIAL TREATMENT IS REQUIRED IF IDEG=2
      IF (IDEG-2) 77,77,78
77    IF (JUMP) 78,78,79
79    JUMP=0
      ADD=-1.0
      AN1=0.0
      AST=-1.0
      LI1=N1-1
      LI2=N1
      GO TO 76
78    CONTINUE
COMMENT SUMMATION OVER THE CLEBSCH-JORDAN COEFICIENTS
C        PROCEEDS,FIRST FOR XODD,THEN FOR XEVEN
      JSET=1
      N=1
C
3     J=1
      L=JSET
18    M=-L+JSET
      JL=L+1
C
10    K=1
      LPP=JSET
15    MPP=-LPP+JSET
C
4     MPA=IABS(MPP-M)
      LPA=IABS(LPP-L)
      IF (LPA-MPA) 24,24,23
23    MPA=LPA
24    MP1=MPP-M+L2MAX+1
      LP1=L+LPP+1
      ACC=CZERO
C
12    ACC=ACC+CLM(N)*FLM(LP1,MP1)
      N=N+1
C
      LP1=LP1-2
      IF (LP1-1-MPA) 13,12,12
C
13    XEVEN(J,K)=ACC*AF(JL)
      K=K+1
      MPP=MPP+2
      IF (LPP-MPP) 14,4,4
14    LPP=LPP+1
```

```
      IF (LMAX-LPP) 16,15,15
C
16    J=J+1
      M=M+2
      IF (L-M) 17,10,10
17    L=L+1
      IF (LMAX-L) 19,18,18
C
19    IF (JSET) 22,22,20
20    DO 21 J=1,NODD
      DO 21 K=1,NODD
21    XODD(J,K)=XEVEN(J,K)
      JSET=0
      GO TO 3
22    CONTINUE
      RETURN
      END
C
C
COMMENT SPHRM4,GIVEN CT=COS(THETA),ST=SIN(THETA),AND
C       CF=EXP(I*FI),CALCULATES ALL THE YLM(THETA,FI)
C       UP TO L=LMAX.SUBSCRIPTS ARE ORDERED THUS:LM=(00)
C       (1-1),(10),(11),(2-2),....
C       DIMENSIONED FOR LMAX=4
      SUBROUTINE SPHRM4(YLM,CT,ST,CF,LMAX)
      DIMENSION FAC1(5),FAC3(5),FAC2(25),YLM(25)
      COMPLEX YLM
      COMPLEX CT,ST,CF,SF,SA,SX,SY
      COMPLEX CMPLX,CSQRT,CEXP,CSIN,CCOS,CLOG
      PI=3.14159265
      LM=0
      CL=0.0
      A=1.0
      B=1.0
      ASG=1.0
      LL=LMAX+1
COMMENT MULTIPLICATIVE FACTORS REQUIRED
      DO 2 L=1,LL
      FAC1(L)=ASG*SQRT((2.0*CL+1.0)*A/(4.0*PI*B*B))
      FAC3(L)=SQRT(2.0*CL)
      CM=-CL
      LN=L+L-1
      DO 1 M=1,LN
      LO=LM+M
      FAC2(LO)=SQRT((CL+1.0+CM)*(CL+1.0-CM)
     1/((2.0*CL+3.0)*(2.0*CL+1.0)))
1     CM=CM+1.0
      CL=CL+1.0
      A=A*2.0*CL*(2.0*CL-1.0)/4.0
      B=B*CL
      ASG=-ASG
2     LM=LM+LN
COMMENT FIRST ALL THE YLM FOR M=+-L AND M=+-(L-1) ARE
C       ARE CALCULATED BY EXPLICIT FORMULAE
      LM=1
      CL=1.0
      ASG=-1.0
      SF=CF
      SA=CMPLX(1.0,0.0)
      YLM(1)=CMPLX(FAC1(1),0.0)
      DO 3 L=1,LMAX
```

```
      LN=LM+L+L+1
      YLM(LN)=FAC1(L+1)*SA*SF*ST
      YLM(LM+1)=ASG*FAC1(L+1)*SA*ST/SF
      YLM(LN-1)=-FAC3(L+1)*FAC1(L+1)*SA*SF*CT/CF
      YLM(LM+2)=ASG*FAC3(L+1)*FAC1(L+1)*SA*CT*CF/SF
      SA=ST*SA
      SF=SF*CF
      CL=CL+1.0
      ASG=-ASG
3     LM=LN
COMMENT USING YLM AND YL(M-1) IN A RECURRENCE RELATION
C       YL(M+1) IS CALCULATED
      LM=1
      LL=LMAX-1
      DO 5 L=1,LL
      LN=L+L-1
      LM2=LM+LN+4
      LM3=LM-LN
      DO 4 M=1,LN
      LO=LM2+M
      LP=LM3+M
      LQ=LM+M+1
      YLM(LO)=-(FAC2(LP)*YLM(LP)-CT*YLM(LQ))/FAC2(LQ)
4     CONTINUE
5     LM=LM+L+L+1
      RETURN
      END
C
C
COMMENT ZGE IS A STANDARD SUBROUTINE TO PERFORM GAUSSIAN
C       ELIMINATION ON AN NR*NC MATRIX A PRIOR TO
C       INVERSION OF A,DETAILS STORED IN INT
      SUBROUTINE ZGE(A,INT,NR,NC,EMACH)
      COMPLEX A,YR,DUM
      DIMENSION A(NR,NC),INT(NC)
      N=NC
      DO 10 II=2,N
      I=II-1
      YR=A(I,I)
      IN=I
      DO 2 J=II,N
      IF(CABS(YR)-CABS(A(J,I)))1,2,2
1     YR=A(J,I)
      IN=J
2     CONTINUE
      INT(I)=IN
      IF(IN-I)3,5,3
3     DO 4 J=I,N
      DUM=A(I,J)
      A(I,J)=A(IN,J)
4     A(IN,J)=DUM
5     IF(CABS(YR)-EMACH)10,10,6
6     DO 9 J=II,N
      IF(CABS(A(J,I))-EMACH)9,9,7
7     A(J,I)=A(J,I)/YR
      DO 8 K=II,N
8     A(J,K)=A(J,K)-A(I,K)*A(J,I)
9     CONTINUE
10    CONTINUE
      RETURN
      END
```

```
C
C
COMMENT ZSU IS A STANDARD BACK SUBSTITUTION SUBROUTINE
C         USING THE OUTPUT OF ZGE TO CALCULATE A INVERSE
C         TIMES X,RETURNED IN X
      SUBROUTINE ZSU(A,INT,X,NR,NC,EMACH)
      COMPLEX A,X,DUM
      DIMENSION A(NR,NC),X(NC),INT(NC)
      N=NC
      DO 5 II=2,N
      I=II-1
      IF(INT(I)-I) 1.2,1
1     IN=INT(I)
      DUM=X(IN)
      X(IN)=X(I)
      X(I)=DUM
2     DO 4 J=II,N
      IF(CABS(A(J,I))-EMACH) 4,4,3
3     X(J)=X(J)-A(J,I)*X(I)
4     CONTINUE
5     CONTINUE
      DO 10 II=1,N
      I=N-II+1
      IJ=I+1
      IF(I-N)6,8,6
6     DO 7 J=IJ,N
7     X(I)=X(I)-A(I,J)*X(J)
8     IF(CABS(A(I,I))-EMACH*1.0E-5) 9,10,10
9     A(I,I)=EMACH*1.0E-5*(1.0,1.0)
10    X(I)=X(I)/A(I,I)
      RETURN
      END
C
C
COMMENT BMMAT CALCULATES THE 2N*2N MATRIX BM WHOSE
C         EIGENVALUES ARE EXP(I*K.C)
C         DIMENSIONS ARE FIXED FOR N=13
      SUBROUTINE BMMAT(N,NN,RPM,RMP,TPP,TMM,BM,EMACH)
      DIMENSION RPM(N,N),RMP(N,N),TPP(N,N),TMM(N,N),
     1BM(NN,NN),XP(13),XM(13),IPL(13)
      DIMENSION CLM(809),PQ(2,13),AS(3)
      COMPLEX RPM,RMP,TPP,TMM,XP,XM,CZERO
      COMPLEX EKP,CI,BM,EKM
      COMPLEX CMPLX,CSQRT,CEXP,CSIN,CCOS,CLOG
      COMMON E,SPA,THETA,FI,VPI,CLM
      COMMON /X1/TV,PQ,AS
      CZERO=CMPLX(0.0,0.0)
      CI=CMPLX(0.0,1.0)
COMMENT XP AND XM ARE PHASE FACTORS DESCRIBING PROPAGATION
C         BETWEEN LAYERS
      AK=SQRT(2.0*E)*SIN(THETA)
      BK2=AK*COS(FI)
      BK3=AK*SIN(FI)
      DO 1 J=1,N
      AK2=BK2+PQ(1,J)
      AK3=BK3+PQ(2,J)
      A=2.0*E-AK2*AK2-AK3*AK3
      EKP=CMPLX(A,-2.0*VPI+0.000001)
      EKP=CSQRT(EKP)*AS(1)
      EKM=-EKP+AS(2)*AK2+AS(3)*AK3
      EKP=EKM+2.0*EKP
```

```
      XP(J)=CEXP(CI*EKP)
1     XM(J)=CEXP(CI*EKM)
COMMENT ASSEMBLY OF BM BEGINS
      DO 4 J=1,N
      DO 3 K=1,N
      EKP=CZERO
      EKM=CZERO
      DO 2 L=1,N
      EKP=EKP-RPM(J,L)*XP(L)*TPP(L,K)
2     EKM=EKM-RPM(J,L)*XP(L)*RMP(L,K)
      BM(J,K)=XP(J)*TPP(J,K)
      BM(J,K+N)=XP(J)*RMP(J,K)
      BM(J+N,K)=EKP
3     BM(J+N,K+N)=EKM
4     BM(J+N,J+N)=BM(J+N,J+N)+XM(J)
COMMENT GAUSSIAN ELIMINATION PRIOR TO INVERSION OF TMM,
C        DETAILS STORED IN IPL
      CALL ZGE(TMM,IPL,N,N,EMACH)
C
      DO 6 K=1,N
      DO 5 J=1,N
      XP(J)=BM(J+N,K)
5     XM(J)=BM(J+N,K+N)
COMMENT INVERSE OF TMM TIMES XP,XM FORMED AND RETURNED
C        IN XP,XM
      CALL ZSU(TMM,IPL,XP,N,N,EMACH)
      CALL ZSU(TMM,IPL,XM,N,N,EMACH)
      DO 6 J=1,N
      BM(J+N,K)=XP(J)
6     BM(J+N,K+N)=XM(J)
C
      RETURN
      END
C
C
COMMENT REAMP GIVEN BM FROM BMMAT,RETURNS THE REFLECTED
C        AMPLITUDES IN XP.RC AND TC ARE WORKING SPACE
C        DIMENSIONS ARE FIXED FOR N=13
      SUBROUTINE REAMP(N,NN,BM,XP,RC,TC,EMACH)
      DIMENSION BM(NN,NN),H(26,26),IPL(26),ISPON(26),
     1DV(26),WV(26),XM(13),XP(13),RC(N,N),TC(N,N),FV(26)
      DIMENSION CLM(809),PQ(2,13),AS(3)
      COMPLEX CZERO,CI,RC,TC,FV,H
      COMPLEX BM,WV,XM,XP,DV,EKP,EKM
      COMPLEX CMPLX,CSQRT,CEXP,CSIN,CCOS,CLOG
      COMMON E,SPA,THETA,FI,VPI,CLM
      COMMON /X1/TV,PQ,AS
      CZERO=CMPLX(0.0,0.0)
      CI=CMPLX(0.0,1.0)
      EPSD=10.0*EMACH
COMMENT XP AND XM ARE PHASE FACTORS DESCRIBING PROPAGATION
C        FROM A LAYER TO HALF WAY BETWEEN LAYERS
      AK=SQRT(2.0*E)*SIN(THETA)
      BK2=AK*COS(FI)
      BK3=AK*SIN(FI)
      DO 1 J=1,N
      AK2=BK2+PQ(1,J)
      AK3=BK3+PQ(2,J)
      A=2.0*E-AK2*AK2-AK3*AK3
      EKP=CMPLX(A,-2.0*VPI+0.000001)
      EKP=CSQRT(EKP)*AS(1)
```

```
      EKM=-EKP+AS(2)*AK2+AS(3)*AK3
      EKP=EKM+2.0*EKP
      XP(J)=CEXP(0.5*CI*EKP)
1     XM(J)=CEXP(0.5*CI*EKM)
COMMENT THE NEXT SECTION OF PROGRAM DOWN TO LABEL
C       17 STORES THE EIGENVALUES OF BM IN DV.THE FIRST
C       AND SECOND N COMPONENTS OF THE I-TH EIGENVECTOR
C       ARE STORED IN TC(J,I) AND RC(J,I) RESPECTIVELY
      DO 2 J=1,NN
      DO 2 K=1,NN
2     H(J,K)=BM(J,K)
COMMENT H IS TRANSFORMED TO AN UPPER HESSENBERG MATRIX
C       WITH THE SAME EIGENVALUES AS H.DETAILS ARE STORED
C       IN IPL
      CALL COMHES(NN,1,NN,H,IPL,EMACH)
C
      DO 3 J=1,NN
      DO 3 K=1,NN
3     BM(J,K)=H(J,K)
COMMENT COMLR PLACES THE EIGENVALUES OF THE UPPER HESSENBERG
C       MATRIX IN DV. IF=THE NUMBER OF EIGENVALUES IT HAS
C       FAILED TO FIND
      CALL COMLR(NN,NN,EMACH,H,DV,IF)
COMMENT THE EIGENVALUES ARE ORDERED WITH ASCENDING MODULUS
C       ONLY THE FIRST N ARE NEEDED
      J2=NN-1
      DO 6 J=1,J2
      K1=J+1
      DO 6 K=K1,NN
      IF (CABS(DV(J))-CABS(DV(K))) 6,6,5
5     EKP=DV(J)
      DV(J)=DV(K)
      DV(K)=EKP
6     CONTINUE
C
      WRITE (6,7) DV
7     FORMAT(12H EIGENVALUES,20(/,2(1E16.5,1E13.5)))
COMMENT COMVEC FINDS EIGENVECTORS OF THE HESSENBERG MATRIX
C       AND COMBAK TRANSFORMS TO EIGENVECTORS OF THE
C       ORIGINAL MATRIX. THE SETTING OF LC IS IMPORTANT
C       WHEN EIGENVALUES ARE DEGENERATE
      LC=1
      FE=0.0
      DO 17 I=1,N
      TEST=ABS(FE-CABS(DV(I)))
      FE=CABS(DV(I))
      IF (TEST-EPSD) 10,10,9
9     LC=1
10    CALL COMVEC(NN,NN,BM,DV(I),FV,EKP,ISPON,
     1H,WV,RES1,IF,LC,EMACH,EPSD)
      CALL COMBAK(NN,1,NN,BM,IPL,FV,EMACH)
      DO 17 J=1,N
      TC(J,I)=FV(J)
17    RC(J,I)=FV(J+N)
COMMENT FORWARD AND BACKWARD TRAVELLING WAVE AMPLITUDES IN THE
C       BLOCH WAVE,RELATIVE TO AN ORIGIN HALF-WAY
C       BETWEEN THE LAYERS,ARE STORED IN RC AND TC
      DO 11 I=1,N
      DO 19 J=1,N
      TC(J,I)=TC(J,I)/XP(J)
19    RC(J,I)=RC(J,I)*XM(J)/DV(I)
```

```
      EKP=CLOG(DV(I))/CI
      WRITE (6,14) EKP,(TC(J,I),J=1,N)
14    FORMAT(5H K.A=,2E13.5,/,3H BP,/,20(3(1E14.3,1E11.3),/))
      WRITE (6,15) (RC(J,I),J=1,N)
15    FORMAT(3H BM,/,20(3(1E14.3,1E11.3),/))
11    CONTINUE
COMMENT GAUSSIAN ELIMINATION PRIOR TO INVERSION OF TC,
C        DETAILS STORED IN IPL
      CALL ZGE(TC,IPL,N,N,EMACH)
C
      DO 12 J=1,N
12    XM(J)=CZERO
      XM(1)=CMPLX(1.0,0.0)
COMMENT XM,THE INCIDENT WAVE AMPLITUDES,ARE MULTIPLIED BY
C        TC INVERSE TO GIVE AMPLITUDES OF EXCITATION OF
C        BLOCH WAVES,RETURNED IN XM
      CALL ZSU(TC,IPL,XM,N,N,EMACH)
COMMENT XM*RC GIVES THE AMPLITUDES OF REFLECTED WAVES.
C        NOTE:IN THIS PROGRAM NO SURFACE BARRIER IS INCLUDED
      DO 13 J=1,N
      XP(J)=CZERO
      DO 13 K=1,N
13    XP(J)=XP(J)+RC(J,K)*XM(K)
C
      WRITE (6,18) XM
18    FORMAT(22H BLOCH WAVE AMPLITUDES,
     120(/,2(1E16.5,1E13.5)))
      WRITE (6,16) XP
16    FORMAT(21H REFLECTED AMPLITUDES,
     120(/,2(1E16.5,1E13.5)))
C
      RETURN
      END
C
C
COMMENT BMMATR CALCULATES THE N*N MATRIX BM WHOSE EIGENVALUES
C        ARE 2*COS(KZCZ),AND THE N*N MATRIX RM=EXP(I*EKP)*(R+T)
C        NOTE:IT CAN ONLY BE USED WHEN R AND T REFER TO A LAYER
C        LYING ON A MIRROR PLANE
C        DIMENSIONS ARE FIXED FOR N=13
      SUBROUTINE BMMATR(N,R,T,BM,RM,EMACH)
      DIMENSION R(N,N),T(N,N),RM(N,N),
     1BM(N,N),XP(13),XM(13),IPL(13)
      DIMENSION CLM(809),PQ(2,13),AS(3)
      COMPLEX R,T,XP,XM,CZ,RM
      COMPLEX EKP,CI,BM,EKM
      COMPLEX CMPLX,CSQRT,CEXP,CSIN,CCOS,CLOG
      COMMON E,SPA,THETA,FI,VPI,CLM
      COMMON /X1/TV,PQ,AS
      CZ=CMPLX(0.0,0.0)
      CI=CMPLX(0.0,1.0)
COMMENT XP IS A PHASE FACTOR BUT NOTE THAT IT REFERS TO ORIGINS
C        SLIGHTLY DIFFERENT FROM THOSE USED IN BMMAT
      AK=SQRT(2.0*E)*SIN(THETA)
      BK2=AK*COS(FI)
      BK3=AK*SIN(FI)
      DO 1 J=1,N
      AK2=BK2+PQ(1,J)
      AK3=BK3+PQ(2,J)
      A=2.0*E-AK2*AK2-AK3*AK3
      EKP=CMPLX(A,-2.0*VPI+0.000001)
```

```
      EKP=CSQRT(EKP)*AS(1)
      EKM=-EKP+AS(2)*PQ(1,J)+AS(3)*PQ(2,J)
      EKP=EKM+2.0*EKP
      XP(J)=CEXP(CI*EKP)
1     XM(J)=CEXP(CI*EKM)
COMMENT ASSEMBLY OF BM AND RM BEGINS
      DO 3 J=1,N
      DO 2 K=1,N
      BM(J,K)=CZ
      RM(J,K)=XP(J)*(R(J,K)+T(J,K))
      DO 2 L=1,N
2     BM(J,K)=BM(J,K)-R(J,L)*XP(L)*(T(L,K)+R(L,K))
3     BM(J,J)=BM(J,J)+XM(J)
COMMENT GAUSSIAN ELIMINATION PRIOR TO INVERSION OF T,
C        DETAILS STORED IN IPL
      CALL ZGE(T,IPL,N,N,EMACH)
C
      DO 6 K=1,N
      DO 5 J=1,N
5     XM(J)=BM(J,K)
COMMENT INVERSE OF T TIMES XM FORMED AND RETURNED IN XM
      CALL ZSU(T,IPL,XM,N,N,EMACH)
      DO 6 J=1,N
6     BM(J,K)=XM(J)
C
      DO 7 J=1,N
      DO 7 K=1,N
7     BM(J,K)=BM(J,K)+RM(J,K)
      RETURN
      END
C
C
COMMENT REAMPR,GIVEN BM AND RM FROM BMMATR RETURNS THE
C        REFLECTED AMPLITUDES IN XP.NOTE THAT THEY MAY BE
C        REFERRED TO A DIFFERENT ORIGIN IN THE X-Y PLANE
C        FROM THOSE RETURNED BY REAMP.RC AND TC ARE WORKING SPACE
C        DIMENSIONS ARE FIXED FOR N=13
      SUBROUTINE REAMPR(N,BM,RM,XP,RC,TC,EMACH)
      DIMENSION BM(N,N),RM(N,N),H(13,13),IPL(13),ISPON(13),
     1DV(13),WV(13),XM(13),XP(13),RC(N,N),TC(N,N),FV(13)
      DIMENSION CLM(809),PQ(2,13),AS(3)
      COMPLEX CZERO,CI,RC,TC,FV,H,RM
      COMPLEX BM,WV,XM,XP,DV,EKP,EKM,RUB
      COMPLEX CMPLX,CSQRT,CEXP,CSIN,CCOS,CLOG
      COMMON E,SPA,THETA,FI,VPI,CLM
      COMMON /X1/TV,PQ,AS
      CZERO=CMPLX(0.0,0.0)
      CI=CMPLX(0.0,1.0)
      EPSD=10.0*EMACH
COMMENT XP IS A PHASE FACTOR DESCRIBING PROPAGATION FROM A LAYER
C        TO HALF WAY BETWEEN LAYERS.NOTE THAT IT IS DIFFERENT
C        FROM THAT USED IN REAMP
      AK=SQRT(2.0*E)*SIN(THETA)
      BK2=AK*COS(FI)
      BK3=AK*SIN(FI)
      DO 1 J=1,N
      AK2=BK2+PQ(1,J)
      AK3=BK3+PQ(2,J)
      A=2.0*E-AK2*AK2-AK3*AK3
      EKP=CMPLX(A,-2.0*VPI+0.000001)
      EKP=CSQRT(EKP)*AS(1)
```

```
      EKM=-EKP+AS(2)*PQ(1,J)+AS(3)*PQ(2,J)
      EKP=EKM+2.0*EKP
1     XP(J)=CEXP(0.5*CI*EKP)
COMMENT THE NEXT SECTION OF PROGRAM DOWN TO LABEL 22
C         STORES THE EIGENVALUES OF BM IN DV AND THE
C         I-TH EIGENVECTOR IN TC(J,I)
      DO 2 J=1,N
      DO 2 K=1,N
2     H(J,K)=BM(J,K)
COMMENT H IS TRANSFORMED TO AN UPPER HESSENBERG MATRIX WITH
C         THE SAME EIGENVALUES AS H.DETAILS ARE STORED IN IPL
      CALL COMHES(N,1,N,H,IPL,EMACH)
C
      DO 3 J=1,N
      DO 3 K=1,N
3     BM(J,K)=H(J,K)
COMMENT COMLR PLACES THE EIGENVALUES OF THE UPPER
C         HESSENBERG MATRIX H IN DV.IF=THE NUMBER OF
C         EIGENVALUES IT HAS FAILED TO FIND.
      CALL COMLR(N,N,EMACH,H,DV,IF)
COMMENT EIGENVALUES ARE ORDERED:THIS IS IMPORTANT IF THERE
C         IS A DEGENERACY
      J2=N-1
      DO 6 J=1,J2
      K1=J+1
      DO 6 K=K1,N
      IF (CABS(DV(J))-CABS(DV(K))) 6,6,5
5     EKP=DV(J)
      DV(J)=DV(K)
      DV(K)=EKP
6     CONTINUE
C
      WRITE (6,7) DV
7     FORMAT(12H EIGENVALUES,20(/,2(1E16.5,1E13.5)))
COMMENT COMVEC FINDS EIGENVECTORS OF THE HESSENBERG MATRIX
C         AND COMBAK TRANSFORMS TO EIGENVECTORS OF THE
C         ORIGINAL MATRIX.THE SETTING OF LC IS IMPORTANT WHEN
C         EIGENVALUES ARE DEGENERATE
      LC=1
      FE=0.0
      DO 22 I=1,N
      TEST=ABS(FE-CABS(DV(I)))
      FE=CABS(DV(I))
      IF (TEST-EPSD) 10,10,9
9     LC=1
10    CALL COMVEC(N,N,BM,DV(I),FV,RUB,ISPON,
     1H,WV,RES1,IF,LC,EMACH,EPSD)
      CALL COMBAK(N,1,N,BM,IPL,FV,EMACH)
      DO 22 J=1,N
22    TC(J,I)=FV(J)
COMMENT FORWARD AND BACKWARD TRAVELLING WAVE AMPLITUDES
C         IN THE BLOCH WAVE,RELATIVE TO AN ORIGIN HALF WAY
C         BETWEEN LAYERS ARE STORED IN TC AND RC.NOTE AGAIN
C         THE CHOICE OF ORIGIN DEFINED BY XP
      DO 11 I=1,N
      EKP=0.5*DV(I)
      EKM=CSQRT(1.0-EKP*EKP)
      EKM=EKP+CI*EKM
      A=1.0-CABS(EKM)
      A=SIGN(1.0,A)
      IF (A) 17,19,19
```

```
17     EKM=1.0/EKM
19     DO 20 J=1,N
       XM(J)=CZERO
       DO 20 K=1,N
20     XM(J)=XM(J)+RM(J,K)*TC(K,I)
       DO 21 J=1,N
       RC(J,I)=(TC(J,I)-XM(J)/EKM)/XP(J)
21     TC(J,I)=(XM(J)-TC(J,I)/EKM)/XP(J)
C
       EKP=CLOG(EKM)/CI
       WRITE (6,14) EKP,(TC(J,I),J=1,N)
14     FORMAT(7H KZ.AZ=,2E13.5,/,3H BP,/,20(3(1E14.3,1E11.3),/))
       WRITE (6,15) (RC(J,I),J=1,N)
15     FORMAT(3H BM,/,20(3(1E14.3,1E11.3),/))
11     CONTINUE
COMMENT GAUSSIAN ELIMINATION PRIOR TO INVERSION OF TC,
C        DETAILS STORED IN IPL
       CALL ZGE(TC,IPL,N,N,EMACH)
C
       DO 12 J=1,N
12     XM(J)=CZERO
       XM(1)=CMPLX(1.0,0.0)
COMMENT XM,THE INCIDENT WAVE AMPLITUDES,ARE MULTIPLIED BY
C        TC INVERSE TO GIVE AMPLITUDES OF EXCITATION OF
C        BLOCH WAVES,RETURNED IN XM
       CALL ZSU(TC,IPL,XM,N,N,EMACH)
COMMENT XM*RC GIVES THE AMPLITUDES OF REFLECTED WAVES.
C        NOTE:IN THIS PROGRAM NO SURFACE BARRIER IS INCLUDED
       DO 13 J=1,N
       XP(J)=CZERO
       DO 13 K=1,N
13     XP(J)=XP(J)+RC(J,K)*XM(K)
C
       WRITE (6,18) XM
18     FORMAT(22H BLOCH WAVE AMPLITUDES,
      120(/,2(1E16.5,1E13.5)))
       WRITE (6,16) XP
16     FORMAT(21H REFLECTED AMPLITUDES,
      120(/,2(1E16.5,1E13.5)))
C
       RETURN
       END
C
C
COMMENT CDBL RETURNS THE REFLECTED AMPLITUDES,WV,
C        GIVEN THE REFLECTION AND TRANSMISSION MATRICES
C        APM,APP,FOR A SINGLE LAYER,BY STACKING LAYERS
C        TOGETHER TO APPROXIMATE AN INFINITE CRYSTAL.
C        DIMENSIONS FIXED FOR N=13
       SUBROUTINE CDBL(N,APM,APP,WV,EMACH)
       DIMENSION APP(13,13),AMM(13,13),APM(13,13),AMP(13,13),
      1XP(13),XM(13),WV(13),PQ(2,13),AS(3),CLM(809)
       COMPLEX APP,AMM,APM,AMP,XP,XM,WV,EKP,EKM,CI,CZ
       COMPLEX CMPLX,CSQRT,CEXP,CSIN,CCOS,CLOG
       COMMON E,SPA,THETA,FI,VPI,CLM
       COMMON /X1/TV,PQ,AS
       CZ=CMPLX(0.0,0.0)
       CI=CMPLX(0.0,1.0)
COMMENT APP,APM,ETC. ARE REFERRED TO AN ORIGIN HALF-WAY
C        BETWEEN LAYERS
       AK1=SQRT(2.0*E)*SIN(THETA)
```

```
      AK2=AK1*SIN(FI)
      AK1=AK1*COS(FI)
      DO 1 J=1,N
      WV(J)=CZ
      BK1=AK1+PQ(1,J)
      BK2=AK2+PQ(2,J)
      EKP=CMPLX(2.0*E-BK1*BK1-BK2*BK2,-2.0*VPI+0.000001)
      EKP=CSQRT(EKP)*AS(1)
      EKM=-EKP+BK1*AS(2)+BK2*AS(3)
      EKP=EKM+2.0*EKP
      XP(J)=CEXP(0.5*CI*EKP)
1     XM(J)=CEXP(0.5*CI*EKM)
COMMENT THE ORIGINS OF REFLECTION AND TRANSMISSION
C       MATRICES ARE MOVED FROM THE CENTRE OF THE LAYER
C       TO POINTS DEFINED BY XP AND XM
      DO 2 J=1,N
      DO 2 K=1,N
      AMM(J,K)=APP(J,K)/(XM(K)*XM(J))
      APP(J,K)=APP(J,K)*XP(K)*XP(J)
      AMP(J,K)=APM(J,K)*XP(K)/XM(J)
2     APM(J,K)=APM(J,K)*XP(J)/XM(K)
COMMENT STACKING OF LAYERS PROCEEDS BY DOUBLING THE
C       NUMBER OF LAYERS EACH TIME
      ALAY=1.0
3     CALL DBL(APP,APP,AMM,AMM,APM,APM,AMP,AMP,N,EMACH)
      TEST=0.0
COMMENT WV IS THE AMPLITUDE REFLECTED BY 2*ALAY LAYERS
      DO 5 J=1,N
      A=CABS(WV(J)-AMP(J,1))
      WV(J)=AMP(J,1)
5     TEST=TEST+A*A
      ALAY=ALAY+ALAY
      IF (ALAY-2.5) 6,6,7
6     ASC=TEST*0.005
      GO TO 3
7     IF (ALAY-31.0) 10,10,11
10    IF (TEST-ASC) 8,3,3
11    WRITE (6,12)
12    FORMAT(24H ***NO CONVERGENCE AFTER)
8     WRITE (6,9) ALAY,WV
9     FORMAT(1F12.5,8H  LAYERS,//,
     121H REFLECTED AMPLITUDES,50(/,2F12.5))
      RETURN
      END
C
C
COMMENT DBL,GIVEN TRANSMISSION AND REFLECTION COEFICIENTS
C       FOR LAYERS A AND B,RETURNS TRANSMISSION
C       AND REFLECTION MATRICES FOR THE PAIR IN APP,AMM,APM,
C       AND AMP
      SUBROUTINE DBL(APP,BPP,AMM,BMM,APM,BPM,AMP,BMP,
     1N,EMACH)
      DIMENSION APP(N,N),BPP(N,N),AMM(N,N),BMM(N,N),APM(N,N),
     1BPM(N,N),AMP(N,N),BMP(N,N)
      DIMENSION AA(13,13),AB(13,13),AC(13,13),WA(13),WB(13),
     1IPL(13)
      COMPLEX APP,BPP,AMM,BMM,APM,BPM,AMP,BMP,AA,AB,AC,
     1WA,WB,CZ
      COMPLEX CMPLX,CSQRT,CEXP,CSIN,CCOS,CLOG
C
      CZ=CMPLX(0.0,0.0)
```

```
COMMENT CALCULATION OF THE NEW APP AND AMP BEGINS
      DO 2 J=1,N
      DO 1 K=1,N
      AA(J,K)=CZ
      DO 1 L=1,N
1     AA(J,K)=AA(J,K)-APM(J,L)*BMP(L,K)
2     AA(J,J)=AA(J,J)+1.0
COMMENT GAUSSIAN ELIMINATION PRIOR TO INVERSION OF AA,
C        DETAILS STORED IN IPL
      CALL ZGE(AA,IPL,N,N,EMACH)
C
      DO 5 K=1,N
      DO 3 J=1,N
3     WA(J)=APP(J,K)
COMMENT INVERSE OF AA TIMES WA IS RETURNED IN WA
      CALL ZSU(AA,IPL,WA,N,N,EMACH)
COMMENT THE NEW APP IS STORED TEMPORARILY IN AB
      DO 4 J=1,N
      AB(J,K)=CZ
      WB(J)=CZ
      DO 4 L=1,N
      AB(J,K)=AB(J,K)+BPP(J,L)*WA(L)
4     WB(J)=WB(J)+BMP(J,L)*WA(L)
COMMENT THE NEW AMP IS STORED TEMPORARILY IN AC
      DO 5 J=1,N
      AC(J,K)=AMP(J,K)
      DO 5 L=1,N
5     AC(J,K)=AC(J,K)+AMM(J,L)*WB(L)
COMMENT CALCULATION OF THE NEW AMM AND APM BEGINS
      DO 7 J=1,N
      DO 6 K=1,N
      AA(J,K)=CZ
      DO 6 L=1,N
6     AA(J,K)=AA(J,K)-BMP(J,L)*APM(L,K)
7     AA(J,J)=AA(J,J)+1.0
COMMENT GAUSSIAN ELIMINATION PRIOR TO INVERSION OF AA,
C     DETAILS STORED IN IPL
      CALL ZGE(AA,IPL,N,N,EMACH)
COMMENT THE NEW AMP IS TRANSFERRED FROM AC
      DO 8 J=1,N
      DO 8 K=1,N
8     AMP(J,K)=AC(J,K)
C
      DO 10 K=1,N
      DO 9 J=1,N
9     WA(J)=BMM(J,K)
COMMENT THE INVERSE OF AA TIMES WA IS RETURNED IN WA
      CALL ZSU(AA,IPL,WA,N,N,EMACH)
      DO 10 J=1,N
10    AC(J,K)=WA(J)
COMMENT THE NEW AMM IS STORED TEMPORARILY IN AC
      DO 12 K=1,N
      DO 11 J=1,N
      AA(J,K)=CZ
      WB(J)=CZ
      DO 11 L=1,N
      AA(J,K)=AA(J,K)+APM(J,L)*AC(L,K)
11    WB(J)=WB(J)+AMM(J,L)*AC(L,K)
      DO 12 J=1,N
12    AC(J,K)=WB(J)
COMMENT NOW THE NEW APM IS CALCULATED
```

```
      DO 14 K=1,N
      DO 13 J=1,N
      WB(J)=BPM(J,K)
      DO 13 L=1,N
13    WB(J)=WB(J)+BPP(J,L)*AA(L,K)
      DO 14 J=1,N
14    APM(J,K)=WB(J)
C
      DO 15 J=1,N
      DO 15 K=1,N
      APP(J,K)=AB(J,K)
15    AMM(J,K)=AC(J,K)
C
      RETURN
      END
```

Test runs of LEED programs

There follows first a sample set of typical data appropriate to a LEED calculation on the (001) surface of copper. Then follows the output obtained by using these input data by employing PERTA, REAMP, REAMPR and finally DBL to calculate reflected amplitudes. These calculations were made on an IBM 370/165 using the FORTG compiler for PERTA and DBL runs, and a FORTX double precision compiler for REAMP and REAMPR runs to handle to eigenvalue/eigenvector problem with sufficient accuracy.

```
  6.81735
 1.0       0.0       0.0
 0.0       1.0       0.0
 0.0       0.0       1.0
  0.00000       0.50000       0.50000
  0.00000      -0.50000       0.50000
  0.50000      -0.50000       0.00000
 4    12    13
 0.0       0.0
 1.0       0.0
 0.0       1.0
-1.0       0.0
 0.0      -1.0
 1.0       1.0
-1.0       1.0
 1.0      -1.0
-1.0      -1.0
 2.0       0.0
 0.0       2.0
-2.0       0.0
 0.0      -2.0
-0.14700

  2.00000
 -1.25000      -0.35000      -0.20000       0.15000       0.00000
  3.00000
 -1.70000      -0.70000      -0.35000       0.37500       0.07000

  0.50000       0.50000
  2.60000
 -1.00000
  0.00000       0.00000
  2.60000
 -3.00000
```

```
 0.10000E 01   0.10000E 01  -0.94868E 00  -0.94868E 00   0.94868E 00
 0.94868E 00   0.89642E 00   0.89642E 00  -0.98198E 00  -0.98198E 00
 0.89642E 00   0.89642E 00  -0.85391E 00  -0.85391E 00   0.96825E 00
 0.96825E 00  -0.96825E 00  -0.96825E 00   0.85391E 00   0.85391E 00
 0.94868E 00   0.94868E 00  -0.64286E 00   0.35714E 00   0.10000E 01
 0.10714E 01   0.10714E 01   0.47246E 00  -0.66144E 00  -0.11339E 01
-0.86258E 00  -0.24152E 00   0.62106E 00   0.11024E 01   0.11024E 01
-0.36822E 00   0.78905E 00   0.11573E 01   0.69588E 00  -0.17894E 00
-0.87482E 00  -0.97423E 00  -0.53682E 00   0.43741E 00   0.11047E 01
 0.11047E 01  -0.94868E 00  -0.94868E 00   0.10714E 01   0.10714E 01
-0.64286E 00   0.35714E 00   0.10000E 01  -0.11024E 01  -0.11024E 01
 0.86258E 00   0.24152E 00  -0.62106E 00  -0.47246E 00   0.66144E 00
 0.11339E 01   0.11047E 01   0.11047E 01  -0.97423E 00  -0.53682E 00
 0.43741E 00   0.69588E 00  -0.17894E 00  -0.87482E 00  -0.36822E 00
 0.78905E 00   0.11573E 01   0.89642E 00   0.89642E 00  -0.47246E 00
 0.66144E 00   0.11339E 01   0.11024E 01   0.11024E 01   0.28409E 00
-0.71591E 00   0.            0.10000E 01  -0.72615E 00   0.18672E 00
 0.91287E 00   0.11932E 01   0.11932E 01  -0.18735E 00   0.64770E 00
-0.38969E 00  -0.12247E 01   0.49569E 00  -0.55638E 00  -0.58916E 00
 0.46291E 00  -0.89223E 00  -0.15579E 00   0.73645E 00   0.12365E 01
 0.12365E 01  -0.98198E 00  -0.98198E 00   0.86258E 00   0.24152E 00
-0.62106E 00  -0.86258E 00  -0.24152E 00   0.62106E 00  -0.72615E 00
 0.18672E 00   0.91287E 00   0.94697E 00   0.61364E 00   0.66667E 00
 0.10000E 01  -0.72615E 00   0.18672E 00   0.91287E 00   0.61570E 00
-0.45151E 00  -0.10672E 01  -0.90499E 00  -0.46321E 00  -0.40337E 00
-0.84515E 00   0.90499E 00   0.46321E 00   0.40337E 00   0.84515E 00
-0.61570E 00   0.45151E 00   0.10672E 01   0.89642E 00   0.89642E 00
-0.11024E 01  -0.11024E 01   0.47246E 00  -0.66144E 00  -0.11339E 01
 0.11932E 01   0.11932E 01  -0.72615E 00   0.18672E 00   0.91287E 00
 0.28409E 00  -0.71591E 00   0.            0.10000E 01  -0.12365E 01
-0.12365E 01   0.89223E 00   0.15579E 00  -0.73645E 00  -0.49569E 00
 0.55638E 00   0.58916E 00  -0.46291E 00   0.18735E 00  -0.64770E 00
 0.38969E 00   0.12247E 01   0.85391E 00   0.85391E 00  -0.36822E 00
 0.78905E 00   0.11573E 01   0.11047E 01   0.11047E 01   0.18735E 00
-0.64770E 00   0.38969E 00   0.12247E 01  -0.61570E 00   0.45151E 00
 0.10672E 01   0.12365E 01   0.12365E 01  -0.10708E 00   0.48295E 00
-0.63724E 00  -0.22727E 00   0.10000E 01   0.36425E 00  -0.67647E 00
-0.26761E 00   0.77311E 00  -0.80136E 00   0.13529E 00   0.93665E 00
 0.13125E 01   0.13125E 01  -0.96825E 00  -0.96825E 00   0.69588E 00
-0.17894E 00  -0.87482E 00  -0.97423E 00  -0.53682E 00   0.43741E 00
-0.49569E 00   0.55638E 00   0.58916E 00  -0.46291E 00   0.90499E 00
 0.46321E 00   0.40337E 00   0.84515E 00  -0.89223E 00  -0.15579E 00
 0.73645E 00   0.36425E 00  -0.67647E 00  -0.26761E 00   0.77311E 00
-0.74956E 00  -0.28409E-01   0.27310E 00   0.55195E 00   0.10000E 01
 0.96372E 00   0.59659E 00   0.60689E 00   0.97403E 00  -0.80136E 00
 0.13529E 00   0.93665E 00   0.96825E 00   0.96825E 00  -0.97423E 00
-0.53682E 00   0.43741E 00   0.69588E 00  -0.17894E 00  -0.87482E 00
 0.89223E 00   0.15579E 00  -0.73645E 00  -0.90499E 00  -0.46321E 00
-0.40337E 00  -0.84515E 00   0.49569E 00  -0.55638E 00  -0.58916E 00
 0.46291E 00  -0.80136E 00   0.13529E 00   0.93665E 00   0.96372E 00
 0.59659E 00   0.60689E 00   0.97403E 00  -0.74956E 00  -0.28409E-01
 0.27310E 00   0.55195E 00   0.10000E 01   0.36425E 00  -0.67647E 00
-0.26761E 00   0.77311E 00  -0.85391E 00  -0.85391E 00   0.11047E 01
 0.11047E 01  -0.36822E 00   0.78905E 00   0.11573E 01  -0.12365E 01
-0.12365E 01   0.61570E 00  -0.45151E 00  -0.10672E 01  -0.18735E 00
 0.64770E 00  -0.38969E 00  -0.12247E 01   0.13125E 01   0.13125E 01
-0.80136E 00   0.13529E 00   0.93665E 00   0.36425E 00  -0.67647E 00
-0.26761E 00   0.77311E 00  -0.10708E 00   0.48295E 00  -0.63724E 00
-0.22727E 00   0.10000E 01   0.10000E 01  -0.12247E 01   0.12247E 01
 0.13693E 01  -0.11180E 01   0.13693E 01  -0.14790E 01   0.11456E 01
-0.11456E 01   0.14790E 01   0.15687E 01  -0.11859E 01   0.11250E 01
```

-0.11859E 01	0.15687E 01	0.12247E 01	-0.50000E 00	0.10000E 01
0.15000E 01	0.33541E 00	-0.13416E 01	-0.82158E 00	0.54772E 00
0.16771E 01	-0.25877E 00	0.15526E 01	0.60134E 00	-0.80178E 00
-0.10022E 01	0.40089E 00	0.18114E 01	0.21348E 00	-0.17078E 01
-0.48412E 00	0.96825E 00	0.76547E 00	-0.61237E 00	-0.11296E 01
0.32275E 00	0.19213E 01	-0.12247E 01	0.15000E 01	-0.50000E 00
0.10000E 01	-0.16771E 01	0.82158E 00	-0.54772E 00	-0.33541E 00
0.13416E 01	0.18114E 01	-0.10022E 01	0.40089E 00	0.60134E 00
-0.80178E 00	-0.25877E 00	0.15526E 01	-0.19213E 01	0.11296E 01
-0.32275E 00	-0.76547E 00	0.61237E 00	0.48412E 00	-0.96825E 00
-0.21348E 00	0.17078E 01	0.13693E 01	-0.33541E 00	0.13416E 01
0.16771E 01	0.16071E 00	-0.71429E 00	0.10000E 01	-0.65611E 00
0.87482E 00	0.18750E 01	-0.96440E-01	0.54006E 00	-0.13887E 01
0.37351E 00	-0.83666E 00	0.35857E 00	-0.87152E 00	0.69722E 00
0.20252E 01	0.65093E-01	-0.44635E 00	0.16366E 01	-0.24603E 00
0.75918E 00	-0.61859E 00	0.54461E 00	-0.80024E 00	0.19562E 00
-0.10333E 01	0.59047E 00	0.21481E 01	-0.11180E 01	0.82158E 00
-0.54772E 00	-0.82158E 00	0.54772E 00	-0.65611E 00	0.87482E 00
0.96429E 00	0.71429E 00	0.10000E 01	-0.65611E 00	0.87482E 00
0.55120E 00	-0.11024E 01	-0.91491E 00	-0.51235E 00	-0.87831E 00
0.91491E 00	0.51235E 00	0.87831E 00	-0.55120E 00	0.11024E 01
-0.47834E 00	0.12756E 01	0.84371E 00	0.27550E 00	0.75761E 00
-0.95287E 00	-0.65340E 00	-0.95831E 00	0.84371E 00	0.27550E 00
0.75761E 00	-0.47834E 00	0.12756E 01	0.13693E 01	-0.16771E 01
0.33541E 00	-0.13416E 01	0.18750E 01	-0.65611E 00	0.87482E 00
0.16071E 00	-0.71429E 00	0.10000E 01	-0.20252E 01	0.87152E 00
-0.69722E 00	-0.37351E 00	0.83666E 00	-0.35857E 00	0.96440E-01
-0.54006E 00	0.13887E 01	0.21481E 01	-0.10333E 01	0.59047E 00
0.54461E 00	-0.80024E 00	0.19562E 00	-0.24603E 00	0.75918E 00
-0.61859E 00	0.65093E-01	-0.44635E 00	0.16366E 01	0.14790E 01
-0.25877E 00	0.15526E 01	0.18114E 01	0.96440E-01	-0.54006E 00
0.13887E 01	-0.55120E 00	0.11024E 01	0.20252E 01	-0.47348E-01
0.30682E 00	-0.83333E 00	0.10000E 01	0.25673E 00	-0.79220E 00
0.64550E 00	-0.77020E 00	0.92423E 00	0.21875E 01	0.27042E-01
-0.20397E 00	0.67497E 00	-0.14142E 01	-0.14309E 00	0.57821E 00
-0.76534E 00	0.26726E 00	0.40725E 00	-0.85328E 00	0.40337E 00
-0.94441E 00	0.80949E 00	0.23202E 01	-0.11456E 01	0.60134E 00
-0.80178E 00	-0.10022E 01	0.40089E 00	-0.37351E 00	0.83666E 00
-0.35857E 00	0.91491E 00	0.51235E 00	0.87831E 00	-0.87152E 00
0.69722E 00	0.25673E 00	-0.79220E 00	0.64550E 00	-0.71023E 00
0.10227E 00	0.50000E 00	0.10000E 01	0.99432E 00	0.68182E 00
0.10000E 01	-0.77020E 00	0.92423E 00	-0.18852E 00	0.73732E 00
-0.87138E 00	0.55419E 00	-0.42797E 00	-0.65870E 00	-0.10351E 01
-0.87626E 00	-0.34621E 00	-0.10415E 00	0.65465E 00	0.99755E 00
0.62703E 00	0.98805E 00	-0.69124E 00	0.11060E 01	0.11456E 01
-0.10022E 01	0.40089E 00	0.60134E 00	-0.80178E 00	0.87152E 00
-0.69722E 00	-0.91491E 00	-0.51235E 00	-0.87831E 00	0.37351E 00
-0.83666E 00	0.35857E 00	-0.77020E 00	0.92423E 00	0.99432E 00
0.68182E 00	0.10000E 01	-0.71023E 00	0.10227E 00	0.50000E 00
0.10000E 01	0.25673E 00	-0.79220E 00	0.64550E 00	0.69124E 00
-0.11060E 01	-0.99755E 00	-0.62703E 00	-0.98805E 00	0.87626E 00
0.34621E 00	0.10415E 00	-0.65465E 00	-0.55419E 00	0.42797E 00
0.65870E 00	0.10351E 01	0.18852E 00	-0.73732E 00	0.87138E 00
-0.14790E 01	0.18114E 01	-0.25877E 00	0.15526E 01	-0.20252E 01
0.55120E 00	-0.11024E 01	-0.96440E-01	0.54006E 00	-0.13887E 01
0.21875E 01	-0.77020E 00	0.92423E 00	0.25673E 00	-0.79220E 00
0.64550E 00	-0.47348E-01	0.30682E 00	-0.83333E 00	0.10000E 01
-0.23202E 01	0.94441E 00	-0.80949E 00	-0.40725E 00	0.85328E 00
-0.40337E 00	0.14309E 00	-0.57821E 00	0.76534E 00	-0.26726E 00
-0.27042E-01	0.20397E 00	-0.67497E 00	0.14142E 01	0.15687E 01
-0.21348E 00	0.17078E 01	0.19213E 01	0.65093E-01	-0.44635E 00

```
 0.16366E 01  -0.47834E 00   0.12756E 01   0.21481E 01  -0.27042E-0
 0.20397E 00  -0.67497E 00   0.14142E 01   0.18852E 00  -0.73732E 00
 0.87138E 00  -0.69124E 00   0.11060E 01   0.23202E 01   0.13385E-01
-0.11364E 00   0.42483E 00  -0.90909E 00   0.10000E 01  -0.91063E-01
 0.45098E 00  -0.80284E 00   0.51541E 00   0.31676E 00  -0.85568E 00
 0.59239E 00  -0.86814E 00   0.99216E 00   0.24609E 01  -0.11859E 01
 0.48412E 00  -0.96825E 00  -0.11296E 01   0.32275E 00  -0.24603E 00
 0.75918E 00  -0.61859E 00   0.84371E 00   0.27550E 00   0.75761E 00
-0.10333E 01   0.59047E 00   0.14309E 00  -0.57821E 00   0.76534E 00
-0.26726E 00  -0.55419E 00   0.42797E 00   0.65870E 00   0.10351E 01
 0.99755E 00   0.62703E 00   0.98805E 00  -0.94441E 00   0.80949E 00
-0.91063E-01   0.45098E 00  -0.80284E 00   0.51541E 00   0.37478E 00
-0.62500E 00  -0.33379E 00   0.25974E 00   0.10000E 01  -0.76189E 00
 0.            0.35185E 00   0.92404E 00   0.10601E 01   0.71591E 00
 0.10621E 01  -0.86814E 00   0.99216E 00   0.11250E 01  -0.76547E 00
 0.61237E 00   0.76547E 00  -0.61237E 00   0.54461E 00  -0.80024E 00
 0.19562E 00  -0.95287E 00  -0.65340E 00  -0.95831E 00   0.54461E 00
-0.80024E 00   0.19562E 00  -0.40725E 00   0.85328E 00  -0.40337E 00
 0.87626E 00   0.34621E 00   0.10415E 00  -0.65465E 00  -0.87626E 00
-0.34621E 00  -0.10415E 00   0.65465E 00   0.40725E 00  -0.85328E 00
 0.40337E 00   0.31676E 00  -0.85568E 00   0.59239E 00  -0.76189E 00
 0.            0.35185E 00   0.92404E 00   0.93695E 00   0.56818E 00
 0.54620E 00   0.64935E 00   0.10000E 01  -0.76189E 00   0.
 0.35185E 00   0.92404E 00   0.31676E 00  -0.85568E 00   0.59239E 00
-0.11859E 01   0.11296E 01  -0.32275E 00  -0.48412E 00   0.96825E 00
-0.10333E 01   0.59047E 00   0.84371E 00   0.27550E 00   0.75761E 00
-0.24603E 00   0.75918E 00  -0.61859E 00   0.94441E 00  -0.80949E 00
-0.99755E 00  -0.62703E 00  -0.98805E 00   0.55419E 00  -0.42797E 00
-0.65870E 00  -0.10351E 01  -0.14309E 00   0.57821E 00  -0.76534E 00
 0.26726E 00  -0.86814E 00   0.99216E 00   0.10601E 01   0.71591E 00
 0.10621E 01  -0.76189E 00   0.            0.35185E 00   0.92404E 00
 0.37478E 00  -0.62500E 00  -0.33379E 00   0.25974E 00   0.10000E 01
-0.91063E-01   0.45098E 00  -0.80284E 00   0.51541E 00   0.15687E 01
-0.19213E 01   0.21348E 00  -0.17078E 01   0.21481E 01  -0.47834E 00
 0.12756E 01   0.65093E-01  -0.44635E 00   0.16366E 01  -0.23202E 01
 0.69124E 00  -0.11060E 01  -0.18852E 00   0.73732E 00  -0.87138E 00
 0.27042E-01  -0.20397E 00   0.67497E 00  -0.14142E 01   0.24609E 01
-0.86814E 00   0.99216E 00   0.31676E 00  -0.85568E 00   0.59239E 00
-0.91063E-01   0.45098E 00  -0.80284E 00   0.51541E 00   0.13385E-01
-0.11364E 00   0.42483E 00  -0.90909E 00   0.10000E 01
```

```
SPA=      6.81735


SURF ASM1 ASM2
   1.0     0.0     0.0
   0.0     1.0     0.0
   0.0     0.0     1.0
SURFACE VECS
    0.0           0.50000      0.50000
    0.0          -0.50000      0.50000
LAYER DISPLACEMENT
    3.40867    -3.40867      0.0
IDEG=     4 N1=    12 N=    13
BEAMS INCLUDED

  PQF1    PQF2
   0.0     0.0
   1.0     0.0
   0.0     1.0
  -1.0     0.0
   0.0    -1.0
   1.0     1.0
  -1.0     1.0
   1.0    -1.0
  -1.0    -1.0
   2.0     0.0
   0.0     2.0
  -2.0     0.0
   0.0    -2.0
VPI=     -0.14700
E1=      2.00000
PHS
   -1.25000     -0.35000     -0.20000      0.15000      0.0
E1=      3.00000
PHS
   -1.70000     -0.70000     -0.35000      0.37500      0.07000


ORDER    1      CRIT = 0.003
IMAX =   8      BNORM =    0.2961E-02
   WVR          WVI
 -0.7788E-01  0.2596E-01
  0.3106E-01  0.1184E 00
 -0.7854E-01  0.2037E-01
 -0.4696E-01  0.1272E-02
  0.4469E-01  0.1323E-01
 -0.1658E-01  0.9827E-02
  0.4810E-02  0.1310E 00
  0.3357E-01 -0.2527E-01
 -0.4873E-02  0.4016E-01
 -0.4449E-02  0.2734E-02
 -0.4973E-02  0.5176E-02
  0.6307E-02  0.1092E 00
 -0.1181E-01  0.3386E-01
ANORM =    0.6696E-01


ORDER    2      CRIT = 0.003
IMAX =   6      BNORM =    0.3705E-03
```

```
   WVR          WVI
 -0.7917E-01  0.2601E-01
  0.3078E-01  0.1185E 00
 -0.7827E-01  0.1984E-01
 -0.4667E-01  0.2669E-03
  0.4403E-01  0.1271E-01
 -0.1662E-01  0.9749E-02
  0.5500E-02  0.1301E 00
  0.3375E-01 -0.2510E-01
 -0.5699E-02  0.4158E-01
 -0.4462E-02  0.2706E-02
 -0.4953E-02  0.5175E-02
  0.7941E-02  0.1066E 00
 -0.1206E-01  0.3400E-01
ANORM =    0.6698E-01

E=     2.6000 THETA=     0.5000 FI=     0.5000
    BEAM              INTENSITY
   0.0      0.0     0.69443E-02
   1.0      0.0     0.0
   0.0      1.0     0.39946E-02
  -1.0      0.0     0.24437E-02
   0.0     -1.0     0.18499E-02
   1.0      1.0     0.0
  -1.0      1.0     0.13497E-01
   1.0     -1.0     0.0
  -1.0     -1.0     0.17918E-02
   2.0      0.0     0.0
   0.0      2.0     0.0
  -2.0      0.0     0.93430E-02
   0.0     -2.0     0.0



ORDER    1      CRIT = 0.003
IMAX =   6      BNORM =    0.1867E-02
   WVR          WVI
  0.5191E-01  0.4688E-01
  0.1066E-02  0.2442E-01
  0.1066E-02  0.2442E-01
 -0.1065E-02 -0.2442E-01
 -0.1065E-02 -0.2442E-01
  0.5153E-01 -0.1748E 00
 -0.5153E-01  0.1748E 00
 -0.5153E-01  0.1748E 00
  0.5154E-01 -0.1748E 00
 -0.1603E-01  0.2973E-01
 -0.1603E-01  0.2973E-01
 -0.1603E-01  0.2973E-01
 -0.1603E-01  0.2973E-01
ANORM =    0.1446E 00

ORDER    2      CRIT = 0.003
IMAX =   6      BNORM =    0.3450E-02
   WVR          WVI
  0.4670E-01  0.5826E-01
 -0.2100E-02  0.2797E-01
```

```
 -0.2100E-02  0.2797E-01
  0.2101E-02 -0.2797E-01
  0.2101E-02 -0.2797E-01
  0.5881E-01 -0.1724E 00
 -0.5881E-01  0.1724E 00
 -0.5881E-01  0.1724E 00
  0.5881E-01 -0.1724E 00
 -0.1459E-01  0.2983E-01
 -0.1459E-01  0.2983E-01
 -0.1459E-01  0.2983E-01
 -0.1459E-01  0.2983E-01
ANORM =    0.1451E 00


ORDER    3      CRIT = 0.003
IMAX =   4      BNORM =    0.3095E-03
   WVR           WVI
  0.4575E-01  0.5791E-01
 -0.1852E-02  0.2815E-01
 -0.1852E-02  0.2815E-01
  0.1853E-02 -0.2815E-01
  0.1853E-02 -0.2815E-01
  0.5823E-01 -0.1724E 00
 -0.5823E-01  0.1724E 00
 -0.5823E-01  0.1724E 00
  0.5823E-01 -0.1724E 00
 -0.1469E-01  0.2981E-01
 -0.1469E-01  0.2981E-01
 -0.1469E-01  0.2981E-01
 -0.1469E-01  0.2981E-01
ANORM =    0.1451E 00

E=     2.6000 THETA=     0.0     FI=     0.0
    BEAM             INTENSITY
   0.0     0.0     0.54470E-02
   1.0     0.0     0.65312E-03
   0.0     1.0     0.65312E-03
  -1.0     0.0     0.65312E-03
   0.0    -1.0     0.65312E-03
   1.0     1.0     0.19493E-01
  -1.0     1.0     0.19493E-01
   1.0    -1.0     0.19493E-01
  -1.0    -1.0     0.19493E-01
   2.0     0.0     0.0
   0.0     2.0     0.0
  -2.0     0.0     0.0
   0.0    -2.0     0.0
```

```
SPA=      6.81735



SURF ASM1 ASM2
   1.0     0.0     0.0
   0.0     1.0     0.0
   0.0     0.0     1.0
SURFACE VECS
    0.0           0.50000      0.50000
    0.0          -0.50000      0.50000
LAYER DISPLACEMENT
    3.40867     -3.40867      0.0
IDEG=     4 N1=    12 N=    13
BEAMS INCLUDED

  PQF1   PQF2
   0.0    0.0
   1.0    0.0
   0.0    1.0
  -1.0    0.0
   0.0   -1.0
   1.0    1.0
  -1.0    1.0
   1.0   -1.0
  -1.0   -1.0
   2.0    0.0
   0.0    2.0
  -2.0    0.0
   0.0   -2.0
VPI=     -0.14700
E1=      2.00000
PHS
   -1.25000     -0.35000     -0.20000      0.15000      0.0
E1=      3.00000
PHS
   -1.70000     -0.70000     -0.35000      0.37500      0.07000

EIGENVALUES
   -0.11847D-03 -0.16700D-04    -0.32787D-03 -0.35422D-03
   -0.26623D-02  0.62728D-03    -0.18881D-01 -0.77988D-02
   -0.27263D-01 -0.12247D-01     0.50271D-01  0.35475D-01
   -0.16969D+00  0.52942D+00    -0.39333D+00 -0.43925D+00
    0.25954D+00  0.54704D+00    -0.45444D+00  0.48402D+00
    0.60691D+00 -0.31287D+00    -0.68164D+00 -0.11319D+00
   -0.59499D+00 -0.37873D+00    -0.96271D+00  0.10409D+01
   -0.13202D+01  0.59290D+00     0.14297D+01  0.31737D+00
   -0.12767D+01 -0.79926D+00     0.30453D+00 -0.16232D+01
   -0.77227D+00  0.15100D+01    -0.96717D+00 -0.15165D+01
    0.10455D+02 -0.12444D+02    -0.26022D+02  0.21032D+02
   -0.38992D+02  0.29596D+02    -0.36549D+03  0.95661D+01
   -0.97363D+03  0.18288D+04    -0.77062D+04  0.32362D+04

K.A= -0.30016D+01  0.90310D+01
BP
    0.138D+00 -0.181D+00    -0.109D+01 -0.730D+00     0.505D+00  0.160D+0
    0.697D-01 -0.516D+00     0.465D-01  0.288D+00    -0.203D+00 -0.461D+0
    0.115D+00  0.531D-03    -0.132D+01 -0.554D+00     0.664D-01 -0.299D+0
   -0.179D+02  0.289D+02    -0.141D+02 -0.603D+01     0.499D+00 -0.730D+0
   -0.297D+01  0.142D+00
```

```
M
  0.338D+00  0.152D+00     0.326D+00  0.423D+00     -0.241D+00 -0.220D+00
 -0.547D-01 -0.239D+00    -0.143D+00 -0.340D+00     -0.146D+00  0.227D-01
 -0.220D+00  0.167D+00     0.145D+00 -0.184D+00     -0.143D+00  0.328D+00
 -0.320D-01  0.377D-01    -0.106D+00 -0.446D-01      0.271D-01  0.128D+00
 -0.118D+00 -0.320D-01
.A= -0.23176D+01  0.76362D+01
P
  0.379D+00  0.817D-01    -0.944D+00 -0.759D+00      0.488D+00 -0.494D-01
  0.203D+00 -0.422D+00     0.338D+00  0.492D+00      0.269D+01 -0.428D+01
 -0.151D+00 -0.219D+00    -0.220D+01 -0.142D+01      0.402D+00 -0.370D+00
 -0.140D+01  0.796D+01    -0.239D+02 -0.856D+01      0.275D+00 -0.668D+00
 -0.563D+01  0.112D+01
M
  0.206D+00  0.502D-01     0.789D-01 -0.673D-01     -0.413D+00 -0.493D+00
  0.819D-01 -0.233D+00    -0.223D+00 -0.182D+00     -0.927D-01 -0.745D-01
 -0.260D+00  0.424D+00     0.132D-01 -0.445D-01     -0.196D+00  0.294D+00
  0.205D-02  0.833D-03    -0.147D+00 -0.774D-01     -0.109D+00  0.385D+00
 -0.200D+00 -0.630D-01
.A=  0.29102D+01  0.59016D+01
P
  0.663D+00  0.395D+00    -0.157D+01 -0.843D+00     -0.709D+00  0.447D+00
  0.541D+00  0.346D-01     0.377D+00  0.308D+00      0.369D+01 -0.202D+02
 -0.100D+00  0.395D+00    -0.976D+00 -0.149D+00      0.708D+00  0.572D-02
 -0.115D+01 -0.427D+00    -0.272D+01 -0.205D+01      0.310D+00  0.534D+00
 -0.270D+01  0.875D+00

 -0.502D+00  0.208D+00    -0.307D+00 -0.126D+00     -0.323D+00 -0.611D+00
  0.380D+00  0.138D+00     0.391D+00  0.105D-01      0.424D-01 -0.248D+00
 -0.203D+00  0.457D+00     0.199D-01  0.212D-01     -0.333D+00 -0.186D+00
  0.105D-01 -0.549D-01    -0.419D-02 -0.108D+00     -0.245D+00  0.245D+00
 -0.642D-01 -0.329D-01
.A= -0.27499D+01  0.38909D+01
P
  0.341D+00 -0.225D-02    -0.784D+00  0.275D+00     -0.116D+00 -0.134D+00
  0.199D+00  0.334D-01    -0.257D-01 -0.340D+00     -0.268D+00  0.844D-01
 -0.376D+00 -0.184D+00     0.211D+01  0.683D+01     -0.283D-01 -0.129D+00
 -0.316D-01 -0.247D-02    -0.455D-01 -0.584D+00      0.233D+00  0.442D+00
 -0.982D+00  0.124D+01

 -0.118D+00  0.171D+00    -0.960D-01  0.403D-02     -0.189D-01  0.187D+00
  0.141D+00 -0.136D+00     0.223D+00  0.126D+00     -0.612D-02 -0.145D-01
  0.217D+00 -0.937D-01     0.178D-01  0.162D+00     -0.960D-01 -0.174D+00
 -0.766D-02 -0.236D-01     0.286D-02 -0.506D-02     -0.770D-02  0.259D+00
 -0.380D-01  0.561D-02
 A= -0.27194D+01  0.35103D+01

 -0.148D+00 -0.118D+00     0.142D+00 -0.269D+00     -0.490D+00 -0.345D-02
 -0.986D-01 -0.133D+00    -0.489D-01  0.454D+00     -0.220D+00  0.438D+00
  0.757D-01  0.681D+00     0.854D+00 -0.133D+01      0.285D+00  0.260D-01
  0.180D-01  0.821D-01    -0.704D+00 -0.308D+00     -0.204D+00 -0.439D+00
  0.224D+01  0.643D+01

 -0.165D-01 -0.165D+00    -0.117D+00  0.583D-01     -0.161D+00 -0.212D+00
 -0.208D+00  0.152D+00    -0.185D+00 -0.186D+00     -0.150D-01 -0.395D-02
 -0.187D+00  0.179D+00     0.397D-01 -0.597D-01     -0.117D+00  0.217D+00
 -0.324D-02  0.502D-02    -0.115D-01 -0.281D-02      0.328D+00 -0.165D+00
 -0.441D-02  0.178D+00
 A=  0.61452D+00  0.27883D+01

  0.962D-01 -0.278D+00    -0.184D+00  0.345D+01      0.743D-01  0.395D+00
```

```
    -0.379D-01 -0.117D+00    -0.144D+00 -0.225D+00    -0.113D+00  0.116D-0
     0.780D-01  0.119D-01     0.262D+00 -0.417D-01    -0.753D-01 -0.470D-0
     0.152D-01  0.164D-02    -0.236D-01 -0.386D-01     0.225D+00 -0.122D-0
    -0.129D-01  0.774D-01
BM
     0.963D-01  0.207D+00     0.573D-01 -0.563D-02    -0.381D-01  0.266D+0
     0.623D-01 -0.122D+00     0.140D+00 -0.154D+00    -0.211D-01  0.123D-0
     0.132D+00 -0.518D-01     0.660D-02 -0.156D-01    -0.103D+00  0.176D-0
    -0.143D-01 -0.378D-03    -0.130D-02  0.184D-02     0.232D-01  0.284D-0
     0.346D-01  0.170D-01
K.A=  0.18810D+01  0.58707D+00
BP
     0.192D+00  0.511D-01    -0.542D-01 -0.819D-01    -0.141D+00  0.553D-0
     0.327D+00 -0.329D+00     0.725D-01  0.686D-01    -0.531D-02  0.139D-0
     0.839D+00  0.113D+00     0.189D-01 -0.116D+00    -0.129D+00 -0.452D-0
    -0.501D-03  0.836D-02    -0.161D-02  0.636D-02    -0.629D+00  0.634D-0
    -0.591D-01  0.143D+00
BM
    -0.186D-01  0.513D-02     0.717D-01  0.725D-01    -0.253D+00 -0.101D-0
     0.428D-01  0.312D-01    -0.395D-01  0.590D-02    -0.102D-01  0.108D-0
     0.361D-02  0.158D+00    -0.768D-02 -0.294D-01    -0.855D-02  0.480D-0
    -0.607D-03  0.260D-02    -0.733D-02  0.260D-02    -0.137D-01 -0.164D+0
     0.658D-02  0.215D-01
K.A= -0.23011D+01  0.52828D+00
BP
     0.220D+00  0.328D+00    -0.186D+00 -0.202D-01    -0.833D-02  0.122D+0
     0.349D-02 -0.715D-01     0.477D-01 -0.229D-01     0.245D-01  0.115D-0
     0.822D-01  0.118D+00     0.605D-02  0.168D-01     0.164D+00 -0.570D-0
     0.178D-02  0.235D-02     0.188D-01 -0.245D-03    -0.215D-01  0.189D-0
     0.692D-01 -0.830D-01
BM
    -0.385D-01  0.558D-01     0.106D+00  0.115D+00    -0.129D+00 -0.780D-0
     0.101D+00 -0.494D-01     0.195D-01 -0.435D-01    -0.205D-01  0.569D-0
    -0.617D-01  0.171D+00    -0.799D-02 -0.144D-01    -0.191D-01  0.622D-0
    -0.295D-02  0.184D-02    -0.124D-01 -0.137D-02    -0.133D+00 -0.858D-0
     0.210D-01 -0.235D-01
K.A=  0.11278D+01  0.50172D+00
BP
     0.289D-01 -0.167D+00     0.861D-02  0.346D-01    -0.388D-02  0.178D+0
    -0.705D+00  0.670D+00     0.775D-01 -0.192D+00     0.225D-01 -0.166D-0
    -0.376D-01 -0.372D+00    -0.103D+00 -0.279D-02    -0.289D-01  0.140D+0
     0.967D-02 -0.302D-02     0.118D-01 -0.180D-01     0.732D-01  0.320D+0
     0.120D+00  0.299D-01
BM
    -0.364D-01  0.642D-01     0.421D-01 -0.112D+00     0.155D+00  0.144D+0
     0.573D-02  0.163D-01     0.304D-01  0.655D-01     0.205D-01 -0.200D-0
     0.122D+00 -0.296D-01    -0.408D-01  0.362D-01     0.635D-01 -0.509D-0
     0.367D-02 -0.268D-02     0.113D-01 -0.145D-02    -0.129D+00  0.350D-0
     0.477D-01 -0.444D-01
K.A=  0.23247D+01  0.40959D+00
BP
     0.273D-02  0.481D-01    -0.126D+00  0.536D-01     0.122D-01 -0.112D+0
     0.253D-01 -0.315D-01     0.110D-01 -0.488D-02     0.326D-01  0.211D-0
    -0.520D+00  0.362D+00     0.116D-01 -0.518D-01    -0.550D-01  0.568D-0
     0.264D-02  0.793D-02     0.224D-01 -0.767D-02    -0.733D+00  0.398D+0
     0.877D-02 -0.204D-01
BM
    -0.829D-01 -0.125D+00    -0.627D-01  0.202D-01     0.240D-01  0.101D+0
    -0.390D-03  0.150D-02    -0.153D-01  0.120D-01     0.215D-03 -0.158D-0
     0.566D-01 -0.114D-01     0.193D-01  0.742D-02    -0.486D-01 -0.103D+0
    -0.250D-02 -0.241D-02     0.851D-02 -0.381D-02     0.496D-01 -0.436D-0
```

```
  0.194D-01  0.200D-01
A= -0.47598D+00  0.38153D+00

 -0.895D-01 -0.641D-01      0.100D+00 -0.159D+00      0.672D-01  0.574D-01
  0.804D-01  0.106D+00      0.578D+00  0.998D+00     -0.529D-03 -0.210D-01
 -0.198D-01 -0.326D-01      0.698D-01 -0.431D-02      0.991D-01  0.671D-01
 -0.606D-02 -0.251D-02      0.191D-01 -0.752D-02      0.217D-01  0.262D-01
 -0.771D-01 -0.223D-02

  0.603D-02 -0.377D-01     -0.977D-01 -0.941D-01      0.251D-02 -0.112D+00
  0.613D-01  0.234D-01     -0.829D-01  0.732D-01     -0.917D-02 -0.156D-01
  0.450D-01  0.101D-01     -0.352D-01  0.351D-01     -0.944D-03  0.294D-01
  0.207D-02 -0.398D-02     -0.110D-01  0.132D-02     -0.573D-01 -0.318D-01
  0.296D-01 -0.424D-01
A= -0.29770D+01  0.36966D+00

 -0.112D+01  0.149D+00     -0.933D-01  0.317D+00      0.466D+00 -0.254D+00
 -0.172D+00  0.795D-01     -0.151D+00  0.730D-01      0.275D-01  0.578D-02
  0.374D+00 -0.119D+00     -0.991D-01 -0.101D+00      0.577D+00 -0.110D+00
  0.102D-01  0.437D-02     -0.706D-02  0.106D-01     -0.261D+00  0.455D-01
  0.398D-01  0.106D+00

  0.116D+00 -0.606D-02     -0.676D-01 -0.265D+00      0.462D-01  0.890D-01
  0.665D-01 -0.156D-01     -0.139D+00 -0.152D-02      0.245D-01 -0.171D-01
  0.908D-01 -0.431D-01     -0.712D-01  0.681D-01     -0.447D-01 -0.291D-01
  0.698D-02 -0.655D-02     -0.767D-03 -0.425D-02     -0.853D-01 -0.604D-01
  0.403D-01 -0.651D-01
A= -0.25747D+01  0.34913D+00

 -0.545D+00  0.174D+00      0.128D+00  0.158D+00      0.250D-01 -0.433D+00
  0.191D-01 -0.205D-02     -0.211D-02 -0.324D-01      0.361D-01 -0.250D-01
 -0.149D-01 -0.803D-01     -0.124D-01  0.469D-01     -0.976D+00  0.503D+00
  0.546D-02 -0.842D-02      0.250D-01 -0.377D-02      0.120D+00 -0.761D-01
 -0.972D-01  0.371D-01

  0.452D-01 -0.815D-01      0.939D-02 -0.986D-01      0.554D-01 -0.160D-01
 -0.672D-01  0.115D-01      0.270D-01  0.690D-02      0.240D-01 -0.320D-02
 -0.780D-01 -0.213D+00     -0.806D-02 -0.115D-01      0.641D-01 -0.634D-01
  0.413D-02 -0.397D-03      0.167D-01 -0.552D-02     -0.173D-01 -0.904D-01
 -0.386D-01  0.311D-01
OCH WAVE AMPLITUDES
  0.17703D-03 -0.28199D-04     -0.34693D-03 -0.74883D-03
 -0.77324D-03  0.21637D-02     -0.72569D-02  0.92402D-02
  0.78457D-02 -0.11039D-02      0.55329D-01  0.40189D-01
  0.19549D+00  0.79752D-01     -0.21209D+00 -0.38486D+00
  0.14252D+00  0.71705D-01     -0.44598D-01  0.55014D-02
 -0.19921D-01  0.93377D-01     -0.54731D+00 -0.15296D+00
 -0.31593D+00 -0.24360D+00
FLECTED AMPLITUDES
 -0.79224D-01  0.26035D-01      0.30772D-01  0.11853D+00
 -0.78337D-01  0.19952D-01     -0.46777D-01  0.28657D-03
  0.43917D-01  0.12708D-01     -0.16616D-01  0.97479D-02
  0.54683D-02  0.13003D+00      0.33743D-01 -0.25099D-01
 -0.58337D-02  0.41460D-01     -0.44619D-02  0.27056D-02
 -0.49512D-02  0.51767D-02      0.79258D-02  0.10661D+00
 -0.12072D-01  0.33998D-01
    2.6000 THETA=     0.5000 FI=     0.5000
  BEAM             INTENSITY
 0.0      0.0     0.69542D-02
 1.0      0.0     0.0
 0.0      1.0     0.40040D-02
```

```
 -1.0     0.0     0.24549D-02
  0.0    -1.0     0.18414D-02
  1.0     1.0     0.0
 -1.0     1.0     0.13488D-01
  1.0    -1.0     0.0
 -1.0    -1.0     0.17832D-02
  2.0     0.0     0.0
  0.0     2.0     0.0
 -2.0     0.0     0.93468D-02
  0.0    -2.0     0.0




EIGENVALUES
   -0.88733D-02 -0.61429D-03     -0.30103D-01  0.91150D-02
    0.39618D-01  0.10976D-01      0.39618D-01  0.10976D-01
   -0.21800D+00 -0.37823D+00     -0.32874D+00  0.34974D+00
    0.30582D+00  0.54121D+00      0.51866D-01 -0.63541D+00
   -0.66869D+00 -0.11756D-01     -0.66869D+00 -0.11756D-01
   -0.22659D+00 -0.62977D+00     -0.22659D+00 -0.62977D+00
   -0.73718D+00 -0.18062D+00     -0.12797D+01  0.31354D+00
   -0.50584D+00  0.14059D+01     -0.50584D+00  0.14059D+01
   -0.14950D+01  0.26283D-01     -0.14950D+01  0.26282D-01
    0.12761D+00  0.15634D+01      0.79140D+00 -0.14005D+01
   -0.14269D+01 -0.15180D+01     -0.11438D+01  0.19846D+01
    0.23442D+02 -0.64942D+01      0.23442D+02 -0.64943D+01
   -0.30430D+02 -0.92139D+01     -0.11216D+03  0.77667D+01

K.A= -0.30725D+01  0.47223D+01
BP
    0.148D-05 -0.792D-06     0.123D+00  0.466D-01     -0.123D+00 -0.466D-0
   -0.123D+00 -0.466D-01     0.123D+00  0.466D-01     -0.352D-05  0.385D-0
    0.307D-05 -0.439D-06     0.134D-05 -0.629D-06     -0.851D-06  0.633D-0
   -0.285D+00 -0.215D+01     0.285D+00  0.215D+01     -0.285D+00 -0.215D+0
    0.285D+00  0.215D+01
BM
    0.682D-05  0.177D-05     0.739D-01 -0.652D-01     -0.739D-01  0.652D-0
   -0.739D-01  0.652D-01     0.739D-01 -0.652D-01      0.161D-05  0.720D-0
    0.977D-05  0.348D-06    -0.160D-05 -0.563D-06     -0.988D-06  0.352D-0
   -0.440D-02 -0.390D-01     0.440D-02  0.390D-01     -0.440D-02 -0.390D-0
    0.440D-02  0.390D-01
K.A=  0.28476D+01  0.34593D+01
BP
    0.455D+00  0.579D+00    -0.482D+00  0.381D+00     -0.482D+00  0.381D+0
    0.482D+00 -0.381D+00     0.482D+00 -0.381D+00     -0.461D+00 -0.800D+0
    0.461D+00  0.800D+00     0.461D+00  0.800D+00     -0.461D+00 -0.800D+0
    0.182D+01 -0.347D+01     0.182D+01 -0.347D+01      0.182D+01 -0.347D+0
    0.182D+01 -0.347D+01
BM
   -0.588D+00  0.356D+00    -0.380D+00 -0.409D+00     -0.380D+00 -0.409D+0
    0.380D+00  0.409D+00     0.380D+00  0.409D+00      0.344D+00 -0.571D+0
   -0.344D+00  0.571D+00    -0.344D+00  0.571D+00      0.344D+00 -0.571D+0
   -0.246D-01  0.803D-01    -0.246D-01  0.803D-01     -0.246D-01  0.803D-0
   -0.246D-01  0.803D-01
K.A=  0.27026D+00  0.31915D+01
BP
   -0.113D-04 -0.141D-04     0.133D+00 -0.270D+00      0.133D+00 -0.270D+0
    0.133D+00 -0.270D+00     0.133D+00 -0.270D+00      0.626D-01  0.599D+0
   -0.925D-05 -0.228D-04    -0.935D-05 -0.208D-04     -0.626D-01 -0.599D+0
```

```
 0.376D+00  0.315D+01      0.376D+00  0.315D+01      -0.376D+00 -0.315D+01
-0.376D+00 -0.315D+01

-0.651D-05  0.959D-05      0.129D+00  0.186D+00       0.129D+00  0.186D+00
 0.129D+00  0.186D+00      0.129D+00  0.186D+00      -0.406D-01  0.394D+00
-0.663D-05 -0.662D-05      0.382D-06 -0.665D-05       0.406D-01 -0.394D+00
 0.632D-02  0.843D-01      0.632D-02  0.843D-01      -0.632D-02 -0.843D-01
-0.631D-02 -0.843D-01
A=  0.27026D+00  0.31915D+01

 0.179D-06 -0.194D-05     -0.112D+00  0.112D+00      -0.140D+00  0.354D+00
-0.112D+00  0.112D+00     -0.140D+00  0.354D+00      -0.346D-01 -0.528D+00
 0.105D+00  0.220D+00     -0.105D+00 -0.220D+00       0.346D-01  0.528D+00
 0.348D+00 -0.163D+01     -0.794D+00 -0.393D+01      -0.348D+00  0.163D+01
 0.794D+00  0.393D+01

-0.265D-06 -0.830D-07     -0.323D-01 -0.115D+00      -0.181D+00 -0.222D+00
-0.323D-01 -0.115D+00     -0.181D+00 -0.222D+00       0.491D-01 -0.345D+00
 0.381D-01  0.156D+00     -0.381D-01 -0.156D+00      -0.491D-01  0.345D+00
 0.112D-01 -0.430D-01     -0.165D-01 -0.106D+00      -0.112D-01  0.430D-01
 0.165D-01  0.106D+00
A= -0.20936D+01  0.82883D+00

-0.146D+00 -0.218D+00      0.393D+00  0.114D+00       0.393D+00  0.114D+00
-0.393D+00 -0.114D+00     -0.393D+00 -0.114D+00       0.347D+00  0.705D+00
-0.347D+00 -0.705D+00     -0.347D+00 -0.705D+00       0.347D+00  0.705D+00
 0.540D-01  0.171D+00      0.540D-01  0.171D+00       0.540D-01  0.171D+00
 0.540D-01  0.171D+00

 0.358D+00 -0.936D-01     -0.265D-01  0.289D+00      -0.265D-01  0.289D+00
 0.265D-01 -0.289D+00      0.265D-01 -0.289D+00      -0.231D+00  0.154D+00
 0.231D+00 -0.154D+00      0.231D+00 -0.154D+00      -0.231D+00  0.154D+00
 0.672D-01 -0.342D-01      0.672D-01 -0.342D-01       0.672D-01 -0.342D-01
 0.672D-01 -0.342D-01
A=  0.23253D+01  0.73400D+00

 0.315D+00  0.107D+01     -0.487D+00  0.846D+00      -0.487D+00  0.846D+00
 0.487D+00 -0.846D+00      0.487D+00 -0.846D+00      -0.636D-01 -0.555D+00
 0.636D-01  0.555D+00      0.636D-01  0.555D+00      -0.636D-01 -0.555D+00
 0.198D+00 -0.442D-01      0.198D+00 -0.442D-01       0.198D+00 -0.442D-01
 0.198D+00 -0.442D-01

-0.221D+00  0.539D-01     -0.251D+00 -0.350D+00      -0.251D+00 -0.350D+00
 0.251D+00  0.350D+00      0.251D+00  0.350D+00       0.540D+00 -0.336D+00
-0.540D+00  0.336D+00     -0.540D+00  0.336D+00       0.540D+00 -0.336D+00
-0.835D-01  0.964D-01     -0.835D-01  0.964D-01      -0.835D-01  0.964D-01
-0.835D-01  0.964D-01
A=  0.10565D+01  0.47540D+00

 0.111D+01 -0.113D+00     -0.310D+00  0.200D+00      -0.310D+00  0.200D+00
 0.310D+00 -0.200D+00      0.310D+00 -0.200D+00       0.307D+00 -0.261D+00
-0.307D+00  0.261D+00     -0.307D+00  0.261D+00       0.307D+00 -0.261D+00
-0.244D-01  0.132D-01     -0.244D-01  0.132D-01      -0.244D-01  0.132D-01
-0.244D-01  0.132D-01

 0.199D-01 -0.748D-01      0.461D-01 -0.117D+00       0.461D-01 -0.117D+00
-0.461D-01  0.117D+00     -0.461D-01  0.117D+00       0.271D+00 -0.215D+00
-0.271D+00  0.215D+00     -0.271D+00  0.215D+00       0.271D+00 -0.215D+00
-0.571D-01  0.504D-01     -0.571D-01  0.504D-01      -0.571D-01  0.504D-01
-0.571D-01  0.504D-01
A= -0.14894D+01  0.45016D+00
```

```
BP
     0.343D-07  0.328D-07     -0.134D-06 -0.709D-07     -0.128D-06 -0.966D-
     0.630D-07  0.332D-07      0.357D-07  0.123D-07      0.577D+00 -0.120D+
     0.577D+00 -0.120D+00      0.577D+00 -0.120D+00      0.577D+00 -0.120D+
     0.396D-07 -0.108D-06      0.601D-07 -0.828D-07     -0.129D-07 -0.349D-
    -0.201D-06  0.714D-06
BM
    -0.191D-06  0.262D-07     -0.312D-07 -0.687D-07     -0.157D-07 -0.607D-
     0.433D-08  0.148D-07      0.256D-07  0.179D-07     -0.135D+00  0.518D-
    -0.135D+00  0.518D-01     -0.135D+00  0.518D-01     -0.135D+00  0.518D-
    -0.226D-07  0.227D-07     -0.204D-07  0.147D-07     -0.750D-08  0.262D-
    -0.135D-07  0.376D-07
K.A= -0.31240D+01  0.40228D+00
BP
     0.158D-06 -0.122D-06     -0.423D+00 -0.324D+00     -0.423D+00 -0.324D+
    -0.423D+00 -0.324D+00     -0.423D+00 -0.324D+00      0.217D+00  0.478D-
     0.107D-07 -0.395D-06      0.845D-07 -0.186D-06     -0.217D+00 -0.478D-
    -0.951D-01  0.228D-01     -0.951D-01  0.228D-01      0.951D-01 -0.228D-
     0.951D-01 -0.228D-01
BM
     0.391D-07  0.457D-07      0.116D+00 -0.116D+00      0.116D+00 -0.116D+
     0.116D+00 -0.116D+00      0.116D+00 -0.116D+00      0.778D-01 -0.648D-
     0.833D-07  0.993D-07      0.597D-07  0.616D-07     -0.778D-01  0.648D-
     0.234D-01 -0.343D-01      0.234D-01 -0.343D-01     -0.234D-01  0.343D-
    -0.234D-01  0.343D-01
K.A= -0.31240D+01  0.40228D+00
BP
    -0.405D-07  0.596D-07      0.286D+00  0.106D+00      0.472D+00  0.247D+
     0.286D+00  0.106D+00      0.472D+00  0.247D+00     -0.174D+00  0.191D-
     0.473D-01  0.103D-01     -0.473D-01 -0.103D-01      0.174D+00 -0.192D-
     0.483D-01 -0.285D-01      0.898D-01 -0.386D-01     -0.483D-01  0.285D-
    -0.898D-01  0.386D-01
BM
    -0.237D-07  0.491D-08     -0.442D-01  0.830D-01     -0.948D-01  0.134D+
    -0.442D-01  0.830D-01     -0.948D-01  0.134D+00     -0.585D-01  0.181D-
     0.170D-01 -0.147D-02     -0.170D-01  0.147D-02      0.585D-01 -0.181D-
    -0.701D-02  0.227D-01     -0.172D-01  0.378D-01      0.701D-02 -0.227D-
     0.172D-01 -0.378D-01
K.A= -0.19162D+01  0.40154D+00
BP
    -0.486D-06  0.152D-06      0.811D-01  0.311D-01      0.811D-01  0.311D-
     0.811D-01  0.311D-01      0.811D-01  0.311D-01      0.458D+00  0.391D+
    -0.143D-05 -0.199D-06     -0.149D-05 -0.140D-06     -0.458D+00 -0.391D+
    -0.325D-01  0.297D-01     -0.325D-01  0.297D-01      0.325D-01 -0.297D-
     0.325D-01 -0.297D-01
BM
     0.513D-06 -0.629D-06      0.187D-01  0.232D-01      0.187D-01  0.232D-
     0.187D-01  0.232D-01      0.187D-01  0.232D-01      0.555D-02  0.269D-
    -0.182D-07 -0.325D-06     -0.253D-07 -0.284D-06     -0.555D-02 -0.269D-
     0.276D-01 -0.827D-02      0.276D-01 -0.827D-02     -0.276D-01  0.827D-
    -0.276D-01  0.827D-02
K.A= -0.19162D+01  0.40154D+00
BP
     0.327D-07 -0.177D-07      0.178D-01 -0.858D-01      0.191D-01 -0.839D-
     0.178D-01 -0.858D-01      0.191D-01 -0.839D-01      0.317D+00 -0.512D+
    -0.207D-02 -0.760D-02      0.207D-02  0.760D-02     -0.317D+00  0.512D+
     0.349D-01  0.275D-01      0.337D-01  0.277D-01     -0.349D-01 -0.275D-
    -0.337D-01 -0.277D-01
BM
    -0.459D-07  0.468D-07      0.200D-01 -0.224D-01      0.201D-01 -0.216D-
     0.200D-01 -0.224D-01      0.201D-01 -0.216D-01      0.257D-01 -0.956D-
```

```
   0.138D-03 -0.332D-03     -0.138D-03  0.332D-03     -0.257D-01  0.956D-02
  -0.127D-01 -0.262D-01     -0.120D-01 -0.260D-01      0.127D-01  0.262D-01
   0.120D-01  0.260D-01
.A= -0.29013D+01  0.27577D+00
P
   0.283D-07  0.252D-07      0.263D+00  0.111D+00     -0.263D+00 -0.111D+00
  -0.263D+00 -0.111D+00      0.263D+00  0.111D+00     -0.109D-06 -0.665D-08
  -0.266D-07 -0.803D-07      0.868D-07  0.137D-07      0.483D-07  0.730D-07
  -0.118D-01 -0.243D-01      0.118D-01  0.243D-01     -0.118D-01 -0.243D-01
   0.118D-01  0.243D-01
M
   0.882D-08  0.269D-07      0.387D-01 -0.229D-01     -0.387D-01  0.229D-01
  -0.387D-01  0.229D-01      0.387D-01 -0.229D-01     -0.412D-07 -0.439D-07
  -0.808D-08  0.286D-07      0.293D-07  0.461D-07      0.210D-07 -0.327D-07
  -0.153D-01  0.106D-02      0.153D-01 -0.106D-02     -0.153D-01  0.106D-02
   0.153D-01 -0.106D-02
LOCH WAVE AMPLITUDES
  -0.31221D-11  0.30087D-12     -0.24297D-02  0.32418D-01
   0.17911D-11 -0.26569D-10     -0.10177D-10 -0.49246D-11
  -0.16214D+00  0.14251D+00     -0.21609D+00 -0.28801D+00
   0.61922D+00  0.32828D+00      0.0           0.0
   0.0           0.0             0.0           0.0
   0.0           0.0             0.0           0.0
   0.0           0.0
EFLECTED AMPLITUDES
   0.45453D-01  0.58377D-01     -0.19974D-02  0.28270D-01
  -0.19974D-02  0.28270D-01      0.19974D-02 -0.28270D-01
   0.19974D-02 -0.28270D-01      0.58307D-01 -0.17227D+00
  -0.58307D-01  0.17227D+00     -0.58307D-01  0.17227D+00
   0.58307D-01 -0.17227D+00     -0.14665D-01  0.29784D-01
  -0.14665D-01  0.29784D-01     -0.14665D-01  0.29784D-01
  -0.14665D-01  0.29784D-01
=     2.6000 THETA=     0.0     FI=     0.0
   BEAM            INTENSITY
  0.0      0.0     0.54738D-02
  1.0      0.0     0.65905D-03
  0.0      1.0     0.65905D-03
 -1.0      0.0     0.65905D-03
  0.0     -1.0     0.65905D-03
  1.0      1.0     0.19473D-01
 -1.0      1.0     0.19473D-01
  1.0     -1.0     0.19473D-01
 -1.0     -1.0     0.19473D-01
  2.0      0.0     0.0
  0.0      2.0     0.0
 -2.0      0.0     0.0
  0.0     -2.0     0.0
```

```
SPA=      6.81735



SURF ASM1 ASM2
   1.0     0.0     0.0
   0.0     1.0     0.0
   0.0     0.0     1.0
SURFACE VECS
    0.0          0.50000      0.50000
    0.0         -0.50000      0.50000
LAYER DISPLACEMENT
    3.40867     -3.40867      0.0
IDEG=     4 N1=    12 N=    13
BEAMS INCLUDED

  PQF1   PQF2
   0.0    0.0
   1.0    0.0
   0.0    1.0
  -1.0    0.0
   0.0   -1.0
   1.0    1.0
  -1.0    1.0
   1.0   -1.0
  -1.0   -1.0
   2.0    0.0
   0.0    2.0
  -2.0    0.0
   0.0   -2.0
VPI=     -0.14700
E1=      2.00000
PHS
   -1.25000     -0.35000     -0.20000      0.15000      0.0
E1=      3.00000
PHS
   -1.70000     -0.70000     -0.35000      0.37500      0.07000

EIGENVALUES
   -0.69760D+00  0.99486D+00     0.10007D+01  0.11249D+01
    0.12934D+01 -0.91220D+00     0.16298D+01 -0.45664D+00
    0.16763D+01  0.53493D+00    -0.20192D+01 -0.26600D+00
    0.20469D+01 -0.21866D+00    -0.12011D+02  0.10957D+02
    0.28533D+02 -0.17501D+02     0.42488D+02 -0.24334D+02
    0.36369D+03  0.37453D+02     0.12004D+04 -0.16887D+04
    0.80579D+04 -0.22198D+04
KZ.AZ= -0.18850D+01  0.50172D+00
BP
    0.228D+00  0.709D+00    -0.104D+00 -0.117D+00    -0.350D+00 -0.703D+0
    0.138D+01 -0.405D+01     0.910D-01  0.908D+00    -0.538D-01  0.111D+0
    0.908D+00  0.137D+01     0.406D+00 -0.199D+00    -0.174D+00 -0.605D+0
   -0.315D-01  0.315D-01    -0.934D-02  0.942D-01    -0.940D+00 -0.110D+0
   -0.531D+00  0.130D+00
BM
    0.106D-01 -0.325D+00    -0.660D-01 -0.525D+00     0.897D+00  0.245D+0
    0.556D-01  0.517D-01     0.253D+00  0.193D+00    -0.758D-01  0.497D-0
   -0.414D+00  0.365D+00     0.850D-01 -0.225D+00    -0.144D+00  0.328D+0
   -0.883D-02  0.179D-01    -0.410D-01  0.287D-01     0.433D+00 -0.401D+0
   -0.953D-01  0.271D+00
KZ.AZ= -0.11319D+01  0.58707D+00
```

```
P
  0.506D+00  0.435D+00     -0.441D-01 -0.326D+00     -0.507D+00 -0.393D-01
  0.147D+01 -0.516D+00      0.119D+00  0.312D+00     -0.363D-01  0.342D-01
  0.236D+01  0.156D+01      0.226D+00 -0.322D+00     -0.323D+00 -0.324D+00
 -0.137D-01  0.245D-01     -0.141D-01  0.168D-01     -0.191D+01 -0.898D+00
 -0.387D+00  0.345D+00
M
 -0.635D-01 -0.116D-01     -0.110D+00 -0.323D+00      0.747D+00  0.399D+00
 -0.836D-01 -0.156D+00      0.128D+00  0.399D-01     -0.465D-01  0.175D-01
 -0.220D+00  0.483D+00      0.197D-01 -0.999D-01     -0.957D-01  0.132D+00
 -0.562D-02  0.696D-02     -0.259D-01 -0.285D-02      0.198D+00 -0.514D+00
 -0.114D-01  0.743D-01
Z.AZ=  0.96927D+00  0.52828D+00
P
 -0.481D+00  0.526D+00     -0.440D-01 -0.334D+00     -0.215D+01  0.509D+00
  0.127D+00 -0.245D-01      0.604D-01  0.737D-01     -0.960D-02  0.478D-01
 -0.171D+00  0.194D+00     -0.268D-01  0.178D-01      0.170D+00  0.263D+00
 -0.335D-02  0.412D-02      0.848D-02  0.329D-01     -0.422D-01 -0.296D-01
  0.175D+00  0.856D-01
M
 -0.114D+00 -0.437D-01      0.157D+00 -0.235D+00     -0.813D-01  0.260D+00
 -0.130D+00 -0.156D+00     -0.845D-01 -0.156D-01     -0.187D-01 -0.335D-01
 -0.325D+00 -0.351D-01      0.218D-01 -0.202D-01     -0.191D-01 -0.309D-01
 -0.448D-02 -0.438D-02     -0.290D-02 -0.223D-01      0.934D-01 -0.270D+00
  0.501D-01  0.268D-01
Z.AZ=  0.69563D+00  0.34913D+00
P
 -0.446D+00 -0.709D+00     -0.177D+00  0.239D+00      0.623D+00 -0.124D+00
  0.992D-02  0.263D-01      0.452D-01 -0.149D-01      0.487D-01  0.420D-01
  0.108D+00 -0.507D-01     -0.710D-01 -0.301D-03     -0.107D+01 -0.120D+01
  0.140D-01  0.465D-02      0.145D-01  0.340D-01      0.152D+00  0.142D+00
 -0.884D-01 -0.124D+00
M
  0.132D+00  0.342D-01     -0.143D+00  0.229D-01     -0.430D-01 -0.727D-01
  0.410D-01  0.910D-01     -0.141D-03 -0.409D-01      0.133D-01  0.328D-01
  0.273D+00 -0.189D+00      0.134D-01 -0.157D-01      0.113D+00  0.675D-01
  0.208D-02  0.571D-02      0.140D-01  0.217D-01      0.122D+00 -0.578D-01
 -0.582D-01 -0.434D-01
Z.AZ= -0.68813D+00  0.40959D+00
P
  0.359D-01  0.833D-01     -0.188D+00  0.176D+00     -0.504D-01 -0.205D+00
  0.243D-01 -0.719D-01      0.163D-01 -0.157D-01      0.711D-01  0.162D-01
 -0.685D+00  0.976D+00     -0.128D-01 -0.990D-01     -0.606D-01  0.136D+00
  0.977D-02  0.123D-01      0.347D-01 -0.280D-01     -0.104D+01  0.118D+01
  0.236D-02 -0.417D-01
M
 -0.227D+00 -0.167D+00      0.977D-01 -0.761D-01     -0.107D+00 -0.163D+00
 -0.281D-03 -0.291D-02      0.192D-01 -0.310D-01     -0.980D-02 -0.281D-01
  0.928D-01 -0.566D-01      0.388D-01  0.695D-03     -0.152D+00 -0.151D+00
 -0.596D-02 -0.265D-02      0.126D-01 -0.122D-01      0.596D-01 -0.109D+00
  0.472D-01  0.228D-01
Z.AZ= -0.27944D+01  0.38153D+00
P
  0.111D+00 -0.663D-02      0.220D-01  0.188D+00     -0.892D-01 -0 231D-02
 -0.130D+00 -0.318D-01     -0.109D+01 -0.415D+00      0.138D-01  0.[illegible]61D-01
  0.363D-01  0.130D-01     -0.520D-01  0.478D-01     -0.120D+00  0.104D-01
  0.635D-02 -0.188D-02     -0.102D-01  0.180D-01     -0.337D-01 -0.682D-02
  0.619D-01 -0.473D-01
M
  0.193D-01  0.335D-01     -0.136D+00 -0.117D-01     -0.692D-01 -0.894D-01
  0.630D-01 -0.206D-01     -0.185D-01  0.110D+00      0.171D-01  0.639D-02
```

```
    -0.418D-01  0.207D-01      0.530D-02 -0.499D-01     -0.179D-01 -0.236D-
     0.906D-03  0.445D-02      0.781D-02 -0.805D-02      0.652D-01 -0.115D-
     0.375D-02  0.521D-01
KZ.AZ=  0.29332D+00  0.36965D+00
BP
     0.124D+00 -0.115D+01     -0.292D+00 -0.169D+00      0.139D+00  0.522D+0
    -0.372D-01 -0.190D+00     -0.358D-01 -0.167D+00     -0.124D-01  0.258D-
     0.277D-01  0.399D+00      0.124D+00 -0.737D-01     -0.300D-01  0.598D+0
    -0.680D-02  0.909D-02     -0.878D-02 -0.955D-02      0.179D-01 -0.270D+0
    -0.114D+00  0.139D-01
BM
    -0.221D-01  0.117D+00     -0.279D+00  0.295D-02      0.993D-01 -0.243D-0
     0.558D-03 -0.696D-01     -0.350D-01  0.137D+00      0.110D-01  0.284D-0
     0.208D-01  0.100D+00     -0.502D-01 -0.869D-01      0.396D-01 -0.372D-
     0.480D-02  0.849D-02      0.439D-02  0.266D-03      0.804D-01 -0.699D-0
     0.547D-01  0.556D-01
KZ.AZ= -0.23983D+01  0.27883D+01
BP
     0.472D+01  0.215D+01     -0.602D+02 -0.897D+01     -0.706D+01  0.645D+0
     0.212D+01 -0.470D+00      0.418D+01 -0.216D+01     -0.147D-01 -0.201D+0
    -0.338D+00  0.135D+01      0.295D+00  0.467D+01      0.950D+00 -0.124D+0
    -0.542D-01  0.265D+00      0.716D+00 -0.349D+00     -0.160D+00  0.397D+0
    -0.134D+01 -0.355D+00
BM
    -0.379D+01  0.134D+01     -0.339D-02 -0.102D+01      0.460D+01  0.111D+0
    -0.203D+01 -0.130D+01     -0.247D+01 -0.272D+01      0.137D-01 -0.373D+0
     0.688D+00  0.241D+01      0.263D+00  0.142D+00     -0.137D+00 -0.184D+0
     0.305D-01 -0.251D+00     -0.302D-01 -0.259D-01     -0.537D+00  0.360D+0
    -0.356D+00  0.579D+00
KZ.AZ=  0.55098D+00  0.35103D+01
BP
    -0.446D+01 -0.328D+01      0.385D+01 -0.806D+01     -0.144D+02  0.478D+0
    -0.305D+01 -0.377D+01     -0.897D+00  0.134D+02     -0.593D+01  0.131D+0
     0.302D+01  0.199D+02      0.234D+02 -0.399D+02      0.837D+01  0.425D+0
     0.614D+00  0.240D+01     -0.210D+02 -0.820D+01     -0.649D+01 -0.126D+0
     0.732D+02  0.186D+03
BM
    -0.680D+00 -0.481D+01      0.337D+01 -0.185D+01      0.498D+01  0.603D+0
     0.592D+01 -0.469D+01      0.564D+01  0.524D+01     -0.444D+00 -0.981D-0
    -0.526D+01  0.546D+01      0.109D+01 -0.180D+01     -0.317D+01  0.650D+0
    -0.890D-01  0.151D+00     -0.340D+00 -0.687D-01      0.940D+01 -0.524D+0
     0.808D-01  0.521D+01
KZ.AZ=  0.52048D+00  0.38908D+01
BP
    -0.115D+02  0.126D+02      0.166D+02 -0.383D+02      0.889D+01  0.310D+0
    -0.802D+01  0.620D+01      0.134D+02  0.107D+02      0.604D+01 -0.128D+0
     0.196D+02 -0.756D+01     -0.324D+03 -0.155D+03      0.572D+01  0.336D+0
     0.117D+01 -0.108D+01      0.231D+02  0.183D+02     -0.242D+02 -0.649D+0
    -0.123D+02 -0.786D+02
BM
    -0.227D+01 -0.102D+02     -0.313D+01  0.368D+01      0.625D+01  0.708D+0
    -0.198D+00 -0.982D+01      0.123D+02 -0.393D+01      0.742D+00  0.268D+0
    -0.393D+01  0.112D+02     -0.659D+01 -0.488D+01      0.971D+01  0.241D+0
     0.113D+01  0.522D+00      0.891D-01  0.278D+00     -0.929D+01 -0.912D+0
     0.109D+01 -0.159D+01
KZ.AZ= -0.10262D+00  0.59016D+01
BP
     0.326D+03  0.257D+03     -0.778D+03 -0.558D+03     -0.409D+03  0.189D+0
     0.286D+03  0.559D+02      0.180D+03  0.190D+03      0.337D+04 -0.105D+0
    -0.808D+02  0.203D+03     -0.510D+03 -0.147D+03      0.377D+03  0.521D+0
    -0.585D+03 -0.308D+03     -0.131D+04 -0.128D+04      0.128D+03  0.306D+0
```

```
 -0.150D+04  0.280D+03

 -0.282D+03  0.760D+02     0.155D+03  0.884D+02      0.130D+03  0.348D+03
 -0.193D+03 -0.996D+02    -0.208D+03 -0.327D+02      0.398D+02 -0.129D+03
 -0.140D+03  0.229D+03     0.911D+01  0.127D+02     -0.165D+03 -0.122D+03
  0.940D+01 -0.285D+02     0.529D+01 -0.581D+02     -0.148D+03  0.114D+03
 -0.319D+02 -0.220D+02
.AZ=  0.95281D+00  0.76362D+01

  0.523D+03  0.784D+03    -0.365D+03 -0.292D+04      0.936D+03  0.740D+03
  0.107D+04 -0.393D+03    -0.244D+03  0.143D+04      0.119D+05 -0.293D+04
  0.107D+03 -0.638D+03    -0.143D+04 -0.621D+04      0.133D+04  0.343D+02
 -0.159D+05  0.115D+05    -0.273D+05 -0.555D+05      0.161D+04 -0.701D+03
 -0.117D+05 -0.759D+04

  0.275D+03  0.437D+03    -0.252D+03 -0.160D+02     -0.115D+03  0.156D+04
 -0.538D+03  0.268D+03     0.803D+02  0.695D+03     -0.358D+02 -0.287D+03
 -0.117D+04  0.299D+03     0.985D+02 -0.554D+02     -0.839D+03  0.182D+03
  0.217D+01  0.492D+01    -0.126D+03 -0.385D+03     -0.842D+03  0.488D+03
 -0.243D+03 -0.449D+03
..AZ=  0.26881D+00  0.90310D+01

  0.174D+04  0.670D+03     0.324D+04 -0.102D+05     -0.966D+02  0.434D+04
  0.422D+04 -0.643D+03    -0.216D+04  0.103D+04      0.358D+05 -0.122D+05
  0.261D+03  0.905D+03     0.131D+04 -0.117D+05      0.251D+04 -0.168D+03
 -0.269D+06 -0.741D+05     0.150D+05 -0.125D+06      0.690D+04  0.225D+04
 -0.797D+04 -0.230D+05

 -0.413D+03  0.301D+04     0.257D+04 -0.354D+04     -0.118D+04  0.240D+04
 -0.175D+04  0.982D+03    -0.235D+04  0.191D+04     -0.515D+03 -0.109D+04
 -0.182D+04 -0.135D+04     0.179D+04  0.717D+03     -0.291D+04 -0.367D+03
 -0.371D+03 -0.165D+03     0.107D+03 -0.936D+03     -0.944D+03  0.508D+03
 -0.205D+02 -0.100D+04
OCH WAVE AMPLITUDES
 -0.21097D-01 -0.29456D-01     0.62914D-01 -0.39681D-02
 -0.23526D+00  0.63680D-01    -0.21508D+00  0.16707D+00
 -0.23273D-01 -0.53779D-02     0.73539D-01 -0.59408D-01
 -0.18673D-01  0.55733D+00     0.19730D-02 -0.33400D-02
  0.26887D-03 -0.26830D-04     0.23317D-03 -0.18921D-04
 -0.90762D-06  0.41774D-05    -0.31665D-06 -0.12187D-06
  0.27716D-08 -0.21649D-07
FLECTED AMPLITUDES
 -0.79223D-01  0.26035D-01    -0.30772D-01 -0.11853D+00
  0.78335D-01 -0.19952D-01     0.46776D-01 -0.28721D-03
 -0.43917D-01 -0.12707D-01    -0.16616D-01  0.97479D-02
  0.54687D-02  0.13003D+00     0.33743D-01 -0.25099D-01
 -0.58329D-02  0.41460D-01    -0.44619D-02  0.27056D-02
 -0.49512D-02  0.51767D-02     0.79268D-02  0.10661D+00
 -0.12072D-01  0.33998D-01
=    2.6000 THETA=    0.5000 FI=    0.5000
  BEAM            INTENSITY
 0.0      0.0     0.69542D-02
 1.0      0.0     0.0
 0.0      1.0     0.40039D-02
-1.0      0.0     0.24548D-02
 0.0     -1.0     0.18414D-02
 1.0      1.0     0.0
-1.0      1.0     0.13488D-01
 1.0     -1.0     0.0
-1.0     -1.0     0.17832D-02
 2.0      0.0     0.0
```

```
  0.0     2.0     0.0
 -2.0     0.0     0.93469D-02
  0.0    -2.0     0.0
```

```
EIGENVALUES
     0.17948D+00  0.92795D+00     -0.73243D+00  0.77613D+00
    -0.73243D+00  0.77613D+00      0.10972D+01 -0.85931D+00
    -0.20169D+01  0.13292D+00     -0.13618D+01  0.16064D+01
    -0.17556D+01 -0.11683D+01     -0.21637D+01  0.14526D-01
    -0.21637D+01  0.14526D-01      0.23482D+02 -0.64833D+01
     0.23482D+02 -0.64833D+01     -0.30460D+02 -0.92049D+01
    -0.11217D+03  0.77661D+01
KZ.AZ= -0.14894D+01   0.45016D+00
BP
    -0.153D-09 -0.509D-11      0.165D-09  0.142D-08      -0.504D-09  0.743D-
    -0.664D-09  0.820D-09      0.120D-10  0.141D-08       0.579D+00 -0.116D+
     0.579D+00 -0.116D+01      0.579D+00 -0.116D+01       0.579D+00 -0.116D+
    -0.143D-08 -0.769D-09     -0.670D-09 -0.985D-09       0.772D-09  0.980D-
     0.141D-08  0.561D-09
BM
    -0.338D-09 -0.253D-09     -0.188D-09 -0.157D-10       0.101D-09 -0.170D-
     0.222D-09 -0.504D-09     -0.211D-09 -0.230D-09      -0.939D-01  0.302D+
    -0.939D-01  0.302D+00     -0.939D-01  0.302D+00      -0.939D-01  0.302D+
    -0.232D-09  0.600D-09     -0.307D-09  0.724D-09       0.340D-09 -0.699D-
     0.163D-09 -0.464D-09
KZ.AZ= -0.19162D+01   0.40154D+00
BP
     0.228D-09 -0.461D-09     -0.819D-01  0.132D+00      -0.870D-01  0.178D+
    -0.819D-01  0.132D+00     -0.870D-01  0.178D+00      -0.912D+00  0.819D+
     0.705D-01 -0.146D+00     -0.705D-01  0.146D+00       0.912D+00 -0.819D+
    -0.413D-01 -0.671D-01     -0.612D-01 -0.799D-01       0.413D-01  0.671D-
     0.612D-01  0.799D-01
BM
     0.321D-08  0.131D-08      0.470D-01 -0.251D-01       0.565D-01 -0.380D-
     0.470D-01 -0.251D-01      0.565D-01 -0.380D-01      -0.557D-01  0.400D-
     0.661D-02 -0.328D-02     -0.661D-02  0.328D-02       0.557D-01 -0.400D-
     0.526D-02  0.514D-01      0.133D-01  0.646D-01      -0.526D-02 -0.514D-
    -0.133D-01 -0.646D-01
KZ.AZ= -0.19162D+01   0.40154D+00
BP
     0.741D-10  0.694D-10      0.692D-01 -0.184D+00      -0.685D-01  0.138D+
     0.692D-01 -0.184D+00     -0.685D-01  0.138D+00       0.559D-01 -0.150D+
     0.824D+00 -0.894D+00     -0.824D+00  0.894D+00      -0.559D-01  0.150D+
     0.679D-01  0.732D-01     -0.471D-01 -0.623D-01      -0.679D-01 -0.732D-
     0.471D-01  0.623D-01
BM
    -0.448D-09  0.806D-10     -0.522D-01  0.428D-01       0.441D-01 -0.292D-
    -0.522D-01  0.428D-01      0.441D-01 -0.292D-01       0.622D-02 -0.386D-
     0.547D-01 -0.919D-02     -0.547D-01  0.919D-02      -0.622D-02  0.386D-
    -0.192D-01 -0.625D-01      0.100D-01  0.502D-01       0.192D-01  0.625D-
    -0.100D-01 -0.502D-01
KZ.AZ=  0.10565D+01   0.47540D+00
BP
     0.209D+01  0.481D+00     -0.688D+00  0.171D+00      -0.688D+00  0.171D+
     0.688D+00 -0.171D+00      0.688D+00 -0.171D+00       0.720D+00 -0.283D+
    -0.720D+00  0.283D+00     -0.720D+00  0.283D+00       0.720D+00 -0.283D+
    -0.525D-01  0.898D-02     -0.525D-01  0.898D-02      -0.525D-01  0.898D-
```

```
 -0.525D-01  0.898D-02
M
  0.826D-01 -0.124D+00    -0.156D+00  0.185D+00    -0.156D+00  0.185D+00
  0.156D+00 -0.185D+00     0.156D+00 -0.185D+00     0.626D+00 -0.223D+00
 -0.626D+00  0.223D+00    -0.626D+00  0.223D+00     0.626D+00 -0.223D+00
 -0.135D+00  0.562D-01    -0.135D+00  0.562D-01    -0.135D+00  0.562D-01
 -0.135D+00  0.562D-01
Z.AZ= -0.29013D+01  0.27577D+00
P
  0.598D-12  0.317D-11     0.206D-01 -0.384D+00    -0.206D-01  0.384D+00
 -0.206D-01  0.384D+00     0.206D-01 -0.384D+00    -0.176D-09 -0.876D-10
  0.335D-08 -0.715D-08    -0.375D-08  0.695D-08    -0.209D-09 -0.856D-10
 -0.254D-01  0.261D-01     0.254D-01 -0.261D-01    -0.254D-01  0.261D-01
  0.254D-01 -0.261D-01
M
  0.111D-10  0.685D-12     0.467D-01  0.386D-01    -0.467D-01 -0.386D-01
 -0.467D-01 -0.386D-01     0.467D-01  0.386D-01     0.124D-10  0.618D-10
  0.235D-08 -0.217D-08    -0.230D-08  0.230D-08     0.145D-10  0.762D-10
  0.831D-02  0.189D-01    -0.831D-02 -0.189D-01     0.831D-02  0.189D-01
 -0.831D-02 -0.189D-01
Z.AZ= -0.20936D+01  0.82883D+00
P
 -0.845D+00 -0.349D+00     0.136D+01 -0.428D+00     0.136D+01 -0.428D+00
 -0.136D+01  0.428D+00    -0.136D+01  0.428D+00     0.237D+01  0.137D+01
 -0.237D+01 -0.137D+01    -0.237D+01 -0.137D+01     0.237D+01  0.137D+01
  0.486D+00  0.390D+00     0.486D+00  0.390D+00     0.486D+00  0.390D+00
  0.486D+00  0.390D+00
M
  0.855D+00 -0.962D+00    -0.482D+00 -0.889D+00    -0.482D+00 -0.889D+00
  0.482D+00  0.889D+00     0.482D+00  0.889D+00    -0.373D+00  0.893D+00
  0.373D+00 -0.893D+00     0.373D+00 -0.893D+00    -0.373D+00  0.893D+00
  0.129D+00 -0.229D+00     0.129D+00 -0.229D+00     0.129D+00 -0.229D+00
  0.129D+00 -0.229D+00
Z.AZ=  0.23253D+01  0.73400D+00
P
 -0.244D+01  0.859D+00    -0.202D+01 -0.102D+01    -0.202D+01 -0.102D+01
  0.202D+01  0.102D+01     0.202D+01  0.102D+01     0.128D+01 -0.215D+00
 -0.128D+01  0.215D+00    -0.128D+01  0.215D+00     0.128D+01 -0.215D+00
  0.127D+00  0.454D+00     0.127D+00  0.454D+00     0.127D+00  0.454D+00
  0.127D+00  0.454D+00
M
 -0.152D+00 -0.505D+00    -0.779D+00  0.624D+00    -0.779D+00  0.624D+00
  0.779D+00 -0.624D+00     0.779D+00 -0.624D+00     0.843D+00  0.121D+01
 -0.843D+00 -0.121D+01    -0.843D+00 -0.121D+01     0.843D+00  0.121D+01
 -0.233D+00 -0.182D+00    -0.233D+00 -0.182D+00    -0.233D+00 -0.182D+00
 -0.233D+00 -0.182D+00
Z.AZ= -0.31240D+01  0.40228D+00
P
  0.363D-09  0.463D-09    -0.825D-01 -0.275D+00    -0.212D-01  0.189D+00
 -0.825D-01 -0.275D+00    -0.212D-01  0.189D+00     0.271D-01  0.714D-02
  0.523D-01  0.820D-01    -0.523D-01 -0.820D-01    -0.271D-01 -0.714D-02
 -0.487D-01 -0.200D-01     0.244D-01  0.248D-01     0.487D-01  0.200D-01
 -0.244D-01 -0.248D-01
M
 -0.130D-09  0.443D-09    -0.873D-01  0.139D-01     0.567D-01  0.142D-01
 -0.873D-01  0.139D-01     0.567D-01  0.142D-01     0.986D-02 -0.428D-03
  0.261D-01  0.221D-01    -0.261D-01 -0.221D-01    -0.986D-02  0.428D-03
  0.210D-01 -0.762D-02    -0.148D-01 -0.829D-03    -0.210D-01  0.762D-02
  0.148D-01  0.829D-03
Z.AZ= -0.31240D+01  0.40228D+00
P
```

```
    -0.238D-08  0.305D-08      0.233D+00 -0.279D+00       0.159D+00 -0.526D+0
     0.233D+00 -0.279D+00      0.159D+00 -0.526D+00      -0.320D-02  0.186D+0
    -0.358D-01 -0.401D-01      0.358D-01  0.401D-01       0.320D-02 -0.186D+0
    -0.127D-01 -0.655D-01     -0.566D-01 -0.834D-01       0.127D-01  0.655D-0
     0.566D-01  0.834D-01
BM
    -0.306D-08 -0.200D-09     -0.755D-01 -0.826D-01      -0.154D+00 -0.701D-0
    -0.755D-01 -0.826D-01     -0.154D+00 -0.701D-01       0.183D-01  0.629D-0
    -0.162D-01 -0.974D-02      0.162D-01  0.974D-02      -0.183D-01 -0.629D-0
     0.227D-01  0.169D-01      0.416D-01  0.101D-01      -0.227D-01 -0.169D-0
    -0.416D-01 -0.101D-01
KZ.AZ=  0.27026D+00   0.31915D+01
BP
     0.441D-07 -0.490D-07     -0.625D+01  0.483D+01      -0.349D+01  0.584D+0
    -0.625D+01  0.483D+01     -0.349D+01  0.584D+01       0.256D+01 -0.142D+0
     0.288D+01 -0.623D+00     -0.288D+01  0.623D+00      -0.256D+01  0.142D+0
     0.274D+02 -0.785D+02     -0.273D+01 -0.715D+02      -0.274D+02  0.785D+0
     0.273D+01  0.715D+02
BM
     0.270D-07  0.504D-07      0.881D+00  0.587D+01       0.255D+01  0.442D+0
     0.881D+00  0.587D+01      0.255D+01  0.442D+01       0.356D+01 -0.880D+0
     0.194D+01 -0.129D-01     -0.194D+01  0.129D-01      -0.356D+01  0.880D+0
     0.823D+00 -0.206D+01      0.113D-01 -0.191D+01      -0.823D+00  0.206D+0
    -0.113D-01  0.191D+01
KZ.AZ=  0.27026D+00   0.31915D+01
BP
    -0.620D-08 -0.103D-07     -0.478D+01  0.147D+01       0.579D+01  0.448D+0
    -0.478D+01  0.147D+01      0.579D+01  0.448D+00      -0.188D+01 -0.108D+0
     0.840D+01 -0.650D+01     -0.840D+01  0.650D+01       0.188D+01  0.108D+0
     0.338D+02 -0.404D+02     -0.537D+02  0.293D+02      -0.338D+02  0.404D+0
     0.537D+02 -0.293D+02
BM
     0.855D-08 -0.204D-09     -0.786D+00  0.368D+01       0.241D+01 -0.364D+0
    -0.786D+00  0.368D+01      0.241D+01 -0.364D+01      -0.106D+01 -0.952D+0
     0.629D+01 -0.305D+01     -0.629D+01  0.305D+01       0.106D+01  0.952D+0
     0.947D+00 -0.104D+01     -0.146D+01  0.717D+00      -0.947D+00  0.104D+0
     0.146D+01 -0.717D+00
KZ.AZ=  0.28476D+01   0.34593D+01
BP
     0.259D+01 -0.297D+02      0.248D+02  0.210D+01       0.248D+02  0.210D+0
    -0.248D+02 -0.210D+01     -0.248D+02 -0.210D+01      -0.852D+01  0.364D+0
     0.852D+01 -0.364D+02      0.852D+01 -0.364D+02      -0.852D+01  0.364D+0
    -0.150D+03  0.523D+02     -0.150D+03  0.523D+02      -0.150D+03  0.523D+0
    -0.150D+03  0.523D+02
BM
     0.272D+02  0.575D+01      0.915D-01 -0.226D+02       0.915D-01 -0.226D+0
    -0.915D-01  0.226D+02     -0.915D-01  0.226D+02      -0.259D+02  0.735D+0
     0.259D+02 -0.735D+01      0.259D+02 -0.735D+01      -0.259D+02  0.735D+0
     0.295D+01 -0.169D+01      0.295D+01 -0.169D+01       0.295D+01 -0.169D+0
     0.295D+01 -0.169D+01
KZ.AZ= -0.30725D+01   0.47223D+01
BP
    -0.127D-07 -0.839D-08      0.251D+02 -0.157D+02      -0.251D+02  0.157D+0
    -0.251D+02  0.157D+02      0.251D+02 -0.157D+02       0.114D-06 -0.810D-0
     0.777D-07  0.147D-07      0.151D-06 -0.206D-06       0.119D-06 -0.792D-0
    -0.424D+03 -0.242D+03      0.424D+03  0.242D+03      -0.424D+03 -0.242D+0
     0.424D+03  0.242D+03
BM
     0.174D-08 -0.589D-08      0.162D+01  0.221D+02      -0.162D+01 -0.221D+0
    -0.162D+01 -0.221D+02      0.162D+01  0.221D+02      -0.118D-07 -0.520D-0
    -0.222D-07  0.383D-07      0.395D-07 -0.672D-07      -0.415D-08  0.346D-0
```

```
 -0.758D+01 -0.452D+01      0.758D+01  0.452D+01     -0.758D+01 -0.452D+01
  0.758D+01  0.452D+01
LOCH WAVE AMPLITUDES
  0.24786D-10 -0.62057D-10      0.26796D-08 -0.10128D-08
 -0.31216D-09  0.27189D-09      0.36048D+00  0.57906D-01
 -0.53978D-11  0.14594D-11     -0.61420D-01  0.82144D-02
 -0.12895D+00  0.86561D-01     -0.35315D-09 -0.52125D-10
 -0.57532D-09 -0.16080D-09      0.45148D-11  0.26524D-11
 -0.10674D-11 -0.14873D-11     -0.50352D-03 -0.62547D-03
  0.61728D-14  0.34926D-13
EFLECTED AMPLITUDES
  0.45454D-01  0.58377D-01      0.19973D-02 -0.28271D-01
  0.19973D-02 -0.28271D-01     -0.19973D-02  0.28271D-01
 -0.19973D-02  0.28271D-01      0.58306D-01 -0.17227D+00
 -0.58306D-01  0.17227D+00     -0.58306D-01  0.17227D+00
  0.58306D-01 -0.17227D+00     -0.14665D-01  0.29784D-01
 -0.14665D-01  0.29784D-01     -0.14665D-01  0.29784D-01
 -0.14665D-01  0.29784D-01
=     2.6000 THETA=     0.0     FI=     0.0
  BEAM            INTENSITY
  0.0     0.0     0.54740D-02
  1.0     0.0     0.65909D-03
  0.0     1.0     0.65909D-03
 -1.0     0.0     0.65909D-03
  0.0    -1.0     0.65909D-03
  1.0     1.0     0.19473D-01
 -1.0     1.0     0.19473D-01
  1.0    -1.0     0.19473D-01
 -1.0    -1.0     0.19473D-01
  2.0     0.0     0.0
  0.0     2.0     0.0
 -2.0     0.0     0.0
  0.0    -2.0     0.0
```

```
SPA=      6.81735



SURF ASM1 ASM2
   1.0     0.0     0.0
   0.0     1.0     0.0
   0.0     0.0     1.0
SURFACE VECS
    0.0          0.50000      0.50000
    0.0         -0.50000      0.50000
LAYER DISPLACEMENT
    3.40867    -3.40867      0.0
IDEG=    4 N1=    12 N=    13
BEAMS INCLUDED

  PQF1   PQF2
   0.0    0.0
   1.0    0.0
   0.0    1.0
  -1.0    0.0
   0.0   -1.0
   1.0    1.0
  -1.0    1.0
   1.0   -1.0
  -1.0   -1.0
   2.0    0.0
   0.0    2.0
  -2.0    0.0
   0.0   -2.0
VPI=     -0.14700
E1=      2.00000
PHS
   -1.25000    -0.35000    -0.20000     0.15000     0.0
E1=      3.00000
PHS
   -1.70000    -0.70000    -0.35000     0.37500     0.07000

    8.00000  LAYERS

REFLECTED AMPLITUDES
   -0.07916     0.02593
    0.03079     0.11858
   -0.07832     0.01991
   -0.04669     0.00031
    0.04374     0.01270
   -0.01662     0.00975
    0.00559     0.13000
    0.03375    -0.02512
   -0.00590     0.04133
   -0.00446     0.00271
   -0.00495     0.00518
    0.00800     0.10655
   -0.01209     0.03401

E=     2.6000 THETA=    0.5000 FI=     0.5000
    BEAM            INTENSITY
   0.0     0.0     0.69393E-02
   1.0     0.0     0.0
   0.0     1.0     0.40010E-02
```

```
  -1.0     0.0     0.24455E-02
   0.0    -1.0     0.18276E-02
   1.0     1.0     0.0
  -1.0     1.0     0.13483E-01
   1.0    -1.0     0.0
  -1.0    -1.0     0.17732E-02
   2.0     0.0     0.0
   0.0     2.0     0.0
  -2.0     0.0     0.93366E-02
   0.0    -2.0     0.0



    8.00000  LAYERS

REFLECTED AMPLITUDES
    0.04556      0.05846
   -0.00196      0.02826
   -0.00196      0.02826
    0.00196     -0.02826
    0.00196     -0.02826
    0.05833     -0.17227
   -0.05833      0.17227
   -0.05833      0.17227
    0.05834     -0.17227
   -0.01466      0.02978
   -0.01466      0.02978
   -0.01466      0.02978
   -0.01466      0.02978

E=     2.6000 THETA=     0.0     FI=     0.0
    BEAM              INTENSITY
   0.0     0.0     0.54937E-02
   1.0     0.0     0.65853E-03
   0.0     1.0     0.65853E-03
  -1.0     0.0     0.65853E-03
   0.0    -1.0     0.65853E-03
   1.0     1.0     0.19475E-01
  -1.0     1.0     0.19476E-01
   1.0    -1.0     0.19476E-01
  -1.0    -1.0     0.19476E-01
   2.0     0.0     0.0
   0.0     2.0     0.0
  -2.0     0.0     0.0
   0.0    -2.0     0.0
```

2. Generation of CLM and ELM

CLM and ELM are required in subroutine XMAT, CLM being required when the real space summation is employed, as in our LEED programs, and ELM being required when a Kambe transformation of the summation has been made. They are defined by

$$C^{\ell}(\ell'm', \ell''m'') = 4\pi(-1)^{\frac{1}{2}(\ell-\ell'-\ell'')}(-1)^{m'+m''}\, Y_{\ell'-m'}(\tfrac{1}{2}\pi, 0) \times B^{\ell}(\ell'm', \ell''m''), \tag{A.55}$$

$$E^{\ell}(\ell'm', \ell''m'') = 4\pi(-1)^{\frac{1}{2}(\ell''-\ell'-\ell)}(-1)^{m''}\, B^{\ell}(\ell'm', \ell''m''), \tag{A.56}$$

$$B^{\ell}(\ell'm', \ell''m'') = \int Y_{\ell m}(\Omega)\, Y_{\ell'm'}(\Omega)\, Y_{\ell''-m''}(\Omega)\, d\Omega. \tag{A.57}$$

The quantity B is evaluated by means of the formula

$$B^{\ell}(\ell'm', \ell''m'') = 0, \qquad \text{unless} \qquad m + m' - m'' = 0,$$

$$= c(-1)^{\frac{1}{2}(|m|+|m'|+|m''|)}\, \tfrac{1}{2}\int_{-1}^{+1} P_{\ell}^{|m|}(\mu)\, P_{\ell'}^{|m'|}(\mu)\, P_{\ell''}^{|m''|}(\mu)\, d\mu, \tag{A.58}$$

otherwise. $c = 4\pi$

$$\left[\frac{(2\ell+1)(\ell-|m|)!}{4\pi(\ell+|m|)!}\cdot\frac{(2\ell'+1)(\ell'-|m'|)!}{4\pi(\ell'+|m'|)!}\cdot\frac{(2\ell''+1)(\ell''-|m''|)!}{4\pi(\ell''+|m''|)!}\right]^{\frac{1}{2}}, \tag{A.59}$$

and from Gaunt's formula (Slater, 1960)

$$\tfrac{1}{2}\int_{-1}^{+1} P_{\ell}^{|m|}(\mu)\, P_{\ell'}^{|m'|}(\mu)\, P_{\ell''}^{|m''|}(\mu)\, d\mu = 0 \qquad \text{unless}$$

$$|\ell' - \ell''| \leqslant \ell \leqslant \ell' + \ell''$$

$$\text{and} \qquad \ell' + \ell'' + \ell = \text{even}$$

$$= a \times b \qquad \text{otherwise.}$$

$$a = (-1)^{s-\ell_2-m_3}\frac{(\ell_2+m_2)!\,(\ell_3+m_3)!\,(2s-2\ell_3)!\,s!}{(\ell_2-m_2)!\,(s-\ell_1)!\,(s-\ell_2)!\,(s-\ell_3)!\,(2s+1)!}$$

$$b = \sum_t (-1)^t \frac{(\ell_1+m_1+t)!\,(\ell_2+\ell_3-m_1-t)!}{t!\,(\ell_1-m_1-t)!\,(\ell_2-\ell_3+m_1+t)!\,(\ell_3-m_3-t)!}$$

where $$2s = \ell_1 + \ell_2 + \ell_3 \tag{A.60}$$

The sum over t is from $\max[0, \ell_3 - \ell_2 - m_1]$ to $\min[\ell_2 + \ell_3 - m_1, \ell_1 - m_1, \ell_3 - m_3]$. The pairs $(\ell_1 m_1)$, $(\ell_2 m_2)$, $(\ell_3 m_3)$ are obtained by permuting the pairs $(\ell|m|)$, $(\ell'|m'|)$, $(\ell''|m''|)$, until the following relations hold

$$m_1 = m_2 + m_3$$

$$\ell_2 \geqslant \ell_3$$

The values of NLM, NN, and N for various values of LMAX are shown below

LMAX	NLM	NN	N
2	76	25	5
3	284	49	7
4	809	81	9
5	1925	121	11
6	4032	169	13
7	7680	225	15

```
COMMENT ROUTINE TO TABULATE THE CLEBSCH-GORDON TYPE COEFFICIENTS
C       CLM AND ELM, FOR USE WITH THE SUBROUTINE XMAT.
C       THE NON-ZERO VALUES ARE TABULATED FIRST FOR (L2+M2) AND
C       (L3+M3) ODD; AND THEN FOR (L2+M2) AND (L3+M3) EVEN -
C       THE SAME SCHEME AS THAT BY WHICH THEY ARE RETRIEVED IN
C       SUBROUTINE XMAT.
C       OUTPUT FROM THIS ROUTINE MAY NOT BE REQUIRED AND CAN BE
C       SUPPRESSED
      SUBROUTINE CELMG(CLM,ELM,NLM,YLM,FAC2,NN,FAC1,N,LMAX)
      DIMENSION CLM(NLM),ELM(NLM),YLM(NN),FAC2(NN),FAC1(N)
      PI=3.14159265
      L2MAX=LMAX+LMAX
C
COMMENT THE ARRAY YLM IS FIRST LOADED WITH SPHERICAL
C       HARMONICS, ARGUMENTS THETA=PI/2.0, FI=0.0
      LM=0
      CL=0.0
      A=1.0
      B=1.0
      ASG=1.0
      LL=L2MAX+1
COMMENT MULTIPLICATIVE FACTORS REQUIRED
      DO 2 L=1,LL
      FAC1(L)=ASG*SQRT((2.0*CL+1.0)*A/(4.0*PI*B*B))
      CM=-CL
      LN=L+L-1
      DO 1 M=1,LN
      LO=LM+M
      FAC2(LO)=SQRT((CL+1.0+CM)*(CL+1.0-CM)
     1/((2.0*CL+3.0)*(2.0*CL+1.0)))
1     CM=CM+1.0
      CL=CL+1.0
      A=A*2.0*CL*(2.0*CL-1.0)/4.0
      B=B*CL
      ASG=-ASG
2     LM=LM+LN
COMMENT FIRST ALL THE YLM FOR M=+-L AND M=+-(L-1) ARE
C       ARE CALCULATED BY EXPLICIT FORMULAE
      LM=1
      CL=1.0
      ASG=-1.0
      YLM(1)=FAC1(1)
      DO 3 L=1,L2MAX
      LN=LM+L+L+1
      YLM(LN)=FAC1(L+1)
      YLM(LM+1)=ASG*FAC1(L+1)
      YLM(LN-1)=0.0
      YLM(LM+2)=0.0
      CL=CL+1.0
      ASG=-ASG
3     LM=LN
COMMENT USING YLM AND YL(M-1) IN A RECURRENCE RELATION
C       YL(M+1) IS CALCULATED
      LM=1
      LL=L2MAX-1
      DO 5 L=1,LL
      LN=L+L-1
      LM2=LM+LN+4
      LM3=LM-LN
      DO 4 M=1,LN
      LO=LM2+M
```

```
      LP=LM3+M
      LQ=LM+M+1
      YLM(LO)=-(FAC2(LP)*YLM(LP))/FAC2(LQ)
4     CONTINUE
5     LM=LM+L+L+1
C
      WRITE(6,6)
6     FORMAT(1H1,2X,1HK,6X,17HL1  M1  L2  M2  L3  M3,10X,6HCLM(K),
     1 9X,6HELM(K),/)
      K=1
      II=0
7     LL=LMAX+II
      DO 11 IL2=1,LL
      L2=IL2-II
      M2=-L2+1-II
      DO 11 I2=1,IL2
      DO 10 IL3=1,LL
      L3=IL3-II
      M3=-L3+1-II
      DO 10 I3=1,IL3
      LA1=MAX0(IABS(L2-L3),IABS(M2-M3))
      LB1=L2+L3
      LA11=LA1+1
      LB11=LB1+1
      M1=M3-M2
      DO 9 L11=LA11,LB11,2
      L1=LA11+LB11-L11-1
      L=(L3-L2-L1)/2+M3
      M=L1*(L1+1)-M1+1
      ALM=((-1.0)**L)*4.0*PI*BLM(L1,M1,L2,M2,L3,-M3,LMAX)
      ELM(K)=ALM
      CLM(K)=YLM(M)*ALM
      WRITE(6,8) K,L1,M1,L2,M2,L3,M3,CLM(K),ELM(K)
8     FORMAT(1H ,I3,5X,6I3,5X,2E15.5)
9     K=K+1
10    M3=M3+2
11    M2=M2+2
      IF(II) 12,12,13
12    II=1
      GOTO 7
13    CONTINUE
      WRITE(7,14) ELM
14    FORMAT(200(5E14.5,/))
      RETURN
      END
C
C
COMMENT FUCTION BLM PROVIDES THE INTEGRAL OF THE PRODUCT
C        OF THREE SPHERICAL HARMONICS, EACH OF WHICH CAN BE
C        EXPRESSED AS A PREFACTOR TIMES A LEGENDRE FUNCTION.
C        THE THREE PREFACTORS ARE LUMPED TOGETHER AS FACTOR
C        'C'; AND THE INTEGRAL OF THE THREE LEGENDRE FUNCTIONS
C        FOLLOWS GAUNT'S SUMMATION SCHEME SET OUT BY SLATER:
C        ATOMIC STRUCTURE, VOL1, 309,310
      FUNCTION BLM(L1,M1,L2,M2,L3,M3,LMAX)
      DIMENSION FAC(20)
      PI=3.14159265
      NN=4*LMAX+1
      FAC(1)=1.0
      DO 1 I=1,NN
1     FAC(I+1)=FLOAT(I)*FAC(I)
```

```
      IF(M1+M2+M3) 8,21,8
21    IF(L1-LMAX-LMAX) 2,2,19
2     IF(L2-LMAX) 3,3,19
3     IF(L3-LMAX) 4,4,19
4     IF(L1-IABS(M1)) 19,5,5
5     IF(L2-IABS(M2)) 19,6,6
6     IF(L3-IABS(M3)) 19,7,7
7     IF(MOD(L1+L2+L3,2)) 8,9,8
8     BLM=0.0
      RETURN
9     NL1=L1
      NL2=L2
      NL3=L3
      NM1=IABS(M1)
      NM2=IABS(M2)
      NM3=IABS(M3)
      IC=(NM1+NM2+NM3)/2
      IF(MAX0(NM1,NM2,NM3)-NM1) 13,13,10
10    IF(MAX0(NM2,NM3)-NM2) 11,11,12
11    NL1=L2
      NL2=L1
      NM1=NM2
      NM2=IABS(M1)
      GOTO 13
12    NL1=L3
      NL3=L1
      NM1=NM3
      NM3=IABS(M1)
13    IF(NL2-NL3) 14,15,15
14    NTEMP=NL2
      NL2=NL3
      NL3=NTEMP
      NTEMP=NM2
      NM2=NM3
      NM3=NTEMP
15    IF(NL3-IABS(NL2-NL1)) 16,17,17
16    BLM=0.0
      RETURN
C
C      CALCULATION OF FACTOR 'A'
C
17    IS=(NL1+NL2+NL3)/2
      IA1=IS-NL2-NM3
      IA2=NL2+NM2
      IA3=NL2-NM2
      IA4=NL3+NM3
      IA5=NL1+NL2-NL3
      IA6=IS-NL1
      IA7=IS-NL2
      IA8=IS-NL3
      IA9=NL1+NL2+NL3+1
      AN=((-1.0)**IA1)*FAC(IA2+1)*FAC(IA4+1)*FAC(IA5+1)*FAC(IS+1)
      AD=FAC(IA3+1)*FAC(IA6+1)*FAC(IA7+1)*FAC(IA8+1)*FAC(IA9+1)
      A=AN/AD
C
C      CALCULATION OF SUM 'B'
C
      IB1=NL1+NM1
      IB2=NL2+NL3-NM1
      IB3=NL1-NM1
      IB4=NL2-NL3+NM1
```

```
      IB5=NL3-NM3
      IT1=MAX0(0,-IB4)+1
      IT2=MIN0(IB2,IB3,IB5)+1
      B=0.
      SIGN=(-1.0)**(IT1)
      IB1=IB1+IT1-2
      IB2=IB2-IT1+2
      IB3=IB3-IT1+2
      IB4=IB4+IT1-2
      IB5=IB5-IT1+2
      DO 18 IT=IT1,IT2
      SIGN=-SIGN
      IB1=IB1+1
      IB2=IB2-1
      IB3=IB3-1
      IB4=IB4+1
      IB5=IB5-1
      BN=SIGN*FAC(IB1+1)*FAC(IB2+1)
      BD=FAC(IT)*FAC(IB3+1)*FAC(IB4+1)*FAC(IB5+1)
18    B=B+(BN/BD)
C
C      CALCULATION OF FACTOR 'C'
C
      IC1=NL1-NM1
      IC2=NL1+NM1
      IC3=NL2-NM2
      IC4=NL2+NM2
      IC5=NL3-NM3
      IC6=NL3+NM3
      CN=FLOAT((2*NL1+1)*(2*NL2+1)*(2*NL3+1))*FAC(IC1+1)*
     1FAC(IC3+1)*FAC(IC5+1)
      CD=FAC(IC2+1)*FAC(IC4+1)*FAC(IC6+1)
      C=CN/(PI*CD)
      C=(SQRT(C))/2.
C
C
      BLM=((-1.0)**IC)*A*B*C
      RETURN
19    WRITE(6,20)L1,L2,M2,L3,M3
20    FORMAT(28H INVALID ARGUMENTS FOR BLM: ,5(I2,1H,))
      RETURN
      END
```

3. Kambe's method for the layer sums

If it is desired to use parts of the preceding programs to calculate conventional band-structures near the Fermi surface, or in a surface-state calculation then

$$\text{VPI} = 0,$$

and subroutine XMAT included with the LEED programs, does not work as it relies on absorption for convergence of the series it sums. There follows an alternative XMAT subroutine based on the work of Kambe described in Section IVD, which will work even when VPI is zero. For all LEED calculations VPI is such that the previous XMAT is the preferred one.

We have from formula (4.46) after some manipulation

$$X_{\ell m, \ell'' m''} = -\exp(i\delta_\ell)\sin(\delta_\ell) \times \left[i\delta_{\ell\ell''}\,\delta_{mm''} + \kappa^{-1} a_{\ell'' m'', \ell m}\right], \tag{A.61}$$

$$a_{\ell'' m'', \ell m} = \sum_{\ell'} E^{\ell'}(\ell m, \ell'' m'')\, D_{\ell' m'}, \tag{A.62}$$

$$E^{\ell'}(\ell m, \ell'' m'') = 4\pi(-1)^{\frac{1}{2}(\ell'' - \ell' - \ell)}(-1)^{m''} \int Y_{\ell m}(\Omega)\, Y_{\ell' m'}(\Omega)\, Y_{\ell'', -m''}(\Omega)\, d\Omega. \tag{A.63}$$

Kambe's expressions for $D_{\ell m}$ were given in equations (4.69)–(4.72)

$$D_{\ell m} = \left[D_{\ell m}^{(1)} + D_{\ell m}^{(2)} + D_{\ell m}^{(3)}\right],$$

$$D_{\ell m}^{(1)} = -(A\kappa 2^{\ell})^{-1} i^{1-m}\left[(2\ell + 1)(\ell + m)!\,(\ell - m)!\right]^{\frac{1}{2}}$$

$$\times \sum_{\mathbf{g}} \exp\left[-im\phi(\mathbf{k}_{0\|} + \mathbf{g})\right] \sum_{n=0}^{\frac{1}{2}(\ell - |m|)} \frac{(K_{gz}^{+}/\kappa)^{2n-1}\,(|\mathbf{k}_{0\|} + \mathbf{g}|/\kappa)^{\ell - 2n}}{n!\,[\frac{1}{2}(\ell - m - 2n)]!\,[\frac{1}{2}(\ell + m - 2n)]!}$$

$$\times \Gamma\left[\tfrac{1}{2}(1 - 2n), \exp(-i\pi)\,\alpha\,(K_{gz}^{+}/\kappa)^2\right], \tag{A.64}$$

$$D_{\ell m}^{(2)} = -\frac{\kappa}{4\pi}\frac{(-1)^{\ell}(-1)^{\frac{1}{2}(\ell + m)}}{2^{\ell}\,[\frac{1}{2}(\ell - m)]!\,[\frac{1}{2}(\ell + m)]!}\left[(2\ell + 1)(\ell - m)!\,(\ell + m)!\right]^{\frac{1}{2}}$$

$$\sum_j{}' \exp\left[-i\mathbf{k}_{0\|}\cdot\mathbf{R}_j - im\phi(\mathbf{R}_j)\right]\left[\tfrac{1}{2}\kappa|\mathbf{R}_j|\right]^{\ell}$$

$$\times \int_0^{\infty} u^{-\frac{3}{2} - \ell} \exp\left[u - (\kappa|\mathbf{R}_j|)/(4u)\right] du, \tag{A.65}$$

$$D_{\ell m}^{(3)} = -\delta_{\ell 0}\,\delta_{m0}\,\kappa(2\pi)^{-1}\left[2\int_0^{\alpha^{1/2}} \exp(t^2)\,dt - \alpha^{-\frac{1}{2}}\exp(\alpha)\right]. \tag{A.66}$$

where Γ is an incomplete gamma function, $\phi(\mathbf{R}_j)$ denotes the azimuthal angle of vector $\mathbf{R}_j$. α is a parameter which can be chosen to divide the summation equally between $D^{(1)}$ and $D^{(2)}$:

$$\alpha = \frac{\kappa^2 A}{4\pi} \tag{A.67}$$

This choice of α gave trouble at higher energies because of rounding errors, and above a certain energy in the program, α is given a constant value. The other symbols take their usual meaning in the text.

Dimensions of variables are a little complicated and a table of dimensions appropriate to various values of LMAX precedes the program itself

DIMENSIONS REQUIRED FOR ARRAYS USED IN XMAT

LMAX = 2	ARRAY	DIMENSION
	XODD	3
	XEVEN	6
	ELM	76
	DENOM	14
	DLM	15
	PREF	9
	FAC	9
	AGK	5
	XPM	5
	PHS	3
	AF	3
	GKN	3

LMAX = 3	ARRAY	DIMENSION
	XODD	6
	XEVEN	10
	ELM	284
	DENOM	30
	DLM	28
	PREF	16
	FAC	13
	AGK	7
	XPM	7
	PHS	4
	AF	4
	GKN	4

LMAX = 4	ARRAY	DIMENSION
	XODD	10
	XEVEN	15
	ELM	809
	DENOM	55
	DLM	45
	PREF	25
	FAC	17
	AGK	9
	XPM	9
	PHS	5
	AF	5
	GKN	5

LMAX = 5	ARRAY	DIMENSION
	XODD	15
	XEVEN	21
	ELM	1925
	DENOM	91
	DLM	66
	PREF	36
	FAC	21
	AGK	11
	XPM	11
	PHS	6
	AF	6
	GKN	6

LMAX = 6	ARRAY	DIMENSION
	XODD	21

```
                         XEVEN                     28
                           ELM                   4032
                         DENOM                    140
                           DLM                     91
                          PREF                     49
                           FAC                     25
                           AGK                     13
                           XPM                     13
                           PHS                      7
                            AF                      7
                           GKN                      7

LMAX = 7                 ARRAY              DIMENSION
                          XODD                     28
                         XEVEN                     36
                           ELM                   7680
                         DENOM                    204
                           DLM                    120
                          PREF                     64
                           FAC                     29
                           AGK                     15
                           XPM                     15
                           PHS                      8
                            AF                      8
                           GKN                      8
```

```
C
C
COMMENT XMAT CALCULATES THE MATRIX DESCRIBING MULTIPLE
C         SCATTERING WITHIN A LAYER, RETURNING IT AS: XODD,
C         CORRESPONDING TO ODD L+M, WITH LM=(10),(2-1),(21),
C         ....; AND XEVEN, CORRESPONDING TO EVEN L+M, WITH
C         LM=(00),(1-1),(11),(2-2),....
C         THE PROGRAM ASSUMES THAT THE LAYER IS A BRAVAIS LATTICE.
C         THE SUMMATION OVER THE LATTICE FOLLOWS THE EWALD SPLIT
C         METHOD SUGGESTED BY KAMBE. DIMENSIONS ARE FIXED FOR
C         LMAX=4. EMACH IS THE MACHINE ACCURACY: E.G. SET TO
C         1.0E-6 FOR AN IBM 370/165 SINGLE PRECISION COMPILER
      SUBROUTINE XMAT(XODD,XEVEN,LMAX,APQ,BPQ,EMACH)
      DIMENSION XODD(10,10),XEVEN(15,15),ELM(809)
      DIMENSION DENOM(55),DLM(45),PREF(25),FAC(17),AGK(9),
     1 XPM(9),PHS(5),AF(5),GKN(5)
      DIMENSION R(2),AKPT(2),AK(2),AR1(2),AR2(2),B1(2),B2(2)
      COMPLEX XODD,XEVEN,DLM,PREF,AGK,XPM,GKN,AF
      COMPLEX ALPHA,RTA,RTAI,KAPPA,KAPSQ,KANT,KNSQ,XPK,XPA,
     1 CF,CI,CP,CX,CZ,CZERO,CERF,Z,ZZ,W,WW,A,ACC,GPSQ,GP,BT
      COMPLEX AA,AB,U,U1,U2,GAM,GK,GKK,SD,ALM
      COMPLEX CMPLX,CSQRT,CEXP
      COMMON E,SPA,THETA,FI,VPI,ELM
      COMMON /X2/ PHS,AR1,AR2
C
COMMENT AK(1) AND AK(2) ARE THE X AND Y COMPONENTS OF THE
C         MOMENTUM PARALLEL TO THE SURFACE, MODULO A RECIPROCAL
C         LATTICE VECTOR
      PI=3.14159265
      RTPI=SQRT(PI)
      CZERO=CMPLX(0.0,0.0)
      CI=CMPLX(0.0,1.0)
      KAPSQ=CMPLX(2.0*E,-2.0*VPI+0.000001)
      KAPPA=CSQRT(KAPSQ)
COMMENT THE FOLLOWING THREE LINES SPECIFY AK IN TERMS OF E,
C         THETA AND FI.  FOR E<0, SOME ALTERNATIVE DEFINITION
C         OF AK MAY BE MORE APPROPRIATE
      AG=SQRT(2.0*E)*SIN(THETA)
      AK(1)=AG*COS(FI)+APQ
      AK(2)=AG*SIN(FI)+BPQ
      LIM=LMAX+1
      DO 1 L=1,LIM
1     AF(L)=SIN(PHS(L))*CEXP(CI*PHS(L))
C
COMMENT THE FACTORIAL FUNCTION IS TABULATED IN FAC. THE ARRAY
C         DLM WILL CONTAIN NON-ZERO, I.E. L+M EVEN, VALUES OF
C         KAMBE'S DLM'S = DLM1+DLM2+DLM3, WITH LM=(00),(1-1),(11),
C         (2-2),....
      L2MAX=LMAX+LMAX
      LL2=L2MAX+1
      FAC(1)=1.0
      II=L2MAX+L2MAX
      DO 2 I=1,II
2     FAC(I+1)=FLOAT(I)*FAC(I)
      NNDLM=L2MAX*(L2MAX+3)/2+1
      DO 3 I=1,NNDLM
3     DLM(I)=CZERO
C
COMMENT KAMBE'S FORMULA FOR THE SEPARATION CONSTANT, ALPHA, IS
C         USED, SUBJECT TO A RESTRICTION WHICH IS IMPOSED TO
C         CONTROL LATER ROUNDING ERRORS
```

```
      TV=ABS(AR1(1)*AR2(2)-AR1(2)*AR2(1))
      ALPHA=TV/(4.0*PI)*KAPSQ
      AL=CABS(ALPHA)
      IF(EXP(AL)*EMACH-5.0E-5)5,5,4
4     AL=ALOG(5.0E-5/EMACH)
      ALPHA=CMPLX(SIGN(1.0,E)*AL,0.0)
5     RTA=CSQRT(ALPHA)
C
COMMENT DLM1, THE SUM OVER RECIPROCAL LATTICE VECTORS, IS
C       CALCULATED FIRST. THE PREFACTOR 'P1' IS TABULATED
C       FOR EVEN VALUES OF L+|M|, THUS LM=(00),(11),(20),
C       (22),....(L2MAX,L2MAX). THE FACTORIAL FACTOR 'F1' IS
C       SIMULTANEOUSLY TABULATED IN DENOM, FOR ALL VALUES OF
C       N=0,(L-|M|)/2
      K=1
      KK=1
      AP1=-2.0/TV
      AP2=-1.0
      CF=CI/KAPPA
      DO 8 L=1,LL2
      AP1=AP1/2.0
      AP2=AP2+2.0
      CP=CF
      MM=1
      IF(MOD(L,2))7,6,7
6     MM=2
      CP=CI*CP
7     NN=(L-MM)/2+2
      DO 8 M=MM,L,2
      J1=L+M-1
      J2=L-M+1
      AP=AP1*SQRT(AP2*FAC(J1)*FAC(J2))
      PREF(KK)=AP*CP
      CP=-CP
      KK=KK+1
      NN=NN-1
      DO 8 I=1,NN
      I1=I
      I2=NN-I+1
      I3=NN+M-I
      DENOM(K)=1.0/(FAC(I1)*FAC(I2)*FAC(I3))
8     K=K+1
C
COMMENT THE RECIPROCAL LATTICE IS DEFINED BY B1,B2. THE SUMMATION
C       BEGINS WITH THE ORIGIN POINT OF THE LATTICE, AND CONTINUES
C       IN STEPS OF 8*N1 POINTS, EACH STEP INVOLVING THE PERIMETER
C       OF A PARALLELOGRAM OF LATTICE POINTS ABOUT THE ORIGIN, OF
C       SIDE 2*N1+1. EACH STEP BEGINS AT LABEL 9.
C       AKPT = THE CURRENT LATTICE VECTOR IN THE SUM
      RTV=2.0*PI/TV
      B1(1)=-AR1(2)*RTV
      B1(2)=AR1(1)*RTV
      B2(1)=-AR2(2)*RTV
      B2(2)=AR2(1)*RTV
      TEST1=1.0E6
      II=1
      N1=-1
9     N1=N1+1
      NA=N1+N1+II
      AN1=FLOAT(N1)
      AN2=-AN1-1.0
```

```
      DO 22 I1=1,NA
      AN2=AN2+1.0
      DO 21 I2=1,4
      AN=AN1
      AN1=-AN2
      AN2=AN
      AB1=AN1*B1(1)+AN2*B2(1)
      AB2=AN1*B1(2)+AN2*B2(2)
      AKPT(1)=AK(1)+AB1
      AKPT(2)=AK(2)+AB2
C
COMMENT FOR EVERY LATTICE VECTOR OF THE SUM, THREE SHORT
C        ARRAYS ARE INITIALISED AS BELOW, AND USED AS TABLES:
C        XPM(M) CONTAINS VALUES OF XPK**|M|
C        AGK(I) CONTAINS VALUES OF (AC/KAPPA)**I
C        GKN(N) CONTAINS VALUES OF (GP/KAPPA)**(2*N-1)*GAM<N,Z>
C        WHERE L=0,L2MAX; M=-L,L; N=0,(L-|M|)/2; I=L-2*N
C        GAM IS THE INCOMPLETE GAMMA FUNCTION, WHICH IS
C        CALCULATED BY RECURRENCE FROM THE VALUE FOR N=0,
C        WHICH IN TURN CAN BE EXPRESSED IN TERMS OF THE
C        COMPLEX ERROR FUNCTION CERF
C        AC = MOD(AKPT): NOTE SPECIAL ACTION IF AC=0
      ACSQ=AKPT(1)*AKPT(1)+AKPT(2)*AKPT(2)
      GPSQ=KAPSQ-ACSQ
      AC=SQRT(ACSQ)
      GP=CSQRT(GPSQ)
      XPK=CZERO
      GK=CZERO
      GKK=CMPLX(1.0,0.0)
      IF(AC-EMACH)11,11,10
10    XPK=CMPLX(AKPT(1)/AC,AKPT(2)/AC)
      GK=AC/KAPPA
      GKK=GPSQ/KAPSQ
11    XPM(1)=CMPLX(1.0,0.0)
      AGK(1)=CMPLX(1.0,0.0)
      DO 12 I=2,LL2
      XPM(I)=XPM(I-1)*XPK
12    AGK(I)=AGK(I-1)*GK
      CF=KAPPA/GP
      ZZ=-ALPHA*GKK
      CZ=CSQRT(-ZZ)
      Z=-CI*CZ
      CX=CEXP(-ZZ)
      GAM=RTPI*CERF(CZ,EMACH)
      GKN(1)=CF*CX*GAM
      BT=Z
      B=0.5
      LLL=L2MAX/2+1
      DO 13 I=2,LLL
      BT=BT/ZZ
      B=B-1.0
      GAM=(GAM-BT)/B
      CF=CF*GKK
13    GKN(I)=CF*CX*GAM
C
COMMENT THE CONTRIBUTION TO THE SUM DLM1 FOR A PARTICULAR
C        RECIPROCAL LATTICE VECTOR IS NOW ACCUMULATED INTO
C        THE ELEMENTS OF DLM. NOTE SPECIAL ACTION IF AC=0
      K=1
      KK=1
      DO 19 L=1,LL2
```

```
      MM=1
      IF(MOD(L,2)) 15,14,15
14    MM=2
15    N=(L*L+MM)/2
      NN=(L-MM)/2+2
      DO 19 M=MM,L,2
      ACC=CZERO
      NN=NN-1
      IL=L
      DO 16 I=1,NN
      ACC=ACC+DENOM(K)*AGK(IL)*GKN(I)
      IL=IL-2
16    K=K+1
      ACC=PREF(KK)*ACC
      IF(AC-1.0E-6) 17,17,165
165   DLM(N)=DLM(N)+ACC/XPM(M)
      IF(M-1) 17,18,17
17    NM=N-M+1
      DLM(NM)=DLM(NM)+ACC*XPM(M)
18    KK=KK+1
19    N=N+1
20    IF(II)21,21,22
21    CONTINUE
22    II=0
C
COMMENT AFTER EACH STEP OF THE SUMMATION A TEST ON THE
C        CONVERGENCE OF THE ELEMENTS OF DLM IS MADE
      TEST2=0.0
      DO 23 I=1,NNDLM
      DNORM=CABS(DLM(I))
23    TEST2=TEST2+DNORM*DNORM
      TEST=ABS((TEST2-TEST1)/TEST1)
      TEST1=TEST2
      IF(TEST-0.001) 27,27,24
24    IF(N1-10) 9,25,25
25    WRITE(6,26) N1
26    FORMAT(31H0**DLM1'S NOT CONVERGED BY N1 =,I2)
      GOTO 285
27    WRITE(6,28) N1
28    FORMAT(25H0DLM1'S CONVERGED BY N1 =,I2)
C
C
COMMENT DLM2, THE SUM OVER REAL SPACE LATTICE VECTORS, BEGINS
C        WITH THE ADJUSTMENT OF THE ARRAY PREF, TO CONTAIN
C        VALUES OF THE PREFACTOR 'P2' FOR LM=(00),(11),(20),
C        (22),....
285   KK=1
      AP1=TV/(4.0*PI)
      CF=KAPSQ/CI
      DO 31 L=1,LL2
      CP=CF
      MM=1
      IF(MOD(L,2)) 30,29,30
29    MM=2
      CP=-CI*CP
30    J1=(L-MM)/2+1
      J2=J1+MM-1
      IN=J1+L-2
      AP2=((-1.0)**IN)*AP1
      DO 31 M=MM,L,2
      AP=AP2/(FAC(J1)*FAC(J2))
```

```
      PREF(KK)=AP*CP*PREF(KK)
      J1=J1-1
      J2=J2+1
      AP2=-AP2
      CP=-CP
31    KK=KK+1
C
COMMENT THE SUMMATION PROCEEDS IN STEPS OF 8*N1 LATTICE
C       POINTS AS BEFORE, BUT THIS TIME EXCLUDING THE
C       ORIGIN POINT.
C       R = THE CURRENT LATTICE VECTOR IN THE SUM
C       AR= MOD(R)
      N1=0
32    N1=N1+1
      NA=N1+N1
      AN1=FLOAT(N1)
      AN2=-AN1-1.0
      DO 40 I1=1,NA
      AN2=AN2+1.0
      DO 40 I2=1,4
      AN=AN1
      AN1=-AN2
      AN2=AN
      R(1)=AN1*AR1(1)+AN2*AR2(1)
      R(2)=AN1*AR1(2)+AN2*AR2(2)
      AR=SQRT(R(1)*R(1)+R(2)*R(2))
      XPK=CMPLX(R(1)/AR,R(2)/AR)
      XPM(1)=CMPLX(1.0,0.0)
      DO 33 I=2,LL2
33    XPM(I)=XPM(I-1)*XPK
      AD=AK(1)*R(1)+AK(2)*R(2)
      SD=CEXP(-AD*CI)
C
COMMENT FOR EACH LATTICE VECTOR THE INTEGRAL 'U' IS OBTAINED
C       FROM THE RECURRENCE RELATION IN L SUGGESTED BY
C       KAMBE. U1 AND U2 ARE THE INITIAL TERMS OF THIS
C       RECURRENCE, FOR L=-1 AND L=0, AND THEY ARE EVALUATED
C       IN TERMS OF THE COMPLEX ERROR FUNCTION CERF
      KANT=0.5*AR*KAPPA
      KNSQ=KANT*KANT
      Z=CI*KANT/RTA
      ZZ=RTA-Z
      Z=RTA+Z
      WW=CERF(-ZZ,EMACH)
      W=CERF(Z,EMACH)
      AA=0.5*RTPI*(W-WW)/CI
      AB=0.5*RTPI*(W+WW)
      A=ALPHA-KNSQ/ALPHA
      XPA=CEXP(A)
      U1=AA*XPA
      U2=AB*XPA/KANT
C
COMMENT THE CONTRIBUTION TO DLM2 FROM A PARTICULAR LATTICE
C       VECTOR IS ACCUMULATED INTO THE ELEMENTS OF DLM.
C       THIS PROCEDURE INCLUDES THE TERM (KANT**L), AND
C       THE RECURRENCE FOR THE INTEGRAL 'U'
      KK=1
      AL=-0.5
      CP=RTA
      CF=CMPLX(1.0,0.0)
      DO 39 L=1,LL2
```

```
      MM=1
      IF(MOD(L,2))35,34,35
34    MM=2
35    N=(L*L+MM)/2
      DO 38 M=MM,L,2
      ACC=PREF(KK)*U2*CF*SD
      DLM(N)=DLM(N)+ACC/XPM(M)
      IF(M-1)36,37,36
36    NM=N-M+1
      DLM(NM)=DLM(NM)+ACC*XPM(M)
37    KK=KK+1
38    N=N+1
      AL=AL+1.0
      CP=CP/ALPHA
      U=(AL*U2-U1+CP*XPA)/KNSQ
      U1=U2
      U2=U
39    CF=KANT*CF
40    CONTINUE
C
COMMENT AFTER EACH STEP OF THE SUMMATION A TEST ON THE
C        CONVERGENCE OF THE ELEMENTS OF DLM IS MADE
      TEST2=0.0
      DO 41 I=1,NNDLM
      DNORM=CABS(DLM(I))
41    TEST2=TEST2+DNORM*DNORM
      TEST=ABS((TEST2-TEST1)/TEST1)
      TEST1=TEST2
      IF(TEST-0.001)45,45,42
42    IF(N1-10)32,43,43
43    WRITE(6,44)N1
44    FORMAT(31H0**DLM2'S NOT CONVERGED BY N1 =,I2)
      GOTO 465
45    WRITE(6,46)N1
46    FORMAT(25H0DLM2'S CONVERGED BY N1 =,I2)
C
COMMENT THE TERM DLM3 HAS A NON-ZERO CONTRIBUTION ONLY
C        WHEN L=M=0. IT IS EVALUATED HERE IN TERMS OF THE
C        COMPLEX ERROR FUNCTION CERF
465   XPA=CEXP(-ALPHA)
      RTAI=1.0/(RTPI*RTA)
      ACC=KAPPA*(CI*(XPA-CERF(RTA,EMACH))-RTAI)/XPA
      AP=-0.5/RTPI
      DLM(1)=DLM(1)+AP*ACC
C
COMMENT FINALLY THE ELEMENTS OF DLM ARE MULTIPLIED BY THE
C        FACTOR (-1.0)**((M+|M|)/2)
      DO 47 L=2,LL2,2
      N=L*L/2+1
      DO 47 M=2,L,2
      DLM(N)=-DLM(N)
47    N=N+1
C
C
COMMENT SUMMATION OVER THE CLEBSCH-GORDON TYPE COEFFICIENTS
C        ELM PROCEEDS, FIRST FOR XODD, AND THEN FOR XEVEN.
C        THIS GIVES KAMBE'S ELEMENTS  A(L2,M2;L3,M3)  WHICH,
C        COMBINED WITH THE PHASE SHIFT TERMS IN AF, GIVE THE
C        ELEMENTS  X(L3,M3;L2,M2)  OF XODD AND XEVEN
      K=1
      II=0
```

```
48      LL=LMAX+II
        I=1
        DO 56 IL2=1,LL
        L2=IL2-II
        M2=-L2+1-II
        DO 56 I2=1,IL2
        J=1
        DO 55 IL3=1,LL
        L3=IL3-II
        M3=-L3+1-II
        DO 55 I3=1,IL3
        ALM=CZERO
        LA1=MAX0(IABS(L2-L3),IABS(M2-M3))
        LB1=L2+L3
        N=(LB1*(LB1+2)+M2-M3+2)/2
        NN=2*LB1
        LB11=LB1+1
        LA11=LA1+1
        DO 49 L1=LA11,LB11,2
        ALM=ALM+ELM(K)*DLM(N)
        N=N-NN
        NN=NN-4
49      K=K+1
        ALM=ALM/KAPPA
        IF(I-J)51,50,51
50      ALM=ALM+CI
51      IF(II)52,52,53
52      XODD(J,I)=-AF(L3+1)*ALM
        GOTO 54
53      XEVEN(J,I)=-AF(L3+1)*ALM
54      M3=M3+2
55      J=J+1
        M2=M2+2
56      I=I+1
        IF(II)57,57,58
57      II=1
        GOTO 48
58      CONTINUE
C
        RETURN
        END
C
C
COMMENT CERF, GIVEN COMPLEX ARGUMENT Z, PROVIDES THE COMPLEX
C          ERROR FUNCTION:      W(Z)=EXP(-Z**2)*(1.0-ERF(-I*Z))
C          THE EVALUATION ALWAYS TAKES PLACE IN THE FIRST QUADRANT.
C          ONE OF THREE METHODS IS EMPLOYED DEPENDING ON THE SIZE
C          OF THE ARGUMENT: A POWER SERIES, A RECURRENCE BASED ON
C          CONTINUED FRACTIONS THEORY, OR AN ASYMPTOTIC SERIES.
C          EMACH IS THE MACHINE ACCURACY
        COMPLEX FUNCTION CERF(Z,EMACH)
        COMPLEX Z,ZZ,CI,SUM,ZZS,XZZS
        COMPLEX H1,H2,H3,U1,U2,U3,TERM1,TERM2
        COMPLEX CMPLX,CEXP,CONJG
        EPS=5.0*EMACH
        PI=3.14159265
        API=1.0/PI
        CI=CMPLX(0.0,1.0)
        ABSZ=CABS(Z)
        IF(ABSZ)2,1,2
1       CERF=CMPLX(1.0,0.0)
```

```
      RETURN
C
COMMENT THE ARGUMENT IS TRANSLATED TO THE FIRST QUADRANT FROM
C        THE NN'TH QUADRANT, BEFORE THE METHOD FOR THE FUNCTION
C        EVALUATION IS CHOSEN
2     X=REAL(Z)
      Y=AIMAG(Z)
      AX=ABS(X)
      AY=ABS(Y)
      ZZ=CMPLX(AX,AY)
      ZZS=ZZ*ZZ
      NN=1
      IF(X.NE.AX)NN=2
      IF(Y.NE.AY)NN=5-NN
      IF(ABSZ.GT.10.0)GOTO 13
      IF(AY.GE.1.0.OR.ABSZ.GE.4.0)GOTO 8
C
COMMENT POWER SERIES: SEE ABRAMOWITZ AND STEGUN'S HANDBOOK OF
C        MATHEMATICAL FUNCTIONS, P297
3     Q=1.0
      FACTN=-1.0
      FACTD=1.0
      TERM1=ZZ
      SUM=ZZ
4     DO 5 N=1,5
      FACTN=FACTN+2.0
      FACTD=FACTD+2.0
      FACT=FACTN/(Q*FACTD)
      TERM1=FACT*ZZS*TERM1
      SUM=SUM+TERM1
5     Q=Q+1.0
      ABTERM=CABS(TERM1)
      IF(ABTERM-EPS)7,6,6
6     IF(Q-100.0)4,7,7
7     FACT=2.0*SQRT(API)
      SUM=FACT*CI*SUM
      XZZS=CEXP(-ZZS)
      CERF=XZZS+XZZS*SUM
      GOTO 14
C
COMMENT CONTINUED FRACTIONS THEORY: W(Z) IS RELATED TO THE LIMITING
C        VALUE OF U<N,Z>/H<N,Z>, WHERE U AND H OBEY THE SAME
C        RECURRENCE RELATION IN N, SEE FADDEEVA AND TERENT'EV:
C        TABLES OF VALUES OF W(Z) FOR COMPLEX ARGUMENTS,PERGAMON,
C        N.Y. 1961
8     TERM2=CMPLX(1.E6,0.0)
      Q=1.0
      H1=CMPLX(1.0,0.0)
      H2=2.0*ZZ
      U1=CMPLX(0.0,0.0)
      RTPI=2.0*SQRT(PI)
      U2=CMPLX(RTPI,0.0)
9     TERM1=TERM2
      DO 10 N=1,5
      H3=H2*ZZ-Q*H1
      U3=U2*ZZ-Q*U1
      H1=H2
      H2=2.0*H3
      U1=U2
      U2=2.0*U3
10    Q=Q+1.0
```

```
      TERM2=U3/H3
      TEST=CABS((TERM2-TERM1)/TERM1)
      IF(TEST-EPS)12,11,11
11    IF(Q-60.0)9,9,3
12    CERF=API*CI*TERM2
      GOTO 14
C
COMMENT ASYMPTOTIC SERIES: SEE ABRAMOWITZ AND STEGUN, P328
13    CERF=0.5124242/(ZZS-0.2752551)+0.05176536/(ZZS-2.724745)
      CERF=CI*ZZ*CERF
C
COMMENT SYMMETRY RELATIONS ARE NOW USED TO TRANSFORM THE FUNCTION
C       BACK TO QUADRANT NN
14    GOTO(18,16,17,15),NN
15    CERF=2.0*CEXP(-ZZS)-CERF
16    CERF=CONJG(CERF)
      RETURN
17    CERF=2.0*CEXP(-ZZS)-CERF
18    RETURN
      END
```

4. Calculation of phase shifts

The following programs calculate phase shifts of an ion core in the Hartree Fock approximation, given as input core state wave functions such as those tabulated by Herman and Skillman (1963). They assume a spherically symmetric ion core, averaging over configurations of incomplete shells.

First subroutine CALCA finds the electrostatic or Hartree potential, ATP/X where X are tabulated values of the radius, r. In a free atom the calculation is straightforward, but in a solid some charge may overlap the muffin-tin radius, RMAX. All core-state wavefunctions must be set equal to zero for $X >$ RMAX, and the deficit in electronic charge thus created can be remedied in two ways. *Either* FRAC is increased for the core state that has been truncated, increasing the effective charge density due to that core state (i.e. the charge that would have overlapped RMAX is assumed to be pushed back inside). *Or* electrons overlapping RMAX can be assumed to be forced into a free-electron band, having a uniform density of RHO inside the sphere radius RMAX. RHO is adjusted until electrical neutrality is obtained, which is easily done because $-$ATP(J) for X(J) $>$ RMAX gives the amount of charge remaining inside RMAX. For materials such as aluminium where the $3s$ and $3p$ levels completely lose their identity in the solid, the atomic $3p$ and $3s$ states are ignored, and replaced by ROE corresponding to 3 free electrons inside the sphere. Both of the techniques are crude by conduction-band standards, but as we have seen at LEED energies scattering is sensitive to different components of the potential. At the same time as ATP is calculated, VS is also evaluated. It is given by

$$-3X\left(\frac{3\rho_T(X)}{8\pi}\right)^{\frac{1}{3}}, \tag{A.68}$$

where ρ_T is the total electronic density, and therefore can be used to include exchange in the Slater approximation as an alternative to the Hartree Fock subroutine.

Subroutine CALCFI integrates the radial Schrödinger equation for a given value of L. It follows the method of Hartree (1957) p. 81. If the wavefunction that concerns us is.

$$\phi_\ell(r)\, Y_{\ell m}(\Omega(\mathbf{r})), \tag{A.69}$$

then

$$-\frac{1}{2r^2}\frac{d}{dr}\left[r^2\frac{d\phi_\ell}{dr}\right]+\frac{1}{2}\frac{\ell(\ell+1)}{r^2}\phi_\ell+V\phi_\ell=E\phi_\ell \tag{A.70}$$

(A.70) can be written as two first order differential equations

$$\frac{d[\mathrm{RAD}(1)]}{d[\mathrm{RAD}(3)]} = \mathrm{RAD}(2) + (\ell + 1)\frac{\mathrm{RAD}(1)}{\mathrm{RAD}(3)} \tag{A.71}$$

$$\frac{d[\mathrm{RAD}(2)]}{d[\mathrm{RAD}(3)]} = (2V - 2E)\,\mathrm{RAD}(1) - (\ell + 1)\frac{\mathrm{RAD}(2)}{\mathrm{RAD}(3)} \tag{A.72}$$

where

$$\mathrm{RAD}(1) = r\phi \tag{A.73}$$

$$\mathrm{RAD}(2) = \frac{d}{dr}\,\mathrm{RAD}(1) - \frac{(\ell + 1)}{r}\,\mathrm{RAD}(1). \tag{A.74}$$

$$\mathrm{RAD}(3) = r$$

Starting conditions for the integration are given by a series expansion about the origin if

$$V(r) = -\frac{Z}{r} - VO - \cdots \tag{A.75}$$

then

$$\mathrm{RAD}(1) = \mathrm{const}\; r^{\ell+1}\left[1 - \frac{Zr}{\ell + 1} + \alpha r^2 - \beta r^3 + O(r^4)\right] \tag{A.76}$$

$$\mathrm{RAD}(2) = \mathrm{const}\; r^{\ell+1}\left[-\frac{Z}{(\ell + 1)} + 2\alpha r - 3\beta r^2 + O(r^3)\right] \tag{A.77}$$

where $\alpha = [2Z^2 - (\ell + 1)(2VO + 2E)]/[2(2\ell + 3)(\ell + 1)]$ (A.78)

$$\beta = Z[2Z^2 - (3\ell + 4)(2VO + 2E)]/[6(\ell + 1)(\ell + 2)(2\ell + 3)] \tag{A.79}$$

Integration proceeds by the Runga-Kutta Merson method, in subroutine DAO1A, taken from the Harwell program library.

The Hartree Fock component of V, itself depends on r and the equations must be iterated to self-consistency. Subroutine CALCHF calculates the Hartree-Fock exchange potential:

$$\int V_{ex}^{\ell}(r, r')\,\phi_\ell(r')\,r'^2\,dr' = \frac{\mathrm{HF(J)}}{\mathrm{X(J)}}, \tag{A.80}$$

$$X(\mathrm{J}) = r$$

The Hartree Fock exchange potential was defined in equations (2.39) and (2.41).

$$\mathrm{HF(J)}/X(\mathrm{J}) = -\sum_{\ell'\ell''} \psi_{\ell'}(r) \int \psi_{\ell'}^*(r')(r_<^{\ell''}/r_>^{\ell''+1})\,\phi_\ell(r')\,r'^2\,dr'$$

$$\times\ C(\ell',\ell'',\ell)$$

$$= -\sum_{\ell'\ell''} C(\ell',\ell'',\ell)\,\psi_{\ell'}(r)\,[\textstyle\int_0^r \psi^*_{\ell'}(r')\,\phi_\ell(r')$$

$$\times\ (r'^{\ell''+2}/r^{\ell''+1})\,dr' + \textstyle\int_r^\infty \psi^*_c(r')\,\phi_\ell(r')(r^{\ell''}/r'^{\ell''-1})\,dr'] \qquad \text{(A.81)}$$

The constants 'C' are defined by equation (2.38) and tabulated at the beginning of the program via subroutines ORG and CONS, supplied by B. S. Ing. The integrations in equation (A.81) are performed by Simpson's rule in CALCHF.

Basic variables in the program to calculate phase shifts

Z	atomic number of the atom.
NC	the number of core states to be used (not counting m-values).
LC(J)	angular momentum of the Jth core state.
FRAC	fractional occupation of the core state (e.g. if there are 8 electrons in a d state, FRAC = 0.8).
AW(J,K)	The amplitude of the Jth core state (in fact $r \times \psi_\ell(r)$), the Kth tabulated value. The intervals of tabulation start at r = 0, K = 1, and proceed with ten increments of 0.01∗AMU (defined in the main programmme, label 43) to K = 11, then in steps of 0.02∗AMU to K = 21 etc. Wavefunctions are tabulated in the correct form in Herman Skillman (1963).
RMAX	radius of the inscribed sphere of the unit cell, given in atomic units.
RHO	determined by the number of electrons, n, forced into a a free-electron band: RHO $= n/(\frac{4}{3}\pi \mathrm{RMAX}^3)$.
CN(LPP+1, L+1,LP+1)	$\frac{1}{2}[(2L+1)(2LP+1)]^{\frac{1}{2}} \int_{-1}^{+1} P_{LPP} P_L P_{LP}\, d\mu.$ Sufficient of these are tabulated to carry phase shift calculations to $\ell = 9$, and for core states up to $\ell_c = 2$. Changing dimensions statements and do-loops 15, can increase this.
E	energy at which the phase shift is to be calculated, in Hartrees (27.2 eV), relative to the zero-potential immediately outside the neutral sphere.
L	angular momentum to which the phase shift refers.
AMU	defines the scale of length for the atom.
X	values of the radius at which the incident electron's wavefunction is to be calculated. It follows the same increments as the core states out to a radius X(NF) defined at label 3, and thereafter uses equal increments.

ATP(K) $r \times V_H(r)$, tabulated at $r = X(K)$. If the sphere of radius RMAX is electrically neutral, as it should be, ATP = 0.0 outside RMAX to within the accuracy of the integration routines (typically ± 0.05).

FI(K) Amplitude of the incident electron's wavefunction, energy E, times $X(K)$,

$$\text{FI(K)} = r\phi_\ell(r), \quad r = X(K).$$

VS(K) defined by

$$VS(K) = -3r(3\rho(r)/8\pi)^{\frac{1}{3}}, \quad r = X(K)$$

where ρ is the total density of electrons at radius r. VS is redundant in the current version of the program.

DEL the phase shift for angular momentum L, energy E.

```
COMMENT MAIN PROGRAM FOR HARTREE-FOCK PHASE SHIFTS
C       PENDRY CAVENDISH                         JUNE 1970
C
C       INPUT - DETAILS OF ATOM
C       Z    = ATOMIC NUMBER
C       NC   = NUMBER OF CORE STATES
C       LC   = ANGULAR MOMENTUM OF CORE STATE
C       N    = NUMBER OF TABULATED VALUES OF WAVE FUNCTION
C       FRAC= FRACTIONAL OCCUPATION OF CORE STATE
C              (E.G. IN NICKEL, WITH D-BAND 94% OCCUPIED, FRAC=0.94)
C       AW   = CORE STATE WAVE FUNCTIONS, AS TABULATED BY HERMAN
C              AND SKILLMAN
C       RMAX= RADIUS OF INSCRIBED SPHERE OF UNIT CELL
C       RHO = DENSITY OF CONDUCTION ELECTRONS WITHIN THIS SPHERE
C              REQUIRED TO MAKE THE SPHERE ELECTRICALLY NEUTRAL
C              (I.E. = NO.OF COND.ELECTS./VOL.OF SPHERE)
      DIMENSION AW(10,200),ATP(200),X(200),HF(200),FI(200),
     1 VS(200),HF2(200)
      DIMENSION A(200),B(200),C(200),D(200),F(35),LC(10),
     1 CN(13,10,3),RAD(3)
      COMMON KKJ,JR,X,CL,E,HF,ATP,VS
      NMESH=200
21    READ (5,37) Z,NC
37    FORMAT(1F9.4/1I4/)
      WRITE (6,50) Z,NC
50    FORMAT(3H0Z=,1F9.4,5X,3HNC=,1I4/)
      IF (NC) 20,38,38
38    DO 49 J=1,10
      DO 49 K=1,NMESH
49    AW(J,K)=0.0
      DO 22 J=1,NC
      READ (5,39) LC(J),N,FRAC
39    FORMAT(1I4/1I4/1F9.4/)
      FRAC=SQRT(FRAC)
      READ (5,40) (AW(J,K),K=1,N)
40    FORMAT(100(5F9.4/))
      WRITE (6,46) LC(J),N,FRAC
46    FORMAT(1I4/1I4/1F9.4/)
      WRITE (6,48) (AW(J,K),K=1,N)
48    FORMAT(50(5(5F9.4/)/))
      DO 22 K=1,N
22    AW(J,K)=FRAC*AW(J,K)
      READ (5,41) RMAX,RHO
41    FORMAT(1F9.4/1F9.4)
      WRITE (6,51) RMAX,RHO
51    FORMAT(6H0RMAX=,1F9.4,5X,4HRHO=,1F9.4/)
      DO 15 K1=1,13
      DO 15 L1=1,10
      DO 15 LP1=1,3
      K=K1-1
      L=L1-1
      LP=LP1-1
      CALL ORG(L,LP,K,CN(K1,L1,LP1))
15    CONTINUE
COMMENT SUBROUTINE ORG CALCULATES CLEBSCH-GORDON COEFFICIENTS
C       INPUT - DETAILS OF REQUIRED PHASE SHIFTS
C       E    = ENERGY AT WHICH THE PHASE SHIFTS ARE TO BE
C              CALCULATED
C       L    = ANGULAR MOMENTUM
```

```
24      READ (5,42) E
42      FORMAT(1F9.4)
        IF (E) 20,25,25
25      READ (5,44) L
44      FORMAT(1I4)
        IF (L) 24,43,43
43      AMU=0.88534138/EXP(ALOG(Z)/3.0)
        DX=0.01*AMU
        ES=SQRT(2.0*E)
        X(1)=0.0
        J=2
        NF=NMESH
1       KDL2=J+9
        DO 2 K=J,KDL2
2       X(K)=X(K-1)+DX
        J=J+10
        IF(J-NMESH+10) 3,3,10
3       IF (0.3/ES-DX) 5,5,4
4       DX=DX+DX
        GO TO 1
5       NF=J-1
        DO 7 K=J,NMESH
7       X(K)=X(K-1)+DX
COMMENT AMU DEFINES THE SCALE OF LENGTH USED - SEE HERMAN AND
C         SKILLMAN. THE ARRAY X DEFINES THE POINTS AT WHICH THE
C         INTEGRATED WAVE FUNCTIONS WILL BE TABULATED. NF IS
C         THE LAST VALUE OF X TO BE INCREMENTED ACCORDING TO THE
C         HERMAN AND SKILLMAN SCHEME. THEREAFTER EQUAL INCREMENTS
C         ARE USED
10      CALL CALCA(ATP,AW,X,NF,NC,Z,A,B,C,D,LC,VS,RMAX,RHO)
        WRITE (6,27) (X(J),ATP(J),VS(J),J=1,100)
27      FORMAT(9H X ATP VS,50(/5(/6F10.5)))
COMMENT SUBROUTINE CALCA CALCULATES ATP - THE ELECTROSTATIC
C         POTENTIAL TIMES THE RADIAL DISTANCE - OVER THE GRID X.
C         THE PRINTOUT IS AN ESSENTIAL CHECK THAT ATP BECOMES
C         NEGLIGIBLE OUTSIDE RMAX. IF IT DOES NOT, THE PROGRAM
C         HAS THE WRONG AMOUNT OF CHARGE INSIDE THE SPHERE OF
C         RADIUS RMAX
C         SUBROUTINE CALCFI INTEGRATES THE SCHROEDINGER EQUATION,
C         SETTING VALUES IN THE ARRAY FI. THE ELEMENTS OF THE ARRAY
C         HF ARE ZERO AT THIS STAGE. SUBROUTINE CALCHF CALCULATES
C         THE EXCHANGE POTENTIAL IN THE HARTREE FOCK APPROXIMATION,
C         SETTING VALUES IN THE ARRAY HF. THIS PROCEDURE IS
C         ITERATED, SINCE HF DEPENDS ON FI ITSELF. THE RESULTING
C         PHASE SHIFT APPEARS IN DEL
30      DO 28 J=1,NMESH
        HF2(J)=0.0
        VS(J)=0.0
28      HF(J)=0.0
        DP=1.0E9
        IIT=0
        WRITE (6,29) L
29      FORMAT(11H DEL FOR L=,1I4)
        CALL CALCFI(L,Z,FI,DEL,NF,RMAX)
        WRITE (6,31) DEL
31      FORMAT(1F13.5)
        CALL CALCHF(AW,LC,HF,FI,NF,L,CN,X,NC)
        CALL CALCFI(L,Z,FI,DEL,NF,RMAX)
        IIT=IIT+1
```

```
35      CALL CALCHF(AW,LC,HF2,FI,NF,L,CN,X,NC)
        WRITE (6,31) DEL
        DO 14 J=1,NMESH
14      HF(J)=0.5*(HF(J)+HF2(J))
        CALL CALCFI(L,Z,FI,DEL,NF,RMAX)
        WRITE (6,31) DEL
        TEST=ABS(DP-DEL)
        IF (TEST-5.0E-3) 33,33,34
34      DP=DEL
        IF (IIT-10) 35,33,33
33      WRITE (6,12) (FI(J),J=1,100)
12      FORMAT(3H FI,100(/5(/5E13.5)))
        WRITE (6,26) (HF(J),J=1,100)
26      FORMAT(3H HF,50(/5(/5E13.5)))
        WRITE (6,11) E,L,DEL
11      FORMAT(3H E=,1F10.5,7H      L=,1I4,4H PS=,1F10.5///)
        GO TO 25
20      STOP
        END
C
C
COMMENT SUBROUTINE ORG, WITH SUBROUTINE CONS, EVALUATES THE
C         CLEBSCH-GORDON COEFFICIENTS ACCORDING TO THE SCHEME
C         SET OUT BY SLATER: ATOMIC STRUCTURE, VOL1, 309-10
        SUBROUTINE ORG(IL,ILP,IK,A)
        K=IK
        LP=ILP
        L=IL
        IF (LP-K) 1,2,2
1       K=LP
        LP=IK
2       IF (L-K) 3,16,16
3       L=K
        K=IL
16      IF(LP+K-L) 7,8,8
8       IF(L-LP+K) 7,9,9
9       CALL CONS(L,LP,K,A)
        A=A*SQRT(FLOAT((2*IL+1)*(2*ILP+1)))
        RETURN
7       A=0.0
        RETURN
        END
        SUBROUTINE CONS(L,LP,K,A)
        DIMENSION F(35)
        F(1)=1.0
        DO 1 I=1,30
1       F(I+1)=F(I)*FLOAT(I)
        FL=FLOAT(L+LP+K)/2.0
        IF(FL-FLOAT(INT(FL))-0.2) 4,4,2
2       A=0.
        RETURN
4       SUM=0.0
        IS=(L+LP+K)/2
        IT=-1
        IT2=K+1
        DO 3 ITT=1,IT2
        IT=IT+1
        L1=L+IT+1
        L2=LP+K-IT+1
```

```
      L3=L-IT+1
      L4=LP-K+IT+1
      L5=K-IT+1
3     SUM=SUM+(-1.)**IT*F(L1)*F(L2)/(F(L3)*F(L4)
     1*F(L5)*F(IT+1))
      L1=IS+IS-K-K+1
      L2=IS-L+1
      L3=IS-K+1
      A=(-1.)**(IS-LP)*F(L1)*SUM*F(IS+1)/(F(L2)*F(L3))
      L1=IS-LP+1
      L2=IS+IS+2
      A=A/(F(L1)*F(L2))*F(K+1)
      RETURN
      END
C
C
COMMENT HARTREE AND SLATER-EXCHANGE POTENTIALS
C        PENDRY CAVENDISH                   JUNE 1970
C        SUBROUTINE CALCA CALCULATES THE ELECTROSTATIC POTENTIAL
C        TIMES THE RADIAL DISTANCE, OVER THE GRID X.  N.B. THIS
C        ROUTINE IS QUITE FLEXIBLE WITH RESPECT TO THE CHARGE
C        DENSITY INPUT. IN PARTICULAR, IF IT IS NOT DESIRED TO
C        TREAT THE CONDUCTION ELECTRONS AS A UNIFORM CHARGE DENSITY
C        INSIDE THE INSCRIBED SPHERE, THEN THIS CAN BE CHANGED
      SUBROUTINE CALCA(ATP,AW,X,NF,NC,Z,A,B,C,D,LC,VS,RMAX,RHO)
      DIMENSION AW(10,200),ATP(200),X(200),VS(200)
      DIMENSION A(200),B(200),C(200),D(200),LC(10)
      NMESH=200
      DO 18 K=1,NMESH
      A(K)=0.0
      B(K)=0.0
      C(K)=0.0
      D(K)=0.0
      VS(K)=0.0
18    ATP(K)=0.0
COMMENT TABULATION OF CHARGE DENSITY OVER THE GRID
      J=1
      DO 19 K=1,NC
      A(K)=FLOAT(LC(K))
19    A(K)=2.0*A(K)+1.0
1     KDL1=J+1
      KDL2=J+10
      DO 2 K=KDL1,KDL2
      DO 2 L=1,NC
2     C(K)=C(K)+A(L)*AW(L,K)*AW(L,K)
      J=J+10
      I=J+1
      IF (C(J)-1.0E-8) 9,9,3
3     IF (NF-J) 4,4,1
4     M=4
      AM=4.0
      I=J+1
8     KDL1=J+1
      KDL2=J+9
COMMENT VALUES OF THE CHARGE DENSITY BETWEEN THE TABULATED GRID
C        POINTS ARE OBTAINED BY QUADRATIC INTERPOLATION
      DO 6 K=KDL1,KDL2,2
      A1=0.0
      A2=0.0
```

```
      A3=0.0
      DO 5 L=1,NC
      A1=A1+A(L)*AW(L,K-1)*AW(L,K-1)
      A2=A2+A(L)*AW(L,K)*AW(L,K)
5     A3=A3+A(L)*AW(L,K+1)*AW(L,K+1)
      CC=A1
      AA=2.0*(A3-2.0*A2+CC)/(AM*AM)
      BB=(A3-CC)/AM-AA*AM
      AM1=1.0
      DO 6 M1=1,M
      C(I)=AM1*(AA*AM1+BB)+CC
      I=I+1
6     AM1=AM1+1.0
      M=M+M
      AM=AM+AM
      J=J+10
      IF (C(I-1)-1.0E-8) 9,9,8
9     A(1)=0.0
      B(1)=0.0
      D(1)=0.0
      J=I
      VS(1)=0.0
      DO 20 JJS=2,J
      ADC=0.0
COMMENT THE FOLLOWING OPTION IS REDUNDANT IN THE PRESENT
C        FORMULATION, BUT CAN BE USED TO ADD A (RHO**(1/3))
C        TERM TO THE POTENTIAL
      IF (RMAX-X(JJS)) 24,24,25
25    ADC=6.28318*RHO*X(JJS)**2
24    AR=2.0*(C(JJS)+ADC)/(4.0*3.14159*X(JJS)**2)
      VS(JJS)=0.0
      IF (AR-1.0E-8) 20,20,21
21    VS(JJS)=-3.0*X(JJS)*EXP(ALOG(3.0*AR/
     1(8.0*3.14159))/3.0)
20    CONTINUE
COMMENT CALCULATION OF THE ELECTROSTATIC FIELD BY SIMPSON'S
C        INTEGRATION PROCEDURE. SIMPSON'S RULE INTEGRATES TWO
C        STEPS AT A TIME, AND HERE THE ODD AND EVEN VALUES OF
C        THE INDEX K ARE TREATED SEPARATELY
      DO 10 K=2,J
10    D(K)=C(K)/X(K)
      DX=X(2)-X(1)
      A(2)=DX*0.5*(C(1)+C(2))
      B(2)=DX*0.5*(D(1)+D(2))
      JJ=3
11    KDL2=JJ+8
      DO 12 K=JJ,KDL2,2
      A(K)=(C(K-2)+4.0*C(K-1)+C(K))*DX/3.0
      A(K+1)=(C(K-1)+4.0*C(K)+C(K+1))*DX/3.0
      B(K)=(D(K-2)+4.0*D(K-1)+D(K))*DX/3.0
12    B(K+1)=(D(K-1)+4.0*D(K)+D(K+1))*DX/3.0
      JJ=JJ+10
      IF (NF-JJ) 14,14,13
13    DX=DX+DX
      AS=0.25*DX*(C(JJ-4)+C(JJ-3))
      BS=0.25*DX*(D(JJ-4)+D(JJ-3))
      A(JJ-1)=(C(JJ-4)+4.0*C(JJ-2)+C(JJ-1))*DX/3.0-AS
      B(JJ-1)=(D(JJ-4)+4.0*D(JJ-2)+D(JJ-1))*DX/3.0-BS
14    IF (J-JJ) 15,15,11
```

```
15      SB1=0.0
        SB2=0.0
        SA1=0.0
        SA2=0.0
        KDL2=J-1
        DO 16 K=2,KDL2,2
        SB2=SB2+B(K)
16      SB1=SB1+B(K+1)
        ATP(1)=-Z
        KDL2=J-1
        DO 17 K=2,KDL2,2
        KA1=K+1
        SA1=SA1+A(K+1)
        SA2=SA2+A(K)
        SB1=SB1-B(K+1)
        SB2=SB2-B(K)
        ATP(K)=-Z+2.0*(SA2+SB2*X(K))
17      ATP(K+1)=-Z+2.0*(SA1+SB1*X(K+1))
        DO 28 K=1,KA1
COMMENT NOW THE ELECTROSTATIC FIELD FROM THE CONDUCTION ELECTRONS
C         IS INCLUDED
        IF (RMAX-X(K)) 27,27,26
26      ATP(K)=ATP(K)+3.14159*RHO*2.0*(RMAX**2-X(K)**2/3.0)*X(K)
        GO TO 28
27      ATP(K)=ATP(K)+(4.0*3.14159*RHO*RMAX**3)/3.0
28      CONTINUE
        RETURN
        END
C
C
COMMENT RADIAL INTEGRATION BY RUNGA-KUTTA WITH CHECKS
C         PENDRY CAVENDISH                          JUNE 1970
C         INTEGRATES THE SCHROEDINGER EQUATION
        SUBROUTINE CALCFI(L,Z,STORE,DEL,NF,RHO)
        DIMENSION ATP(200),X(200),HF(200),VS(200),STORE(200)
        DIMENSION A(2),B(2),C(2),D(2),F(2),RAD(3)
        COMMON K,JR,X,CL,E,HF,ATP,VS
        NMESH=200
        DO 29 J=1,NMESH
29      STORE(J)=0.0
1       CL=FLOAT(L)
        M=1
        SPLIT=1.0
4       DX=X(2)-X(1)
COMMENT SPLIT IS THE DEGREE OF SUBDIVISION OF DX, THE BASIC
C         INTERVAL OF TABULATION. THE STARTING CONDITIONS FOR
C         THE INTEGRATION ARE SET UP AS IN HARTREE: CALCULATION
C         OF ATOMIC STRUCTURE, 81
        VO=-(ATP(1)-ATP(2))/DX
        JR=0
        XV=DX/SPLIT
        ALPHA=(2.0*Z*Z-(CL+1.0)*(2.0*VO+2.0*E))/
       1(2.0*(2.0*CL+3.0)*(CL+1.0))
        BETA=Z*(2.0*Z*Z-(3.0*CL+4.0)*(2.0*VO
       1+2.0*E))/(6.0*(CL+1.0)*(CL+2.0)*(2.0*CL+3.0))
        RL=1.0
        KDL2=L+1
        DO 2 K=1,KDL2
2       RL=XV*RL
```

```
      RAD(1)=RL*(1.0-XV*Z/(CL+1.0)+ALPHA*XV*XV
     1-BETA*XV*XV*XV)
      RAD(2)=RL*(-Z/(CL+1.0)+2.0*ALPHA*XV
     1-3.0*XV*XV)
      RAD(3)=XV
      N1=2
      INIT=0
      IF (M-1) 5,6,5
5     IF (INIT) 3,10,3
3     RAD(1)=ST1
      RAD(2)=ST2
      RAD(3)=ST3
10    PRAD=STORE(JR+11)
6     DO 11 K=1,10
      J2=IFIX(SPLIT)
      IF (J2-N1) 21,20,20
COMMENT INTEGRATION PROCEEDS OVER TEN VALUES OF THE BASIC GRID.
C        AFTER EACH RANGE OF TEN VALUES, AND BEFORE PROCEEDING
C        TO THE NEXT, THE STEP SIZE IS HALVED AND THE ACCURACY OF
C        THE CURRENT RANGE IS CHECKED. IF THE REQUIRED ACCURACY HAS
C        NOT BEEN REACHED AFTER HALVING THE STEP LENGTH 4 TIMES,
C        THE PROGRAM PRINTS OUT THE ACCURACY ACHIEVED, AND PROCEEDS.
C        THIS MONITOR OFTEN OCCURS IN THE FIRST RANGE OF TEN, AND
C        CAN BE IGNORED IF TEST IS SMALL. SUBROUTINE DA01A
C        ADVANCES THE INTEGRATION ACROSS ONE GRID INTERVAL
20    DO 7 J=N1,J2
      XT=RAD(3)
      DDX=DX/SPLIT
      CALL DA01A(RAD,A,B,C,D,F,2,XT,DDX)
      RAD(3)=XT
      IF (BIG-ABS(RAD(1))) 12,12,7
12    BIG=ABS(RAD(1))
7     CONTINUE
21    N1=1
      JRK=JR+K+1
11    STORE(JRK)=RAD(1)
      SPLIT=2.0*SPLIT
      M=M+1
      IF (M-5) 28,28,26
26    TEST=ABS((PRAD-RAD(1))/BIG)
      WRITE (6,27) TEST
27    FORMAT(18H MONITOR M>5 TEST=,1E13.5)
      GO TO 16
28    IF (M-2) 13,14,13
14    IF (INIT) 15,4,15
15    IF (INIT-1) 13,5,13
13    IF (ABS((PRAD-RAD(1))/BIG)-1.0E-4) 16,16,17
17    IF (INIT) 18,4,18
18    IF (INIT-1) 16,5,16
16    ST1=RAD(1)
      ST2=RAD(2)
      ST3=RAD(3)
      SPLIT=1.0
      INIT=1
      IF (NF-JR-12) 23,23,22
22    DX=2.0*DX
23    JR=JR+10
      M=1
COMMENT THE PHASE SHIFT IS NOW CALCULATED AND SET IN DEL. DL IS
```

```
C         A LOGARITHMIC DERIVATIVE
       IF (RHO-X(JR+1)) 19,19,6
19     DL=(RAD(2)+(CL+1.0)*RAD(1)/RAD(3))/RAD(1)
      1-1.0/RAD(3)
       ES=SQRT(2.0*E)
       BFARG=RAD(3)*ES
       CALL CALCBF(BFARG,CL,BFJ1,BFN1)
       DCL=CL+1.0
       CALL CALCBF(BFARG,DCL,BFJ2,BFN2)
       DJL=CL*BFJ1/BFARG-BFJ2
       DNL=CL*BFN1/BFARG-BFN2
       A1=ES*DJL-DL*BFJ1
       A2=ES*DNL-DL*BFN1
       DEL=3.14159/2.0
       IF (ABS(A2)-1.0E-8) 25,25,24
24     TD=A1/A2
       DEL=ATAN(TD)
25     RETURN
       END
C
C
COMMENT HARTREE-FOCK EXCHANGE POTENTIAL
C        PENDRY CAVENDISH        JUNE 1970
C        SUBROUTINE CALCHF CALCULATES THE EXCHANGE POTENTIAL
C        IN THE HARTREE-FOCK APPROXIMATION
       SUBROUTINE CALCHF(AW,LC,HF,FI,NF,L,CN,X,NC)
       DIMENSION AW(10,200),X(200),HF(200),FI(200)
       DIMENSION A(200),B(200),C(200),D(200),LC(10),
      1 CN(13,10,3)
       NMESH=200
       DO 17 K=1,NMESH
       A(K)=0.0
       B(K)=0.0
       C(K)=0.0
       D(K)=0.0
17     HF(K)=0.0
1      DO 13 N=1,NC
       LP=LC(N)
       LK1=IABS(LP-L)
       LK2=LP+L
       A(1)=0.0
       B(1)=0.0
       IAJ=2
28     JDL2=IAJ+9
       DO 2 J=IAJ,JDL2
       CX=X(J)**LK1
       A(J)=AW(N,J)*CX
2      B(J)=AW(N,J)/(CX*X(J))
       IAJ=IAJ+10
       IF (NF-IAJ) 26,27,27
27     IF (ABS(AW(N,IAJ))-1.0E-7) 18,18,28
26     M=4
       AM=4.0
       I=NF+1
       J=I
14     KDL2=J+8
COMMENT VALUES OF THE CORE STATE WAVE FUNCTIONS ARE FOUND BY
C        QUADRATIC INTERPOLATION, AND PROVIDE THE FINER MESH
C        REQUIRED FOR THE INTEGRATION OF PHI - THE PRODUCT OF THE
```

```
C          RADIAL DISTANCE AND THE RADIAL PART OF THE SCATTERING
C          WAVE FUNCTION. A AND B THEN CONTAIN THE TWO QUANTITIES
C          TO BE INTEGRATED
          DO 15 K=J,KDL2,2
          CC=AW(N,K-1)
          AA=2.0*(AW(N,K+1)-2.0*AW(N,K)+CC)/(AM*AM)
          BB=(AW(N,K+1)-CC-AA*AM*AM)/AM
          AM1=1.0
          DO 15 M1=1,M
          AWF=AM1*(AA*AM1+BB)+CC
          CX=X(I)**LK1
          A(I)=AWF*CX
          B(I)=AWF/(CX*X(I))
          I=I+1
15        AM1=AM1+1.0
          J=J+10
          IF (ABS(AW(N,J))-1.0E-7) 18,18,16
16        M=M+M
          AM=AM+AM
          GO TO 14
18        W=FLOAT(LP)
          Y=FLOAT(L)
          CT=CN(LK1+1,L+1,LP+1)*SQRT((2.0*W+1.0)/(2.0*Y+1.0))
COMMENT CLEBSCH-GORDON COEFFICIENTS CN. THE INTEGRATION PROCEEDS
C          USING SIMPSON'S RULE, WHICH INTEGRATES TWO STEPS AT A TIME.
C          ODD AND EVEN VALUES OF THE INDEX K ARE TREATED SEPARATELY
3         DX=X(2)-X(1)
          J=1
          DO 9 K=1,NMESH
          C(K)=0.0
9         D(K)=0.0
7         DO 4 K=1,5
          J=J+2
          C(J)=(FI(J-2)*A(J-2)+4.0*FI(J-1)*A(J-1)
         1+FI(J)*A(J))*DX/3.0
          C(J+1)=(FI(J-1)*A(J-1)+4.0*FI(J)*A(J)
         1+FI(J+1)*A(J+1))*DX/3.0
          D(J)=(FI(J-2)*B(J-2)+4.0*FI(J-1)*B(J-1)
         1+FI(J)*B(J))*DX/3.0
          D(J+1)=(FI(J-1)*B(J-1)+4.0*FI(J)*B(J)
         1+FI(J+1)*B(J+1))*DX/3.0
4         CONTINUE
          IF (NF-J) 6,6,5
5         DX=2.0*DX
          CCT=0.25*DX*(FI(J-1)*A(J-1)+FI(J-2)*A(J-2))
          DDT=0.25*DX*(FI(J-1)*B(J-1)+FI(J-2)*B(J-2))
          C(J+1)=((FI(J-2)*A(J-2)+4.0*FI(J)*A(J)+FI(J+1)*A(J+1))
         1*DX/3.0)-CCT
          D(J+1)=((FI(J-2)*B(J-2)+4.0*FI(J)*B(J)+FI(J+1)*B(J+1))
         1*DX/3.0)-DDT
6         IF (ABS(A(J))-1.0E-9) 8,8,7
8         SD1=0.0
          SD2=0.0
          DO 19 K=3,J,2
          SD1=SD1+D(K)
19        SD2=SD2+D(K+1)
          DX=X(2)-X(1)
          SC1=0.0
          SC2=SC1+DX*0.5*(FI(1)*A(1)+FI(2)*A(2))
```

```
      DO 10 K=1,J,2
      ADD=(SC1*B(K)+SD1*A(K))*CT
      HF(K+1)=HF(K+1)-(SC2*B(K+1)+SD2*A(K+1))*CT
      HF(K)=HF(K)-ADD
21    SC1=SC1+C(K+2)
      SC2=SC2+C(K+3)
      SD1=SD1-D(K+2)
10    SD2=SD2-D(K+3)
      LK1=LK1+2
      IF (LK2-LK1) 13,11,11
11    KDL2=J+10
      DO 12 K=2,KDL2
      W=X(K)*X(K)
      A(K)=A(K)*W
12    B(K)=B(K)/W
      GO TO 18
13    CONTINUE
      RETURN
      END
C
C
COMMENT SPHERICAL BESSEL FUNCTIONS 'J' AND 'N'
C        PENDRY CAVENDISH                JUNE 1970
      SUBROUTINE CALCBF(Z,CL,BJ,BN)
      L=IFIX(CL)
      IF (CL-0.5) 10,10,4
10    BJ=1.0
      IF (1.0E-6-ABS(Z)) 11,11,8
11    BJ=SIN(Z)/Z
      GO TO 8
4     IF (1.1*CL-ABS(Z)) 12,12,6
12    BJ=0.0
      IF (1.0E-6-ABS(Z)) 13,13,8
13    BF2=SIN(Z)/Z
      BF1=SIN(Z)/(Z*Z)-COS(Z)/Z
      BFJ=2.0
      IF (L-1) 23,23,22
22    DO 14 J=2,L
      BFS=BF1
      BF1=(2.0*BFJ-1.0)*BF1/Z-BF2
      BF2=BFS
14    BFJ=BFJ+1.0
23    BJ=BF1
      GO TO 8
6     BFP=1.0
      BF=0.0
      BFLL=2.0*CL+3.0
      BFSR=1.0
      BFSX=0.5*Z*Z
7     BF=BF+BFP
      BFP=-BFP*BFSX/(BFSR*BFLL)
      BFSR=BFSR+1.0
      BFLL=BFLL+2.0
      IF (ABS(BFP)-ABS(BF)*1.0E-6) 15,15,7
15    BFLX=1.0
      BFNM=1.0
      DO 16 J=1,L
      BFLX=Z*BFLX
16    BFNM=BFNM+1.0
```

```
      BFZ=1.0
      BFY=2.0*CL+1.0
      LL=IFIX(BFY)+1
      DO 17 JBB=1,LL,2
      BFZ=BFZ*BFY
17    BFY=BFY-2.0
      BJ=BFLX*BF/BFZ
8     IF (CL-0.5) 18,18,9
18    BN=-COS(Z)/Z
      GO TO 5
9     BF2=-COS(Z)/Z
      BF1=-COS(Z)/(Z*Z)-SIN(Z)/Z
      BFJ=2.0
      IF (L-1) 21,21,20
20    DO 19 J=2,L
      BFS=BF1
      BF1=(2.0*BFJ-1.0)*BF1/Z-BF2
      BF2=BFS
19    BFJ=BFJ+1.0
21    BN=BF1
5     RETURN
      END
C
C
COMMENT RUNGE-KUTTA-MERSON WITH ESTIMATED TRUNCATION ERROR E(I).
C       ADVANCES THE INTEGRATION OF THE SET OF DIFFERENTIAL
C       EQUATIONS  DY(I)/DX=F(I),I=1,N  BY ONE STEP FROM X0
C       TO X0+DX. THERE ARE FIVE ENTRIES TO DYBDX TO SET UP
C       F(J)=FUNC(Y(I),X),I=1,N
C       MCVICAR HARWELL                             FEBRUARY 1963
      SUBROUTINE DA01A(Y,E,A,B,C,D,N0,X0,DX)
      DIMENSION Y(1),E(1),A(1),B(1),C(1),D(1)
      N=N0
      Z=X0
      H=DX/3.
      DO 1 I=1,N
      D(I)=Y(I)
1     CONTINUE
      X=Z
      DO 41 J=1,5
      CALL DYBDX(Y,E,N,X)
      DO 21 I=1,N
      GO TO (11,12,13,14,15),J
11    A(I)=H*E(I)
      Y(I)=D(I)+A(I)
      GO TO 21
12    B(I)=H*E(I)
      Y(I)=D(I)+(A(I)+B(I))*0.5
      GO TO 21
13    B(I)=H*E(I)
      Y(I)=D(I)+(A(I)+B(I)*3.)*0.375
      GO TO 21
14    C(I)=H*E(I)
      Y(I)=D(I)+(A(I)-B(I)*3.+C(I)*4.)*1.5
      GO TO 21
15    Z=D(I)
      D(I)=H*E(I)
      Y(I)=Z+(A(I)+C(I)*4.+D(I))*0.5
      E(I)=(A(I)*2.-B(I)*9.+C(I)*8.-D(I))/10.
```

```
21    CONTINUE
      GO TO (31,41,33,34,41),J
31    X=Z+H
      GO TO 41
33    X=Z+DX*0.5
      GO TO 41
34    X=Z+DX
41    CONTINUE
      X0=X
      RETURN
      END
C
C
COMMENT SUBROUTINE DYBDX SUPPORTS THE INTEGRATION SUBROUTINE DA01A
      SUBROUTINE DYBDX(YT,F,N,XT)
      DIMENSION VS(200),YT(2),F(2),HF(200),ATP(200),X(200)
      COMMON K,JR,X,CL,E,HF,ATP,VS
      JRK=JR+K
      A=(XT-X(JRK))/(X(JRK+1)-X(JRK))
      V=(ATP(JRK)+(ATP(JRK+1)-ATP(JRK))*A)/XT
      VHF=(HF(JRK)+(HF(JRK+1)-HF(JRK))*A)
      VSA=(VS(JRK)+(VS(JRK+1)-VS(JRK))*A)/XT
      F(1)=(YT(2)+(CL+1.0)*YT(1)/XT)
      F(2)=((2.0*V+2.0*VSA-2.0*E)*YT(1)+2.0*VHF
     1-(CL+1.0)*YT(2)/XT)
      RETURN
      END
$      EXECUTE
```

test runs of phase shift program

There follows first a sample set of typical data for aluminium, then output from the phase shift program

```
13.0000
 3

 0
60
 1.0000

 0.        0.3252    0.6193    0.8847    1.1236
 1.3377    1.5291    1.6994    1.8502    1.9830
 2.0993    2.2871    2.4231    2.5152    2.5706
 2.5953    2.5945    2.5727    2.5339    2.4813
 2.4178    2.2675    2.0985    1.9217    1.7449
 1.5732    1.4101    1.2577    1.1169    0.9884
 0.8718    0.6731    0.5152    0.3919    0.2967
 0.2238    0.1684    0.1266    0.0951    0.0714
 0.0537    0.0306    0.0176    0.0103    0.0062
 0.0039    0.0025    0.0017    0.0012    0.0009
 0.0006    0.0004    0.0003    0.0002    0.0001
 0.0001    0.0000    0.0000    0.0000    0.0000

 0
70
 1.0000

 0.       -0.0845   -0.1607   -0.2294   -0.2907
-0.3453   -0.3935   -0.4358   -0.4724   -0.5037
-0.5301   -0.5694   -0.5924   -0.6014   -0.5980
-0.5840   -0.5609   -0.5300   -0.4924   -0.4493
-0.4015   -0.2954   -0.1798   -0.0590    0.0632
 0.1844    0.3024    0.4156    0.5230    0.6238
 0.7175    0.8825    1.0177    1.1243    1.2046
 1.2614    1.2974    1.3156    1.3186    1.3091
 1.2892    1.2261    1.1424    1.0477    0.9490
 0.8510    0.7568    0.6684    0.5868    0.5126
 0.4457    0.3334    0.2464    0.1805    0.1312
 0.0949    0.0683    0.0490    0.0351    0.0250
 0.0178    0.0089    0.0043    0.0000    0.0000
 0.0000    0.0000    0.0000    0.0000    0.0000

 1
70
 1.0000

 0.        0.0009    0.0036    0.0080    0.0139
 0.0211    0.0297    0.0395    0.0504    0.0623
 0.0751    0.1032    0.1342    0.1674    0.2025
 0.2390    0.2765    0.3148    0.3536    0.3926
 0.4316    0.5087    0.5839    0.6562    0.7248
 0.7893    0.8493    0.9048    0.9555    1.0015
 1.0428    1.1120    1.1643    1.2012    1.2244
 1.2357    1.2367    1.2290    1.2139    1.1928
 1.1667    1.1036    1.0310    0.9536    0.8748
 0.7972    0.7224    0.6514    0.5849    0.5234
 0.4668    0.3685    0.2885    0.2244    0.1736
 0.1337    0.1026    0.0786    0.0600    0.0458
 0.0350    0.0205    0.0121    0.0000    0.0000
 0.0000    0.0000    0.0000    0.0000    0.0000

 2.6993
```

```
 0.0364
 4.0000
 0
-1
 6.0000
 1
-1
-1.0000
```

```
SNUMB = C0272, ACTIVITY # = 02, REPORT CODE = 06, RECORD COUNT = 00878

Z=  13.0000      NC=    3
  0
 60
  1.0000
  0.        0.3252    0.6193    0.8847    1.1236
  1.3377    1.5291    1.6994    1.8502    1.9830
  2.0993    2.2871    2.4231    2.5152    2.5706
  2.5953    2.5945    2.5727    2.5339    2.4813
  2.4178    2.2675    2.0985    1.9217    1.7449

  1.5732    1.4101    1.2577    1.1169    0.9884
  0.8718    0.6731    0.5152    0.3919    0.2967
  0.2238    0.1684    0.1266    0.0951    0.0714
  0.0537    0.0306    0.0176    0.0103    0.0062
  0.0039    0.0025    0.0017    0.0012    0.0009

  0.0006    0.0004    0.0003    0.0002    0.0001
  0.0001    0.        0.        0.        0.
  0
 70
  1.0000
  0.       -0.0845   -0.1607   -0.2294   -0.2907
 -0.3453   -0.3935   -0.4358   -0.4724   -0.5037
 -0.5301   -0.5694   -0.5924   -0.6014   -0.5980
 -0.5840   -0.5609   -0.5300   -0.4924   -0.4493
 -0.4015   -0.2954   -0.1798   -0.0590    0.0632

  0.1844    0.3024    0.4156    0.5230    0.6238
  0.7175    0.8825    1.0177    1.1243    1.2046
  1.2614    1.2974    1.3156    1.3186    1.3091
  1.2892    1.2261    1.1424    1.0477    0.9490
  0.8510    0.7568    0.6684    0.5868    0.5126

  0.4457    0.3334    0.2464    0.1805    0.1312
  0.0949    0.0683    0.0490    0.0351    0.0250
  0.0178    0.0089    0.0043    0.        0.
  0.        0.        0.        0.        0.
  1
 70
  1.0000
  0.        0.0009    0.0036    0.0080    0.0139
  0.0211    0.0297    0.0395    0.0504    0.0623
  0.0751    0.1032    0.1342    0.1674    0.2025
  0.2390    0.2765    0.3148    0.3536    0.3926
  0.4316    0.5087    0.5839    0.6562    0.7248

  0.7893    0.8493    0.9048    0.9555    1.0015
  1.0428    1.1120    1.1643    1.2012    1.2244
  1.2357    1.2367    1.2290    1.2139    1.1928
  1.1667    1.1036    1.0310    0.9536    0.8748
  0.7972    0.7224    0.6514    0.5849    0.5234

  0.4668    0.3685    0.2885    0.2244    0.1736
  0.1337    0.1026    0.0786    0.0600    0.0458
  0.0350    0.0205    0.0121    0.        0.
  0.        0.        0.        0.        0.

RMAX=    2.6993      RHO=    0.0364
X ATP VS
```

0.	-13.00000	0.	0.00377	-12.83144	-0.06019
0.00753	-12.66397	-0.11651	0.01130	-12.49779	-0.16917
0.01506	-12.33393	-0.21835	0.01883	-12.17237	-0.26422
0.02259	-12.01390	-0.30695	0.02636	-11.85835	-0.34671
0.03012	-11.70631	-0.38363	0.03389	-11.55752	-0.41787
0.03765	-11.41241	-0.44959	0.04518	-11.13258	-0.50592
0.05271	-10.86684	-0.55367	0.06024	-10.61454	-0.59372
0.06777	-10.37499	-0.62698	0.07531	-10.14732	-0.65423
0.08284	-9.93051	-0.67621	0.09037	-9.72377	-0.69361
0.09790	-9.52593	-0.70709	0.10543	-9.33642	-0.71726
0.11296	-9.15405	-0.72469	0.12802	-8.80866	-0.73346
0.14308	-8.48465	-0.73741	0.15814	-8.17810	-0.73992
0.17320	-7.88584	-0.74379	0.18826	-7.60592	-0.75099
0.20332	-7.33639	-0.76265	0.21839	-7.07666	-0.77908
0.23345	-6.82539	-0.79979	0.24851	-6.58282	-0.82401
0.26357	-6.34783	-0.85066	0.29369	-5.90198	-0.90750
0.32381	-5.48631	-0.96391	0.35394	-5.10139	-1.01589
0.38406	-4.74543	-1.06143	0.41418	-4.41880	-1.09978
0.44430	-4.11885	-1.13082	0.47442	-3.84559	-1.15488
0.50455	-3.59568	-1.17239	0.53467	-3.36908	-1.18403
0.56479	-3.16204	-1.19031	0.62504	-2.80351	-1.18934
0.68528	-2.50562	-1.17386	0.74552	-2.25720	-1.14767
0.80577	-2.04676	-1.11401	0.86601	-1.86737	-1.07573
0.92626	-1.71071	-1.03514	0.98650	-1.57343	-0.99425
1.04675	-1.44943	-0.95481	1.10699	-1.33793	-0.91854
1.16723	-1.23412	-0.88658	1.28772	-1.04877	-0.84014
1.40821	-0.88333	-0.82023	1.52870	-0.73448	-0.82577
1.64919	-0.59871	-0.85144	1.76968	-0.47650	-0.89087
1.89017	-0.36645	-0.93878	2.01066	-0.27017	-0.99166
2.13114	-0.18690	-1.04728	2.25163	-0.11861	-1.10444
2.37212	-0.06471	-1.16243	2.49261	-0.02726	-1.22092
2.61310	-0.00575	-1.27956	2.73359	-0.00152	-0.09831
2.85408	-0.00119	-0.08751	2.97456	-0.00151	-0.06398
3.09505	-0.00119	0.	3.21554	-0.00151	0.
3.33603	-0.00119	0.	3.45652	-0.00151	0.
3.57701	-0.00119	0.	3.69750	-0.00151	0.
3.81799	-0.00119	0.	3.93847	-0.00151	0.
4.05896	-0.00119	0.	4.17945	-0.00151	0.
4.29994	-0.00119	0.	4.42043	-0.00151	0.
4.54092	-0.00119	0.	4.66141	-0.00151	0.
4.78190	-0.00119	0.	4.90238	0.	0.
5.02287	0.	0.	5.14336	0.	0.
5.26385	0.	0.	5.38434	0.	0.
5.50483	0.	0.	5.62532	0.	0.
5.74580	0.	0.	5.86629	0.	0.
5.98678	0.	0.	6.10727	0.	0.
6.22776	0.	0.	6.34825	0.	0.
6.46874	0.	0.	6.58923	0.	0.
6.70971	0.	0.	6.83020	0.	0.
6.95069	0.	0.	7.07118	0.	0.

```
DEL FOR L=   0
     1.42533
    -1.52109
    -1.52427
    -1.52427
    -1.52587
FI

 0.            0.35843E-02  0.68209E-02  0.97296E-02  0.12330E-01
 0.14639E-01   0.16675E-01  0.18453E-01  0.19990E-01  0.21299E-01
 0.22396E-01   0.24001E-01  0.24905E-01  0.25194E-01  0.24949E-01
 0.24240E-01   0.23133E-01  0.21684E-01  0.19947E-01  0.17967E-01
 0.15788E-01   0.10975E-01  0.57578E-02  0.37204E-03 -0.50470E-02

-0.10359E-01 -0.15463E-01 -0.20279E-01 -0.24751E-01 -0.28836E-01
-0.32505E-01 -0.38530E-01 -0.42779E-01 -0.45299E-01 -0.46196E-01
-0.45611E-01 -0.43705E-01 -0.40645E-01 -0.36599E-01 -0.31731E-01
-0.26197E-01 -0.13707E-01 -0.18723E-03  0.13462E-01  0.26493E-01
 0.38297E-01  0.48397E-01  0.56434E-01  0.62163E-01  0.65443E-01

 0.66230E-01  0.60577E-01  0.46461E-01  0.26101E-01  0.23323E-02
-0.21762E-01 -0.43209E-01 -0.59458E-01 -0.68648E-01 -0.69771E-01
-0.62764E-01 -0.48487E-01 -0.28616E-01 -0.54515E-02  0.18340E-01
 0.40021E-01  0.57099E-01  0.67508E-01  0.70340E-01  0.64981E-01
 0.52147E-01  0.           0.           0.           0.

 0.           0.           0.           0.           0.
 0.           0.           0.           0.           0.
 0.           0.           0.           0.           0.
 0.           0.           0.           0.           0.
 0.           0.           0.           0.           0.

HF

 0.          -0.31178E-01 -0.58864E-01 -0.83297E-01 -0.10430E 00
-0.12220E 00 -0.13700E 00 -0.14909E 00 -0.15850E 00 -0.16565E 00
-0.17061E 00 -0.17519E 00 -0.17387E 00 -0.16823E 00 -0.15945E 00
-0.14861E 00 -0.13648E 00 -0.12371E 00 -0.11074E 00 -0.97922E-01
-0.85483E-01 -0.62328E-01 -0.41813E-01 -0.24035E-01 -0.87435E-02

 0.43360E-02  0.15531E-01  0.25084E-01  0.33230E-01  0.40122E-01
 0.45903E-01  0.54485E-01  0.59640E-01  0.61822E-01  0.61570E-01
 0.59325E-01  0.55617E-01  0.50848E-01  0.45473E-01  0.39781E-01
 0.34102E-01  0.23459E-01  0.14527E-01  0.76637E-02  0.27571E-02
-0.47515E-03 -0.24144E-02 -0.34035E-02 -0.37523E-02 -0.36924E-02

-0.34008E-02 -0.25620E-02 -0.17494E-02 -0.11340E-02 -0.71202E-03
-0.44349E-03 -0.27619E-03 -0.17524E-03 -0.11271E-03 -0.74581E-04
-0.50178E-04 -0.34842E-04 -0.24018E-04 -0.16793E-04 -0.12019E-04
-0.41528E-05  0.47018E-13  0.11755E-05  0.80045E-13  0.
 0.           0.           0.           0.           0.

 0.           0.           0.           0.           0.
 0.           0.           0.           0.           0.
 0.           0.           0.           0.           0.
 0.           0.           0.           0.           0.
 0.           0.           0.           0.           0.

E=   4.00000      L=   0 PS=  -1.52587
```

X ATP VS

```
0.         -13.00000    0.          0.00377  -12.83144   -0.06019
0.00753    -12.66397   -0.11651     0.01130  -12.49779   -0.16917
0.01506    -12.33393   -0.21835     0.01883  -12.17237   -0.26422
0.02259    -12.01390   -0.30695     0.02636  -11.85835   -0.34671
0.03012    -11.70631   -0.38363     0.03389  -11.55752   -0.41787

0.03765    -11.41241   -0.44959     0.04518  -11.13258   -0.50592
0.05271    -10.86684   -0.55367     0.06024  -10.61454   -0.59372
0.06777    -10.37499   -0.62698     0.07531  -10.14732   -0.65423
0.08284     -9.93051   -0.67621     0.09037   -9.72377   -0.69361
0.09790     -9.52593   -0.70709     0.10543   -9.33642   -0.71726

0.11296     -9.15405   -0.72469     0.12802   -8.80866   -0.73346
0.14308     -8.48465   -0.73741     0.15814   -8.17810   -0.73992
0.17320     -7.88584   -0.74379     0.18826   -7.60592   -0.75099
0.20332     -7.33639   -0.76265     0.21839   -7.07666   -0.77908
0.23345     -6.82539   -0.79979     0.24851   -6.58282   -0.82401

0.26357     -6.34783   -0.85066     0.29369   -5.90198   -0.90750
0.32381     -5.48631   -0.96391     0.35394   -5.10139   -1.01589
0.38406     -4.74543   -1.06143     0.41418   -4.41880   -1.09978
0.44430     -4.11885   -1.13082     0.47442   -3.84559   -1.15488
0.50455     -3.59568   -1.17239     0.53467   -3.36908   -1.18403

0.56479     -3.16204   -1.19031     0.62504   -2.80351   -1.18934
0.68528     -2.50562   -1.17386     0.74552   -2.25720   -1.14767
0.80577     -2.04676   -1.11401     0.86601   -1.86737   -1.07573
0.92626     -1.71071   -1.03514     0.98650   -1.57343   -0.99425
1.04675     -1.44943   -0.95481     1.10699   -1.33793   -0.91854

1.16723     -1.23412   -0.88658     1.28772   -1.04877   -0.84014
1.40821     -0.88333   -0.82023     1.52870   -0.73448   -0.82577
1.64919     -0.59871   -0.85144     1.76968   -0.47650   -0.89087
1.89017     -0.36645   -0.93878     2.01066   -0.27017   -0.99166
2.13114     -0.18690   -1.04728     2.25163   -0.11861   -1.10444

2.37212     -0.06471   -1.16243     2.49261   -0.02726   -1.22092
2.61310     -0.00575   -1.27956     2.73359   -0.00152   -0.09831
2.85408     -0.00119   -0.08751     2.97456   -0.00151   -0.06398
3.09505     -0.00119    0.          3.21554   -0.00151    0.
3.33603     -0.00119    0.          3.45652   -0.00151    0.

3.57701     -0.00119    0.          3.69750   -0.00151    0.
3.81799     -0.00119    0.          3.93847   -0.00151    0.
4.05896     -0.00119    0.          4.17945   -0.00151    0.
4.29994     -0.00119    0.          4.42043   -0.00151    0.
4.54092     -0.00119    0.          4.66141   -0.00151    0.

4.78190     -0.00119    0.          4.90238    0.          0.
5.02287      0.         0.          5.14336    0.          0.
5.26385      0.         0.          5.38434    0.          0.
5.50483      0.         0.          5.62532    0.          0.
5.74580      0.         0.          5.86629    0.          0.

5.98678      0.         0.          6.10727    0.          0.
6.22776      0.         0.          6.34825    0.          0.
6.46874      0.         0.          6.58923    0.          0.
6.70971      0.         0.          6.83020    0.          0.
6.95069      0.         0.          7.07118    0.          0.
```

```
DEL FOR L=   1
     -0.67180
     -0.49812
     -0.50443
     -0.50443
     -0.50749
FI

 0.            0.13799E-04  0.53856E-04  0.11826E-03  0.20520E-03
 0.31295E-03   0.43989E-03  0.58447E-03  0.74525E-03  0.92084E-03
 0.11100E-02   0.15238E-02  0.19779E-02  0.24639E-02  0.29748E-02
 0.35039E-02   0.40456E-02  0.45948E-02  0.51467E-02  0.56974E-02
 0.62432E-02   0.73076E-02  0.83186E-02  0.92597E-02  0.10118E-01

 0.10886E-01   0.11555E-01  0.12123E-01  0.12587E-01  0.12947E-01
 0.13203E-01   0.13412E-01  0.13239E-01  0.12714E-01  0.11876E-01
 0.10762E-01   0.94143E-02  0.78723E-02  0.61753E-02  0.43611E-02
 0.24660E-02  -0.14318E-02 -0.52670E-02 -0.88213E-02 -0.11911E-01
-0.14389E-01  -0.16146E-01 -0.17111E-01 -0.17252E-01 -0.16574E-01

-0.15122E-01  -0.10226E-01 -0.34911E-02  0.38687E-02  0.10566E-01
 0.15455E-01   0.17714E-01  0.16975E-01  0.13375E-01  0.75283E-02
 0.41919E-03  -0.67601E-02 -0.12807E-01 -0.16706E-01 -0.17800E-01
-0.15899E-01  -0.11319E-01 -0.48287E-02  0.24772E-02  0.93637E-02
 0.14664E-01   0.           0.           0.           0.

 0.            0.           0.           0.           0.
 0.            0.           0.           0.           0.
 0.            0.           0.           0.           0.
 0.            0.           0.           0.           0.
 0.            0.           0.           0.           0.

HF

 0.           -0.11397E-03 -0.43383E-03 -0.92708E-03 -0.15611E-02
-0.23047E-02  -0.31337E-02 -0.40218E-02 -0.49479E-02 -0.58923E-02
-0.68397E-02  -0.86885E-02 -0.10418E-01 -0.11969E-01 -0.13323E-01
-0.14467E-01  -0.15411E-01 -0.16165E-01 -0.16756E-01 -0.17197E-01
-0.17520E-01  -0.17877E-01 -0.17992E-01 -0.17970E-01 -0.17896E-01

-0.17802E-01  -0.17717E-01 -0.17633E-01 -0.17548E-01 -0.17438E-01
-0.17297E-01  -0.16850E-01 -0.16140E-01 -0.15138E-01 -0.13886E-01
-0.12432E-01  -0.10864E-01 -0.92402E-02 -0.76431E-02 -0.61097E-02
-0.46987E-02  -0.23040E-02 -0.53708E-03  0.63359E-03  0.13173E-02
 0.16302E-02   0.16929E-02  0.15959E-02  0.14161E-02  0.12002E-02

 0.98434E-03   0.61307E-03  0.35282E-03  0.19357E-03  0.10248E-03
 0.54290E-04   0.29048E-04  0.16454E-04  0.96996E-05  0.61369E-05
 0.39281E-05   0.26639E-05  0.16947E-05  0.10957E-05  0.69124E-06
 0.23546E-06  -0.34894E-14 -0.60744E-07 -0.55584E-14  0.
 0.            0.           0.           0.           0.

 0.            0.           0.           0.           0.
 0.            0.           0.           0.           0.
 0.            0.           0.           0.           0.
 0.            0.           0.           0.           0.
 0.            0.           0.           0.           0.

E=   6.00000      L=    1 PS=   -0.50749
```

5. Calculation of temperature dependent phase shifts

The following programs given the zero temperature phase shifts as input, calculate the phase shifts appropriate to finite temperature calculations. The assumption is made that atoms vibrate isotropically, and the program works by solving equation (6.55), for $\delta_\ell(T)$:

$$\exp(i\delta_\ell(T))\sin(\delta_\ell(T)) = \sum_{L\ell''} i^L \exp(-2\alpha\kappa^2)$$

$$\times\, j_L(-2i\alpha\kappa^2)\exp(i\delta_{\ell''})\sin(\delta_{\ell''})[4\pi(2L+1)(2\ell''+1)(2\ell+1)^{-1}]^{\frac{1}{2}}$$

$$\times \int Y_{\ell''0}\, Y_{L0}\, Y_{\ell 0}\, d\Omega \qquad \text{(A.82)}$$

$\frac{1}{2}\kappa^2$ = energy relative to muffin tin zero

$$\alpha = \frac{3T}{2mk_B\Theta^2} \qquad \text{(A.83)}$$

in the expression for α our units are such that $\hbar$ is unity, T and Θ are measured in °K, the mass of the atom, m, in electron mass units, and k_B in Hartrees/°K.

Basic variables in the program to calculate $\delta_\ell(T)$

EMZERO	mass of the atom in units of the *proton* mass.
DTEMP	Debye temperature in °K.
E	energy relative to the muffin tin zero measured in Hartrees.
PHS	phase shifts at $T = 0$ for energy E, for $\ell = 0, 1, \ldots (N3 - 1)$.
DELP	phase shifts at finite temperature for energy E, for $\ell = 0, 1, \ldots (N2 - 1)$.
TEMP	temperature in °K.
ALFA	the number α defined in (A.83).

```
COMMENT PROGRAM TO CALCULATE PHASE SHIFTS AS A FUNCTION OF
C         TEMPERATURE.  PHASE SHIFTS FOR TEMPERATURE ZERO ARE
C         READ IN, AND THE NON-ZERO PHASE SHIFTS FOR ANY OTHER
C         REQUIRED TEMPERATURE ARE CALCULATED AND PRINTED OUT.
C         INPUT:
C         EMZERO= ATOMIC WEIGHT OF THE ATOM CONCERNED
C         DTEMP = DEBYE TEMPERATURE FOR THE ATOM CONCERNED
C         E     = ENERGY
C         PHS   = PHASE SHIFTS FOR ENERGY E, TEMP. ZERO, AND
C                 L=0,6
C         TEMP  = TEMPERATURE AT WHICH PHASE SHIFTS ARE TO BE
C                 CALCULATED
      DIMENSION PPP(13,7,7),PHS(7)
      COMPLEX BJ(13),DEL(7),SUM(7),CTAB(7)
      COMPLEX Z,FF,FN,FN1,FN2,CI,CS,CC,CL
      COMPLEX CMPLX,CSIN,CCOS,CEXP,CLOG
      PI=3.14159265
      CI=CMPLX(0.0,1.0)
      N3=7
      N2=7
      N1=N2+N3-1
      CALL CPPP(PPP,N1,N2,N3)
1     READ(5,2)EMZERO
2     FORMAT(F9.4)
      IF(EMZERO)27,27,3
3     READ(5,35)DTEMP
35    FORMAT(F9.4)
      WRITE(6,4)EMZERO,DTEMP
4     FORMAT(15H1ATOMIC WEIGHT=,F9.4,5X,11HDEBYE TEMP=,F9.4)
5     READ(5,6)E
6     FORMAT(F9.4)
      IF(E)1,65,65
65    READ(5,66)PHS
66    FORMAT(5F9.4)
7     READ(5,8)TEMP
8     FORMAT(F9.4)
      IF(TEMP)5,9,9
9     ALFA=(1.5*TEMP)/(1836.0*EMZERO*3.169E-6*DTEMP*DTEMP)
      FALFE=-4.0*ALFA*E
      WRITE(6,10)TEMP,E
10    FORMAT(6H0TEMP=,F9.4,5X,7HENERGY=,F9.4,//,
     1 11X,11HPHASE SHIFT,23X,8HABSOLUTE)
C
COMMENT BJ(N1) IS LOADED WITH SPHERICAL BESSEL FUNCTIONS OF
C         THE FIRST KIND; ARGUMENT Z
      Z=FALFE*CI
      LLL=N1-1
      CS=CSIN(Z)
      CC=CCOS(Z)
      BJ(1)=CS/Z
      LP=INT(3.5*CABS(Z))
      IF(LLL-LP)14,14,11
11    FF=CMPLX(1.0,0.0)
      CL=FF
      LPP=LP+1
      DO 12 MM=1,LPP
      CL=CL+(2.0,0.0)
12    FF=FF*Z/CL
      FL=FLOAT(LP)
```

```
      DO 13 J=LPP,LLL
      FL=FL+1.0
      B=1.0/(12.0*FL+42.0)
      FN=1.0-B*Z*Z
      B=1.0/(8.0*FL+20.0)
      FN=1.0-B*Z*Z*FN
      B=1.0/(4.0*FL+6.0)
      FN=1.0-B*Z*Z*FN
      BJ(J+1)=FF*FN
      CL=CL+(2.0,0.0)
13    FF=FF*Z/CL
      LLL=LP
      IF(LP)14,17,14
14    BJ(2)=(BJ(1)-CC)/Z
      FN1=BJ(2)
      FN2=BJ(1)
      AM=3.0
      IF(LP-1)15,17,15
15    ILL=LLL+1
      DO 16 IA=3,ILL
      FN=AM*FN1/Z-FN2
      FN2=FN1
      FN1=FN
      AM=AM+2.0
16    BJ(IA)=FN
17    CS=CMPLX(1.0,0.0)
      FL=1.0
      DO 18 I=1,N1
      BJ(I)=EXP(FALFE)*FL*CS*BJ(I)
      FL=FL+2.0
18    CS=CS*CI
C
      FL=1.0
      DO 19 I=1,N3
      CTAB(I)=(CEXP(2.0*PHS(I)*CI)-(1.0,0.0))*FL
19    FL=FL+2.0
C
      ITEST=0
      LLLMAX=N2
      FL=1.0
      DO 26 LLL=1,N2
      SUM(LLL)=CMPLX(0.0,0.0)
      DO 20 L=1,N3
      LLMIN=IABS(L-LLL)+1
      LLMAX=L+LLL-1
      DO 20 LL=LLMIN,LLMAX
20    SUM(LLL)=SUM(LLL)+PPP(LL,LLL,L)*CTAB(L)*BJ(LL)
      DEL(LLL)=-CI*CLOG(SUM(LLL)+(1.0,0.0))/(2.0,0.0)
      ABSDEL=CABS(DEL(LLL))
      IL=LLL-1
      WRITE(6,21)IL,DEL(LLL),ABSDEL
21    FORMAT(1H ,I3,5X,2E13.5,5X,E13.5)
      IF(ABSDEL-1.0E-2)23,22,22
22    ITEST=0
      GO TO 26
23    IF(ITEST-1)25,24,25
24    LLLMAX=LLL
      GO TO 7
25    ITEST=1
```

```
26      FL=FL+2.0
        GO TO 7
27      STOP
        END
C
C
COMMENT CPPP TABULATES THE FUNCTION PPP(I1,I2,I3), EACH ELEMENT
C         CONTAINING THE INTEGRAL OF THE PRODUCT OF THREE LEGENDRE
C         FUNCTIONS P(I1),P(I2),P(I3). THE INTEGRALS ARE CALCULATED
C         FOLLOWING GAUNT'S SUMMATION SCHEME SET OUT BY SLATER:
C         ATOMIC STRUCTURE, VOL1, 309,310
        SUBROUTINE CPPP(PPP,N1,N2,N3)
        DIMENSION PPP(N1,N2,N3),F(30)
        F(1)=1.0
        DO 1 I=1,29
1       F(I+1)=F(I)*FLOAT(I)
        DO 11 I1=1,N1
        DO 11 I2=1,N2
        DO 11 I3=1,N3
        IM1=I1
        IM2=I2
        IM3=I3
        IF(I1-I2)2,3,3
2       IM1=I2
        IM2=I1
3       IF(IM2-I3)4,7,7
4       IM3=IM2
        IM2=I3
5       IF(IM1-IM2)6,7,7
6       J=IM1
        IM1=IM2
        IM2=J
7       CONTINUE
        A=0.0
        IS=I1+I2+I3-3
        IF(MOD(IS,2)-1)8,11,8
8       IF(IABS(IM2-IM1)-IM3+1)9,9,11
9       SUM=0.0
        IS=IS/2
        SIGN=1.0
        DO 10 IT=1,IM3
        SIGN=-SIGN
        IA1=IM1+IT-1
        IA2=IM1-IT+1
        IA3=IM3-IT+1
        IA4=IM2+IM3-IT
        IA5=IM2-IM3+IT
10      SUM=SUM-SIGN*F(IA1)*F(IA4)/(F(IA2)*F(IA3)*F(IA5)*F(IT))
        IA1=2+IS-IM1
        IA2=2+IS-IM2
        IA3=2+IS-IM3
        IA4=3+2*(IS-IM3)
        A=-(-1.0)**(IS-IM2)*F(IA4)*F(IS+1)*F(IM3)*SUM/
       1(F(IA1)*F(IA2)*F(IA3)*F(2*IS+2))
11      PPP(I1,I2,I3)=A
        RETURN
        END
```

Test run of program to calculate $\delta_\ell(T)$

There follows a data stream for the program, using zero temperature phase shifts for nickel, then the output from the program.

```
  58.7100
 440.0000
   4.0000
   1.0100   -1.0900   -0.5800    0.4200    0.1100
   0.0300    0.0100    0.0000    0.0000    0.0000
 300.0000
 600.0000
-100.0000
-100.0000
-100.0000
```

```
ATOMIC WEIGHT=   58.7100        DEBYE TEMP= 440.0000

TEMP= 300.0000        ENERGY=    4.0000

             PHASE SHIFT                             ABSOLUTE
   0         0.10602E 01   0.80404E-01         0.10632E 01
   1        -0.10750E 01   0.43623E-01         0.10759E 01
   2        -0.56782E 00   0.55128E-01         0.57049E 00
   3         0.38134E 00   0.35907E-01         0.38303E 00
   4         0.11771E 00   0.47956E-02         0.11780E 00
   5         0.32799E-01   0.46002E-03         0.32802E-01
   6         0.10478E-01   0.37301E-04         0.10478E-01

TEMP= 600.0000        ENERGY=    4.0000

             PHASE SHIFT                             ABSOLUTE
   0         0.11161E 01   0.15858E 00         0.11273E 01
   1        -0.10602E 01   0.85856E-01         0.10637E 01
   2        -0.55357E 00   0.10710E 00         0.56384E 00
   3         0.34439E 00   0.66263E-01         0.35070E 00
   4         0.12277E 00   0.94733E-02         0.12313E 00
   5         0.35598E-01   0.11258E-02         0.35616E-01
   6         0.11065E-01   0.10303E-03         0.11066E-01
```

6. Calculation of LEED intensities from overlayer systems

The following set of subroutines calculates intensities from an adsorbate system. Layers of the bulk are assumed identical and each layer has only one atom per unit cell of a bulk layer. The overlayer can have any structure that is simply related to the bulk, and is restricted to one atom per unit overlayer-cell. Like the clean surface programs in Appendix C1 no surface barrier is included, and all energies are relative to the muffin-tin zero. Programming was in standard ASA fortran IV and the subroutines have been tested on an IBM 370/165 under WATFIV and FORTG compilers.

Organisation of calculations is parallel with that in the clean surface programs. Only the significant differences will be described.

Calculation of reflection and transmission coefficients of bulk layers is divided between three subroutines SLIND, FMAT, and MSMAT. In calculating the lattice sum the method of division into sublattices described in Section VIIF is used. FMAT calculates formula (7.53) for

$$
\begin{aligned}
F_{\ell m}(t) &= i\kappa(-1)^{\ell+m}\sum_u{}' h_\ell^{(1)}(\kappa|\mathbf{R}_t^{(0)} + \mathbf{R}_u^{(C)}|) \\
&\quad \times Y_{\ell-m}(\Omega(-\mathbf{R}_t^{(0)} - \mathbf{R}_u^{(C)}))\exp\left[i\mathbf{k}_{0\|}\cdot(\mathbf{R}_u^{(C)} + \mathbf{R}_t^{(0)})\right] \\
&= i\kappa(-1)^m Y_{\ell-m}(\theta = \tfrac{\pi}{2}, \phi = 0)\sum_u{}' h_\ell^{(1)}(\kappa|\mathbf{R}_t^{(0)} + \mathbf{R}_u^{(C)}|) \\
&\quad \times \exp\left(-im\phi(\mathbf{R}_t^{(0)} + \mathbf{R}_u^{(C)})\right) \\
&\quad \times \exp\left[i\mathbf{k}_{0\|}\cdot(\mathbf{R}_t^{(0)} + \mathbf{R}_u^{(C)})\right].
\end{aligned} \tag{A.84}
$$

The latter manipulation follows because we specialise to one atom per unit cell, and all atoms are coplanar. The values of $F/Y_{\ell-m}$ for each sublattice are stored in the vector FLMS. The function of SLIND is as follows: in subroutine FMAT, calculations are speeded up by noting that in (A.84), if the bulk layers have an IDEG-fold axis of symmetry, only the angular factors change for the symmetrically related lattice vectors. Hence they can be added to F without recalculating $h_\ell^{(1)}$. However, the difficulty arises that rotation by 2π/IDEG although it takes the *lattice* into itself, may take one *sublattice* into a different sublattice. In that case the contribution from the new point must be added to the FLMS appropriate to the different sublattice. SLIND calculates JJS such that a rotation of 2πJ/IDEG takes the Ith sublattice into the JJS(I,J)th sublattice, and hence FMAT knows where to put the new contribution.

Subroutine MSMAT called with the parameters set as in the first call, calculates the reflection and transmission matrices of the bulk, by assembling the quantities FLMS in a manner appropriate to whichever of the factorised reflection transmission matrices is being found. A subsequent call of MSMAT with a new set of parameters calculates the overlayer reflection and transmission matrices. Use has been made of the fact that when the bulk and overlayer lattices are simply related, the overlayer lattice coincides with the first sublattice of the bulk. Therefore the sum over the overlayer lattice has already been made in FLMS(1) and need not be recalculated.

The overlayer and layers of the bulk are put together by RFS perturbation theory in subroutine PERTA. Its working is closely parallel to that of the clean surface subroutine except that different matrices must be used to describe transmission and reflection by overlayer and bulk layers, and advantage has been taken of the factorised form of bulk reflection and transmission matrices.

Calls of FMAT + MSMAT are slow relative to calls of PERTA. Therefore in a structure determination it would be advantageous to alter the program slightly so that once reflection and transmission matrices have been found, several calls of PERTA are made for different positions of the overlayer relative to the bulk. Once again we have presented the program in its simplest form. Sophistications can be added if desired.

Basic variables in the main overlayer LEED program

LMAX	maximum value of ℓ for the phase shifts, set equal to 4.
CLM	values of the quantities defined in (4.56) tabulated in the order required by the program. 809 values, appropriate to LMAX = 4, are read in.
SPA	characteristic length in the bulk layers, e.g. in a cubic lattice, SPA is the length of unit cell side in atomic units.
SURF	a 3D vector normal to the surface pointing inwards.
ASM1,ASM2	two orthogonal 3D vectors lying in the plane of the surface, to serve as axes.
AR1, ARS	two 2D vectors forming the sides of unit cell of a bulk layer, referred to axes ASM1 and ASM2. They are read into AS as 3D vectors referred to *xyz* axes *in units of SPA*. Multiplication by SPA and projection onto ASM1 and ASM2 being done in the program.
ASB	a 3D vector describing displacement of successive bulk layers deeper into the crystal. The data stream expresses ASB in units of SPA, referred to axes along SURF, ASM1 and ASM2, in that order.
BR1, BR2	two 2D vectors forming the sides of unit overlayer cell referred to axes ASM1 and ASM. They are read into AS as 3D vectors referred to *xyz* axes *in unit of SPA*. Multiplication by SPA and projection onto ASM1 and ASM2 being done in the program.

ASA	a 3D vector describing displacement of an atom in the first bulk layer relative to an atom in the overlayer. The data stream expresses ASA in units of SPA referred to axes along SURF, ASM1, ASM2.
TVA	area of overlayer unit cell.
RBR1,RBR2	2D vectors forming sides of the unit cell of the overlayer's reciprocal lattice, referred to axes ASM1, ASM2.
TVB	area of unit cell of a bulk layer.
RAR1,RAR2	2D vectors forming sides of the unit cell of a bulk layer's reciprocal lattice, referred to axes ASM1, ASM2.
IDEG	order of the axis of symmetry that passes through the centres of overlayer atoms.
NL1,NL2	are explained by reference to Section VIIB. Because the overlayer and bulk layer lattices are simply related, a general point in a *bulk layer lattice* can by (7.10) be written

$$M\cdot\mathbf{BR1} + N\cdot\mathbf{BR2} + m\cdot\mathbf{AR1} + n\cdot\mathbf{AR2}$$

MN are integers taking any value; **AR1** etc. stand for the two components of the vectors; m and n take a finite number of values as explained in Section VIIB:

$$0 \leqslant m < \mathrm{NL1}$$

$$0 \leqslant n < \mathrm{NL2}$$

NL	the number of sublattices into which the bulk layers are decomposed. From above it follows that

$$\mathrm{NL} = \mathrm{NL1}*\mathrm{NL2}.$$

NSL(I)	The number of vectors considered in the Ith reciprocal sublattice of the overlayer. NSL(1) is assumed to be at least as large as any other member of the set.
PQF(J,K)	The Jth component of the Kth reciprocal lattice vector in units of RAR1 and RAR2 (this corresponds to the conventional naming of beams in terms of the clean surface beams).
PQ(J,K)	The Jth component of the Kth reciprocal lattice vector referred to axes ASM1, ASM2.
NA	total number of reciprocal lattice vectors.
V(J,K)	contains the polar coordinates of the vector $m\mathbf{AR1} + n\mathbf{AR2}$ (see definition of NL1, NL2)

$$V(J,1) = |m\mathbf{AR1} + n\mathbf{AR2}|$$

$$V(J,2) = \arctan\left[(mAR1(2) + nAR2(2)) / (mAR1(1) + nAR2(1))\right].$$

in the order in which they are to be used in the program.

VL(J,K)	is used to identify the sublattice defined by **ADR** = m**AR1** + n**AR2**: $\mathrm{VL}(J,1) = \exp[i(\mathbf{ADR}\cdot\mathbf{RBR1})]$, $\mathrm{VL}(J,2) = \exp[i(\mathbf{ADR}\cdot\mathbf{RBR2})]$.
JJS(I,J)	as explained in the text above, a rotation of 2πJ/IDEG takes the Ith sublattice into the JJS(I,J)th sublattice.
VPI	the imaginary part of the potential in the crystal, V_{0i}, in Hartrees (VPI must always be negative).
E,E,E2, PHS,PHS1, PHS2	phase shifts at energy E, PHS, are found by linear interpolation of phase shifts at energies $E1$ and $E2$. In PHS1 and PHS2, phase shifts of bulk atoms for $\ell = 0, 1\ldots$ LMAX take the first 5 spaces, and phase shifts for the overlayer atoms the next 5. Energies are measured in Hartrees relative to the muffin tin zero.
THETA,FI	THETA is the angle between the normal to the surface and the incident beam, FI the angle between the plane containing the normal and the incident beam and ASM1 (rotations towards ASM2 are positive); angles are in radians.
RCB,TCB	reflection and transmission matrices for a single bulk layer referred to an origin at the centre of a bulk atom. These matrices factorise as explained in Section VIIE. The various factors are stored by the first element such that, e.g. RCB(NSL(I − 1) + J,K) is the (J,K)th element of the Ith factor.
RCA,TCA	reflection and transmission matrices for the overlayer referred to an origin at the centre of an overlayer atom.
WV	reflected amplitudes.
AT	reflected intensities

```
COMMENT MAIN PROGRAM FOR CALCULATION OF LEED INTENSITIES
C        FOR THE GENERAL OVERLAYER PROBLEM, USING RFS
C        PERTURBATION THEORY.
C        DIMENSIONS ARE FIXED FOR N=13,NM=9,NL=2,LMAX=4
      DIMENSION AS(3),AT(13),SURF(3),ASM1(3),ASM2(3),
     1PHS(5),PQF(2,13),PQ(2,13),RCB(13,13),TCB(13,13)
      DIMENSION FLMS(2,45),JFLM(9,17)
      DIMENSION RCA(13,9),TCA(13,9),VL(2,2),V(2,2),JJS(2,6),
     1 AST(2),NSL(2),CLM(809),AF(5),ASA(3),ASB(3)
      DIMENSION PHS1(10),PHS2(10),WV(13),AR1(2),AR2(2),
     1 BR1(2),BR2(2),RAR1(2),RAR2(2),RBR1(2),RBR2(2)
      COMPLEX AF,FLMS,RCA,TCA,RCB,TCB,VL,WV,BM,CM,RM
      COMMON E,SPA,THETA,FI,VPI,CLM
      COMMON /X1/TV,PQ,ASA,ASB
      COMMON /X2/PHS,AR1,AR2
      COMMON /X3/ BR1,BR2,RBR1,RBR2,NL1,NL2
COMMENT EMACH IS THE MACHINE ACCURACY AND HAS BEEN
C        FIXED FOR AN IBM COMPUTER
      EMACH=1.0E-6
      EPSD=10.0*EMACH
      PI=3.14159265
      LMAX=4
      LMAX1=LMAX+1
      READ (2,1) (CLM(J),J=1,809)
1     FORMAT(5E14.5)
COMMENT SPA IS THE CHARACTERISTIC LENGTH OF THE CRYSTAL
      READ (5,3) SPA
      READ (5,2) SURF,ASM1,ASM2
2     FORMAT(3F7.1)
      READ (5,3) AS
3     FORMAT(3F12.5)
      WRITE (6,4) SPA
4     FORMAT(5H SPA=,1F12.5)
      WRITE (6,5)
5     FORMAT(///,15H SURF ASM1 ASM2)
      WRITE (6,2) SURF,ASM1,ASM2
      A=0.0
      B=A
      C=A
      D=A
      DO 6 J=1,3
      A=A+ASM1(J)**2
      B=B+ASM2(J)**2
      C=C+AS(J)*ASM1(J)*SPA
6     D=D+AS(J)*ASM2(J)*SPA
      AR1(1)=C/SQRT(A)
      AR1(2)=D/SQRT(B)
      WRITE (6,7) AS
7     FORMAT(18H BULK SURFACE VECS,/ ,3F12.5)
      READ (5,3) AS
      WRITE (6,3) AS
      C=0.0
      D=C
      DO 8 J=1,3
      C=C+AS(J)*ASM1(J)*SPA
8     D=D+AS(J)*ASM2(J)*SPA
      AR2(1)=C/SQRT(A)
      AR2(2)=D/SQRT(B)
COMMENT AR1,2 DEFINE THE UNIT CELL WITHIN A BULK LAYER
```

```
      READ (5,3) AS
      DO 9 J=1,3
9     ASA(J)=AS(J)*SPA
      WRITE(6,10)ASA
10    FORMAT(24H BULK LAYER DISPLACEMENT,/,3F12.5)
      READ (5,3) AS
      WRITE (6,11) AS
11    FORMAT(23H OVERLAYER SURFACE VECS,/,3F12.5)
      C=0.0
      D=C
      DO 12 J=1,3
      C=C+AS(J)*ASM1(J)*SPA
12    D=D+AS(J)*ASM2(J)*SPA
      BR1(1)=C/SQRT(A)
      BR1(2)=D/SQRT(B)
      READ (5,3) AS
      WRITE (6,3) AS
      C=0.0
      D=C
      DO 13 J=1,3
      C=C+AS(J)*ASM1(J)*SPA
13    D=D+AS(J)*ASM2(J)*SPA
      BR2(1)=C/SQRT(A)
      BR2(2)=D/SQRT(B)
COMMENT BR1,2 DEFINE THE UNIT CELL IN THE OVERLAYER
      READ (5,3) AS
      DO 14 J=1,3
14    ASB(J)=AS(J)*SPA
      WRITE (6,15) ASB
15    FORMAT(23H OVERLAYER DISPLACEMENT,/,3F12.5)
      TVA=ABS(AR1(1)*AR2(2)-AR1(2)*AR2(1))
      ATV=2.0*PI/TVA
      RAR1(1)=AR2(2)*ATV
      RAR1(2)=-AR2(1)*ATV
      RAR2(1)=AR1(2)*ATV
      RAR2(2)=-AR1(1)*ATV
COMMENT RAR1,2 DEFINE THE UNIT CELL OF THE RECIPROCAL LATTICE
C       CORRESPONDING TO A BULK LAYER
      TVB=ABS(BR1(1)*BR2(2)-BR1(2)*BR2(1))
      ATV=2.0*PI/TVB
      RBR1(1)=BR2(2)*ATV
      RBR1(2)=-BR2(1)*ATV
      RBR2(1)=BR1(2)*ATV
      RBR2(2)=-BR1(1)*ATV
COMMENT RBR1,2 DEFINE THE UNIT CELL OF THE RECIPROCAL LATTICE
C       CORRESPONDING TO THE OVERLAYER
      READ(5,16)IDEG,NL1,NL2
16    FORMAT(3I5)
      NL=NL1*NL2
COMMENT IDEG IS THE ORDER OF THE AXIS OF SYMMETRY NORMAL TO
C       THE SURFACE.
C       NL1,2 ARE USED WITH AR1,2 IN SUBROUTINE SLIND TO
C       CONSTRUCT THE ADDING VECTORS WHICH DEFINE THE SUB-
C       LATTICES WITHIN A BULK LAYER.
C       NL IS THE NUMBER OF SUBLATTICES SO DEFINED
      WRITE(6,17)IDEG,NL1,NL2,NL
17    FORMAT(6H IDEG=,I2,5X,5H NL1=,I2,5X,5H NL2=,
     1 I2,5X,4H NL=,I3)
      WRITE(6,18)
```

```
18     FORMAT(15H BEAMS INCLUDED,//,14H    PQF1    PQF2,
      1 5X,14H    PQ1    PQ2)
       N=0
       NM=0
COMMENT PQ ARE THE RECIPROCAL LATTICE VECTORS.
C         NSL RECORDS THE NUMBER OF LATTICE VECTORS INCLUDED
C         FOR EACH RECIPROCAL SUBLATTICE.
C         N IS THE TOTAL NUMBER OF RECIPROCAL LATTICE VECTORS
       DO 21 I=1,NL
       READ(5,19)NSL(I)
       WRITE(6,19)NSL(I)
19     FORMAT(I5)
       NB=NSL(I)
       NM=MAX0(NB,NM)
COMMENT NM IS THE MAXIMUM NUMBER OF RECIPROCAL LATTICE VECTORS
C         INCLUDED FOR ANY ONE SUBLATTICE, AND IS THE VALUE
C         REQUIRED FOR THE SECOND DIMENSION OF RCA,TCA
       DO 21 J=1,NB
       N=N+1
       READ(5,20)PQF(1,N),PQF(2,N)
20     FORMAT(2F7.1)
       PQ(1,N)=PQF(1,N)*RAR1(1)+PQF(2,N)*RAR2(1)
       PQ(2,N)=PQF(1,N)*RAR1(2)+PQF(2,N)*RAR2(2)
21     WRITE(6,22)PQF(1,N),PQF(2,N),PQ(1,N),PQ(2,N)
22     FORMAT(2F7.1,5X,2F7.3)
       WRITE(6,23)N
23     FORMAT(3H0N=,I5)
COMMENT A MAPPING BETWEEN THE SUBLATTICES IN SYMMETRICALLY
C         RELATED SECTORS IS FOUND
       CALL SLIND(V,VL,JJS,AST,NL,IDEG,EPSD)
       READ (5,3) VPI
       WRITE (6,24) VPI
24     FORMAT(5H0VPI=,1F12.5)
25     READ (5,26) E1,PHS1,E2,PHS2
26     FORMAT(2(1F12.5,/,2(5F12.5,/)))
       WRITE (6,27) E1,PHS1,E2,PHS2
27     FORMAT(2(4H E1=,1F12.5,/,4H PHS,/,2(5F12.5,/)))
28     READ (4,3) THETA,FI
29     READ (4,3) E
       IF (E) 30,32,32
30     IF (E+1.5) 31,28,28
31     IF (E+2.5) 35,25,25
32     FAC=(E-E1)/(E2-E1)
COMMENT LINEAR INTERPOLATION OF PHASE SHIFTS FOR BULK
       DO 33 L=1,LMAX1
33     PHS(L)=PHS1(L)+FAC*(PHS2(L)-PHS1(L))
COMMENT THE QUANTITIES FLMS ARE COMPUTED FOR EACH SUBLATTICE
       CALL FMAT(FLMS,JFLM,V,JJS,AST,NL,IDEG,LMAX)
COMMENT REFLECTION AND TRANSMISSION MATRICES FOR
C         A SINGLE BULK LAYER ARE FOUND.  THESE FACTORISE
C         ACCORDING TO RECIPROCAL SUBLATTICES
       TV=TVA
       CALL MSMAT(RCA,TCA,N,NM,FLMS,JFLM,V,NSL,NL,LMAX,0,EPSD)
COMMENT LINEAR INTERPOLATION OF PHASE SHIFTS FOR SURFACE
       DO 34 L=1,LMAX1
       LL=L+LMAX1
34     PHS(L)=PHS1(LL)+FAC*(PHS2(LL)-PHS1(LL))
COMMENT REFLECTION AND TRANSMISSION MATRICES FOR THE
C         OVERLAYER ARE FOUND
```

```
      TV=TVB
      CALL MSMAT(RCB,TCB,N,N,FLMS,JFLM,V,NSL,NL,LMAX,1,EPSD)
COMMENT PERTA CALCULATES REFLECTED AMPLITUDES BY RFS
C         PERTURBATION THEORY
      CALL PERTA(RCA,TCA,N,RCB,TCB,NSL,NL,NM,WV)
COMMENT REFLECTED AMPLITUDES,WV,ARE USED TO CALCULATE
C         REFLECTED INTENSITIES,AT
      CALL RINT2(N,WV,PQF,PQ,AT)
      GO TO 29
35    STOP
      END
C
C
C
COMMENT SLIND SETS UP A MATRIX JJS(JS,J) CONTAINING DETAILS OF
C         HOW THE SUBLATTICES JS ARE TRANSFORMED INTO ONE
C         ANOTHER BY ROTATIONS THROUGH J*2.0*PI/IDEG.  V(JS,2)
C         CONTAINS THE ADDING VECTORS DEFINING THE SUBLATTICES
C         JS IN POLAR FORM.  AST RECORDS WHETHER AN ADDING
C         VECTOR IS COLLINEAR WITH BR1 - THIS INFORMATION IS
C         USED IN CONTROLLING THE SUMMATION SCHEME IN FMAT
      SUBROUTINE SLIND(V,VL,JJS,AST,NL,IDEG,EPSD)
      DIMENSION V(NL,2),JJS(NL,6),VL(NL,2),AST(NL)
      DIMENSION PHS(5),AR1(2),AR2(2),BR1(2),BR2(2),
     1 RBR1(2),RBR2(2)
      COMPLEX VL,VLA,VLB,CI
      COMPLEX CEXP,CMPLX
      COMMON /X2/ PHS,AR1,AR2
      COMMON /X3/ BR1,BR2,RBR1,RBR2,NL1,NL2
C
      CI=CMPLX(0.0,1.0)
      PI=3.14159265
      I=1
      S1=0.0
      DO 4 J=1,NL1
      S2=0.0
      DO 3 K=1,NL2
      ADR1=S1*AR1(1)+S2*AR2(1)
      ADR2=S1*AR1(2)+S2*AR2(2)
      V(I,1)=SQRT(ADR1*ADR1+ADR2*ADR2)
      V(I,2)=0.0
      IF(V(I,1))2,2,1
1     V(I,2)=ATAN2(ADR2,ADR1)
2     VL(I,1)=CEXP(CI*(ADR1*RBR1(1)+ADR2*RBR1(2)))
      VL(I,2)=CEXP(CI*(ADR1*RBR2(1)+ADR2*RBR2(2)))
      AST(I)=0.0
      TEST=ABS(ADR1*BR2(1)+ADR2*BR2(2))
      IF(TEST.LE.EPSD)AST(I)=1.0
      I=I+1
3     S2=S2+1.0
4     S1=S1+1.0
C
      AINC=2.0*PI/FLOAT(IDEG)
      DO 7 I=1,NL
      ADR=V(I,1)
      ANG=V(I,2)
      DO 7 K=1,IDEG
      ANG=ANG+AINC
      CANG=COS(ANG)
```

```
      SANG=SIN(ANG)
      A=RBR1(1)*CANG+RBR1(2)*SANG
      B=RBR2(1)*CANG+RBR2(2)*SANG
      VLA=CEXP(CI*ADR*A)
      VLB=CEXP(CI*ADR*B)
      DO 5 J=1,NL
      TEST=CABS(VLA-VL(J,1))+CABS(VLB-VL(J,2))
      IF (TEST-1.0E-3) 6,6,5
5     CONTINUE
6     JJS(I,K)=J
7     CONTINUE
      WRITE(6,8)
8     FORMAT(4H0JJS)
      DO 9 I=1,NL
9     WRITE(6,10)(JJS(I,J),J=1,IDEG)
10    FORMAT(1H ,6I5)
      RETURN
      END
C
C
C
COMMENT FMAT CALCULATES THE VALUES OF THE SUM FLMS(JS,LM),
C       OVER LATTICE POINTS OF EACH SUBLATTICE JS, WHERE
C       LM=(0,0),(1,-1),(1,1),.....
C       NOTE: FOR ODD (L+M), FLMS IS ZERO
      SUBROUTINE FMAT(FLMS,JFLM,V,JJS,AST,NL,IDEG,LMAX)
      DIMENSION FLMS(NL,45),V(NL,2),JJS(NL,6),AST(NL),CLM(809),
     1 JFLM(9,17)
      DIMENSION SCC(6,17),SA(48),ANC(6),ANS(6),RTAB(4),
     1AK(2),BR1(2),BR2(2),RBR1(2),RBR2(2),R(2)
      COMPLEX FLMS,SCC,SA,RTAB,CZERO,CI,KAPPA,SC,SD,SE,Z,ACS,ACC,RF
      COMPLEX CMPLX,CSQRT,CEXP,CCOS,CSIN
      COMMON E,SPA,THETA,FI,VPI,CLM
      COMMON /X3/ BR1,BR2,RBR1,RBR2,NL1,NL2
COMMENT AK(1) AND AK(2) ARE THE X AND Y COMPONENTS OF THE
C       MOMENTUM PARALLEL TO THE SURFACE,MODULO A RECIPROCAL
C       LATTICE VECTOR
      PI=3.14159265
      CZERO=CMPLX(0.0,0.0)
      CI=CMPLX(0.0,1.0)
      KAPPA=CMPLX(2.0*E,-2.0*VPI+0.000001)
      KAPPA=CSQRT(KAPPA)
      AG=SQRT(2.0*E)*SIN(THETA)
      AK(1)=AG*COS(FI)
      AK(2)=AG*SIN(FI)
C
COMMENT ANC,ANS AND SA ARE PREPARED TO BE USED IN THE SUM
C       OVER SYMMETRICALLY RELATED SECTORS OF THE LATTICE
      L2MAX=LMAX+LMAX
      LIM=L2MAX*IDEG
      ANG=2.0*PI/FLOAT(IDEG)
      D=1.0
      DO 1 J=1,IDEG
      ANC(J)=COS(D*ANG)
      ANS(J)=SIN(D*ANG)
1     D=D+1.0
      D=1.0
      DO 2 J=1,LIM
      SA(J)=CEXP(-CI*D*ANG)
```

```
2       D=D+1.0
        K2=(L2MAX+1)*(L2MAX+2)/2
        DO 3 J=1,NL
        DO 3 K=1,K2
3       FLMS(J,K)=CZERO
COMMENT THE LATTICE SUM STARTS.THE SUM IS DIVIDED INTO ONE
C         OVER A SINGLE SECTOR,THE OTHER (IDEG-1) SECTORS
C         ARE RELATED BY SYMMETRY EXCEPT FOR FACTORS
C         INVOLVING THE DIRECTION OF R.  THE SUMMATION
C         INCLUDES POINTS WITHIN A CIRCLE OF RADIUS DCUT
        DCUT=-4.0*SQRT(2.0*E)/(AMIN1(VPI,-0.05))
        WRITE(6,4) DCUT
4       FORMAT(6H0DCUT=,E13.4)
        ADD=1.0
        ANN=-1.0
        ASST=1.0
5       DO 14 JS=1,NL
        AN1=ANN
        ADR1=V(JS,1)*COS(V(JS,2))
        ADR2=V(JS,1)*SIN(V(JS,2))
        AN2=ASST*AST(JS)
6       AN1=AN1+ADD
COMMENT R=THE CURRENT LATTICE VECTOR IN THE SUM
C         AR=MOD(R)
C         RTAB(1)=-EXP(I*FI(R))
        R(1)=ADR1+AN1*BR1(1)+AN2*BR2(1)
        R(2)=ADR2+AN1*BR1(2)+AN2*BR2(2)
        AR=SQRT(R(1)*R(1)+R(2)*R(2))
        IF(AR-DCUT)7,7,14
7       RTAB(1)=-CMPLX(R(1)/AR,R(2)/AR)
        ABC=1.0
        ABB=0.0
        IF (AG-1.0E-4) 9,9,8
8       ABC=(AK(1)*R(1)+AK(2)*R(2))/(AG*AR)
        ABB=(-AK(2)*R(1)+AK(1)*R(2))/(AG*AR)
9       SC=CI*AG*AR
COMMENT SCC CONTAINS FACTORS IN THE SUMMATION DEPENDENT ON
C         THE DIRECTION OF R. CONTRIBUTIONS FROM SYMMETRICALLY
C         RELATED SECTORS CAN BE GENERATED SIMPLY AND ARE
C         ACCUMULATED FOR EACH SECTOR, INDEXED BY THE SUBSCRIPT J.
C         THE SUBSCRIPT M IS ORDERED:M=(-L2MAX),(-L2MAX+1)....
C         (+L2MAX)
        DO 10 J=1,IDEG
        AD=ABC*ANC(J)-ABB*ANS(J)
        SD=CEXP(SC*AD)
        SCC(J,L2MAX+1)=SD
        MJ=0
        SE=RTAB(1)
        DO 10 M=1,L2MAX
        MJ=MJ+J
        L2P=L2MAX+1+M
        L2M=L2MAX+1-M
        SCC(J,L2P)=SD*SA(MJ)/SE
        SCC(J,L2M)=SD*SE/SA(MJ)
10      SE=SE*RTAB(1)
        Z=AR*KAPPA
        ACS=CSIN(Z)
        ACC=CCOS(Z)
COMMENT RTAB(3)=SPHERICAL HANKEL FUNCTION OF THE FIRST KIND,L=0
```

```
C        RTAB(4)=SPHERICAL HANKEL FUNCTION OF THE FIRST KIND,L=1
      RTAB(3)=(ACS-CI*ACC)/Z
      RTAB(4)=((ACS/Z-ACC)-CI*(ACC/Z+ACS))/Z
      AL=0.0
      LIML=L2MAX+1
COMMENT THE SUMMATION OVER FACTORS INDEPENDENT OF THE
C        DIRECTION OF R IS ACCUMULATED IN FLMS, FOR EACH
C        SUBLATTICE INDEXED BY SUBSCRIPT JSP.  THE SECOND
C        SUBSCRIPT ORDERS L AND M AS: (0,0),(1,-1),(1,1),(2,-2),
C        (2,0),(2,2)...
      JF=1
      DO 13 JL=1,LIML
      RF=RTAB(3)*CI
      JM=L2MAX+2-JL
      DO 12 KM=1,JL
      JFLM(JL,JM)=JF
      DO 11 J=1,IDEG
      JSP=JJS(JS,J)
11    FLMS(JSP,JF)=FLMS(JSP,JF)+SCC(J,JM)*RF
      JF=JF+1
      JM=JM+2
12    CONTINUE
COMMENT SPHERICAL HANKEL FUNCTIONS FOR HIGHER L ARE
C        GENERATED BY RECURRENCE RELATIONS
      ACS=(2.0*AL+3.0)*RTAB(4)/Z-RTAB(3)
      RTAB(3)=RTAB(4)
      RTAB(4)=ACS
      AL=AL+1.0
13    CONTINUE
      R(1)=R(1)+BR2(1)
      R(2)=R(2)+BR2(2)
      AR=SQRT(R(1)*R(1)+R(2)*R(2))
      IF(AR-DCUT)7,7,6
14    CONTINUE
COMMENT SPECIAL TREATMENT IS REQUIRED IF IDEG=2
      IF (IDEG-2) 15,15,17
15    IF (ASST) 17,17,16
16    ADD=-1.0
      ANN=0.0
      ASST=0.0
      GO TO 5
17    CONTINUE
C
      RETURN
      END
C
C
C
COMMENT PERTA, GIVEN REFLECTION AND TRANSMISSION MATRICES
C        FOR BULK AND OVERLAYER, RETURNS THE REFLECTED
C        AMPLITUDES IN WV BY RFS PERTURBATION THEORY.
C        DIMENSIONS ARE FIXED FOR N=13
      SUBROUTINE PERTA(RCA,TCA,N,RCB,TCB,NSL,NL,NM,WV)
      DIMENSION RCA(N,NM),TCA(N,NM),RCB(N,N),TCB(N,N),
     1 WV(N),NSL(NL),AKP(13),AKM(13),BKP(13),BKM(13)
      DIMENSION PQ(2,13),ANEW(13,12),AW(13,3),CLM(809),
     1 ASA(3),ASB(3)
      COMMON E,SPA,THETA,FI,VPI,CLM
      COMMON /X1/TV,PQ,ASA,ASB
```

```
      COMPLEX RCA,TCA,RCB,TCB
      COMPLEX EK,EKP,EKM,ANEW,AKP,AKM,BKP,BKM,CZERO,CI,WV,AW
      COMPLEX CMPLX,CSQRT,CEXP,CSIN,CCOS,CLOG
      CRIT=0.003
      CZERO=CMPLX(0.0,0.0)
      CI=CMPLX(0.0,1.0)
COMMENT AKP AND AKM ARE PHASE FACTORS DESCRIBING THE
C        PROPAGATION OF FORWARD AND BACKWARD TRAVELLING
C        WAVES FROM A LAYER TO HALF WAY BETWEEN LAYERS.
C        BKP AND BKM DESCRIBE PROPAGATION BETWEEN BULK
C        AND OVERLAYER
      AK=SQRT(2.0*E)*SIN(THETA)
      BK2=AK*COS(FI)
      BK3=AK*SIN(FI)
      DO 1 J=1,N
      AK2=BK2+PQ(1,J)
      AK3=BK3+PQ(2,J)
      A=2.0*E-AK2*AK2-AK3*AK3
      EK=CMPLX(A,-2.0*VPI+0.000001)
      EKP=CSQRT(EK)*ASA(1)
      EKM=-EKP+AK2*ASA(2)+AK3*ASA(3)
      EKP=EKM+2.0*EKP
      AKP(J)=CEXP(0.5*CI*EKP)
      AKM(J)=CEXP(-0.5*CI*EKM)
      EKP=CSQRT(EK)*ASB(1)
      EKM=-EKP+AK2*ASB(2)+AK3*ASB(3)
      EKP=EKM+2.0*EKP
      BKP(J)=CEXP(0.5*CI*EKP)
1     BKM(J)=CEXP(-0.5*CI*EKM)
COMMENT ANEW(J,I) IS THE CONTRIBUTION OF THE CURRENT
C        ORDER OF RFS PERTURBATION THEORY TO THE
C        AMPLITUDE OF THE J TH BEAM BETWEEN THE (I-1) TH
C        AND I TH LAYERS.
C        IO IS THE NUMBER OF THE PASS BEING EXECUTED.
      DO 2 I=1,N
      WV(I)=CZERO
      DO 2 J=1,12
2     ANEW(I,J)=CZERO
COMMENT INWARD PASS STARTS HERE
      ANEW(1,1)=CMPLX(1.0,0.0)
      IO=0
      IFAIL=0
      ANORM1=1.0E-6
3     ANORM2=0.0
      ALPHI=1.0
      IO=IO+1
      DO 4 K=1,N
      AW(K,2)=BKP(K)*ANEW(K,1)
4     AW(K,3)=BKM(K)*ANEW(K,2)
      DO 5 IG=1,N
      AW(IG,1)=CZERO
COMMENT TRANSMISSION AND REFLECTION BY THE OVERLAYER
      DO 5 K=1,N
5     AW(IG,1)=AW(IG,1)+TCB(IG,K)*AW(K,2)+RCB(IG,K)*
     1AW(K,3)
COMMENT NEW AMPLITUDES BETWEEN OVERLAYER AND 2ND LAYER
      DO 6 IG=1,N
6     ANEW(IG,2)=BKP(IG)*AW(IG,1)
      I=2
```

```
7       BNORM=0.0
        I=I+1
COMMENT FOR THE SECOND AND SUBSEQUENT LAYERS, THE RECIPROCAL
C         SUBLATTICES ARE TREATED INDIVIDUALLY
        NA=0
        DO 11 NN=1,NL
        NB=NSL(NN)
COMMENT OLD AMPLITUDES AT THE (I-1) TH LAYER
        DO 8 K=1,NB
        KNA=K+NA
        AW(K,2)=AKP(KNA)*ANEW(KNA,I-1)
8       AW(K,3)=AKM(KNA)*ANEW(KNA,I)
        DO 9 IG=1,NB
        IGNA=IG+NA
        AW(IG,1)=CZERO
COMMENT TRANSMISSION AND REFLECTION BY THE (I-1) TH LAYER
        DO 9 K=1,NB
9       AW(IG,1)=AW(IG,1)+TCA(IGNA,K)*AW(K,2)+RCA(IGNA,K)*
       1AW(K,3)
COMMENT NEW AMPLITUDES BETWEEN THE (I-1) TH AND I TH LAYERS
        DO 10 IG=1,NB
        IGNA=IG+NA
        ANEW(IGNA,I)=AKP(IGNA)*AW(IG,1)
        AB=CABS(ANEW(IGNA,I))
10      BNORM=BNORM+AB*AB
        NA=NA+NB
11      CONTINUE
COMMENT TESTS AND CHECKS ON THE DEPTH OF PENETRATION INTO THE
C         CRYSTAL.FOR INCREASED PENETRATION INCREASE THE SECOND
C         DIMENSION OF ANEW AND CHANGE IF STATEMENT 14
        IF(IO-1)12,12,13
12      IF(BNORM-CRIT)17,17,14
13      ALPHI=ALPHI*ALPHA
        IF(SQRT(ALPHI*BNORM)-CRIT)17,17,14
14      IF(I-11)7,15,15
15      WRITE (6,16)
16      FORMAT(23H0**FIRST ORDER TOO DEEP,/)
        IFAIL=1
        GOTO 19
17      WRITE (6,18)IO,CRIT
18      FORMAT(7H0ORDER ,I3,5X,7HCRIT = ,F5.3)
19      IMAX=I-1
        WRITE (6,20)IMAX,BNORM
20      FORMAT(8H IMAX = ,I2,5X,8HBNORM = ,E12.4,/,
       1 4X,3HWVR,9X,3HWVI)
        IMAX=IMAX+1
COMMENT IT IS ASSUMED THAT THE BEAM IS NOW SO STRONGLY
C         ATTENUATED THAT IT HAS EFFECTIVELY ZERO AMPLITUDE
        DO 21 IG=1,N
21      ANEW(IG,IMAX+1)=CZERO
COMMENT OUTWARD PASS STARTS HERE
        DO 25 I=2,IMAX
        II=IMAX-I+2
COMMENT FOR ALL EXCEPT THE OVERLAYER THE RECIPROCAL
C         SUBLATTICES ARE TREATED INDIVIDUALLY
        NA=0
        DO 25 NN=1,NL
        NB=NSL(NN)
COMMENT OLD AMPLITUDES AT THE II TH LAYER
```

```
      DO 22 K=1,NB
      KNA=K+NA
      AW(K,2)=AKM(KNA)*ANEW(KNA,II+1)
22    AW(K,3)=AKP(KNA)*ANEW(KNA,II)
COMMENT TRANSMISSION AND REFLECTION BY THE II TH LAYER
      DO 23 IG=1,NB
      IGNA=IG+NA
      AW(IG,1)=CZERO
      DO 23 K=1,NB
23    AW(IG,1)=AW(IG,1)+TCA(IGNA,K)*AW(K,2)+RCA(IGNA,K)*
     1AW(K,3)
COMMENT NEW AMPLITUDES BETWEEN THE (II-1) TH AND II TH LAYERS
      DO 24 IG=1,NB
      IGNA=IG+NA
24    ANEW(IGNA,II)=AKM(IGNA)*AW(IG,1)
      NA=NA+NB
25    CONTINUE
COMMENT OLD AMPLITUDES AT OVERLAYER
      DO 26 K=1,N
      AW(K,2)=BKM(K)*ANEW(K,2)
26    AW(K,3)=BKP(K)*ANEW(K,1)
COMMENT TRANSMISSION AND REFLECTION BY THE OVERLAYER
      DO 27 IG=1,N
      AW(IG,1)=CZERO
      DO 27 K=1,N
27    AW(IG,1)=AW(IG,1)+TCB(IG,K)*AW(K,2)+RCB(IG,K)*
     1AW(K,3)
COMMENT NEW AMPLITUDES AT SURFACE
      DO 28 IG=1,N
28    ANEW(IG,1)=BKM(IG)*AW(IG,1)
COMMENT CONTRIBUTION OF THE CURRENT PASS TO AMPLITUDES AT THE
C        SURFACE
      DO 29 IG=1,N
      AB=CABS(ANEW(IG,1))
      ANORM2=ANORM2+AB*AB
      WV(IG)=WV(IG)+ANEW(IG,1)
29    ANEW(IG,1)=CZERO
COMMENT CHECKS ON CONVERGENCE OVER PASSES
      DO 31 I=1,N
      WRITE (6,30)WV(I)
30    FORMAT(1H ,2E12.4)
31    CONTINUE
      ANORM1=ANORM1+ANORM2
      WRITE (6,32)ANORM1
32    FORMAT(9H ANORM = ,E12.4,/)
      IF(IFAIL-1)33,39,33
33    IF(ANORM2/ANORM1-0.001)39,39,34
34    IF(IO-1)35,35,36
35    ALPHA=EXP(ALOG(BNORM)/FLOAT(IMAX-1))
      GOTO 3
36    IF (IO-6) 3,37,37
37    WRITE (6,38) IO
38    FORMAT(/,24H ***NO CONVERGENCE AFTER,
     1I4,11H ITERATIONS)
39    RETURN
      END
C
C
C
```

```
COMMENT MSMAT RETURNS THE TRANSMISSION,TC,AND REFLECTION,
C       RC,COEFFICIENTS OF A LAYER OF ATOMS,REFERRED TO AN
C       ORIGIN AT ONE OF THE ATOMIC SITES.
C       THE LAYER MUST BE A BRAVAIS LATTICE.
C       DIMENSIONS ARE FIXED FOR N=13,LMAX=4
      SUBROUTINE MSMAT(RC,TC,N,NM,FLMS,JFLM,V,NSL,NL,LMAX,IO,EPSD)
      DIMENSION RC(N,NM),TC(N,NM),FLMS(NL,45),V(NL,2),NSL(NL),
     1 FLM(45),JFLM(9,17)
      DIMENSION XODD(10,10),XEVEN(15,15),CYLM(13,25),YLM(25),
     1CLM(809),PHS(5),AMULT(13),PQ(2,13),AR1(2),AR2(2),AF(5)
      DIMENSION YLO(10),YLE(15),IPO(10),IPE(15),ASA(3),ASB(3)
      COMPLEX AF,RC,TC,CYLM,YLM,AMULT,XA,YA,CZERO,CI,ST,CF,CT
      COMPLEX FLMS,FLM,XODD,XEVEN,YLO,YLE
      COMPLEX CMPLX,CSQRT,CEXP,CSIN,CCOS,CLOG
      COMMON E,SPA,THETA,FI,VPI,CLM
      COMMON /X1/TV,PQ,ASA,ASB
      COMMON /X2/PHS,AR1,AR2
COMMENT LMMAX=NUMBER OF VALUES OF L AND M IN FLM
C       NEVEN=NUMBER OF EVEN VALUES OF L+M IN XEVEN
C       NODD=NUMBER OF ODD VALUES OF L+M IN XODD
      LMMAX=(LMAX+1)**2
      PI=3.14159265
      CZERO=CMPLX(0.0,0.0)
      CI=CMPLX(0.0,1.0)
      LIM=LMAX+1
      DO 1 L=1,LIM
1     AF(L)=SIN(PHS(L))*CEXP(CI*PHS(L))
      B=FLOAT(LMAX)
      A=0.5*(B+1.0)*(B+2.0)
      NEVEN=IFIX(A)
      A=0.5*B*(B+1.0)
      NODD=IFIX(A)
      AK=SQRT(2.0*E)
      AK2=AK*SIN(THETA)*COS(FI)
      AK3=AK*SIN(THETA)*SIN(FI)
      YA=CMPLX(2.0*E,-2.0*VPI+0.000001)
      YA=CSQRT(YA)
COMMENT FOR EACH DIFFRACTED WAVE BETWEEN LAYERS,WITH X
C       AND Y WAVE VECTORS BK2 AND BK3,THE Z WAVE VECTOR,
C       XA,IS CALCULATED
      DO 4 JG=1,N
      BK2=PQ(1,JG)+AK2
      BK3=PQ(2,JG)+AK3
      C=BK2*BK2+BK3*BK3
      XA=CMPLX(2.0*E-C,-2.0*VPI+0.000001)
      XA=CSQRT(XA)
COMMENT THE ARGUMENTS THETA AND FI OF THE SPHERICAL
C       HARMONIC ARE CALCULATED FOR THE CURRENT WAVE AS
C       CF=EXP(I*FI),CT=COS(THETA),ST=SIN(THETA)
      B=0.0
      CF=CMPLX(1.0,0.0)
      IF (C-1.0E-7) 3,3,2
2     B=SQRT(C)
      CF=CMPLX(BK2/B,BK3/B)
3     CT=XA/YA
      ST=B/YA
      AMULT(JG)=-8.0*PI*PI*CI/(AK*TV*XA)
COMMENT ALL THE SPHERICAL HARMONICS UP TO L=LMAX ARE
C       CALCULATED AND SET IN YLM ORDERED THUS:LM=
```

```
C         (00),(1-1),(10),(11),(2-2).....,AND STORED IN CYLM
      CALL SPHRM4(YLM,CT,ST,CF,LMAX)
      DO 4 K=1,LMMAX
4     CYLM(JG,K)=YLM(K)
COMMENT FOR A BULK LAYER,IO=0, RC AND TC FACTORISE ACCORDING TO
C         RECIPROCAL SUBLATTICES.  IN THIS CASE THE FACTORISED
C         MATRICES ARE COMPUTED INDIVIDUALLY INSIDE THE DO-LOOP
C         LABELLED 100.  FOR THE OVERLAYER,IO=1, THIS DO-LOOP IS
C         EXECUTED ONCE ONLY, WITH NB=N
      L2MAX=LMAX+LMAX
      KK=(L2MAX+1)*(L2MAX+2)/2
      NLL=NL
      IF(IO.EQ.1)NLL=1
      NA=0
      DO 100 NN=1,NLL
      NB=NSL(NN)
      IF(IO.EQ.1)NB=N
COMMENT THE APPROPRIATE FLM'S ARE CALCULATED FOR THE
C         RECIPROCAL LATTICE CONCERNED.  IF IO=1, ONLY
C         THE FLMS(1)'S ARE INCLUDED
      DO 5 I=1,KK
5     FLM(I)=CZERO
      DO 6 JS=1,NLL
      ABR=V(JS,1)*(PQ(1,NA+1)*COS(V(JS,2))+PQ(2,NA+1)*SIN(V(JS,2)))
      DO 6 I=1,KK
6     FLM(I)=FLM(I)+FLMS(JS,I)*CEXP(ABR*CI)
COMMENT THE MATRICES XODD AND XEVEN DESCRIBING MULTIPLE
C         SCATTERING WITHIN A LAYER ARE CALCULATED. THEIR
C         SUBSCRIPTS ARE ORDERED THUS:XODD,LM=(10),(2-1),
C         (21),.....,XEVEN,LM=(00),(1-1),(11),(2-2),....
      CALL XMAT(XODD,XEVEN,NODD,NEVEN,LMAX,FLM,JFLM,CLM,AF)
COMMENT (XODD-1) AND (XEVEN-1) ARE INVERTED USING GAUSSIAN
C         ELIMINATION: N.B. A MINUS SIGN IS INCLUDED IN AMULT
      DO 7 J=1,NODD
7     XODD(J,J)=XODD(J,J)-(1.0,0.0)
      DO 8 J=1,NEVEN
8     XEVEN(J,J)=XEVEN(J,J)-(1.0,0.0)
      CALL ZGE(XODD,IPO,NODD,NODD,EPSD)
      CALL ZGE(XEVEN,IPE,NEVEN,NEVEN,EPSD)
COMMENT YLM(JGP)*AF*(I)**-L IS SET INTO YLO AND YLE WITH
C         PROPER ORDERING OF SUBSCRIPTS
      DO 21 JGP=1,NB
      JGPA=JGP+NA
      ST=CMPLX(1.0,0.0)
9     L=0
      K=1
      J=1
      JL=1
10    M=-L
      JL=JL+L+L
      JLMP=JL+M
      YLE(K)=CYLM(JGPA,JLMP)*AF(L+1)/ST
      K=K+1
      GO TO 12
C
11    JLMP=JL+M
      YLE(K)=CYLM(JGPA,JLMP)*AF(L+1)/ST
      JLMPM1=JL+M-1
      YLO(J)=CYLM(JGPA,JLMPM1)*AF(L+1)/ST
```

```
      J=J+1
      K=K+1
C
12    M=M+2
      IF (L-M) 13,11,11
13    L=L+1
      ST=ST*CI
      IF (LMAX-L) 14,10,10
COMMENT YLO AND YLE ARE MULTIPLIED BY INVERSE XODD,XEVEN,
C        AND THE RESULT RETURNED IN YLO AND YLE
14    CALL ZSU(XODD,IPO,YLO,NODD,NODD,EPSD)
      CALL ZSU(XEVEN,IPE,YLE,NEVEN,NEVEN,EPSD)
COMMENT SCALAR PRODUCT TAKEN AND ACCUMULATED IN XA FOR TC
C        AND YA FOR RC
      DO 21 JG=1,NB
      JGNA=JG+NA
      XA=CZERO
      YA=CZERO
      ST=CMPLX(1.0,0.0)
      L=0
      K=1
      J=1
      JL=1
15    M=-L
      JL=JL+L+L
      CF=CZERO
      JLMM=JL-M
      CT=YLE(K)*CYLM(JGNA,JLMM)/ST
      K=K+1
      GO TO 17
C
16    JLMM=JL-M
      CT=YLE(K)*CYLM(JGNA,JLMM)/ST
      JLMMP1=JL-M+1
      CF=YLO(J)*CYLM(JGNA,JLMMP1)/ST
      K=K+1
      J=J+1
17    XA=XA+(CT-CF)
      YA=YA+(CT+CF)
C
      M=M+2
      IF (L-M) 18,16,16
18    L=L+1
      ST=ST*CI
      IF (LMAX-L) 19,15,15
C
19    CONTINUE
      YA=YA*AMULT(JGPA)
      XA=XA*AMULT(JGPA)
COMMENT RC IS THE REFLECTION MATRIX AND TC THE TRANSMISSION
C        MATRIX
      RC(JGPA,JG)=YA
      TC(JGPA,JG)=XA
      IF(JGP-JG)21,20,21
20    TC(JGNA,JG)=TC(JGNA,JG)+(1.0,0.0)
21    CONTINUE
      NA=NA+NB
100   CONTINUE
      RETURN
```

```
      END
C
C
C
COMMENT RINT2,GIVEN THE REFLECTED AMPLITUDES,WV,
C        RETURNS THE REFLECTED INTENSITIES,AT.
      SUBROUTINE RINT2(N,WV,PQF,PQ,AT)
      DIMENSION WV(N),PQF(2,N),PQ(2,N),CLM(809),AT(N)
      COMMON E,SPA,THETA,FI,VPI,CLM
      COMPLEX WV
1     AK=SQRT(2.0*E)
      BK2=AK*SIN(THETA)*COS(FI)
      BK3=AK*SIN(THETA)*SIN(FI)
      DO 3 J=1,N
      AK2=PQ(1,J)+BK2
      AK3=PQ(2,J)+BK3
      A=2.0*E-AK2**2-AK3**2
      AT(J)=0.0
COMMENT IT IS ESTABLISHED WHETHER THE J TH BEAM EMERGES
C        FROM THE CRYSTAL
      IF (A) 3,3,2
2     A=SQRT(A)
      B=CABS(WV(J))
COMMENT THE REFLECTED INTENSITY IS PROPORTIONAL TO THE
C        Z COMPONENT OF MOMENTUM OF THE WAVE
      AT(J)=B*B*A/(AK*COS(THETA))
3     CONTINUE
      WRITE (6,4) E,THETA,FI
4     FORMAT(3H E=,1F10.4,7H THETA=,1F10.4,4H FI=,1F10.4)
      WRITE (6,5)
5     FORMAT(29H      BEAM              INTENSITY)
      WRITE (6,6) (PQF(1,J),PQF(2,J),AT(J),J=1,N)
6     FORMAT(2F7.1,1E15.5)
      WRITE (6,7)
7     FORMAT(////)
      RETURN
      END
C
C
C
COMMENT XMAT CALCULATES THE MATRIX DESCRIBING MULTIPLE
C        SCATTERING WITHIN A LAYER,RETURNED AS XODD
C        CORRESPONDING TO ODD L+M,WITH LM=(10),(2-1),(21),
C        .....,AND XEVEN CORRESPONDING TO EVEN L+M WITH
C        LM=(00),(1-1),(11),(2-2)....
C        THE PROGRAM ASSUMES THAT THE LAYER IS A BRAVAIS
C        LATTICE.DIMENSIONS ARE FIXED FOR LMAX=4
      SUBROUTINE XMAT(XODD,XEVEN,NODD,NEVEN,LMAX,FLM,JFLM,CLM,AF)
      DIMENSION XODD(NODD,NODD),XEVEN(NEVEN,NEVEN),FLM(45),
     1 JFLM(9,17),AF(5),CLM(809)
      COMPLEX XODD,XEVEN,FLM,AF,ACC,CZERO
      COMPLEX CMPLX
COMMENT SUMMATION OVER THE CLEBSCH-JORDAN COEFICIENTS
C        PROCEEDS,FIRST FOR XODD,THEN FOR XEVEN
      CZERO=CMPLX(0.0,0.0)
      L2MAX=LMAX+LMAX
      JSET=1
      N=1
C
```

```
1       J=1
        L=JSET
2       M=-L+JSET
        JL=L+1
C
3       K=1
        LPP=JSET
4       MPP=-LPP+JSET
C
5       MPA=IABS(MPP-M)
        LPA=IABS(LPP-L)
        IF (LPA-MPA) 7,7,6
6       MPA=LPA
7       MP1=MPP-M+L2MAX+1
        LP1=L+LPP+1
        ACC=CZERO
C
8       JF=JFLM(LP1,MP1)
        ACC=ACC+CLM(N)*FLM(JF)
        N=N+1
C
        LP1=LP1-2
        IF (LP1-1-MPA) 9,8,8
C
9       XEVEN(J,K)=ACC*AF(JL)
        K=K+1
        MPP=MPP+2
        IF (LPP-MPP) 10,5,5
10      LPP=LPP+1
        IF (LMAX-LPP) 11,4,4
C
11      J=J+1
        M=M+2
        IF (L-M) 12,3,3
12      L=L+1
        IF (LMAX-L) 13,2,2
C
13      IF (JSET) 16,16,14
14      DO 15 J=1,NODD
        DO 15 K=1,NODD
15      XODD(J,K)=XEVEN(J,K)
        JSET=0
        GO TO 1
16      CONTINUE
        RETURN
        END
C
C
C
COMMENT SPHRM4,GIVEN CT=COS(THETA),ST=SIN(THETA),AND
C       CF=EXP(I*FI),CALCULATES ALL THE YLM(THETA,FI)
C       UP TO L=LMAX.SUBSCRIPTS ARE ORDERED THUS:LM=(00),
C       (1-1),(10),(11),(2-2),....
C       DIMENSIONED FOR LMAX=4
      SUBROUTINE SPHRM4(YLM,CT,ST,CF,LMAX)
      DIMENSION FAC1(5),FAC3(5),FAC2(25),YLM(25)
      COMPLEX YLM
      COMPLEX CT,ST,CF,SF,SA,SX,SY
      COMPLEX CMPLX,CSQRT,CEXP,CSIN,CCOS,CLOG
```

```
      PI=3.14159265
      LM=0
      CL=0.0
      A=1.0
      B=1.0
      ASG=1.0
      LL=LMAX+1
COMMENT MULTIPLICATIVE FACTORS REQUIRED
      DO 2 L=1,LL
      FAC1(L)=ASG*SQRT((2.0*CL+1.0)*A/(4.0*PI*B*B))
      FAC3(L)=SQRT(2.0*CL)
      CM=-CL
      LN=L+L-1
      DO 1 M=1,LN
      LO=LM+M
      FAC2(LO)=SQRT((CL+1.0+CM)*(CL+1.0-CM)
     1/((2.0*CL+3.0)*(2.0*CL+1.0)))
1     CM=CM+1.0
      CL=CL+1.0
      A=A*2.0*CL*(2.0*CL-1.0)/4.0
      B=B*CL
      ASG=-ASG
2     LM=LM+LN
COMMENT FIRST ALL THE YLM FOR M=+-L AND M=+-(L-1) ARE
C        CALCULATED BY EXPLICIT FORMULAE
      LM=1
      CL=1.0
      ASG=-1.0
      SF=CF
      SA=CMPLX(1.0,0.0)
      YLM(1)=CMPLX(FAC1(1),0.0)
      DO 3 L=1,LMAX
      LN=LM+L+L+1
      YLM(LN)=FAC1(L+1)*SA*SF*ST
      YLM(LM+1)=ASG*FAC1(L+1)*SA*ST/SF
      YLM(LN-1)=-FAC3(L+1)*FAC1(L+1)*SA*SF*CT/CF
      YLM(LM+2)=ASG*FAC3(L+1)*FAC1(L+1)*SA*CT*CF/SF
      SA=ST*SA
      SF=SF*CF
      CL=CL+1.0
      ASG=-ASG
3     LM=LN
COMMENT USING YL(M) AND YL(M-1) IN A RECURRENCE RELATION
C        YL(M+1) IS CALCULATED
      LM=1
      LL=LMAX-1
      DO 5 L=1,LL
      LN=L+L-1
      LM2=LM+LN+4
      LM3=LM-LN
      DO 4 M=1,LN
      LO=LM2+M
      LP=LM3+M
      LQ=LM+M+1
      YLM(LO)=-(FAC2(LP)*YLM(LP)-CT*YLM(LQ))/FAC2(LQ)
4     CONTINUE
5     LM=LM+L+L+1
      RETURN
      END
```

```
C
C
C
COMMENT ZGE IS A STANDARD SUBROUTINE TO PERFORM GAUSSIAN
C         ELIMINATION ON AN NR*NC MATRIX A PRIOR TO
C         INVERSION OF A,DETAILS STORED IN INT
      SUBROUTINE ZGE(A,INT,NR,NC,EMACH)
      COMPLEX A,YR,DUM
      DIMENSION A(NR,NC),INT(NC)
      N=NC
      DO 10 II=2,N
      I=II-1
      YR=A(I,I)
      IN=I
      DO 2 J=II,N
      IF(CABS(YR)-CABS(A(J,I)))1,2,2
1     YR=A(J,I)
      IN=J
2     CONTINUE
      INT(I)=IN
      IF(IN-I)3,5,3
3     DO 4 J=I,N
      DUM=A(I,J)
      A(I,J)=A(IN,J)
4     A(IN,J)=DUM
5     IF(CABS(YR)-EMACH)10,10,6
6     DO 9 J=II,N
      IF(CABS(A(J,I))-EMACH)9,9,7
7     A(J,I)=A(J,I)/YR
      DO 8 K=II,N
8     A(J,K)=A(J,K)-A(I,K)*A(J,I)
9     CONTINUE
10    CONTINUE
      RETURN
      END
C
C
C
COMMENT ZSU IS A STANDARD BACK SUBSTITUTION SUBROUTINE
C         USING THE OUTPUT OF ZGE TO CALCULATE A INVERSE
C         TIMES X,RETURNED IN X
      SUBROUTINE ZSU(A,INT,X,NR,NC,EMACH)
      COMPLEX A,X,DUM
      DIMENSION A(NR,NC),X(NC),INT(NC)
      N=NC
      DO 5 II=2,N
      I=II-1
      IF(INT(I)-I)1,2,1
1     IN=INT(I)
      DUM=X(IN)
      X(IN)=X(I)
      X(I)=DUM
2     DO 4 J=II,N
      IF(CABS(A(J,I))-EMACH)4,4,3
3     X(J)=X(J)-A(J,I)*X(I)
4     CONTINUE
5     CONTINUE
      DO 10 II=1,N
      I=N-II+1
```

```
      IJ=I+1
      IF(I-N)6,8,6
6     DO 7 J=IJ,N
7     X(I)=X(I)-A(I,J)*X(J)
8     IF(CABS(A(I,I))-EMACH*1.0E-5)9,10,10
9     A(I,I)=EMACH*1.0E-5*(1.0,1.0)
10    X(I)=X(I)/A(I,I)
      RETURN
      END
```

REFERENCES

Abramowitz, M., and Stegun, I. A. (1964), "Handbook of Mathematical Functions", N.B.S.; Washington.
Allen, R. E., and Dewette, F. W. (1969a), *Phys. Rev.,* **179,** 873–86.
Allen, R. E., and Dewette, F. W. (1969b), *Phys. Rev.,* **188,** 1320–3.
Allen, R. E., Dewette, F. W., and Rahman, A. (1969), *Phys. Rev.,* **179,** 887–92.
Andersson, S. (1969), *Surface Science,* **18,** 325–40.
Andersson, S. (1970), *Surface Science,* **19,** 21–8.
Andersson, S., and Kasemo, B. (1970), *Solid State Comm.,* **8,** 961–4.
Andersson, S., and Kasemo, B. (1971), *Surface Science,* **25,** 273–88.
Andersson, S., and Kasemo, B. (1972), *Surface Science*, **32,** 79–99.
Andersson, S., and Pendry, J. B. (1972), *J. Phys. C.,* **5,** L41–5.
Andersson, S., and Pendry, J. B. (1973), *J. Phys. C.,* **6,** 601–20.
Appelbaum, J. A., and Hamann, D. R. (1972), *Phys. Rev.,* **B6,** 1122–30.
Baker, J. M., Strozier, J. A., and Blakely, J. M. (1969), *report 1241,* Mat. Sci. Centre, Cornell University.
Baker, J. M. (1969), proceedings of the conference on "Structure et Propriétés des Surfaces des Solides", pp. 69–79, Centre National de la Recherche Scientifique, Paris.
Baker, J. M. (1970), *Ph.D. Thesis; Cornell University.*
Barnes, R. F., Lagally, M. G., and Webb, M. B. (1968), *Phys. Rev.,* **171,** 627–33.
Beeby, J. L. (1968), *J. Phys. C.,* **1,** 82–7.
Bennett, A. J., McCarroll, B., and Messmer, R. P. (1971a), *Surface Science,* **24,** 191–208.
Bennett, A. J., McCarroll, B., and Messmer, R. P. (1971b), *Phys. Rev.,* **B3,** 1397–406.
Bethe, H. (1928), *Annalen der Physik,* **87,** 55–129.
Blount, E. I. (1962), *Solid State Physics,* **13,** 362–73.
Born, M., and Huang, K. (1954), "Dynamical Theory of Crystal Lattices", Clarendon Press, Oxford.
Boudreaux, D. S., and Heine, V. (1967), *Surface Science,* **8,** 426–44.
Burdick, G. A. (1963), *Phys. Rev.,* **129,** 138–50.
Burton, J. J., and Jura, G. (1967), *J. Phys. Chem.,* **71,** 1937–9.

Capart, G. (1969), *Surface Science,* **13,** 361–76.
Capart, G. (1971), *Surface Science,* **26,** 429–53.
Clark, B. C., Herman, R., and Wallis, R. F. (1965), *Phys. Rev.,* **139,** A860–7.
Clementi, E. (1965), *IBM Journal of Research and Development,* **9,** 2–19.
Cohen, M. H., and Heine, V. (1961), *Phys. Rev.,* **122,** 1821–6.
Cochran, W. (1963), *Reps. Prog. Phys.,* **26.**
Cyrot Lackmann, F. (1970), *J. Physique,* **31,** C1–67.
Davison, S. G., and Levine, S. D. (1970), *Solid State Physics,* **25,** 1–149.
Dicke, R. H., and Wittke, J. P. (1960), "Introduction to Quantum Mechanics", Addison-Wesley, Reading.
Dirac, P. A. M. (1930), *Proc. Camb. Phil. Soc.,* **26,** 376–85.
Duke, C. B., Anderson, J. R., and Tucker, C. W. (1970), *Surface Science,* **19,** 117–58.
Duke, C. B., and Bennet, A. (1967a), *Phys. Rev.,* **160,** 541–53.
Duke, C. B., and Bennet, A. (1967b), *Phys. Rev.,* **162,** 578–88.
Duke, C. B., and Tucker, C. W. (1969), *Surface Science,* **15,** 231–56.
Duke, C. B., and Laramore, G. E. (1970), *Phys. Rev.,* **B2,** 4765–82.
Estrup, P. J. (1970), in "Modern Diffraction and Imaging Techniques in Materials Science" (S. Amelinckx *et al.* eds.), North Holland, Amsterdam.
Ewald, P. P. (1921), *Annalen der Physik,* **64,** 253–7.
Fadeeva, V. N., and Terent'ev, N. M. (1961), "Tables of Values of the Function $\omega(z)$ for Complex Argument", Pergamon Press New York.
Falicov, L. M. (1964), "Group Theory and its Physical Applications", University of Chicago, Chicago.
Farnsworth, H. E. (1964), *Advances in Catalysis,* **15,** 31–63.
Feder, R. (1972), *Phys. Stat. Sol. (b),* **49,** 699–710.
Feinstein, L. G. (1970), *Surface Science,* **19,** 366–70.
Forstmann, F., Berndt, W., and Büttner, P. (1973), *Phys. Rev. Lett.,* **30,** 17–9.
Forstmann, F., and Berndt, W. (1973), *to be published.*
Gadzuk, W. (1969), *J. Phys. Chem. Solids,* **30,** 2307–19.
Gerlach, R. L., and Rhodin, T. N. (1970), *Surface Science,* **19,** 403–26.
Gersten, J. I., and McRae, E. G. (1972), *Surface Science,* **29,** 483–500.
Glauber, R. J. (1955), *Phys. Rev.,* **98,** 1692–8.
Goldstein, H. (1950), "Classical Mechanics", Addison Wesley, Reading Mass. U.S.A.
Goodman, R. M., Farrell, H. H., and Somorjai, G. A. (1968), *J. Chem. Phys.,* **49,** 692–700.
Grimley, T. B. (1971), *Z. für Electrochemie,* **75,** 1003–8.
Hartree, D. R. (1957), "The Calculation of Atomic Structures", Chapman and Hall, London.
Hedin, L., and Lundqvist, S. (1969), *Solid State Physics,* **23,** 1–181.
Heine, V. (1960), "Group Theory in Quantum Mechanics: an Introduction to its Present Usage", Pergamon, New York.
Heine, V. (1963), *Proc. Phys. Soc.,* **81,** 300–10.
Heppel, T. A. (1967), *J. Sci. Inst.,* **44,** 686–8.
Herman, F., and Skillman, S. (1963), "Atomic Structure Calculations", Prentice Hall, Englewood Cliffs, N.J.
Herring, C. (1940), *Phys. Rev.,* **57,** 1169–77.
Hirsch *et al.* (1965), "Electron Microscopy of Thin Crystals", Butterworths, London.
Holland, B. W. (1971), *Surface Science,* **28,** 258–66.
Holland, B. W., Hannum, R. W., Gibbons, A. M., and Woodruff, D. P. (1971), *Surface Science,* **25,** 576–86.
Huang, C. H., and Estrup, P. J. (1973), private communication.

Ibach, H. (1970), *Phys. Rev. Lett.,* **24,** 1416–8.
Ignatjevs, A., Jones, A. V., and Rhodin, T. N. (1972), *Surface Science,* **30,** 573–91.
Ignatjevs, A., Pendry, J. B., and Rhodin, T. N. (1971), *Phys. Rev. Lett.,* **26,** 189–91.
Ing, B. S. (1972), Ph.D. thesis, University of Cambridge.
Inkson, J. C. (1971), *Surface Science,* **28,** 69–76.
Iyengar, P. K., Venkataraman, G., Vijarayaghaven, P. R., and Roy, A. P. (1965), in "Inelastic Neutron Scattering, Vol. 1", pp. 153–79, International Atomic Energy Agency, Vienna.
James, R. W. (1962), "The Optical Principles of the Diffraction of X-rays", Bell, London.
Jenkins, F. A., and White, H. E. (1957), "Fundamentals of Optics", McGraw Hill, New York.
Jennings, P. J. (1970), *Surface Science,* **20,** 18–26.
Jennings, P. J., and McRae, E. G. (1970), *Surface Science,* **23,** 368–88.
Jennings, P. J., and Sim, B. K. (1972), *Surface Science,* **33,** 1–10.
Jepsen, D. W., and Marcus, P. M. (1971), in "Computational Methods in Band Theory", Marcus, P. M., Janak, J. F., and Williams, A. R., eds. Plenum Press, New York.
Jepsen, D. W., Marcus, P. M., and Jona, F. (1971), *Phys. Rev. Lett.,* **26,** 1365–8.
Jepsen, D. W., Marcus, P. M., and Jona, F. (1972), *Phys. Rev.,* **B5,** 3933–52.
Jona, F. (1970), *IBM Journal of Research and Development,* **14,** 444–52.
Jones, A. V. (1968), Ph.D. thesis: University of Cambridge.
Jones, E. R., McKinney, J. T., and Webb, M. B. (1966), *Phys. Rev.,* **151,** 476–83.
Kambe, K. (1967a), *Z. Naturf.,* **22a,** 322–30.
Kambe, K. (1967b), *Z. Naturf.,* **22a,** 422–31.
Kambe, K. (1968), *Z. Naturf.,* **23a,** 1280–94.
Lagally, M., Ngoc, T. C., and Webb, M. B. (1971), *Phys. Rev. Lett.,* **26,** 1557–60.
Lagally, M., Ngoc, T. C., and Webb, M. B. (1972), to be published.
Lander, J. J. (1965), *Progress in Solid State Chemistry,* **2,** 26–116.
Lander, J. J., and Morrison, J. (1962), *J. Chem. Phys.,* **37,** 729–46.
Lang, N. D. (1971), *Sol. St. Comm.,* **9,** 1015–9.
Lang, N. D., and Kohn, W. (1971), *Phys. Rev.,* **B3,** 1215–23.
Laramore, G. E., and Duke, C. B. (1970), *Phys. Rev.,* **B2,** 4783–95.
Lonsdale, K. (1959), "International Tables for X-ray Crystallography", Kynock Press, Birmingham.
Loucks, T. (1967), "Augmented Plane Wave Method", Benjamin, New York.
Lundqvist, B. I. (1969), *Physica Status Solidii,* **32,** 273–80.
Maradudin, A. A., and Melnagailis, J. (1964), *Phys. Rev.,* **133,** A1188–93.
Marcus, P. M., and Jepsen, D. W. (1968), *Phys. Rev. Lett.,* **20,** 925–9.
Mattheis, L. F. (1964), *Phys. Rev.,* **134,** 970–3.
McKinney, J. T., Jones, E. R., and Webb, M. B. (1967), *Phys. Rev.,* **160,** 523–30.
McRae, E. G. (1966), *J. Chem. Phys.,* **45,** 3258–76.
McRae, E. G. (1968a), *Surface Science,* **11,** 479–91.
McRae, E. G. (1968b), *Surface Science,* **11,** 492–507.
McRae, E. G. (1971), *Surface Science,* **25,** 491–512.
McRae, E. G., and Caldwell, C. W. (1967), *Surface Science,* **7,** 41–67.
Molière, K. (1939), *Annalen der Physik,* **34,** 461–72.
Morabito, J. M., Steiger, R. F., and Somorjai, G. A. (1969), *Phys. Rev.,* **179,** 638–44.
Newns, D. M. (1969), *Phys. Rev.,* **178,** 1123–35.
Newton, R. G. (1966), "Scattering Theory of Waves and Particles", McGraw Hill, New York.
Nozière, P. (1964), "The Theory of Interacting Fermi Systems", Benjamin, New York.

Palmberg, P. W., and Rhodin, T. N. (1968), *J. Chem. Phys.*, **49**, 134–46.
Pendry, J. B. (1968), *J. Phys. C.*, **1**, 1065–74.
Pendry, J. B. (1969a), *J. Phys. C.*, **2**, 1215–21.
Pendry, J. B. (1969b), *J. Phys. C.*, **2**, 2273–82.
Pendry, J. B. (1969c), *J. Phys. C.*, **2**, 2283–9.
Pendry, J. B. (1971a), *J. Phys. C.*, **4**, 427–34.
Pendry, J. B. (1971b), *J. Phys. C.*, **4**, 2501–13.
Pendry, J. B. (1971c), *J. Phys. C.*, **4**, 2514–23.
Pendry, J. B. (1971d), *J. Phys. C.*, **4**, 3095–106.
Pendry, J. B. (1972), *J. Phys. C.*, **5**, 2567–78.
Pendry, J. B., and Capart, G. (1969), *J. Phys. C.*, **2**, 841–51.
Phillips, J. C., and Kleinman, L. (1959), *Phys. Rev.*, **116**, 287–94.
Pines, D. (1963), "Elementary Excitations in Solids", Benjamin, New York.
Pynn, R., and Squires, G. L. (1968), in "Neutron Inelastic Scattering, Vol. **1**" pp. 215–22, International Atomic Energy Agency, Vienna.
Quinn, J. J., and Ferrell, R. A. (1958), *Phys. Rev.*, **112**, 812–27.
Reid, R. J. (1971), *Phys. Stat. Sol (a)*, **4**, K211–6.
Reid, R. J. (1972), *Surface Science*, **29**, 623–43.
Ritchie, R. H. (1957), *Phys. Rev.*, **106**, 874–81.
Ritchmeyer, F. K., Kennard, E. H., and Lauritsen, T. (1955), "Introduction to Modern Physics", McGraw Hill, New York.
Schiff, L. J. (1968), "Quantum Mechanics", McGraw Hill, New York.
Schilling, J. S., and Webb, M. B. (1970), *Phys. Rev.*, **B2**, 1665–76.
Schlier, R. E., and Farnsworth, H. E. (1959), *J. Chem. Phys.*, **30**, 917–26.
Shaw, R. W., and Pynn, R. (1969), *J. Phys. C.*, **2**, 2071–88.
Shen, A. P. (1971), *Phys. Rev.*, **B4**, 382–93.
Sherwood, P. M. A. (1972), "Vibrational Spectroscopy of Solids", Cambridge University Press, Cambridge.
Slater, J. C. (1937), *Phys. Rev.*, **51**, 840–6.
Slater, J. C. (1960), "Quantum Theory of Atomic Structure", McGraw Hill, New York.
Slater, J. C. (1967), "Insulators Semiconductors and Metals", McGraw Hill, New York.
Snow, E. C. (1968), *Phys. Rev.*, **172**, 708–11.
Somorjai, G. A., and Farrell, H. H. (1971), *Advances in Chemical Physics*, **20**, 215–339.
Sondhi, P. (1972), Ph.D. thesis, University of Oxford.
Squires, G. L. (1966), *Proc. Phys. Soc.*, **88**, 919–28.
Stern, E. A., and Ferrell, R. A. (1960), *Phys. Rev.*, **120**, 130–6.
Stern, R. M., and Howie, A. (1972), to be published.
Strozier, J. A., and Jones, R. O. (1971), *Phys. Rev.*, **B3**, 3228–43.
Tabor, D., and Wilson, J. M. (1970), *Surface Science*, **20**, 203–8.
Tait, R. H., Tong, S. Y., and Rhodin, T. N. (1972), *Phys. Rev. Lett.*, **28**, 553–6.
Taylor, N. J. (1966), *Surface Science*, **4**, 161.
Tick, R. A., and Witt, A. F. (1971), *Surface Science*, **26**, 165–83.
Tinkham, M. (1964), "Group Theory and Quantum Mechanics", McGraw Hill, New York.
Tucker, C. W., and Duke, C. B. (1970), *Surface Science*, **23**, 411–6.
Tucker, C. W., and Duke, C. B. (1971), *Surface Science*, **24**, 31–60.
Tucker, C. W., and Duke, C. B. (1972), *Surface Science*, **29**, 237–64.
Watts, C. M. K. (1968), *J. Phys. C.*, **1**, 1237–45.
Wilkinson, J. H. (1965), "The Algebraic Eigenvalue Problem", Clarendon Press, Oxford.
Wilson, J. M. (1971), *Ph.D. thesis, University of Cambridge.*

Wood, E. A. (1964), *J. Appl. Phys.*, **35,** 1306–12.
Woodruff, D. P. (1971), *Surface Science,* **25,** 576–86.
Zehner, D. M., and Farnsworth, H. E. (1972), *Surface Science,* **30,** 335–62.
Hirsch, P.B., Howie, A., Nicholson, R. B., Pashley, D. W., and Whelan M. J. (1965), "Electron Microscopy of Thin Crystals", Butterworths, London.
Ziman, J. M. (1964), "Principles of the Theory of Solids", C.U.P., Cambridge.
Ziman, J. M. (1971), *Solid State Physics,* **26,** 1–101.

Authors Index

A

Abramowitz, M., 138, *395*
Allen, R. E., 200, *395*
Anderson, J. R., 154, *396*
Andersson, S., 21, 33, 71, 84, 95, 104, 112, 113, 183, 208, 251, 260, 262, 263, *395*
Appelbaum, J. A., 73, *395*

B

Baker, J. M., 34, 157, *395*
Barnes, R. F., 189, *395*
Beeby, J. L., 130, *395*
Bennet, A., 73, *396*
Bennett, A. J., 222, *395*
Berndt, W., 265, *396*
Bethe, H., 109, *395*
Blakely, J. M., 34, *395*
Blount, E. I., 91, *395*
Born, M., 193, *395*
Boudreaux, D. S., 109, 144, *395*
Burdick, G. A., 95, *395*
Burton, J. J., 10, 70, *395*
Buttner, P., 265, *396*

C

Caldwell, C. W., 21, 112, *397*
Capart, G., 44, 109, 144, 146, *396*, *398*
Clark, B. C., 193, 200, *396*
Clementi, E., 39, 52, *396*
Cochran, W., 193, *396*
Cohen, M. H., 54, *396*
Cyrot-Lackman, F., 222, *396*

D

Davison, S. G., 122, *396*
Dewette, F. W., 200, *395*
Dicke, R. H., 156, *396*
Dirac, P. A. M., 43, *396*
Duke, C. B., 73, 131, 154, 175, 201, 243, *396*, *397*, *398*

E

Estrup, P. J., 6, 18, *396*
Ewald, P. P., 136, *396*

F

Fadeeva, V. N., 138, *396*
Falicov, L. M., 147, *396*
Farnsworth, H. E., 6, 222, 227, *396*, *398*, *399*
Farrell, H. H., 3, 189, 221, 223, *396*, *398*
Feder, R., 144, *396*
Feinstein, L. G., 144, *396*
Ferrell, R. A., 63, 72, *398*
Forstmann, F., 265, *396*

G

Gadzuk, W., 73, *396*
Gerlach, R. L., 251, *396*
Gersten, J. I., 109, *396*
Gibbons, A. M., 155, *396*
Glauber, R. J., 187, *396*
Goldstein, H., 88, *396*
Goodman, R. M., 189, *396*
Grimley, T. B., 222, *396*

H

Hamann, D. R., 73, *395*
Hannum, R. W., 155, *396*

Hartree, D. R., 345, *396*
Hedin, L., 60, *396*
Heine, V., 54, 91, 109, 144, 147, *396*
Heppel, T. A., 4, *396*
Herman, F., 39, 52, 345, 347, *396*
Herman, R., 193, 200, *396*
Herring, C., 54, *396*
Hirsch, P. B., 114, 146, 195, *396*, *399*
Holland, B. W., 155, 201, *396*
Howie, A., 30, *398*
Huang, C. H., 18, *396*
Huang, K., 193, *395*

I

Ibach, H., 29, *397*
Ignatjevs, A., 83, 160, 222, *397*
Ing, B. S., 66, 217, *397*
Inkson, J. C., 73, *397*
Iyengar, P. K., 195, *397*

J

James, R. W., 2, 195, *397*
Jenkins, F. A., 78, *397*
Jennings, P. J., 109, 138, *397*
Jepsen, D. W., 44, 96, 109, 124, *397*
Jona, F., 44, 109, 167, *397*
Jones, A. V., 222, *397*
Jones, E. R., 188, 192, *397*
Jones, R. O., 44, 156, 157, *398*
Jura, G., 10, 70, *395*

K

Kambe, K., 130, 137, *397*
Kasemo, B., 84, 95, 113, 251, 262, *395*
Kennard, E. H., 1, *398*
Kleinman, L., 54, *398*
Kohn, W., 73, *397*

L

Lagally, M. G., 157, 175, 177, 189, 242, *395*, *397*
Lander, J. J., 5, 222, *397*
Lang, N. D., 73, 238, *397*
Laramore, G. E., 201, *396*, *397*
Lauritsen, T., 1, *398*
Levine, S. D., 122, *396*
Lonsdale, K., 10, 198, 217, *397*
Loucks, T., 39, 44, *397*
Lundqvist, B. I., 65, 160, *397*
Lundqvist, S., 60, *396*

M

McCarroll, B., 222, *395*
McKinney, J. T., 188, 192, *397*
McRae, E. G., 21, 96, 109, 112, 124, 130, 154, 164, *396*, *397*
Maradudin, A. A., 192, 193, 200, *397*
Marcus, P. M., 44, 96, 109, 124, *397*
Mattheiss, L. F., 38, *397*
Melngailis, J., 192, 193, 200, *397*
Messmer, R. P., 222, *395*
Molière, K., 93, *397*
Morabito, J. M., 221, *397*
Morrison, J., 222, *397*

N

Newns, D. M., 222, *397*
Newton, R. G., 51, *397*
Ngoc, T. C., 157, 175, 177, 242, *397*
Nozière, P., 63, *397*

P

Palmberg, P. W., 25, *398*
Pendry, J. B., 44, 54, 71, 83, 96, 109, 125, 130, 146, 157, 158, 160, 166, 171, 177, 178, 183, 251, 260, 263, *395*, *397*, *398*
Phillips, J. C., 54, *398*
Pines, D., 60, *398*
Pynn, R., 195, *398*

Q

Quinn, J. J., 63, *398*

R

Rahman, A., 200, *395*
Reid, R. J., 189, 192, 217, *398*
Rhodin, T. N., 25, 83, 160, 166, 222, 251, *396*, *397*, *398*
Ritchie, R. H., 72, *398*
Ritchmeyer, F. K., 1, *398*
Roy, A. P., 195, *397*

S

Schiff, L. J., 51, *398*
Schilling, J. S., 30, 34, *398*
Schlier, R. E., 222, *398*
Shaw, R. W., 195, *398*
Shen, A. P., 144, *398*
Sherwood, P. M. A., 193, *398*
Sim, B. K., 138, *397*
Skillman, S., 39, 52, 345, 347, *396*
Slater, J. C., 43, 93, 326, *398*
Snow, E. C., 39, *398*
Somorjai, G. A., 3, 189, 221, 223, *396*, *397*, *398*
Sondhi, P., 144, *398*
Squires, G. L., 195, *398*
Stegun, I. A., 138, *395*
Steiger, R. F., 221, *397*
Stern, E. A., 72, *398*
Stern, R. M., 30, *398*
Strozier, J. A., 34, 44, 156, 157, *395*, *398*

T

Tabor, D., 25, 186, *398*
Tait, R. H., 166, *398*
Taylor, 233
Terent'ev, N. M., 138, *396*
Tick, R. A., 71, *398*
Tinkham, M., 147, *398*
Tong, S. Y., 166, *398*
Tucker, C. W., 131, 154, 175, 243, *396*, *398*

V

Venkataraman, G., 195, *397*
Vijarayahaven, P. R., 195, *397*

W

Wallis, R. F., 193, 200, *396*
Watts, C. M. K., 122, *398*
Webb, M. B., 30, 34, 157, 175, 177, 188, 189, 192, 242, *395*, *397*, *398*
White, H. E., 78, *397*
Wilkinson, J. H., 125, 145, *398*
Wilson, J. M., 25, 28, 186, 191, 215, *398*
Witt, A. F., 71, *398*
Wittke, 156, *396*
Wood, E. A., 9, *399*
Woodruff, D. P., 155, *396*, *399*

Z

Zehner, D. M., 227, *399*
Ziman, J. M., 91, 144, *399*

Subject Index

A

absorption, see also 'penetration' and 'imaginary potential', 82, 92–4, 106, 112, 131, 135, 174, 203, 217
activation energy, 3
aluminium, 36, 52–3, 56, 68, 71, 160, 166–7, 198, 224
angle of incidence, 11
angular resolution, 6
annealing, 8
anode, 4
antimony, 221
antiphase domains, see 'domains'
argon, 71
atomic scattering factor, 30
atomic units, 12
Auger effect, 8, 28–9
averaging schemes, 157, 174–8
——— for surface structure, 240–5
———, theory of, 178–85
axis of symmetry, 148–9

B

backward scattering, 31–7, 84, 89, 155, 168, 174, 204, 210
band gaps, 91–2, 109
band structure, 3, 38, 89–93, 95, 102, 107, 108, 121–2, 124–5, 135, 253–5
beam current, 4
beryllium, 34–5, 156–7, 198
binding energy, 222
bismuth, 221
Bloch theorem, 13, 30
Bloch waves, see 'normal modes'
bond lengths, 3, 267
Born Oppenheimer approximation, 187
boundary conditions, 193
Bragg reflection, 23, 34, 82, 89–90, 175, 189, 213–42, 255, 260
———, multiple scattering generalisation, 34, 92, 107–8, 174, 255
Bravais lattice, 9, 14, 194
Brillouin zone, 194

C

cadmium sulphide, 221
calcium, 71
cancellation theorem, 40, 54–5
carbon, 224
carbon dioxide, 224
carbon monoxide, 224, 227–8
catalysis, 3
cathode, 4
characteristic losses, 27–9
chemisorption, 222
chromium, 29, 198
cleaning of surfaces, 2, 7
cleaving, 7
coherence length, 5–6, 236, 252
coincidence lattice, 228–36, 250
computation times, 133, 141–2, 147, 151–2, 166, 172, 250, 258, 276
convergence of calculations over beams, 102, 122, 124, 141, 258
——— over partial waves, 102, 122
copper, 21, 23–4, 33, 36, 40, 68, 71, 95–6, 102–4, 105, 113, 115–6, 167, 170–1, 189, 191, 198, 208–18, 224, 233
core levels, 39, 41–7, 52
———, excitation of, 27, 61, 66–8
corrosion, 3

D

d-band levels, 113

Debye temperature, 188–9, 192, 198 (table), 210, 217, 212
———, effective, 192, 213–7
Debye spectrum, 195–6
Debye-Waller factor, 2, 30, 188–9, 204, 207, 219, 239
density of vibrational modes, 193
deuterium, 224
differential cross section, 51
diamond, 198
diffraction pattern, 3, 6, 11–9
dilation of the last atomic layer, 71, 260–2
dipole moment of the surface, 223, 237, 262
domains, 226
———, antiphase, 233–6, 252

E

eigenvalues, 121, 125, 127, 142, 145, 160, 174
eigenvectors, 121, 125, 173
Einstein frequency, 195
electron gun, 3
electronic differentiation, 26
electronic excitations, 27–9
electrostatic field in a solid, 43
energy selecting grids, 26
error function, 137
evanescent beams, 17
Ewald split, 136
exchange potential, 43–4
expansions of Green's function, 272
——— of Legendre polynomials, 271
——— of plane waves, 272
——— of spherical waves, 272–3
experimental apparatus, 3–7

F

Fabry-Perôt interferometer, 255, 260
faceting, 220–1
Faraday cup, 6
Fermions, 41
field ion microscope, 2
flashing, 2, 7
focusing, 4
forward scattering, 31–7, 52, 84, 155, 168, 190, 204, 218–9
Fourier expansion, 13–5, 78, 97, 122, 136, 141, 144–7

G

gallium antimonide, 221
gallium arsenide, 221
gamma function, 137
germanium, 198, 221, 224
giant matrix method, 142–3, 152
gold, 25, 198, 221
graphite, 21–2, 122
grid, 4, 6

H

Hartree approximation, 43
Hartree-Fock approximation, 43
high index surfaces, 220–1
Huyghen's construction, 78
hydrogen, 224

I

imaginary potential, see also 'absorption' and 'penetration of beams', 31–7, 57, 66, 82, 92, 94, 117–8, 138, 160, 177, 178, 182–3, 217, 219, 236–7
——— spatial variation, 68, 218
incident beam conventions, 11
incoherent scattering, 26–30
indium antimonide, 221
inelastic scattering, 26–30, 31–7
inner potential, 23, 32
intensity measurements, 3, 8, 17–26, 32, 84
———, absolute, 6
intensity reflected, calculation from amplitude, 81
iodine, 224, 264–6
ion-bombardment, 2, 7
ion-core potential, 37–47, 115, 128, 145, 220
iridium, 221, 222
iron, 198

irrationally related lattices, 229, 232, 249
isotropic scattering model, 154

K

Kambe's method for planar scattering, 135–8
Kikuchi pattern, 30
kinematic theory, 75–84, 94–5, 99, 114–5, 117, 157, 174–8, 187, 192, 201, 207, 240–5
———, comparison with experiment, 84

L

lattice summation, 132, 136–8, 249–51
lattice vibrations, 193–201
Laue condition, 82
layered compounds, 122
layers of atoms, 10, 11, 96, 123
lead, 71, 189
Legendre polynomials, 268–9
Lennard-Jones potential, 71
lifetime of an electron, 218
logarithmic derivative, 48

M

magnesium, 194–5
mass spectrometer, 7
matching of amplitudes at a surface, 87, 109–10
many body corrections, 41, 57–69
matrix doubling method for intensities, 141–3, 152
mercury, 34, 198
mirror plane symmetry, 125, 148-50
molybdenum, 18–9, 25, 190–1
molybdenum disulphide, 122
Morse potential, 71
muffin tin approximation, 38–9, 128, 137
muffin tin zero of potential, 179
multiple scattering, 75, 85, 92, 99, 100, 106, 108, 128–38, 155, 157, 161, 178–9, 180–2, 190, 198, 202, 207, 213, 260

N

nickel, 53–4, 56, 71, 84, 113–4, 119–20, 160–1, 177–8, 183–4, 200, 223, 224, 225, 238, 251–64
niobium, 25–6, 186
nitrogen, 29, 224
non-locality of potentials, 43
normal modes (or Bloch waves)—electronic, 84, 88–9, 100, 105–9, 121–8, 144, 151, 155–6, 174, 190, 192, 253–5
———, resonance in excitation of, 109–11
——— —vibrational—, 193

O

one dimensional problem, 84–96
optical potential, see also 'imaginary potential', 57–69, 76, 145
orientation of crystals, 7
oxidation, 222
oxygen, 25, 224

P

paladium, 221
partial wave expansion, 44, 77–8, 128–32
———, convergence of, 51–2
pendulums, analogues for normal modes, 88
penetration of beams, 1, 23, 30, 76, 105–6, 191, 217
——— at finite temperatures, 218
perturbation theory, 50, 153–85
———, application to interplanar scattering, 162–74
———, application to planar scattering, 157–61
———, convergence of, 156, 159–61, 166, 168, 172, 174
———, Rayleigh-Schrödinger, 155–6
———, R.F.S. scheme, 152, 168–74, 208, 212, 249
———, rules for calculating interplanar scattering, 164–5
phase shifts, 48, 52, 56 (table), 129, 154,

217, 236–7, 252, 262, 266
———, derivatives of, 57
——— at finite temperatures, 206–7, 210
phonons, 27, 29, 187, 194–5
physisorption, 222
plasmons, 27, 61, 65
platinum, 189, 221, 224
post acceleration grid, 2
potassium, 225
primary peaks in intensity, 23
propagators, 169
pseudopotential, 54, 179
pseudopotential methods for intensities, 143–7, 151, 156

Q

quasi-elastic electrons, 29
quasi-kinematic intensities, 180

R

rationally related lattices, 229
rearrangement of surfaces, 2, 10, 221–2
reciprocal lattice, 14, 17, 77, 79, 136, 231–6, 253
reflected amplitudes from a surface, 81
——— intensities from a surface, 81
relativistic effects in planar scattering, 138
renormalisation, 168
resolution in angle, 4, 6
resolution in energy, 4, 29
RFS scheme, see 'perturbation theory'
rhenium, 227

S

scattering by ion cores, 1, 30, 32, 47–57, 75, 115–6, 129, 154, 156, 157, 175, 177, 178, 236, 252
——— by a plane of ion cores, 77–80, 98, 128–35, 140, 157–8, 245, 249–51
——— by a layer of ion cores, 138–41
——— by ion cores at finite temperature, 188, 202, 204
———, s-wave, 154, 244
screen, fluorescent, 2
screening, 39–40, 64–5
secondary electrons, 26–8
secondary peaks in intensity, 23
selenium, 224
self energy, see 'optical potential'
semiconductors, 221
sensitivity of reflected intensities
to surface structure, 256–62
to the potential, 115–20
silicon, 221, 224
silver, 71, 188, 192, 198, 221, 264–6
single particle excitations, 27
single scattering theory, see 'kinematic theory'
Slater exchange, 43
sodium, 223, 238, 251–64
sound waves, 193
spherical Bessel and Hankel functions, 270–1
spherical harmonics, 269–70
spot photometer, 6
structure factor, 79
sublattice, 231–2, 250, 252
sulphur, 29, 224
surface barrier, 70–4, 97, 99, 110–2, 119–20, 238–9, 249, 253
surface plasmon, 72
surface state resonance, 2, 109–12, 119, 122, 264
surface structures, unit cells of, irrationally related, 229, 249
——— rationally related, 229
——— simply related, 229
———, convention for description of, 229–30
———, determinations of, 260–7
surface structives of adsorbates, 224 (table)
——— of clean surfaces, 221 (table), 222
symmetry, 8–15, 123, 125, 147–51

T

tellurium, 221
temperature, influence of, 25–6, 29–30, 161, 186–219
thermal diffuse scattering, 188, 218
thermal vibrations, 168, 186, 193–201,

239, 264
———, correlation between atoms', 198–200, 203–4
thermal vibrations of surface atoms, 2, 190, 192, 200, 207–8, 214–7
thermal spread of energies, 4
t-matrix, 50, 53–4, 80, 115–6, 156, 178, 252
——— at finite temperatures, 204–5, 210, 217–8
tungsten, 224, 233

U

ultra high vacuum, 2, 7
uncertainty principle, 37, 82, 92, 112

V

vacuum, 7

W

water, 224
wave length of electrons, 23
widths of peaks, 21, 23–4, 37, 69, 82–3, 112–5, 179–80, 218–9, 261
work function, 36, 223, 237, 262

X

xenon, 57, 83–4, 160, 222

Z

zero angle scattering, 182
zero point motion, 197
zinc oxide, 29